GW01458990

Audubon to Xántus

Photographs were supplied by, or are reproduced by, kind permission of the following:

United Kingdom—*H.S.Gladstone collection*: Bell, Bulwer, Cassin, Lawrence, LeConte, Verreaux. *National Portrait Gallery, London*: Barrow (Reg. No. 886), Bewick (Reg. No. 319), Cook (Reg. No. 26), Franklin (Reg. No. 2515 (81)), Henslow (Reg. No. 42266), McCormick (Reg. No. 1216), Sabine (Reg. No. 907), Strickland (Reg. No. 30646). *Trustees of the National Library of Scotland, Edinburgh*: Buller, Palmer, Sclater, C.H.Townsend. *The Natural History Museum, London*: Gray, Swainson (the latter by permission of Mrs Isabel Warre). *Edinburgh University Library, Edinburgh*: Whitney. *National Maritime Museum, Greenwich*: J.C.Ross (Neg. No. A7326). *Linacre Professor, Department of Zoology, University of Oxford*: Thomson. *The Board of Trustees of the National Museums and Galleries on Merseyside (Walker Art Gallery, Liverpool)*: Traill (Neg. No. 10726/5). *Warrington Library, Warrington*: Blackburne family. *Sir Douglas Elphinstone*: Tristram. *B. and R. Mearns*: Cory, Masséna, Nuttall, Xántus.
France—*Musée National du Chateau de Malmaison, Rueil-Malmaison*: C.L.Bonaparte, Z.Bonaparte. *Marquise Costa de Beauregard, Douvaine*: Costa. *Musée d'Art et d'Histoire de Rochefort*: Lesson.
Denmark—*Det Nationalhistoriske Museum på Frederiksborg*: Hornemann.
Italy—*Museo Civico di Storia Naturale, Bra*: Craveri.
Austria—*Bild-Archiv der Österreichischen Nationalbibliothek, Vienna*: Forster.
Germany—*Stadtarchiv, Mainz:* Kittlitz.
C.I.S.—*Zoology Museum of the Moscow Lomonosov State University, Moscow*: Fischer.
U.S.A.—*Ruthven Deane collection, Library of Congress, Washington, D.C.*: Abert, Allen, Brandt, Brewer, Cooper, Gundlach, Lincoln, B.R.Ross, Thayer. *Smithsonian Institution, Washington, D.C.*: Baird family (Neg. No. 10714), S.F.Baird (Neg. No. 64746), Ridgway (Smithsonian Archives photographic collection, Record Unit 95, Neg. No. 91–2158). *Graphics File, National Portrait Gallery, Smithsonian Institution, Washington, D.C.*: J.H.Clark. *U.S. Army Military History Institute, Carlisle Barracks, Pennsylvania*: Bendire, Couch. *U.S. Army Military History Institute and Massachusetts Commandery, Military Order of the Loyal Legion*: Hammond, Scott, Williamson. *Library of The Academy of Natural Sciences of Philadelphia*: Botteri, Say, Vaux. *Delaware Valley Ornithological Club*: Harris, Heermann, J.K. Townsend. *Independence National Historical Park Collection, Philadelphia, Pennsylvania*: Bartram, W.Clark, Lewis. *American Philosophical Society Library, Philadelphia*: Wilson. *Archives of the Cleveland Museum of Natural History, Ohio*: Kirtland, Oberholser. *Bernice P. Bishop Museum, Honolulu, Hawaii*: Bishop (Bradley and Rulofson (S.F.), Neg. No. CP50644), Dole (J.A.Gonsalves, Neg. No. CB23310). *Department Library Services, American Museum of Natural History, New York*: Audubon (Neg. No. 335470, Photo by Logan). *Ottenheimer Library Archives and Special Collections, University of Arkansas at Little Rock, Arkansas*: McCown. *The Charleston Museum, Charleston, South Carolina*: Bachman (Neg. No. Mk 1935c). *The Grove National Historic Park, Glenview Park District, Illinois*: Kennicott. *Gray Herbarium Archives, Harvard University, Cambridge, Massachusetts*: Wright. *Indiana University Archives, Bloomington, Indiana*: McKay. *Dean Amadon*: Munro. *Mrs Richard K. Anderson*: Anderson. *Emmanuel D. Rudolph*: Sprague.
Canada—*British Columbia Archives and Record Service, Victoria*: Tolmie (Cat. No. HP16040/ Neg. No. G-4990).
Australia—*R.P.Truman*: MacGillivray.

The publishers and authors have made every attempt to contact the copyright holders of the portraits. In the few instances where they have been unsuccessful, they invite the copyright holders to contact them direct.

The artist is grateful to the Western Foundation of Vertebrate Zoology for the loan of specimens to make the drawings, and for other valuable assistance.

Audubon to Xántus

The Lives of Those Commemorated in North American Bird Names

Barbara and Richard Mearns

Illustrated by Dana Gardner

ACADEMIC PRESS
Harcourt Brace Jovanovich, Publishers
London · San Diego · New York · Boston · Sydney · Tokyo · Toronto

ACADEMIC PRESS LIMITED
24-28 Oval Road
London NW1 7DX

United States Edition published by
ACADEMIC PRESS INC.
San Diego, CA 92101

Copyright © 1992 by
ACADEMIC PRESS LIMITED

All Rights Reserved
No part of this book may be reproduced in any form by photostat, microfilm, or any other
means, without written permission from the publishers

This book is printed on acid free paper

A catalogue record for this book is available from the British Library

ISBN 0-12-487423-1

Typeset by Photo·graphics, Honiton, Devon, England
Printed and bound in Great Britain
by Mackays of Chatham PLC,
Chatham, Kent

Contents

Maps

Foreword

Some years ago a British husband and wife team, Barbara and Richard Mearns undertook to write biographies of individuals after whom European birds were named. They went to great lengths to also provide a portrait or, for more recent times, a photograph of each of their subjects. The resulting volume *Biographies for Birdwatchers* has been widely praised. The accounts are not superficial thumb-nail sketches, but scholarly set-pieces, sometimes running to several thousand words. They are written in an engaging, at times sprightly, style.

The Mearnses have now produced the present companion volume of biographies of those for whom American (north of Mexico) birds were named. One hundred and one individuals are included, ranging, to quote the title, from *Audubon to Xántus*. While Audubon is known to all, Xántus, an obscure Hungarian, at times miscast as a soldier in the United States Army, was eccentric enough to deserve his position as 'odd-man out', number 101 in the Mearns's list.

In selecting their entries the authors have limited themselves to names currently recognized in the official (1983) 'Check-List' of the American Ornithologists' Union. To have included all those that have fallen by the wayside as synonyms or varieties would have tripled their task. Nevertheless they have added an appendix containing short notices of some of those who nearly made it to the main list. Antonio Vallisnieri, for example, an Italian botanist, is associated with a favorite food of the Canvasback Duck. Knowing this, Alexander Wilson named this waterfowl *Aythya valisineria*, hence providing an indirect link between the bird, its habits and a fellow scientist.

This is a volume that will always repay the browser. Others will turn to it for specific information. Is there, for example, a clue here to the curious name, *Icterus parisorum*, of a beautiful oriole of the southwest? *Parisorum*? Yes, we learn that the first specimen to reach Europe and there to be made known to science, was in a shipment of odds and ends from some traders then residing in Mexico—the 'Paris brothers'. The English name of this bird—Scott's Oriole—was chosen by a US Army lieutenant who wanted to honour his hero, Major General Winfield Scott.

Thus of those who have been immortalized by having a bird named for them, one might say, as in other fields of human recognition, that "some were born great, some achieved greatness and some had greatness thrust upon them". But famous or obscure, most of us prefer such a name as *Icterus parisorum* to *Icterus pustulatus* or Bachman's Warbler to Worm-eating Warbler.

Dean Amadon

Preface

The origins of our interest in avian eponyms were outlined in *Biographies for Birdwatchers: The Lives of Those Commemorated in Western Palearctic Bird Names* (1988). Briefly stated, they perhaps began in 1959 when Richard received a copy of E. Thomas Gilliard's *Living Birds of the World* (1958) for his ninth birthday. Within its glossy pages was a reference to Mearns's Quail and he wondered if Mearns could have been a relative. Many years later, on one of our trips to Alaska, we were struck by the surprising number of birds we were seeing, such as Baird's Sandpiper, Barrow's Goldeneye, Bonaparte's Gull, Sabine's Gull, Say's Phoebe, Steller's Eider and Wilson's Warbler, that were named after people. Our curiosity aroused, we tried to find a book that would tell us about all these characters at length but soon discovered that we would have to write it ourselves. We duly commenced on the birds of the Western Palearctic (Europe, North Africa and the Middle East), fortuitously beginning our task by reading Leonhard Stejneger's biography of Steller, which is surely one of the most remarkable tales of adventure and scientific endeavour that has ever been recorded. In retrospect, a couple of memorable days spent storm-bound in our tent on the Seward Peninsula must also have whetted our appetite: we had no light reading except for an enthralling recent reprint of Joseph Grinnell's birding and prospecting experiences at Nome and on the Kobuk River in the 1890s (*Gold Hunting in Alaska—The Alaska Journal*, Spring 1983).

Our second volume is the result of our unabated interest in early naturalists and it seemed an obvious choice for us to turn to the birds of North America (Canada and the United States). Two later birding trips (California–Arizona and Texas–New Mexico–Arizona) were made all the more enjoyable because of our knowledge of the discovery and naming of many of the birds we saw and we hope this book will give others an added dimension to their birding. We also hope that it will provide a useful addition to the biohistorical literature, particularly with regard to those lesser known personalities for whom we have uncovered new material or presented biographical details in English for the first time. We also hope that it will stimulate further research into the activities of the more obscure individuals.

Our biographees have been chosen for us by current nomenclature, but our degree of respect for them is often reflected in the length of the chapters—provided there is sufficient information available. Our greatest challenges were some of the European naturalists who worked in Mexico and we must admit to being completely defeated by Dr Alexandre, the Paris brothers and Wollweber. Within the United States our search into Charles L. McKay's background proved to be one of the most satisfying enquiries but we found nothing new on William

Hutton. We remain completely baffled by the origins and peculiar fate of Henry C. Palmer, who disappeared shortly after collecting in Hawaii for Lord Rothschild.

Regrettably, there is no chapter on Edgar Alexander Mearns since Mearns's Quail is now usually known as the Montezuma Quail—Mearns is therefore relegated to the Appendix. As for any family connection with this great American ornithologist of the past we can only say that, though no positive link has been found, both families did originate from Aberdeen on the north-east coast of Scotland, where Mearns is a local and still uncommon name.

Once again we must acknowledge the help, forbearance and patience of a great many library and museum staff, particularly Linda Birch (Alexander Library, Edward Grey Institute, Oxford), Geoff Creamer and other staff (Ewart Public Library, Dumfries), Ann Datta, Effie Warr and staff (British Museum (Natural History), London and Tring), Bill Deiss (Smithsonian Institution, Washington, D.C.), Mary Elizabeth Flahive (Cleveland Natural History Museum, Cleveland, Ohio), and Bill Harper (Scottish Ornithologists' Club Library, Edinburgh).

Too numerous to thank individually were the staff of the following institutions: National Library of Scotland, Edinburgh; Cambridge University Library, Cambridge; British Library, British Newspaper Library, Royal Botanic Gardens at Kew, Royal Entomological Society, Royal Geographical Society, London; Royal Meteorological Society, Bracknell; Warrington Library, Warrington; Weybridge Public Library, Weybridge; Bibliothèque National, Ministère des Affaires Étrangeres, Musée d'Histoire Naturelle, Paris; Bibliothèque Municipale, Musée d'Histoire Naturelle, Bordeaux; Bibliothèque Municipale, La Châtre; Société de Sciences Naturelles Loire-Forez, St Etienne; Archives de la Marine, Vincennes; Musée National du Chateau de Malmaison, Rueil-Malmaison, France; Craveri Museum, Bra, Italy; Hessisches Landesmuseum, Darmstadt, Germany; Australian Museum Library, Sydney; State Library of Victoria, Melbourne; Auckland Institute and Museum, New Zealand; Birmingham Public Library, Alabama; Anchorage Municipal Library, Alaska; University of Arkansas at Little Rock, Arkansas; California Historical Society at San Francisco, California State Library at Sacramento, University of California at Berkeley, California; Peabody Museum of Natural History, New Haven, Connecticut; Bernice P. Bishop Museum, Hawaii State Library, Honolulu, Hawaii; Illinois Audubon Society, Wayne, Illinois; Bangor Public Library, Maine; Enoch Pratt Free Library at Baltimore, Maryland Historical Society at Ann Arundel, Maryland; The Commonwealth of Massachusetts State Library, Boston, Massachusetts; Museum of Comparative Zoology, Harvard, Massachusetts; New York Zoological Society Library, New York Academy of Sciences, United States Military Academy Library, West Point, New York; American Philosophical Society at Philadelphia, Dickinson College at Carlisle, Historical Society of Pennsylvania at Philadelphia, Philadelphia Academy of Natural Sciences, Wistar Institute, Philadelphia, Pennsylvania; South Caroliniana Library at the University of South Carolina, South Carolina; National Archives and Record Service, Smithsonian Institution, Washington D.C.; Appleton Public Library, Wisconsin; British Columbia Archives and Record Service, Victoria, British

Columbia; Archives Départementales de la Guadeloupe, Basse-Terre, Guadeloupe.

Special thanks must be reserved for Joe and Nesta Ewan, C. Stuart Houston, Arthur Smith and Gordon Sauer for spontaneous help and encouragement as well as for responding to our specific enquiries so enthusiastically. For a second time Christian Jouanin fulfilled a most valuable role in providing information on French subjects and their portraits. Pietro Passerin d'Entrèves was our Italian connection who provided assistance and put us in touch with Ettore Molinaro at the Craveri Museum. Robert Ralph tracked down the MacGillivray portrait in Australia for his own biography of him but generously allowed us to use it first. And we could not have managed without the help of the following who were each essential for the completeness of one or more chapters: Dean Amadon, Simon Baatz, Alan Brady, Michael J. Brodhead, Anna K. Bullard, David Clugston, John B. Cox, Guillermo N. Falcon, Hans Feustel, Ross Galbreath, Robert Gladstone, Ronnie and Lesley Graham, Brina Kessel, Andrew Kitchener, Joyce Kinney, Norah F. McMillan, Mick Marquiss, Gordon M. Meade, Andy Mitchell, Ian Newton, H. Douglas Pratt, Miriam Rothschild, Emmanuel D. Rudolph and Ann Zwinger.

It is a privilege to thank Le Comte d'Albufera, Mary White Anderson, Kay Cobbett, Marquese Costa de Beauregard, Sir Douglas Elphinstone, Geoffrey Swainson, R.P. Truman and Isabel Warre for providing biographical details or portraits of their relatives.

For translation work we are indebted to Christopher and Rosamund Mearns (French), Sue Carroll (Italian), Benjamin Engel, Jean Muir and Joanna Spencer (German), and Benjamin Engel and Margaret Gatherer (Spanish). Bobby Smith gave some photographic assistance.

Jill Smith and Steve Tilley provided us with hospitality on some of our library visits, as did Andy and Jo Richford—Andy being our trusty and valued Editor. Lastly, heartfelt thanks to Barbara's brother and sister-in-law, Colin Kinloch and Joan Curzio, for being our computer advisers and American bankers.

Introduction

The History of Nomenclature

For centuries the traditional bird names were perfectly adequate for the use of country people, the majority of whom were both sedentary and illiterate. Inevitably, a multitude of names arose in different localities for a relatively small number of easily distinguishable birds and, conversely, groups of similar species were often referred to together under single names. In the eighteenth century, as the study of birds gathered momentum, the need arose for the creation of new names for the many newly discovered species and for the standardization of the existing folk names.

Just as the standardization of vernacular names became essential for people within a country to understand each other quickly and easily, so it became necessary for naturalists from different countries to understand each other, hence the importance of the Latinized names for every species. Many naturalists were not classicists and so names were formed from a type of technical Latin, Latinized Greek or even just bad Latin; it is, therefore, preferable to call them scientific names even though there has been nothing scientific about their formulation. The relatively stable nomenclature of the present day was preceded by years of ambiguity and inconsistency. In 1753 Carl Linné, or Linnaeus, of Sweden introduced a binomial system of classification which was later adopted throughout the scientific community. All botanical names in use before 1753 and all zoological names in use before 1758 are now judged to be Pre-Linnean and have been discarded. This even applies to many names which Linnaeus himself formulated before these dates—they too are Pre-Linnean!

The Linnean system was not accepted at once. Well into the nineteenth century several authors on natural history used their own system of classification and nomenclature or another of their choice. Some bird species thereby acquired long lists of different names. For example, in Elliott Coues's *Birds of the Northwest* (1874) the Golden Eagle and Osprey have over twenty synonyms each and require almost a page of text for each list.[1]

The proliferation of synonyms arose partly because of the desire of some authors to have their own system universally accepted but also because of the limited contact between naturalists. Many of them worked in isolation without access to either foreign literature or large skin collections so that a single species was often described as new, over and over again. Sometimes this could have been avoided by a little more research, but on other occasions, repetition was inevitable. Who, for instance, would have suspected that Wilson's Phalarope, obtained in Canada and named by Joseph Sabine in 1823, was the same species that had been described in 1819 by L.J.P. Vieillot from a Paraguayan specimen? Or that Cassin's Auklet, found in California by William Gambel and named in 1845, had been described in Russia in 1811 by P.S. Pallas from specimens from the northern Pacific?

It soon became obvious that a set of rules was necessary to control the number of synonyms and other problems that existed. Unfortunately, such a variety of systems were proposed that in time:

> "English systematists were following the Stricklandian Code; French systematists were following the International Code; German systematists were following the German Code; American systematists were divided between the Stricklandian, the A.O.U., the Dall and the International Code; systematists in special groups were in some cases following special or even personal codes; and systematists of Italy, Russia and some other countries were following either the International or some other code."[2]

All these have now, thankfully, been superseded by the single International Code of Zoological Nomenclature.

The Rule of Priority

In addition to the general decision to adopt the Linnean system of nomenclature, one of the earliest and most fundamental rules was the Rule of Priority: the oldest published specific name had priority over any other. Even this is not as simple as it sounds, as some early descriptions are barely recognizable, being so vague that it could not be agreed whether they constituted valid descriptions or not. Moreover, descriptions were sometimes discovered in the very old literature pre-dating those which had been accepted for decades. To obviate change under these circumstances, a fifty-year rule was proposed, whereby any name that had been out of use for fifty years or more should not be allowed. Somewhat predictably, even this rule is sometimes debated.

The Naming of New Species

The scientific name of each species now consists of two words, as introduced by Linnaeus. The first, generic name tells us to which genus the species belongs, and it always has an initial capital letter. The second word is the specific name and does not, now, begin with a capital letter, even if it refers to a place or person, e.g. Canada Goose *Branta canadensis*. If the species can be separated into races, then this binomial system can be extended to a trinomial one, e.g. *Branta canadensis hutchinsii*. The first race to be described is known as the nominate race, i.e. *B.c.canadensis*, regardless of whether it is the most widespread or the most typical of the species.

When a new species is named a single specimen, thereafter known as the type specimen, must have a name and description published in sufficient detail for it to be distinguished from any other closely allied species. The type locality is the term given to the location where the specimen was obtained.

When the full scientific name of a bird is formally given, it is followed by the name of the person who gave the original description of it and the date of the publication, e.g. Black-capped Chickadee *Parus atricapillus* Linnaeus, 1766. Revisions in classification mean that sometimes the generic name has been changed, in which case the describer's name is placed in brackets, e.g. Northern Cardinal *Cardinalis cardinalis* (Linnaeus, 1758)—or (Linnaeus) 1758; Linnaeus originally named it *Loxia cardinalis*. Occasionally, as in the chapter headings of this book, the full reference for the publication and the type locality is given.

Different authorities have different opinions about how some birds should be classified. This may affect the family or genus into which a bird is placed but the problem more often lies in whether a bird should be a full species or a subspecies. Some birds are notorious for swinging to and from full specific status depending upon whether lumping or splitting is in vogue. In this book we have usually followed the classification used in the A.O.U. *Check-list of North American Birds* (1983) and its various supplements up to 1989. Charles G. Sibley and Burt L. Munroe's *Distribution and Taxonomy of Birds of the World* (1990) has introduced some revolutionary changes based on biochemical methods of measuring genetic similarity between species. This type of work may help to bring stability to avian taxonomy but for ordinary birders it heralds a period of change that will be hard to keep up with.

The Origins of Bird Names

Most birds are named after some obvious distinguishing feature. This may be their colour (e.g. Blue Jay, Vermilion Flycatcher, *Selasphorus rufus*); the colour of a specific part (e.g. Yellow-billed Magpie, Ruby-crowned Kinglet, *Xanthocephalus xanthocephalus*); their structure (e.g. Broad-winged Hawk, Scissor-tailed Flycatcher, *Toxostoma curvirostre*); their size (e.g. Elf Owl, Great Cormorant, *Sitta pygmaea*); their call or song (e.g. Bobolink, Dickcissel, *Melospiza melodia*); or some aspect of their behaviour (e.g. Burrowing Owl, Parasitic Jaeger, *Turdus migratorius*). Alternatively a bird may be named after a location where it was discovered or where it often occurs, either in a geographical sense (e.g. Canada Warbler, California Quail, *Vireo philadelphicus*) or with reference to its supposed or actual preferred habitat (e.g. Orchard Oriole, Canyon Wren, *Carduelis pinus*). In desperation some common names have been taken directly from their present or former generic names (e.g. Phainopepla, Pyrrhuloxia and phalarope).

Combination names can sometimes be a bit unwieldy (e.g. Black-throated Blue Warbler, Northern Beardless-Tyrannulet). When one considers that the warblers of the world include the Green, Greenish, Olive, Olivaceous, Olive-backed, Yellow, Yellow-green, Yellow-headed, Yellow-browed, Yellow-throated, Icterine, Citrine, Golden-cheeked, Golden-winged and Golden-bellied, one can appreciate that simple descriptive titles can become totally inadequate. It is not surprising that many birds have been named after people.

Birds Named After People—Avian Eponyms

The early botanists started to name species after people and the idea was adopted by Linnaeus and extended to the naming of birds. By the beginning of the nineteenth century the practice had become widespread in both the scientific and common names of birds. It was not just the original discoverers, describers and eminent naturalists who were honoured in this way, but also their wives, daughters, patrons and friends. It is this varied group of people—naturalists, explorers, soldiers, sailors, doctors, clergymen, artists and businessmen—who form the subject of this book.

We have included all the individuals who have one or more North American bird species named after them in either the scientific or vernacular English name **(taking Canada and the United States as our definition of North America)**. An appendix covers people commemorated by birds accidental to the area, birds of uncertain status, a selection of subspecies and some recently obsolete names. Archilochus, King Canute and Montezuma are also relegated to the appendix as they seemed out of place in the main text.

In ninety-eight chapters, biographical sketches of one hundred and one persons are presented. Twenty-four of these individuals appeared in *Biographies for Birdwatchers: The Lives of Those Commemorated in Western Palearctic Bird Names* but we have included them again for the sake of completeness. Most of these chapters have been rewritten with the emphasis on those aspects of particular interest to North Americans, using different portraits wherever possible. In eight cases (Adams, Barrow, Bulwer, Hornemann, Leach, McDougall, Mauri and Sabine) where the subject had little or no connection with the New World, we have slightly condensed the original biography to allow more space for fresh material in other chapters.

About half of our main subjects were citizens of the United States or Canada, either by birth or by naturalization. For those from elsewhere, it was not always possible to assign a nationality, but two (perhaps three) originated in Australia and New Zealand and the remainder were Europeans: over twenty were British, ten French, seven German and four Italian, reflecting, as might be expected, the amount of work done in the New World by non-American naturalists during the eighteenth and nineteenth centuries.

There are eight women, most of them included by virtue of their status as the wife, daughter or sister of an eminent male naturalist. None of them was an outstanding ornithologist in her own right, but several were certainly knowledgeable contributors to the subject. Amongst the men, the commonest profession was that of 'naturalist', with nearly forty falling into this category as collectors, natural history dealers, artist-naturalists, professors or museum curators, while some had sufficient independent means to devote their lives to zoology. Medical and military careers are also well represented: twelve men served, at least briefly, in the U.S. Army, but quite a few of them joined up only because the army offered them a good opportunity to study the wildlife of the remoter regions.

Many of the early naturalists were great travellers and between them our subjects explored almost every corner of North America. If some sketches are read in groups, the reader will get an idea of the development of ornithology in some of the more frequently mentioned states, e.g. Alaska—Adams, Brandt, Fischer, Kennicott, Kittlitz, McKay, Steller, C.H. Townsend; California—Bell, Gambel, Heermann, Hutton, Néboux, Nuttall, Whitney, Williamson, Xántus; Hawaii—Ballieu, Bishop, Dole, Munro, Palmer; Texas—J.H. Clark, Couch, Heermann, McCown, Oberholser, Wright. The famous search for the Northwest Passage is well represented in the accounts of Adams, Barrow, Franklin, Ross and Sabine. Details of Audubon's Missouri River expedition of 1843 can be found in Bell, Harris, Smith and Sprague. Those who collected on the Columbia River could also be read together: W. Clark, Lewis, Tolmie, J.K. Townsend and Nuttall.

The survival of eponyms has been largely a matter of chance. The old literature is full of names such as Maximilian's Jay, Mandt's Guillemot, Maynard's Cuckoo, Parkman's Wren and many others which mean nothing to young ornithologists. Other names have only recently been changed, such as Coues's Flycatcher, Lichtenstein's Oriole, Rivoli's Hummingbird and Wied's Crested Flycatcher. As each succeeding generation becomes more distanced from these early naturalists less and less is remembered about most of them. But whether one approves of such eponyms or not it cannot be denied that many have now stood the test of time and their names remain in constant use by birders. It is our sincere hope that the names which still honour the work of outstanding naturalists will always be used.

ABERT. *TOPOGRAPHICAL ENGINEER, NATURALIST AND ARTIST WHO EXPLORED IN OKLAHOMA, NORTHERN TEXAS AND NEW MEXICO.*

James William Abert
(1820-1897)

ABERT'S TOWHEE *Pipilo aberti* Baird

Pipilo aberti Baird, 1852. In S.F.Baird and H.Stansbury, *Exploration and Survey of the Valley of the Great Salt Lake of Utah*, p. 325: New Mexico [= Gila Bend, Maricopa County, Arizona]

During the middle years of the nineteenth century the key role of reconnaissance, so essential to the American dream of western expansion, was played by the Topographical Engineers of the U.S. Army. This Corps of energetic and intelligent officers, the cream of West Point Military Academy, considered themselves to be the intellectual élite of the army. They were responsible not just for providing tactical maps for cavalry and infantry officers, but for seeking out postal and wagon train routes across the West and undertaking the numerous Pacific Railroad Surveys.

Colonel John James Abert (1788–1863), was for many years the head of the Topographical Engineers and partly responsible for the high quality of army exploration in the south-western regions. He was also a naturalist and had taken pleasure in assisting Audubon in 1837 by making the Revenue cutter *Campbell* available to him for his trip along the shores of the Gulf of Mexico. The Tassel-eared, or Abert Squirrel *Sciurus aberti* which inhabits the high Yellow Pine forests of the south-western states was discovered by Dr Samuel Woodhouse in 1851 and named by him for Colonel Abert.

Colonel Abert's eldest son, James William, the subject of this chapter, was born at Mount Holly, New Jersey, on 18 November 1820, nine years before his father took charge of the Topographical Bureau in Washington D.C. He studied at a classical seminary in the city before entering Princeton as a sophomore in 1835. Three years later he joined the U.S. Military Academy at West Point and graduated from there at the age of twenty-one. After a year with the Fifth Infantry, in garrison at Detroit, he was made an assistant Topographical Engineer in the survey of the northern lakes from 1834 to 1844. By this time he was married, to Jane (Stone) Abert, and had one small son.

On his first journey to the West, in 1845, Abert was led by the most famous of all the Topographical Engineers, John Frémont, who was now making his third expedition. This time he planned to explore the Great Basin and reconnoitre the high passes of the Sierras. In Missouri they were joined by Joseph Walker, "the man who didn't follow trails, but made them". Walker already knew the mountains and their passes better than Frémont ever would, but Abert was not destined to accompany them all the way to California. After travelling up the Arkansas River to Bent's Fort, a trading post one hundred miles east of the Rocky Mountains, Abert was placed in command of a detachment which was to map the Canadian River from its source in the eastern foothills of the Rockies to its debouchement in eastern Oklahoma. His report contained many natural history observations and included a number of watercolour sketches.

The next year, 1846, saw Abert as one of a small group of engineers sent westwards to join the forces of General Kearney in the war against Mexico. On 27 June, led by Lieutenant William Emory, they left Fort Leavenworth, on the Missouri, and headed across the prairies towards Bent's Fort. For the next year Abert kept a diary in which he recorded his observations in both words and pictures. He frequently mentioned birds, insects and flowers that he collected along the way and the following entry, made on 29 June, was typical:

> "This morning at the Kansas River we saw the first flock of parroquets [Carolina Parakeets]; they lit in a large cottonwood tree directly over our heads. Amongst the birds we noticed doves, flickers, bluebirds, towhee buntings [Rufous-sided Towhees], [and] crows; the latter were flying near a large cornfield and were doubtless watching with intense interest the ripening of the grain. Those friends of prairie voyageurs the cowbirds made their appearance and quietly installed themselves on the backs of the mules. The elder is yet in bloom. Amongst the undergrowth near the timber we noticed too the brown butterfly which flits around the Asclepias, and the common yellow butterfly."[1]

He also recorded the frustrations of life on the move: mosquitos that made every moment a torment for both man and beast, a swollen creek that swept away one of their number as they tried to cross and persistent problems with the mules. Then on 21 July Abert mentioned that he was suffering from fever, with attacks of vomiting and stomach cramps. For the next week his diary was laid aside while he continued the journey lying in the wagon and he had to be left behind at Bent's Fort while the others carried on to Mexico.

Throughout August he recuperated at the fort, a square adobe building on a level bluff on the northern side of the Arkansas River. Once he had recovered a little Abert made good use of his time by drawing the Cheyennes who traded there and he tried to learn something of their language. He also collected and sketched plants, birds and animals and read the two books he carried with him, a Greek New Testament and a copy of Horace's works.

By September he was fit enough to make the three hundred mile journey south to Santa Fe. He hoped to catch up with the army there, but by the time he arrived they had gone. With Lieutenant Peck (who had been sick at Santa Fe), he was to survey part of the province of New Mexico, which General Kearney had claimed as a possession of the United States on 15 August.

During October, November and December Abert and Peck with an escort of troopers explored the mountains and valleys around Albuquerque, visiting the pueblos and collecting data on their population size and the political organization of the province. They were continually in danger and on 1 December Abert wrote:

"The month came in today with much wind, scattering the leaves and rustling through the boughs of the cottonwood trees. Everyone seems to be on the alert. We at one time had slight alarm from someone's supposing a herd of mules were a troop of Mexicans. Threatened by danger on all sides, from Mexicans, from Navajoes, from Apaches, and with no pleasant anticipations with reference to the manner in which we shall spend this winter, it should not be wondered [at] if some of us wished ourselves fairly clear of all these annoyances."[2]

Abert was unable to do much hunting but one morning he shot a quail that he could not identify. A few days later he shot five more, but lamented that "as the arsenic that we had obtained in St Louis had been taken to California I was obliged to fill the skins with corn meal." He also shot some Red-shafted Flickers, Western Meadowlarks and Horned Larks.

The party spent Christmas in safety at Santa Fe, but then had to begin the return journey to Fort Leavenworth, which proved to be an even worse ordeal than Abert had anticipated. In Raton Pass the snow was five feet deep and the mules constantly lost their footing, plunging men and stores into the snow or icy rivers. Seven of the soldiers suffered from measles and others were attacked by frostbite. Beyond Bent's Fort they had to struggle through snowstorms and avoid Arapaho and Pawnee warriors who were ambushing travellers. When a band of Pawnees stole their mules they had to harness the able men to the wagons to drag along their weaker companions. Abert at last reached the haven of Fort Leavenworth on 1 March and the following Sunday he made the following entry:

"March 7: Spent much of my time reading and in musing over the events I had gone through. They seem like the realization of some romance, and I wonder at myself that I could have borne so much."[3]

Abert's Towhee

Within days of Abert arriving back home with his small collection, his father was writing to Audubon about some of the specimens, in particular the quail that they could not find in the new octavo edition of *The Birds of America* (1840–44). Colonel Abert wondered whether the birds were undescribed: but they were Scaled Quail, named by Nicholas Vigors seventeen years earlier, though this was the first record within the United States.[4]

The type specimen of Abert's Towhee is supposed to have been part of this collection. When Professor Baird named it *Pipilo aberti* five years later he acknowledged that by "the kind permission of Colonel J.J. Abert" he had been permitted to examine "a small but exceedingly interesting collection of birds and mammals, procured by his son, Lieutenant J.W. Abert, in New Mexico. Among them we found a new species of bird." Baird dedicated the new towhee "to its accomplished discoverer, Lieutenant Jas. W. Abert". He also listed eight other species which Abert had collected in New Mexico: American Kestrel, Yellow-headed Blackbird, Yellow-bellied Sapsucker, White-winged Dove, Scaled Quail, Gambel's Quail, Upland Sandpiper and American Avocet.

However, the exact type locality of Abert's Towhee presents something of a mystery. Baird could only say that it was from "New Mexico" but the birds have not been recorded in any part of New Mexico that Abert visited. The type specimen, with perhaps some further clue on the label, was thrown out by the U.S. National Museum more than a hundred years ago because it was in poor condition.

The problem was considered by John Davis in *Distribution and Variation of the Brown Towhees* (1951). Noting that Abert travelled as far west as Laguna and as far south as Valverde (about twenty miles south of Socorro), Davis speculated that one of Abert's men might have collected the towhee at Cliff, or Redrock, or some other locality on the Gila River about a hundred miles to the west. Another possibility is that the towhee's range was slightly more extensive in those days, but if the type specimen was really brought back from New Mexico by Abert, then it seems strange that both he and his father failed to realize that it was undescribed, since they wasted no time in seeking Audubon's opinion about the Scaled Quail. It is also surprising that Abert omitted Abert's Towhee from a paper he wrote many years later, published in 1882 in the *Journal of the Cincinnati Society of Natural History*. Under locational headings, Abert listed the birds he had observed throughout the course of his 1846–47 expedition and gave dates for each sighting.

It therefore seems likely that Abert's Towhee was never seen alive by Abert. It could have been given to him by Emory—for they travelled as far as Bent's Fort together and Emory continued westwards to the Gila and Colorado Rivers, where Abert's Towhees are common but shy inhabitants of the valley bottoms. Emory stated in his report that he collected some birds during the expedition but the zoology for this portion was never written up. Emory might well have passed the towhee on to Abert at a later date.

The exact nature of the link between Abert and his towhee will probably never be known.

After his adventures in New Mexico Abert was content to serve on the staff at West Point for two years, teaching drawing and painting, English literature and moral philosophy. His wife died at around this time. In 1850 he left the Academy and for the next six years travelled again with the Topographical Engineers, being involved in the 'improvement' of western rivers. Promoted to Captain, he then served in Florida for two years during the third Seminole war. In 1860 the army sent him to Europe to study military affairs and he took every opportunity to visit art galleries wherever he went. When he realized that civil war was imminent he cut short his itinerary and returned to the United States to fight for the north, as did his brother William Stretch Abert.

During 1861–62 J.W. Abert took part in the Shenandoah campaign on the staff of General Banks and displayed great bravery in the retreat of 24 May 1862. On the advance to Frederick, Maryland, in September, while approaching the battlefield of South Mountain he was thrown off his horse and seriously injured. A year later, by then a Major in the Corps of Engineers, he joined General Gillmore in South Carolina but his injuries still troubled him and he resigned from the army in June 1864, receiving the brevet rank of Lieutenant-Colonel "for faithful and meritorious service in the valley of the Shenandoah from June, 1861, to September, 1862".

Abert embarked on his civilian career at the age of forty-three; he was a merchant in Cincinnati from 1864 to 1869, then an examiner of patents in Washington from 1869 to 1871 and lastly, a lecturer at Missouri State University at Rolla from 1871 to 1878. After his retiral he returned to the Cincinnati area, making his home at Newport on the Kentucky side of the Ohio. He continued to paint in oils and watercolours, experimented with sculpture and lectured and wrote on various subjects. He had married for a second time and when he died, on 10 August 1897, in Newport, he was survived by his widow and four children.

Edward Adams†
(1824-1856)

YELLOW-BILLED LOON *Gavia adamsii* (Gray)

Colymbus adamsii G.R.Gray, 1859. *Proceedings of the Zoological Society of London*, p. 167: Alaska

"This fine species is named after the late Mr. Adams Surgeon of H.M.S. *Enterprise*, commanded by Captain Collinson, in the voyage made by him through Behring's Straits. Mr. Adams employed his pencil in producing beautiful drawings of the remarkable birds obtained during the voyage."[1]

These words by George Gray relate to his friend Edward Adams,[2] an unfortunate naturalist who survived two rigorous and exciting expeditions to the Arctic only to die soon afterwards. Because of his early death, an ornithological paper written by Adams in western Alaska between October 1850 and July 1851 was overlooked for many years. It was eventually found by his brother William among other manuscripts and drawings and was published under the title 'Notes on the Birds of Michalaski, Norton Sound' in *The Ibis* of 1878. But by then Robert Kennicott, William Dall and others had visited the same area and Adams's labours were of less consequence. Nevertheless Edward Adams remains a pioneer Alaskan ornithologist.

He was born at Great Barton, near Bury St Edmunds in Suffolk, England, on 24 February 1824 and spent a great part of his youth pursuing his interest in natural history. He entered the medical profession and qualified as a surgeon in April 1847, at the age of twenty-three. In August of the same year he obtained a commission in the Royal Navy as assistant surgeon and was appointed to Haslar Hospital, Gosport, but after three months he was transferred to the Naval Hospital at Devonport. His service here was also brief because after four months he volunteered to join James Clark Ross's expedition to the Arctic to search for Sir John Franklin. He was appointed assistant surgeon and naturalist to the *Investigator* under Captain Bird; Ross commanding the *Enterprise*.

The two ships left England on 8 May 1848 and sailed northwards through Baffin Bay and Lancaster Sound to winter near Somerset Island. They found no

†No portrait traced

trace of Franklin, but Adams collected birds and geological specimens and sketched a number of Arctic scenes. Ironically, Adams and Ross were close to the Boothia Peninsula where, in the 1830s, Ross had spent three winters trapped in the ice and returned with three skins of the Yellow-billed Loon! Edward Sabine, however, had persuaded Ross "against his own better judgement" that they were only very old males of the similar Common Loon.[3]

After eighteen months the two ships arrived back in England and were immediately prepared for a second search expedition that was to approach the islands of the Canadian Arctic from the west, by way of the Bering Straits. This time Adams was appointed assistant surgeon to the *Enterprise* under Captain Richard Collinson, while the *Investigator* was commanded by Robert McClure. They left Plymouth on 20 January 1850, after less than three months in England. The *Investigator* was the slower of the two ships but, in true tortoise-and-hare fashion, she arrived in the Arctic first, and choosing a route north of Victoria Island tried to force her way eastward. The *Enterprise*, with young Adams on board, had been becalmed north of Hawaii and did not arrive in the Bering Straits until August. (For a map of Alaska see p. 318.) They were astonished to hear that McClure had gone on ahead but by then it was too late for them to follow because winter was already closing in. Putting into Port Clarence, north of Nome, Collinson heard several stories from the captain of the *Plover* concerning possible survivors from Franklin's expedition. The most promising account told of two officers and eight men in a "distressed state" who had bartered guns for food at a place 550 miles upriver from the mouth of the Yukon. Collinson determined to investigate this report and Lieutenant Barnard, Adams and a seaman were put ashore at St Michael, just north of the Yukon Delta. Barnard waited until winter before setting off so that he could travel with the Russian Governor and his interpreter, believing that this would speed up his enquiries. Adams remained behind, but on 25 February he received a note saying that the Russian trading post of Nulato had been attacked by Indians and that Barnard was "dangerously wounded". The doctor left at once but by the time he arrived Barnard was dead.

Yellow-billed Loon

The Russians enjoyed good relations with the Athabascan Indians and had therefore not been on their guard. Some Indians from further north had surprised the Russians as they slept, killed the Governor, fled after a struggle with Barnard and then killed a number of local Indians. The interpreter had been wounded but was saved by Adams's surgical skills. There was no sign of the white men whom Barnard had been looking for; they were probably trappers who had arrived on the Yukon overland from the Mackenzie.

On his return to St Michael, Adams continued his observations, enjoying especially the brief period between the arrival of spring in mid-May and the arrival of the *Plover*, which took Adams and the sailor back to Port Clarence to rejoin their ship on 4 July. By then he had made notes on many of the typical birds, such as Red-necked Phalaropes, Sandhill Cranes, Sabine's Gulls, King Eiders, Spectacled Eiders and Lapland Longspurs. He was the first to record the Bluethroat in America[4] and also saw the Yellow Wagtail, another North American 'rarity' whose breeding range only just extends from Siberia across the Bering Sea into Alaska. Concerning the Yellow-billed Loon, Adams noted that the "Natives kill numbers of these birds at sea during the autumn. They have plenty of skins, both old and young, which they convert into bags for their tools. I saw none of the birds myself." Although he saw plenty of these loons later, he was not likely to have seen them on this part of the expedition because they are at sea from October to May and then move far to the north of Norton Sound to nest.

After leaving St Michael the previous autumn Collinson had taken the *Enterprise* southwards to the Russian capital of Sitka, then crossed the Pacific to overwinter at Hong Kong. With Adams on board once again they took a more southerly route than the *Investigator*, below Banks and Victoria Islands, and succeeded in penetrating further east than any ship had done before. In the spring of 1853 some of the crew reached Gateshead Island by sledge and although they found not the slightest trace of the missing men they were only a few miles from the position where Franklin had perished—as long ago as 1847.

The senior surgeon and Adams made worthwhile collections of insects, flowers and birds, although more often than not any shooting was for the pot. Eighty-eight birds of various kinds were later presented to the British Museum but no account of the expedition's natural history discoveries was ever prepared.

The *Enterprise* did not meet up with McClure (who was trapped in the ice, well to the north) and in all they spent three years in complete isolation within the Arctic before returning homewards. Their welcome by the Admiralty was less than enthusiastic. McClure had already claimed the credit and been given the prize money for completing the North-west Passage, even though he had not actually sailed his ship through; the *Investigator* had been crushed and lost in the ice and the crew had been transferred to another ship which had approached from the opposite direction. Four other ships had also been lost in the Arctic, causing the Admiralty to lose interest in the search for Franklin and they made little of Collinson's achievements.

In the course of a two-month rest Adams appeared to recover from a lung disorder contracted on the return voyage. He passed his full naval surgeon's exams, was appointed to the steam sloop *Hecla*, joined her at Devonport in

November 1855 and went out to the coast of West Africa in the following May. After only a month in the region he was set ashore as an invalid while the *Hecla* carried out her duties elsewhere. When she returned three weeks later Adams was taken on board in the hope of returning him to England but an attack of typhus was too much for him and he died on 12 November, aged thirty-two. He was buried at a small cemetery in Sierra Leone.

Dr Alexandre†
(fl. 1846)

BLACK-CHINNED HUMMINGBIRD *Archilochus alexandri* (Bourcier and Mulsant)

Trochilus Alexandri Bourcier and Mulsant, 1846. *Annales des Sciences Physiques et Naturelles de la Société Impériale d'Agriculture de Lyon* 9, p. 330: Sierra Madre [Occidental], Mexico

The Black-chinned Hummingbird was first described by the French naturalists Jules Bourcier and Martial Étienne Mulsant. The latter was primarily an entomologist, but like Bourcier he developed a special interest in the Trochilidae. In 1846, when they wrote the original description of 'L'Alexandre' *Trochilus Alexandri* they gave anatomical measurements, the type locality and full details

Black-chinned Hummingbird

†No portrait traced

of both male and female plumages. However, their dedication to the bird's discoverer tells us very little about that gentleman:

"Nous avons dédié cette espèce au naturaliste qui l'a découverte, M. le docteur Alexandre, de Mexico."

Biohistorians on both sides of the Atlantic—Owen Wynne and T.S. Palmer among them—have failed to learn anything more about the identity of Dr Alexandre and our own enquiries have been equally unsuccessful.

ALLEN. COLLECTOR BASED AT NICASIO, CALIFORNIA. THE PHOTOGRAPH DATES FROM ABOUT 1895.

Charles Andrew Allen
(1841-1930)

ALLEN'S HUMMINGBIRD *Selasphorus sasin* (Lesson)

Ornismya Sasin Lesson, 1829. *Histoire Naturelle des Oiseaux-Mouches,* p.xxx; 1830, p. 190, Pl. 66, 67: La Californie, la côte N.-O. d'Amérique [= San Francisco, California]

In the summertime Allen's Hummingbird is common in the chaparral, brushy hillsides and open coniferous forests of coastal California. Its name commemorates the collector Charles Allen who lived in Marin County, California, for fifty-seven years.

Allen's interest in birds sprang from a childhood experience on the east coast. He was born on 21 August 1841, at Milton, Massachusetts, and at the age of about eleven he visited Cape Cod with James Gatly, a taxidermist from Boston. While they were walking along the shore Allen found a dead Cory's Shearwater and gave the corpse to his friend who made a mounted specimen of it. The boy was so delighted that Gatly began to instruct him in taxidermy and bird lore, thus directing the course of his life.

On leaving school Allen worked as a carpenter for four years, but when the Civil War broke out he enlisted with the First Massachusetts Regiment and served for two years. Afterwards he worked in the Baker Chocolate Mills at Lowell and spent his evenings and holidays collecting and preparing bird specimens. He was then employed at a planing mill but the dust affected his lungs so badly that he would leave for short spells and work with Gatly until he recovered. Wanting to learn more about seabirds he spent three years as a fisherman on the Banks of Newfoundland, where the men often used the flocks of diving birds to help them locate the shoals of cod.

After he returned to work at the planing mills his health suffered so badly that early in 1873 he went to California and collected there for several months. On being offered a job as timber guard on a ranch near Nicasio in Marin County, just to the north of San Francisco, Allen decided to stay on. He and his wife Abby, whom he had married in 1870, set up home in a small cabin on the edge of a Redwood forest and their three daughters, Jessie, Hattie and Ruth, were all born there.

Allen augmented his guard's income by mounting deer heads for hunters and by collecting and dealing in bird skins. He was in an ideal location since the Pacific coast had not yet been systematically explored by naturalists, although Collie, Douglas, Botta, Nuttall, Townsend, Gambel, Bell, Hutton, Heermann and Craveri had made brief visits. Ornithologists living in the eastern states remained eager to obtain specimens from the West and for the rest of his life Allen hunted the resident and migratory birds of California. He sent them to William Brewster in Massachusetts, Charles K. Worthen in Illinois, C. Hart Merriam in Washington D.C. and many others.

In 1877 Allen sent a batch of hummingbirds to Brewster, drawing his attention to differences in the tail feathers and suggesting that this might indicate that they belonged to two separate species. Brewster handed the specimens over to his friend Henry Henshaw who had collected in California with the Wheeler survey, and Henshaw named the new species *Selasphorus alleni* in the *Bulletin of the Nuttall Ornithological Club* (1877). He made the following comparison with the very similar Rufous Hummingbird:

> "I am in possession of but few notes bearing upon the habits of this Hummer. Mr. Allen remarks incidentally in a letter that the Green-backs [Allen's] are much the livelier and more active of the two, keeping constantly in the open, and always perching upon the most prominent dead twigs they can find. Their extreme shyness, as contrasted with the unsuspicious nature of the Rufous-backed, is quite remarkable. They seem to possess a larger share than usual of the courage and pugnacity which is so constantly displayed in birds of this family. Not only do they always come off the victors when chance encounters take place between them and the Rufous-backs, but Mr. Allen has seen a pair attack and put to rout a Red-tailed Hawk; while, as he remarks, "Sparrow-Hawks [American Kestrels] have no chance at all with them." He has often seen the little fellows in hot chase after these latter birds, and their only care seemed to be to get out of the way as soon as possible of foes so determined."[1]

Henshaw did not realize that Allen's Hummingbird had already been described by René Primevère Lesson, whose scientific name takes priority. Many years

Allen's Hummingbird

later, when Allen had come to know the species better, he wrote to Major Bendire telling him that:

"Allen's Hummingbird arrives in the vicinity of Nicasio, California, about the middle of February, and commences to nest soon after arrival. The earliest date on which I found one was February 27, 1879; this was then about half finished, when a heavy storm set in which lasted about five days, and I did not visit the locality again until March 8, when the nest was completed and contained two fresh eggs. I have taken their nests as late as July 3, and am well convinced that two broods are raised in a season, at least by all of the earlier breeding birds. They select all sorts of situations and various kinds of trees and bushes to nest in. I have found their nests as low as 10 inches and again as high as 90 feet from the ground."[2]

Allen sent other observations to Major Bendire and the notes were incorporated into A.C. Bent's *Life Histories of North American Birds* (1919–1968). Most of Allen's information was about the nesting habits of birds, among them the White-headed, Hairy and Downy Woodpeckers, Red-breasted Sapsucker, Vaux's Swift and Hermit Warbler. Allen was the first naturalist to find a Hermit Warbler nest and he took three altogether, in Blue Cañon; all were built in pine trees, high up on thick, scraggy limbs and so well hidden that he could only locate them by patiently watching the birds back to their nests.

His notes on bird behaviour reveal a deep prejudice against the corvids, especially the Scrub Jay. He blamed it for robbing the nests of small birds, stealing his hens' eggs as fast as they could lay them, destroying vast quantities of fruit and even pulling up young wheat. Allen shot them on sight.

In the early 1880s Allen was given a cottage and about an acre of land on the Mailliard's San Geronimo Ranch. Here he continued his collecting and took on various seasonal jobs, sometimes as a fire guard in the summer and patrolman during the deer season. The brothers John and Joseph Mailliard were amateur ornithologists who benefited, as youths, from Allen's considerable experience. In an autobiographical paper Joseph later offered a few remarks on the old collector:

"Allen could hardly be called an ornithologist and he had not had much opportunity for education. He was not a student, but an excellent collector who became well versed in the habits of the birds and other animals with which he came in contact. He certainly was a handy and interesting man to be with in camp, and a pleasant companion in the field.

The labels which Allen attached to his specimens are apt to be misleading as regards the exact locality in which the specimens were obtained, especially as anything taken in the township of that name was labeled 'Nicasio' in the old days, although possibly collected in an entirely different valley from that in which the town is situated. Allen dried his bird skins in a paper cylinder, to make which he invariably heated up a pot of glue for pasting together the edges of the paper. On this cylinder were written merely the sex of the specimen and the day of the month, but no label was attached to the skin until it was sold. The disposition of it might be years after it was put away, so that the exact locality and the year of taking became only matters of memory on his part."[3]

Towards the end of his life Allen became very frail and forgetful. On one occasion Joseph Mailliard visited him and was appalled to discover that the old man had destroyed all his correspondence with other naturalists, probably not realizing that it would be of interest in the future. During his last few years he was confined to bed and died at the age of eighty-eight on 29 June 1930.

Mary Virginia Anderson (1833-1912)

VIRGINIA'S WARBLER *Vermivora virginiae* (Baird)

Helminthophaga virginiae Baird, 1860. In S.F.Baird, J.Cassin and G.N.Lawrence, *Birds of North America*, p. vi, Atlas, Pl. 79, Fig. 1: Cantonment [= Fort] Burgwyn, N.M.

The original description of Virginia's Warbler appeared in *Birds of North America* (1860) and was given by Professor Baird who wrote:

> "*Helminthophaga Virginiae* BAIRD.—Similar in general appearance to *H. ruficapilla* [Nashville Warbler]. Top and sides of head, back and wings light ashy plumbeous, with an almost imperceptible wash of olivaceous green; quills and tail feathers brown, edged with pure ashy plumbeous; the latter indistinctly and narrowly margined with whitish internally and at the end. Rump and upper and lower tail coverts bright yellow (with a greenish tint above) in vivid contrast to the rest of the body. Crown with a concealed patch of orange brown. Rest of under parts brownish white with indications of yellow along the median region (perhaps entirely yellow when mature). Inside of wings and axillars whitish. A white ring around the eye. Length 5 inches; extent 7¼; wing 2½. No. 10719. Fort Burgwyn, N.M. Collected by Dr. W.W. Anderson, U.S.A."

Baird had received the type specimen in 1858 from Dr William Wallace Anderson, Assistant Surgeon in the U.S. Army, with the request that it be named after his wife Virginia.

Virginia Anderson was no stranger to army life, for she was the daughter of General Thomas Childs of Pittsfield, Massachusetts, who had served with considerable distinction in the war with Mexico but had died of yellow fever in 1853 while commanding military operations in East Florida. Though Virginia was one of nine children, only two others survived to adulthood; her sister Catharine (who became the wife of General P. Woodbury) and her brother Frederick.

Virginia was twenty-two when she married William Anderson at Wilmington, North Carolina, on 27 December 1855. Her thirty-one-year-old husband had served with the army in Texas prior to the wedding and soon afterwards he was

ANDERSON. *WIFE OF DR W.W. ANDERSON WHO DISCOVERED VIRGINIA'S WARBLER. SHE ACCOMPANIED HER HUSBAND TO FORT BURGWYN, NEW MEXICO, WHERE HE COLLECTED BIRDS FOR THE SMITHSONIAN INSTITUTION. THE PHOTO-GRAPH IS FROM A GOLD LOCKET IN THE POSSESSION OF THE FAMILY.*

again ordered West, to New Mexico, travelling there with his bride via New Orleans and the Arkansas River. As Virginia was an accomplished singer and musician she took her piano with her and though it suffered considerable shaking in a covered wagon and a soaking in a ford Dr Anderson was somehow able to restore it almost to its original state after they arrived at Fort Burgwyn.

Virginia's musical talents must have ensured her popularity at this isolated post in the Sangre de Cristo Mountains (ten miles south of Taos). Their remote location held great attractions for her husband whose interests were both botanical and ornithological. As well as the historical Virginia's Warbler, he sent Baird other birds, among them Steller's Jay, Scrub Jay, Clark's Nutcracker, White-tailed Ptarmigan, Townsend's Warbler and Broad-tailed Hummingbird. Some were collected at Fort Burgwyn and others originated in the Cochetopa Pass, Colorado, where he travelled with Captain R.B. Marcy. In the summer of 1860 Dr Anderson was sent to Texas where he continued to supply Baird, who wrote to him on 15 October 1860 with typical enthusiasm:

"My dear Doctor:
    On my return home the beginning of this month I found the two boxes from you: one from Burgwyn, and the other containing your Texas things. I found a great many rare birds and other specimens in both, but the Texas box was especially important in having, among other specimens, one of the only species of hawk wanting to our collection, namely the Falco femoralis [Aplomado Falcon]. Think of that! There were also eggs of two hawks new to our collection, Buteo swainsoni [Swainson's Hawk] and Accipiter mexicanus [Gray Hawk]. Think of that again!
    In the Burgwyn box were the second and third specimens of the Colorado Raven besides such varieties as Cassin's Purple Finch, Maximilian's Jay [Pinyon Jay], etc.
    I really don't see how we are to get along without you at Burgwyn. It is certainly the greatest centre of rare birds in the United States. I would write more but I do not know whether or how soon you will get this letter. I will send it to the Surgeon General's Office. Let me know where you are and I will write again."[1]

Dr Anderson's western sketchbook, which has been preserved, contains the only known picture of Fort Burgwyn, drawings of a number of birds and some exquisite miniatures of his children. The eldest, Woodbury, was born in 1857, Thomas in 1858 and Elizabeth and William in 1859. Years later, Virginia entertained the family with stories about the Indian squaws who called at her quarters and sat around her in a circle, watching in amazement as she bathed, oiled and powdered her little white babies.

In 1861, when South Carolina seceded from the Union, Dr Anderson was based at San Antonio under the command of Robert E. Lee. The Doctor immediately resigned from the U.S. Army and returned East with his family in order to serve with the Confederacy. Initially commissioned as a Surgeon, his responsibilities rapidly increased until he was made 'Medical Inspector and Superintendant of Vaccination of the Armies, Hospitals and Camps of Instruction of the Confederate States'—a position second in importance only to that of the Surgeon General. Virginia and her children moved into Borough House in Statesburg, South Carolina, the home of her elderly father-in-law and his second wife, who was paralysed.

Virginia's Warbler

The war years brought much grief to the Anderson family; in 1862 William's younger brother, Edward, was killed at the battle of Williamsburg and in 1863 his sister Mary died. Mary Anderson had married Virginia's brother, Frederick Childs, after a romance that began at William and Virginia's wedding. Dr Anderson Senior died in May 1864, leaving the house occupied solely by women and children. Virginia, then pregnant with her fifth child, went to Fayetteville Arsenal in North Carolina where her brother Frederick was the Commanding Officer so that she could care for his small children. Her own baby, Ann Catherine, was born there in October and shortly afterwards the arsenal was successfully besieged by General Sherman. Before the war Sherman had been a close friend of both the Childs and Anderson families and he was overheard to remark in a resigned tone of voice, "If I catch him, I suppose I'll have to hang him." He was spared the dilemma for Frederick made good his escape as soon as he realized defeat was inevitable. Virginia was too weak to travel at the time, but after a few weeks she asked for a safe conduct pass from Sherman, who told her she was a "damn fool" but gave in to her request. She walked out of the building with her baby in her arms and most of her brother's silverware under her hooped skirt and returned safely to Borough House.

At the end of the war each Confederate soldier who still owned a mount was allowed to keep it—Dr Anderson was allotted an old ambulance and during the lean years of the Reconstruction period, when the Andersons fared forth it was in this dilapidated carriage. The old house in the Santee Hills, which had stood for over a century, survived the war, though it had been rifled and ransacked by General Potter's Negro army. Mrs Anderson Senior, who had refused to leave her house, died soon after peace was made and so William and Virginia made their home there, the Doctor taking up his father's old practice. In that impoverished community he was rarely paid in hard cash. He more often received a pig, a chicken, eggs, corn, sweet potatoes or some other payment in kind—or nothing at all. Under the new laws he was forced to sharecrop the plantation with the one hundred remaining ex-slaves. Virginia gave birth to another four children; Frederick in 1866, William in 1869 (the first William Junior having died in infancy), Mary Virginia in 1872 and Benjamin in 1875.

In 1875 Virginia and her husband were instrumental in founding the Stateburg Literary and Musical Society, which brought the neighbours together once a month to enjoy musicals, plays, poetry readings, dancing and feasting. There is an account of one meeting which began at ten in the morning and finished after midnight, suggesting that the hardship of the times had not completely quenched their capacity for enjoyment. For over thirty years Virginia was the organist at the nearby Episcopal Church of the Holy Cross where she and Dr Anderson were both much involved.

From the end of the war until his death in 1911 Dr Anderson made regular meteorological observations, which he sent to the Weather Bureau, continuing a practice that his father had begun. He also took an interest in his superb garden with its wide sloping lawns, tree-shaded walks and glorious vista across the fields to the Wateree Swamp. Some of the plants and young trees he cherished with particular care, having sent the seeds back to his family while stationed in New Mexico and Texas, but there is no indication that he actively pursued his former interest in birds.

In William and Virginia's old age the house was often full of visiting grandchildren. Of their own seven children who reached adulthood, only Ann Catherine remained in Stateburg after marriage. When her parents became too feeble to live alone at Borough House she and her family moved in to care for them. In the early 1900s Virginia had to have a gallbladder operation but recovered her health and lived on until 1912, leaving her family with precious memories of a very gracious and loving old lady.

AUDUBON. ARTIST AND NATURALIST FAMED FOR THE BIRDS OF AMERICA. THIS PICTURE WAS PAINTED IN 1848 BY HIS SONS, VICTOR AND JOHN.

John James Audubon (1785-1851)

AUDUBON'S SHEARWATER *Puffinus lherminieri* Lesson

Pufflnus [sic] *Lherminieri* Lesson, 1839. *Revue Zoologique* [Paris] 2, p. 102: ad ripas Antillarum [= Straits of Florida]

AUDUBON'S ORIOLE *Icterus graduacauda* Lesson

Icterus graduacauda Lesson, 1839. *Revue Zoologique* [Paris] 2, p. 105: Mexico

Since Audubon's death his accomplishments have been eulogized and his life romanticized more than that of any other American naturalist. One of his many skills was the ability to sell not only his books, but himself, and the image he promoted has caught the imagination of succeeding generations. But along with the paintings, the myths and the ornithological discoveries we have inherited another legacy in the names of the North American birds. Twenty of the species illustrated in this book—nearly one sixth of the total—owe their scientific or vernacular names to Audubon: and thus have Bachman, Baird, Bell, Bewick, Brewer, Harris, Henslow, LeConte, Lincoln, MacGillivray, Nuttall, Smith, Sprague, Swainson, Townsend and Traill been 'immortalised' in eponyms.

The tradition of naming birds after people was well established by Audubon's day and many ornithologists had taken the opportunity to flatter wives, sweethearts, friends, patrons and employers as well as deserving naturalists. But no other ornithologist ever indulged in the habit as indiscriminately as Audubon. Audubon proposed ninety-one new species of bird[1] and named fifty-seven of them after people. He often honoured the same people more than once: Townsend nine times, Bachman four times, Harris three times and Brewer, MacGillivray, Morton, Nuttall, Stanley and Trudeau twice. Only about one-third of his proposed names are now in use, as many of his 'new species' had already been described by other naturalists or were birds in female or immature plumage that he failed to associate with the very different males, while some have since proved to be hybrids, of doubtful status, or to have come from outside North America.

If all of Audubon's eponyms *had* survived then they would celebrate another nine Americans (Dr William O. Ayres, Major Joseph Delafield, Dr Richard Harlan, Miss Maria Martin, Dr Samuel Morton, Dr George Parkman, Dr George Shattuck, Dr James Trudeau and George Washington) and another thirteen Europeans (Charles Bonaparte, John George Children, Baron Cuvier, Dr Meredith Gairdner, Robert Havell, Dr Benjamin Phillips, the Rathbone family, Sir John Richardson, William Roscoe, Prideaux J. Selby, Lord Stanley, Nicholas Vigors and William Yarrell). Only a few of these people made a significant contribution to American ornithology; most were European naturalists or simply friends or acquaintances to whom Audubon owed favours.

Audubon's Shearwater

The story of Audubon's life began on 26 April 1785, when Jeanne Rabine, a servant girl from Nantes, gave birth at Captain Jean Audubon's plantation home. Six months later she succumbed to the fevers and infections prevalent around the port of Les Cayes on the Caribbean island of San Domingo (Haiti). At the age of six the boy and his half-sister sailed to Nantes, on the west coast of France, where they spent the rest of their childhood with their father, Captain Audubon, and his childless wife Anne, who adopted them as her own.

In 1803 Captain Audubon sent his son to oversee his plantation at Mill Grove, near Valley Forge, Pennsylvania, but the youth showed little flair for farming or any other conventional method of earning a living. Audubon's enthusiasm was reserved for watching and sketching birds and courting a young neighbour, Lucy Bakewell of Fatland Ford, whom he married in April 1808. They set up home in Louisville, Kentucky, a port on the Ohio where the first of their two sons, Victor Gifford Audubon, was born in June 1809.

One day in March of the following year, Audubon was behind the counter of his store when Alexander Wilson walked in, carrying the first two volumes of his *American Ornithology*. Audubon chose not to subscribe but he showed Wilson his own crayon drawings and two days later they went out hunting together. Wilson then continued on his way to New Orleans, leaving Audubon fired up with the ambition to see his own artwork published.

Soon after this memorable encounter Audubon moved downriver to Henderson, Kentucky, where he continued storekeeping for another nine years, but he often left a partner in charge while he went out hunting or was engrossed in stuffing and drawing his birds. In November 1812 John Woodhouse Audubon was born and two girls followed, in 1815 and 1819, but both died in infancy. In the summer of 1819 Audubon was forced to leave Henderson after unwise investments made him bankrupt. During the next few years he had to grab at any chance of earning a living; he practised taxidermy at the Western Museum in Cincinnati, drew chalk portraits in New Orleans, taught art, music and dancing to young ladies in Louisiana, painted signs and murals in Shippingport and built up his portfolio of lifesize bird paintings at every opportunity.

In 1824 he decided that the time was ripe for him to seek scientific recognition and a publisher. Leaving Lucy to support herself and the boys by teaching at *Beech Grove* plantation, Louisiana, he set off for the east coast. In Philadelphia, Charles Bonaparte introduced him to the Academy of Sciences, but most of that powerful fraternity rejected Audubon and his art. In New York, publishers responded with such indifference that he realized he would have to find support in Europe. However, he made two important contacts before leaving the United States: Edward Harris, who became a loyal friend and benefactor and Dr Richard Harlan, a young zoologist, who offered to exchange specimens and information.

Between the summer of 1826 and the autumn of 1839 Audubon spent over eight and a half years in Great Britain during four prolonged visits. The first occasion was marked by his triumphant exhibition in Liverpool and the favour of many eminent and influential individuals: Lord Stanley, Dr Traill and the Roscoes and the Rathbones on Merseyside; Sir William Jardine and Prideaux J. Selby in Edinburgh; Thomas Bewick in Newcastle; Professor Henslow in Cambridge; and William Swainson, J.G. Children and Nicholas Vigors in London. Swainson was particularly helpful. An accomplished bird artist himself, he wrote a glowing review of Audubon's work for *Loudon's Magazine*, provided a lamb from his farm for the oil painting 'Eagle and Lamb' and went with his friend to Paris, where they met Baron Cuvier and the Duke and Duchess of Rivoli. Within the year Audubon had made a contract with Robert Havell and his son, in London, who engraved the plates for *The Birds of America* between 1827 and 1838.

During Audubon's second visit from April 1830 to August 1831 he engaged William MacGillivray at Edinburgh to help him in writing the *Ornithological Biography* (1831–39) which was the text to *The Birds of America* but published separately to keep down expenses.

When Audubon returned to the States his pressing need was for more birds and so in the fall he optimistically headed for Florida. On the way south, in Charleston, he met the Reverend John Bachman who quickly became one of his most supportive and intimate friends. After ascending the St John's River in a naval cutter Audubon wrote despondently about the stinking river, the lack of birds, the dreary sand barrens and the impenetrable, alligator-infested swamps. The expedition became more worthwhile in the spring when he cruised through the Florida Keys and the Dry Tortugas, marvelling at the turtles and the breeding seabirds.

The Audubon family spent the summer of 1832 in New England and New Brunswick. At Boston, Audubon sought out Thomas Nuttall, author of *A Manual of the Ornithology of the United States and of Canada* (1832–34) and they enjoyed an instant rapport. The Audubons then sailed to Eastport, Maine, and stayed at nearby Dennysville in August. Their host was Judge Lincoln whose son, Thomas, led Audubon, Victor and John on some exciting hunts into the surrounding forests. Winter was spent in Boston, preparing for a major expedition to the coast of Quebec the following spring. Dr George Parkman, a leading Bostonian, became a firm friend and lent Audubon some of the funds needed to charter a schooner; Audubon responded with gratitude by naming a western House Wren *Troglodytes Parkmanii*. Another of Parkman's debtors was less grateful; in 1849 the doctor was murdered by Professor John White Webster, to whom he had lent a substantial sum of money.

On the expedition to Canada Audubon painted many of the northern birds for the first time but discovered just one species new to science, Lincoln's Sparrow. He named it for young Tom Lincoln, who had come along with John Audubon, George Shattuck and two other youths (see Lincoln).

In the fall of 1836, after returning from his third trip to Britain, Audubon struck gold when the Philadelphia Academy parted with a selection of west coast birds collected by Thomas Nuttall and John Kirk Townsend. Audubon drew the skins during the winter while staying with Bachman's hospitable family in Charleston. He was aided by Maria Martin, the pastor's sister-in-law, who worked on other parts of the compositions, providing scenery, flowers and insects. Identifying and describing birds that he had never seen alive posed considerable problems for Audubon and one species that he found especially confusing was the Hairy Woodpecker. He presented it as five species, three of which he named himself: Maria's Woodpecker *Picus Martini* to please his skilful assistant; Harris's Woodpecker, *Picus Harrisi* for his New Jersey friend; and Phillips's Woodpecker *Picus Phillipsi* after his family physician in London.

In the spring Audubon sailed along the Gulf coast from New Orleans to Galveston, accompanied by John again and by Edward Harris (see Harris).

Audubon's last two years in Britain saw the fulfilment of his artistic vision, but his longing for some reasonable financial reward was frustrated until profits accrued from his octavo edition (1840–44). By then he had started another ambitious venture, *The Viviparous Quadrupeds of North America* (1845–54), for which the text was provided by Bachman.

At the end of 1841 Audubon purchased twenty-four wooded acres on the north of Manhattan Island and had a wooden house built on a slope overlooking the Hudson River; *Minnie's Land* was the family home for the rest of his life. In the spring of 1843 he made his last expedition, to the Upper Missouri, with the aim of collecting more mammals and information about their habits. He set off from Philadelphia in March, with Edward Harris, John G. Bell and Isaac Sprague. At Baltimore he visited Gideon B. Smith, his agent in that city, who afterwards received lengthy letters regarding their progress (see also Harris, Bell and Sprague). As on his other expeditions, Audubon kept the overland travelling to a minimum and used river steamers between Pittsburgh and Fort Union. Though the party returned in the fall with fewer new mammals than Bachman had hoped for, Audubon painted some more bird species for the first time that summer, among them Harris's Sparrow, Bell's Vireo, Sprague's Pipit, Smith's

Longspur, Le Conte's Sparrow, Western Meadowlark, Brewer's Blackbird and Common Poorwill, all of which he included in his last octavo volume.

He illustrated a large portion of the mammals collected on his Missouri River trip, but by 1846 his sight was failing so badly that he had to completely hand over the remaining artwork to John, while Victor continued to manage the business affairs and the editing of the manuscript. The following spring, around the time of his sixty-second birthday, Audubon suffered a slight stroke and thereafter his faculties rapidly diminished. When Bachman visited in May 1848 he was shocked by his friend's senility. Less than three years later Audubon died at *Minnie's Land*, on 27 January 1851.

Audubon's propensity for eponyms was such that it would be strangely ironic if no-one had honoured him in this way. Audubon's Shearwater and Audubon's Oriole were first described by the French ornithologist R.P. Lesson but Audubon's name was not associated with either until some years later. In 1884 Baird, Brewer and Ridgway suggested that *Puffinus lherminieri* should be called Audubon's Dusky Shearwater; Audubon had watched the petrel in the Gulf of Mexico but illustrated it under the name of *Puffinus obscurus*. His name was first attached to Audubon's Oriole in 1841, in a paper in the *Annals* of the New York Lyceum of Natural History written by the treasurer of the society, Jacob P. Giraud.

The Crested Caracara was formerly known as Audubon's Caracara, owing to John Cassin's appellation *Polyborus Audubonii* (1865), but both the scientific and the English names have been revised since then. However, another Audubonian name, that of Audubon's Warbler, is still much used. It distinguishes the western race of the Yellow-rumped Warbler from the northern and eastern race, otherwise known as the Myrtle Warbler. *Sylvia Auduboni* was Townsend's invention for a warbler he collected in the forests along the Columbia River while the 'American Backwoodsman' endured two long years in British cities.

Audubon's Oriole

BACHMAN. *LUTHERAN PASTOR AND NATURALIST. HE WAS THE FIRST TO DESCRIBE MANY NORTH AMERICAN MAMMALS AND WROTE THE TEXT OF* THE VIVIPAROUS QUADRUPEDS *(1845–54). THE PORTRAIT IS BY HIS SON-IN-LAW, JOHN WOODHOUSE AUDUBON.*

John Bachman
(1790-1874)

BLACK OYSTERCATCHER *Haematopus bachmani* Audubon

Haematopus Bachmani Audubon, 1838. *The Birds of America* (folio), Vol. 4, Pl. 427, Fig. 1 (1839, *Ornithological Biography*, Vol. 5, p. 245): Mouth of the Columbia River

BACHMAN'S WARBLER *Vermivora bachmanii* (Audubon)

Sylvia Bachmanii Audubon, 1833. *The Birds of America* (folio), Vol. 2, Pl. 185, (1834, *Ornithological Biography*, Vol. 2, p. 483): a few miles from Charleston [= Edisto River], in South Carolina

BACHMAN'S SPARROW *Aimophila aestivalis* (Lichtenstein)

Fringilla aestivalis Lichtenstein, 1823. *Verzeichniss der Doubletten des Zoologischen Museums der Koenigliche Universitaet zu Berlin ...*, p. 25: Georgia

The Reverend Doctor John Bachman's association with the above-named birds came about through his friendship with John James Audubon who inspired him to devote more of his time to ornithology. As a result, Bachman discovered two new warblers in the woods around his home at Charleston, South Carolina, and they were added to Audubon's *The Birds of America* (1827–1838). Bachman, in turn, instructed Audubon in the study and identification of mammals and provided the text for *The Viviparous Quadrupeds of North America* (Audubon and Bachman, 1845–1854). *The Viviparous Quadrupeds* became a family enterprise since Audubon's two sons, who assisted their father in its production, each married a daughter of Bachman's: in 1837 John Woodhouse Audubon married Maria Bachman and in 1839 Victor Gifford Audubon married her younger sister, Eliza Bachman.

The Bachmans were of Swiss-German descent. John Bachman was born at Rhinebeck, New York, on 4 February 1790, the son of a prosperous farmer who owned numerous slaves. Bachman later attributed his interest in natural history to one of the slaves who often took him, as a small boy, into the woods to trap birds and animals and later helped him to trap enough furs to pay for his first natural history books. Thus from the very beginning Bachman approached his study of nature through both field craft and book learning.

Bachman chose a legal career but tuberculosis interfered with his plans and during a spell of illness he felt called to enter the ministry. During his theological training he had to teach to earn his living and spent a year at a school in Pennsylvania. On a visit to Philadelphia he met up with Audubon's forerunner, Alexander Wilson, who persuaded him to move to his old school at Milestown. There, Bachman taught classes in Latin, French and German for a year and sometimes spent his Saturdays hunting for birds with Wilson. The next year Bachman taught in Philadelphia and in his spare time collected some of the less common birds for his friend, but in 1813 Wilson suddenly died, with his *American Ornithology* (1808–1814) incomplete. That same year, Bachman was licensed as a pastor in the Lutheran Church and returned to Rhinebeck, where he served for the next two years, until called to St. John's Church in Charleston, South Carolina.

His new congregation in Charleston, composed of both blacks and whites, met in a small wooden church three times every Sunday and once during the week. One of the parishioners described the new minister: "His height was medium; his figure slender; his complexion fair; features regular and eyes blue … His voice was strong, clear and sweet." First impressions were apparently favourable, but over the years it was Bachman's unshakable faith, his compassion and very likely his unquenchable sense of humour that endeared him to his flock. It was a rule with him that his duty to the church and the needs of his family were given priority; his scientific work had to be fitted into his limited hours of leisure and was never allowed to become an obsession.

Soon after his arrival in Charleston he married Harriet Martin and their first child, Maria, was born within the year. By the autumn of 1831, when Audubon met the family for the first time, there were six daughters and two sons, five other children having died in infancy. An important member of the family was Harriet's sister, Maria Martin (whom Bachman referred to as his sister) for she not only helped to look after the children but also shared her brother-in-law's interests and was a botanical and zoological artist.

Audubon's visit marked the beginning of a joyful friendship between the two men and the two families. He arrived in Charleston on 16 October, on his way to Florida, accompanied by George Lehman, a landscape painter and Henry Ward, a taxidermist. Audubon gave an account of his first meeting with Bachman in a letter to his wife Lucy, relating how on his second day in the town he went searching for cheaper lodgings accompanied by a local clergyman to whom he had been given a letter of introduction. They met Bachman in the street— and Bachman immediately insisted that the travellers should come and stay with him for the next three weeks.

"We removed to his house in a crack," wrote Audubon, "found a room already arranged for Henry to skin our birds—another for me and Lehman to draw, and a third for thy husband to rest his bones in on an excellent bed! Mr. Bachman!! Why Lucy, Mr. Bachman … would have us to make free there … as if his heart had been purposely made of the most benevolent materials granted to man by the Creator to render all about him most happy."[1]

During the day Bachman took his guest all around the countryside in his carriage and in the evenings they worked companionably side by side in a study overflowing with books and journals, stuffed birds and animals, pet songsters and caged rodents. Audubon resumed his journey in the middle of the month and Bachman wrote to Mrs Audubon:

Bachman's Sparrow

"The last has been one of the happiest months of my life. I was an enthusiastic admirer of nature from my boyhood, and fond of every branch of Natural History. Ornithology is, as a science, pursued by very few persons—and by no one in this city. How gratifying was it, then, to become acquainted with a man, who knew more about birds than any man now living—and who, at the same time, was communicative, intelligent, and amiable, to an extent seldom found associated in the same individual. He has convinced me that I was but a novice in the study; and besides receiving many lessons from him in Ornithology, he has taught me how much can be accomplished by a single individual, who will unite enthusiasm with industry. For the short month he remained with my family, we were inseparable. We were engaged in talking about Ornithology—in collecting birds— in seeing them prepared, and in laying plans for the accomplishment of that great work which he has undertaken. Time passed rapidly away, and it seems but as yesterday since we met, and now, alas! he is already separated from me—and in all human probability we shall never meet again."[2]

Bachman was quite wrong in this gloomy prediction, for Audubon was back at Charleston within two years. In the meantime they corresponded enthusiastically and a year later, on 20 October 1832, Bachman wrote:

"Dear Audubon … My sister Maria, has made several drawings, which she thinks of sending you; but I am anxious to retain them for awhile, in hopes that you may be tempted to come for them yourself. Ever since you left us, I have been studying up my Ornithology, in order to be useful to you, and, if I am spared, I hope to be so. A month in your society would afford me a greater treat than the highest prize in a lottery. I cannot, I find, feel myself at home with new birds without having the skins to refer to. My cabinet is enlarging every day. Henry Ward now prepares the skins–a pair of each ... I am afraid that our Northern Sylvias do not pass near our sea-coast; I rather think that they follow the mountains; the "Henslow's Bunting" is not rare here; I killed three yesterday, and saw, at least, a half dozen. I shall, next week, write all I know about the *Fringilla* I found last spring."[3]

The new '*Fringilla*' is now known as Bachman's Sparrow, though Audubon called it Bachman's Pinewood-Finch, not realizing that it had already been described by Hinrich Lichtenstein from a Georgian specimen. However, two warblers that Bachman shot were indeed hitherto unknown to science. One was Swainson's Warbler, which he found in a muddy swamp near the Edisto River after being attracted by an unfamiliar whistle. The other was Bachman's Warbler. In his account of the Bachman's Warbler, quoted here, Audubon gave the discoverer full credit:

"My friend BACHMAN has the merit of having discovered this pretty little species of Warbler, and to him I have the pleasure of acknowledging my obligations for the pair which you will find represented in the plate, accompanied with a figure of one of the most beautiful of our southern flowers, originally drawn by my friend's sister, Miss MARTIN. I myself have never had the good fortune to meet with any individuals of this interesting Sylvia, respecting which little is as yet known, its discoverer having only procured a few specimens of both sexes, without being able to find a nest. The first obtained was found by him a few miles from Charleston, in South Carolina, in July 1833, while I was rambling over the crags of Labrador. According to my amiable friend, it was 'a lively active bird, gliding among the branches of thick bushes, occasionally mounting on the wing and seizing insects

Bachman's Warbler

in the air in the manner of a Flycatcher. It was an old female that had to all appearance just reared a brood of young.' Shortly after, several were seen in the same neighbourhood; and we may still expect an account of its manners, migration, and breeding, although not yet discovered."[4]

It was not in Bachman's nature to be content just with collecting birds, however useful that might be to Audubon. He investigated the diet, seasonal movements, nests and eggs of many different species and his observations are liberally scattered throughout the *Ornithological Biography*. In 1833 he wrote an essay on a subject which fascinated him, 'The Migration of the Birds of North America', which was published by the *American Journal of Science and the Arts* in April 1836. In it he demonstrated that his understanding of the phenomenon was unusually advanced. He was aware of the ability of birds to fly for long periods and to cover immense distances, sometimes over sea, and realized that the young of some species migrate independently of their parents. He knew that birds have some innate sense which guides them and pondered the nature of instinct. He also suspected, correctly, that migration is prompted more by the necessity for an abundant food supply than by the avoidance of severe weather.

Through his spiritual eye he perceived birds as a marvellous part of creation, their beauty and complexity inspiring him to worship and praise their Creator. His philosophy was in harmony with the age in which he lived, and in some ways, ahead of his time, for he had, in essence, a modern concept of ecology. As a horticulturalist he endeavoured to teach others the importance of having a broad understanding of ornithology, chemistry, entomology and botany because all "are closely allied to, and inseparably connected with, the science of horticulture". He realized that ignorant interference with the balance of nature could do more harm than good and by way of example he cited the following stories:

> "The Purple [Common] Grackle, in New England, was destroyed in consequence of the Governor's offering three-pence per head; and the result was, that the insects multiplied so rapidly, that the herbage was destroyed, and the inhabitants were obliged to obtain hay from Pennsylvania and England. The poor Woodpecker is shot by every idle boy, because he is said to extract the juices of apple trees; when, in most cases, he is attracted there by the worm which is perforating the tree; and thus the bird on which the sentence of death is pronounced as an enemy, has come to save the tree by feeding on its destroyer."[5]

In the late summer of 1833, after his return from the Gulf of St. Lawrence, Audubon travelled overland to Charleston, taking his wife with him. Their son John was sent on ahead by sea with all of the baggage, and arrived first. When they were reunited at the beginning of October it was obvious that John was already a keen admirer of sixteen year old Maria Bachman and by the spring, he and Maria considered themselves to be engaged although no formal agreement had been reached between their parents. During the winter Audubon made good progress with his bird plates, with Maria Martin and John both helping to provide the backgrounds. The visitors departed in March, taking with them four Northern Mockingbirds and two vultures from the Bachman aviary. In April, Audubon and Lucy sailed to Britain, where they stayed for the next two and a half years overseeing the tedious and frustrating process of publishing the third volume of *The Birds of America*, and writing the text (which was published separately as the *Ornithological Biography*) in collaboration with William MacGillivray.

During these years, from 1834 to 1836, Thomas Nuttall and John Kirk Townsend were making exciting ornithological discoveries in the Rocky Mountains and on the west coast. As soon as Audubon returned to the States he sought out his friend Nuttall, now back in Boston, and persuaded him to part with some of his bird collection. With typical generosity, Nuttall also gave him some butterflies to pass on to Bachman, but insisted that the Philadelphia

Academy, who had sponsored the expedition, would have to decide whether or not Audubon could have access to the other specimens.

With the help of his wealthy friend Edward Harris, Audubon was able to purchase ninety-three bird skins and triumphantly headed south to spend another winter in Charleston, taking John with him again. They found Bachman preoccupied with the needs of his congregation, following an epidemic during the summer and fall. His strength had been taxed to the utmost in ministering to the sick and the bereaved and worse was still to come. During the winter, fire swept through the city, leaving many of his people homeless. The visit of his friends was therefore a much needed tonic. Once again, Audubon painted intensively, working through Nuttall and Townsend's collection. The Black Oystercatcher he named Bachman's Oystercatcher, *Haematopus Bachmani*, in honour of his host. Maria Martin and young John again helped with the floral or scenic settings and by mid-February seventy-six drawings had been finished.

Audubon's next ploy was an expedition to the Gulf coast, accompanied by his son and Edward Harris, who arrived in time for them all to leave together on 17 February. The Bachmans waved farewell to their friends knowing that the parting would be brief, since the wedding of John and Maria was to take place in the summer.

That spring, Bachman made one of his infrequent journeys and wrote to his wife on 25 May 1837, telling her about some of his activities while in Philadelphia:

"I spent most of my time at the Academy of Natural Sciences and the Philosophical Hall. I found Nuttall friendly—after all there is as much in the manner of men to attract interest, as there is in the spices that render food palatable. In the Philosophical Hall, I frequently took a seat in the old chair of Dr. Franklin; I could not avoid thinking, if knowledge could be communicated in this short way, by touch or sympathy, what a world of Philosophers Franklin's old chair would have produced!

I was invited yesterday, to meet old General Clarke, the companion of Lewis. He is now very aged and in failing health; he is on a visit to Philadelphia, accompanied by an interesting and beautiful family. I was quite pleased with him. He is intimately acquainted with Mr. and Mrs. Audubon, and spoke in raptures

Black Oystercatcher

of her talents, and his beautiful taste. He seemed to retain his recollection of past events in a remarkable way. Old Dr. Kurtz and his family send kind regards. Love to your mother, to the children, and to all dear friends."[6]

Bachman returned to Charleston in time for the wedding, following which the young couple left for a honeymoon at Niagara before travelling on to England with Audubon. Bachman was saddened by his daughter's departure and was in such poor health that he neglected natural history throughout the whole summer. He had still not recovered by the following spring and was encouraged by the church to take a long period of leave and relaxation. Invited to address an international gathering of naturalists at the German town of Freyburg in September, he decided to spend his vacation in Europe and take the opportunity of visiting his daughter.

He arrived at the Audubon's London home on the day that Maria gave birth to his first grandchild, but after a short rest he went on to Edinburgh for three weeks, where Audubon showed him the sights. He then returned to London for the christening of little Lucy and a week of study at the British Museum and the Zoological Garden. In mid-August he continued on his way to the congress, gave his address in German on the progress of the study of natural history in the United States and greatly enjoyed meeting so many other naturalists. He also appreciated the honour of being made a Doctor of Philosophy by Berlin University. In January 1839 he returned to South Carolina after an absence of eight months, stimulated and mentally refreshed by his travels but still so lacking in vigour that the church gave him an assistant.

Soon after Bachman's return Audubon's elder son, Victor, arrived for a visit and fell in love with Eliza. In common with others in her family Eliza was consumptive and her delicate health worried her father. Nevertheless, he allowed the couple to marry at the end of the year, though Eliza's departure caused him considerable worry and grief.

His misgivings were not untoward. In February John brought Maria home to Charleston with their second daughter, Harriet; young Lucy being left with Grandmother Audubon. Maria's health had deteriorated rapidly since the birth of her second baby and she needed more than the tender loving care of her family to restore her strength. She died in September, aged twenty-three, and the condition of her sister Eliza worsened at around that time; eight months later Eliza died in New York. To make matters worse, a breach developed between Audubon and Bachman. Audubon had visited Charleston with the expectation that the household would provide the balm and cheerfulness which he always sought there and instead found the family in anticipatory mourning for Maria. By resorting to the whiskey bottle for consolation, Audubon distressed the whole family and deeply offended his old friend, who had a perfect hatred of drunkenness. Fortunately, Victor was able to heal the rift and their old, easy relationship seems to have been restored. It was as well that it was, for their collaboration on *The Viviparous Quadrupeds* was just beginning.

After studying mammals for all of his life, Bachman, now at the age of fifty, was acutely aware of the difficulties that lay ahead. He warned the Audubons: "the birds are a mere trifle compared with this". He knew only too well that many of the American quadrupeds had never been studied and that books would be of very limited help to them. Already he had described twelve new mammals in his preparatory research: Southeastern Shrew, Townsend's Mole, Marsh Rabbit, Nuttall's Cottontail, Swamp Rabbit, White-tailed Jack Rabbit, Least Chipmunk, Townsend's Chipmunk, Townsend's Ground Squirrel, Douglas's Squirrel, Townsend's Vole and Creeping Vole. Bachman foresaw that the taxonomic status of certain species and subspecies, especially the deer and the squirrels, would cause enormous difficulties. Procuring skins and illustrating them would be easy in comparison with the problem of gaining accurate information about the behaviour and distribution of the many secretive and nocturnal small mammals. He stressed the need to collect extensively, to visit foreign museums and to obtain publications. As his pastoral work prevented him making more than an occasional journey himself, Audubon and the boys would have to provide him with specimens, books and papers.

Prepared though he was for endless setbacks and difficulties, the next fifteen years of scientific labour were to be even more exhausting and frustrating than he had anticipated. Though the illustrations always evoked his unrestrained praise, he often had to upbraid his colleagues for the sad lack of factual information accompanying the skins which they sent him. The Missouri River expedition of 1843 was a particular disappointment to Bachman. He eagerly awaited the return of his friends with a hoard of animals, great and small, but it was some time before he was allowed to realize that the party had squandered most of their time hunting Buffalo—when he visited *Minnie's Land* in the summer of 1845, the telltale journal was kept out of his hands. An expedition that John W. Audubon made to Texas was equally unproductive. Nevertheless, between 1842 and 1851 he authored, or co-authored with Audubon, another nine original descriptions in *The Viviparous Quadrupeds* or scientific journals. These were of the Broad-footed Mole, Hairy-tailed Mole, California Myotis, Yellowbelly Marmot, Harris's Antelope Squirrel, Ring-tailed Ground Squirrel, Eastern Harvest Mouse, Red Wolf and Black-footed Ferret.

The folio plates were published in Philadelphia between 1845 and 1848, about half of the illustrations being the work of John W. Audubon. Bachman's text, which was edited by Victor, appeared in three parts during 1846, 1851 and 1854. In 1846, as Audubon's sight began to fail, his sons increasingly took over the work and as senility crept up on him, their burden increased.

Bachman's troubles multiplied also. In July 1846 his wife died after a painful illness and during the following spring another beloved daughter, Julia, was threatened by the old enemy, tuberculosis. In July he took her to Red Sulphur Springs, Virginia, hoping that she would respond to the waters and the healthier climate, but she died there just a month before her twenty-second birthday. More than her elder sisters, Maria and Eliza, Julia had tended to keep her deeper feelings to herself and Bachman was partly consoled, during these sad months, by a new openness in their relationship and spiritual depths to his daughter's character which he had not expected. Her confident expectation of a fuller life beyond the grave and her acceptance of death both surprised and comforted him.

Nevertheless, during the winter following this bereavement Bachman was glad of the discipline which his zoological work imposed on his time. In December, a sudden accident brought all his labour to a temporary halt and on the 13th he wrote:

"My dear Victor—I have been unable to write to you for ten days. I had returned from the meeting of the Synod in Georgia, and had but two days at home, when a sad accident befell me, which, but for God's providence, might have rendered me for the rest of my days like Milton, blind and sad.

I had prepared a mixture of gunpowder, sulphur and lard to anoint a mangy dog, and gave it to Sam, our little servant to carry to the yard. I was intently engaged in writing, seated by the fire with my feet on the fender. In his wisdom, Sam supposed that the lard should be melted, and he clapped it on the fire, about eighteen inches from my nose—an explosion took place something like that of a cannon—it was nearly half a pound of powder. I was knocked over—saved twenty-five cents in hair-cutting, lost my eye lashes and eyelids, and was laid on my back for ten days, with grated Irish potato poultices as a remedy. Nothing but my spectacles, (bless them), saved my eyes from total blindness. I have now a new skin from forehead to chin. Yesterday I left the dark room, and looked again upon the light of heaven, and my eyes are so much better to-day, that I have been able to show you the scratch of my pen."[7]

On 1 January 1848 he was visited by Professor Louis Agassiz but was disappointed to discover that the famous naturalist knew nothing about mammals. May brought a trip to New York for the General Synod of the Lutheran Church and he took with him his daughters Jane and Lynch. They stayed at *Minnie's Land* but it was painful to see Audubon with his mind and memory gone, playing silly pranks and pestering all the family. There was pleasure though in the company of Victor and John, both now remarried. Each had named the first daughter of their second marriages after their first wives, and Bachman became as fond of little Maria and Eliza as of his own grandchildren.

On his return to Charleston he resumed his labours, but his eye troubles continued and he relied heavily on his faithful sister-in-law, Maria Martin, as his amanuensis. This remarkable lady, who lent her talents to both Audubon and Bachman over many years, married the old pastor on 28 December. It was a great relief to both of them when their work on *The Viviparous Quadrupeds* came to an end in 1854, leaving them some leisure until their lives were disrupted by the War Between the States.

The hostilities began on 12 April 1861 at Charleston, when Fort Sumter was bombarded by the Confederates. Bachman sent his family inland to Columbia for safety but he remained in the city, often spending seven hours a day at the hospital, which was crammed full of sick and wounded men. One of his two sons, William Bachman, commanded the German artillery at Charleston and fought in numerous battles in Virginia. In 1863 John Bachman joined his family for Christmas and was with his second wife when she died on 26 December, two days before their fifteenth anniversary. After the funeral he returned to his work, but following the fall of Columbia in 1865 he had to leave Charleston with the last evacuation trains in February. In April, General Lee's army surrendered and he was able to return to his home in June, but in such straightened circumstances that a timely gift of $50 from Lucy Audubon was greatly appreciated.

On 1 January 1870, fifty-five years after his arrival at St. John's Church, Bachman preached there for the last time, but he continued his visiting until the end of his life, which came on 24 February 1874, twelve days after a stroke.

During Bachman's long lifetime, no other naturalist saw Bachman's Warbler within the United States; the only ornithologist to observe the species was Johannes Gundlach, who found it wintering in Cuba. When the birds were again reported in the States, in 1886, they were being slaughtered by Charles S. Galbraith for the millinery trade. He took one skin in the spring of 1886, six in 1887 and thirty-one in 1888. All were shot on migration in March and one of them is now in the American Museum of Natural History, stuffed and bent into a contorted pose for decorating the top of a bonnet.

The first nest was found in 1897 by Otto Widmann, in the St Francis River country of Missouri and Arkansas. Other breeding sites have since been tracked down in the swamps of South Carolina, Kentucky and Alabama, but no individuals have been sighted on breeding territory since the early 1960s and a report of a female in Cuba, in 1980, was the only wintering record for fifteen years. Extinction seems inevitable, but so little is known about this distinctive warbler that the reasons for its decline can only be a matter of speculation. No living Bachman's Warbler has ever been captured, banded or weighed. Audubon's confident expectation, that "we may still expect an account of its manners, migration and breeding" will never be realized.

Lucy Hunter Baird
(1848-1913)

LUCY'S WARBLER *Vermivora luciae* (Cooper)

Helminthophaga luciae J.G.Cooper, 1861. *Proceedings of the California Academy of Sciences*, Ser. 1, 2, p. 120: Fort Mojave, near lat. 35° [N.] in the Colorado Valley [Arizona]

With its trills and whistles the tiny Lucy's Warbler enlivens many mesquite thickets and cottonwood creeks across the south-western deserts. It builds in cavities in trees, or sometimes uses an old woodpecker hole in a Saguaro cactus or the bulky, abandoned nest of a Verdin. The first nest was found at Tucson, Arizona, by Charles Bendire and the first bird was collected eleven years earlier, during the spring of 1861, by Dr James G. Cooper. While working with Whitney's Geological Survey, based at Fort Mohave in the Colorado Valley, Cooper discovered two new birds: the Elf Owl and Lucy's Warbler. After describing the first and naming it Whitney's Owl in the *Proceedings of the California Academy of Sciences* (1861) he went on to say that "the next, which is undoubtedly a new bird, I have dedicated to the interesting little daughter of my kind friend, Prof. S.F. Baird". Cooper added that the warblers first appeared around Fort Mohave about 25 March and were still common there when he had to leave at the end of May, but he made no mention of behaviour or habitat. He collected five males, noting their ashy-grey upperparts, chestnut rump and crown-patch, and he also took one female.

Lucy Baird was only thirteen years old when Dr Cooper 'gave' her the warbler, but she was already used to receiving strange presents from her remarkable father and his friends. Professor Baird supplied his only child (born 8 February 1848) with harmless snakes to play with so that she would be free of irrational fears about them or other creepy-crawlies. When she was four, John H. Clark sent a collection to her father from Texas and included a pet squirrel for the little girl, and years later, before he departed for Hungary, John Xántus sent her the autographs of General Scott and Lajas Kossuth.

During her childhood Lucy met most of the young men who were inspired by her father since they frequently gathered in the home on Sunday evenings in wintertime. Though Mrs Baird would invariably say "Now you must not

THE BAIRD FAMILY. *PROFESSOR BAIRD AND HIS WIFE MARY, WITH THEIR DAUGHTER LUCY.*

talk shop in my parlor", she was invariably disregarded because her husband was as eager to hear of each collector's exploits as they were to recount them. Lucy listened eagerly to their adventures and helped her mother to serve them refreshments.

Amongst the many naturalists who were always welcome at the Bairds' home were Elliott Coues, Henry Bryant, John S. Newberry, William Stimpson and James G. Cooper. When Lucy was nine years old Robert Kennicott became a special friend of the household. Like most of the other students then working at the Smithsonian Institution he had very little money, but when Stimpson rented a cottage the youths all moved in together and called themselves the Megatherium Club. To reduce the cost of living they kept hens and whenever there was a surplus of eggs they celebrated with a bowl of egg nog and held a noisy party. Soon rumours of terrible dissipation reached the Bairds, and Lucy, already loyal and protective towards her friends and admirers, warned them of the reports. At the next meeting of the club it was decided that alcoholic concoctions would have to be sacrificed and the following morning Kennicott appeared on the doorstep bearing a basket of live hens for Miss Lucy—so that the temptation to indulge in egg nog would not recur.

When Lucy was fifteen the family spent the summer at Wood's Hole, Massachusetts, since the Professor needed to rest and Lucy was considered to be 'delicate'. By the time she was in her early twenties Mrs Baird's health was declining and soon both her parents were causing her concern. Though there was a continual stream of interesting young men calling at the house, Lucy never married; perhaps she was too devoted to her mother and father to leave them. (Baird's journal, which recorded all visitors to the house, showed that during a five-year period there were less than half a dozen days when no-one came to see him.) Baird died in the summer of 1887 when Lucy was thirty-nine, and for the next few years her life revolved around caring for her ailing mother, until the old lady passed away on 27 December 1891.

Now that she had the time, Lucy considered writing a biography of her father in collaboration with Professor G. Brown Goode, who had succeeded him at the Smithsonian. She began to collate his letters and journals with reminiscences of her own but progress was delayed by Goode's early death and her own poor health. Gradually Lucy realized that she would never finish the task and towards

Lucy's Warbler

the end of her life she instructed the executor of her will to ensure that the memoir was completed by a suitable person. She died at her home in Philadelphia on 19 June 1913 at the age of sixty-five.

The work of producing the biography was then taken up by William H. Dall who had first met Baird at the age of seventeen and who had been taught and influenced by him for more than twenty years. *Spencer Fullerton Baird, A Biography* was published in 1915 and included many of the letters sent between Baird and other naturalists such as Audubon, Agassiz, Dana, Cassin and Kennicott. It was dedicated by Dall 'To the memory of a devoted daughter Lucy Hunter Baird'.

Spencer Fullerton Baird
(1823-1887)

BAIRD'S SANDPIPER *Calidris bairdii* (Coues)

Actodromas Bairdii Coues, 1861. *Proceedings of the Academy of Natural Sciences of Philadelphia* [13], p. 194: Fort Resolution [Great Slave Lake, Mackenzie]

BAIRD'S SPARROW *Ammodramus bairdii* (Audubon)

Emberiza Bairdii Audubon, 1844. *The Birds of America* (octavo edn), Vol. 7, p. 359, Pl. 500: Prairie of the upper Missouri [= near Old Fort Union, North Dakota]

KAUAI CREEPER *Oreomystis bairdi* (Stejneger)

Oreomyza bairdi Stejneger, 1887. *Proceedings of the United States National Museum* 10, p. 99: Kauai

Baird's Sparrow was the last bird ever described, named and figured by John James Audubon. It was also the first species to be named after Spencer Fullerton Baird.

On Wednesday 26 July 1843 Audubon spent a whole day riding hard across the prairie, hunting Buffalo near the junction of the Missouri and Yellowstone Rivers. With him were two men from Fort Union and three travelling companions from back east, Edward Harris, John G. Bell and Lewis Squires. Bell, who was employed as hunter and taxidermist, shot a couple of 'grouse' and some other birds but did not get around to skinning them until the next day. Amongst the trophies were two small, brown, streaked sparrows, a male and female, which proved to be new to science. Audubon did not even mention them in his journal, either because at first he failed to realize that they were new or because he was too carried away by the excitement of the Buffalo hunt to remember them.

In the following year when the seventh volume of the octavo edition of *The Birds of America* was published, Audubon named the discovery Baird's Bunting "after my young friend Spencer F. Baird, of Carlisle, Pennsylvania". A rather uninspiring painting of the male bird made up the completed series

BAIRD. *FROM A DAGUERREOTYPE TAKEN ABOUT 1850, WHEN HE BEGAN HIS LONG CAREER AT THE SMITHSONIAN INSTI-TUTION. HE ENCOURAGED THE ZOOLOGICAL EXPLO-RATION OF NORTH AMERICA BY PLACING NATURALISTS AT STRATEGIC LOCATIONS AND ON VARIOUS GOVERNMENT SURVEYING EXPEDITIONS.*

of five hundred plates. Baird's Bunting has since been renamed Baird's Sparrow.

Ironically, Baird could have discovered the new species for himself. Though only twenty years old he showed such promise as a naturalist that Audubon had invited him to join his expedition. After much deliberation he had turned down the offer because of his heart condition and the anxious pleading of relatives who feared for his safety. From subsequent events, it is easy to imagine that there would have been more scientific success and less pointless Buffalo slaughter if Baird had been amongst them exerting his industrious and sobering influence.

Baird's Sparrow

As a youth Baird greatly admired the older generation of American naturalists; in addition to Audubon he became acquainted with Jacob P. Giraud, James De Kay, Thomas Nuttall, Titian Peale and John Kirk Townsend. He was also friendly with many younger men, particularly Thomas Mayo Brewer, John Cassin and George Newbold Lawrence, and knew all the other prominent men of science on the east coast. He learnt a lot from them and in return he encouraged scores of naturalists to help elucidate the nation's natural history in an extraordinary burst of collecting activity. In his first twenty-eight years at the Smithsonian Institution over 400,000 specimens were redistributed to other museums and individuals, yet the general holdings still increased from around 6000 specimens to several hundred thousand. By the time of his death the museum held over two and a half million items.

Baird's character is worth examining at this early stage since it was in a large part the key to his success. At the American Fisheries Congress of 1898 Livingstone Stone declared that:

"Prof. Baird lived on a higher plane of life and breathed a purer atmosphere than most men. Quiet and unassuming, with a nature as gentle as a child's, his natural superiority never failed to show itself when he was with other men; not even when among the distinguished men who gathered in the winter at the national capital... Prof. Baird had the enviable gift not only of endearing every one to him who came in contact with him, but of inspiring them with his own enthusiasm and energy."[1]

Daniel G. Elliot believed that "no naturalist that ever lived possessed to such a degree the power to imbue others with his own enthusiasm, and to attract them to become devotees of the study of Nature's Kingdom by the irresistible magnetism of his own personality."[2]

The ornithologist and ethnologist Henry W. Henshaw looked back fondly on his contacts with Baird when he wrote some autobiographical notes for *The Condor*:

> "Whatever his errand to Washington, no bird lover in those days ever visited the city without calling on Prof. Baird, in his little office in the north tower of the Smithsonian. Baird's love of young men was as great as his modesty and urbanity, and no one ever left his presence without the impression that he had seen and talked with a really great man. So great was his personal magnetism that no one could long know without loving him. His wisdom and sound judgment were ever at the disposal of those in need, especially the young man struggling to gain a foothold on the ladder of science, and it has been given to few men in this country to wield the personal influence in science that he did. Though he had given up all active work in ornithology and other branches of natural science, before I saw him in 1872, in favor of administrative work, he was as keenly interested in the labors of others as ever, and his first question to the man just returned from the field was, 'Well, what about the season's work? Tell me all about it.'"[3]

Young Robert Kennicott had earlier expressed his admiration much more simply: "Prof. Baird is just about the best and most wonderful man I ever did see."[4]

The third of seven children, Spencer Fullerton Baird was born at Reading, Pennsylvania, on 3 February 1823. After early schooling at a Quaker establishment in Maryland and at Reading Grammar School, he entered Dickinson College at Carlisle at the age of thirteen and remained there for the next four years. In his spare time, alone or with his eldest brother William, he spent hours and hours hunting for birds until he had combed the entire district, spring, summer, fall and winter. He travelled hundreds of miles on foot, sometimes completing over fifty miles in a day, his pack becoming increasingly heavy during his more successful journeys. Spencer and his brother formed a joint collection of birds, each of them being equally enthusiastic and proficient with their guns. When Spencer was seventeen they found a flycatcher that they could not identify. Spencer wrote for the first time to Audubon, whose reply assured them that they had indeed discovered a new species. Further correspondence and subsequent meetings led to Spencer's invitation to join the Missouri River expedition, which had to be declined for the reasons already stated. Instead Spencer continued working locally and together with William published a review of the flycatchers in the *Proceedings of the Academy of Natural Sciences of Philadelphia*. By the time it was published in 1843 it contained the descriptions of two new birds, the Yellow-bellied and Least Flycatchers, both discovered near Carlisle.

William began to do less fieldwork in order to pursue a law career but he maintained a strong interest in Spencer's activities and, for a few years at least, carried out small tasks and helped fund the build-up of the collection. After graduating from college in 1840 Spencer went to New York to study medicine, a subject he soon abandoned in order to continue his usual pursuits. He bought important natural history reference books when he could afford them, he studied foreign languages so that he could read them all and he enjoyed visiting other naturalists. Writing to his brother from New York in 1841 he outlined some of his recent successes:

"I have made a great haul lately ... I have obtained skins of Black bellied Plover, Piping & Ring Plover, Turnstone; Both Species of Godwits, Oyster Catcher, Buff Breasted Sandpiper, Semipalmated Sandpiper, Sanderling, Red Phalarope, Red Breasted Snipe, Long Billed & Esquimaux Curlew, Young Ibis, Ivory Heron, Eider Duck, & Two species of Tern, Three species of Gulls, one Jager, Two species of Guillemots, Red throated Diver, Red Necked Grebe, and the Little Grebe ... Some of these I obtained from a Young man Named Brashear, of Brooklyn who has a good many water birds. Some I got from Mr. Giraud who has, as I told you before, the best collection of American Birds I ever saw. But the most I procured from a young man named Peale, son of Peale's Museum in New York ...

I showed Mr. Audubon the birds I brought with me ... The old man still continues to be as clever as ever; he even offered the other day to teach me to paint & draw after his own peculiar manner, on condition of telling no one, and I have already commenced with him. I have drawn (from his originals) Fox Colored Sparrow, Cedar Bird, and am now at the feet of Harris' Buzzard ... He gave me to-day a copy of his letter press 5 vols. a pretty clever present, and is going to give me some rare bird skins.

I have just finished the other day looking over Major Leconte's Entomological drawings, of which he has about 9000 sheets, a species to each sheet; they are most beautifully executed. The Major is a first Cousin of Grandmother's.

They have some very good books here. The second vol. of Swainson's Fishes, Amphibians and Reptiles, which you know contains all the Generic descriptions, for $1.75. All the rest of his works at $1.75 per vol. Constable's Edition of Jardine's Wilson's Ornithology in 3 duodecimo vols. for $4.00, English Edition. I wish you would send me some money soon, as I can about this time get many interesting small things in market, Say's duck, Velvet duck, Brant, and many small winter birds."[5]

Moving back home to Carlisle again, Baird frequently went to Philadelphia to see John Cassin, Isaac Lea, George Ord and Samuel Morton, and met William Gambel and Adolphus Heermann when they were in the east between collecting trips. After a few years of self-education Baird accepted the Professorship of Natural History at Dickinson College where he taught for four years. Two of his students proved to be first-rate collectors: Caleb Kennerly with the North West Boundary Survey and John H. Clark with the U.S. and Mexican Boundary Survey.

In 1844 the family of Colonel Sylvester Churchill moved into Carlisle close to the Bairds and soon Spencer was courting the daughter Mary, a year older than himself. The marriage was a most fortuitous partnership because Baird's new father-in-law had become Inspector General of the Army in 1841 and was able to help him considerably in placing naturalists on the frontier.

Around this time Baird heard that the newly established Smithsonian Institution would be looking for an Assistant Secretary. He applied to the Secretary, Professor Joseph Henry, armed with letters of recommendation from James Dana, Samuel Morton, Major LeConte, Asa Gray and Audubon. His chum Cassin sent him a testimonial with a note saying that if it was not good enough then Baird should rewrite it and he would sign it. Cassin then urged him to burn the note "as it wouldn't look well in our published correspondence a hundred years from this"![6]

It was three years before Baird could be officially offered the position, partly because the buildings were still incomplete. In the meantime he rejoiced in the birth of his one and only child, Lucy Hunter Baird, born on 8 February 1848. He continued to teach at the College, making the best use of the summer recesses for fieldwork. In 1848 he went on a brief fish collecting foray with Dr Jared Kirtland near Cleveland, Ohio, and in the following year he travelled with Kennerly and another student to western Virginia. The year after that he took Mary, Lucy and Kennerly on a vacation-cum-collecting trip to Lake Champlain, afterwards going on alone with Kennerly to southern Ontario and Quebec.

On his return he had just a few weeks to pack up his collections and in October loaded up two railroad cars with boxes, kegs and packages of mammal remains, bird skins, nests and eggs, reptiles, fish, amphibians, insects, minerals and fossils. He

Baird's Sandpiper

then took the whole caboodle to the Smithsonian, where it formed the nucleus of the nation's natural history collection. He was now twenty-seven years old and one way or another he devoted much of the rest of his life to augmenting it with specimens from home and abroad.

Baird's fieldwork was now drastically reduced because of his official duties, but he more than made up for it in other ways. He gave fatherly help and advice to a motley assortment of students who were allowed to study in the recesses of the new buildings not yet taken up by the rapidly expanding collections. Amongst the earliest to arrive were Henry Bryant, James G. Cooper, Elliott Coues, Robert Kennicott, Fielding Meek, John S. Newberry and William Stimpson. Coues proved sufficiently brilliant for Baird to save him some of the best fieldwork and choicest study material. After returning from Labrador Coues was given some Canadian specimens to work through and found a new bird species which gave him the impetus to write 'A Monograph on the Tringeae of North America' for the *Proceedings of the Academy of Natural Sciences of Philadelphia* (1861):

"In the latter part of 1860, during the examination at the Smithsonian Institution of an extensive and valuable collection of birds made by Messrs. Robert Kennicott and Bernard R. Ross in the vicinity of Great Slave Lake and McKenzie's River, my attention was directed to a Sandpiper, nearly allied to *Actodromus maculata*

[Pectoral Sandpiper] and *Bonapartei* [White-rumped Sandpiper], but differing from both in many important features … "

After naming it "ACTODROMAS BAIRDII Coues—Baird's Sandpiper" he presented a glowing dedication to his mentor:

"In presenting to the scientific world this my *first* new species, I should do violence to my feelings, did I give it any other name than the one chosen. To SPENCER F. BAIRD, I dedicate it, as a slight testimonial of respect for scientific acquirements of the highest order, and in grateful remembrance of the unvarying kindness which has rendered my almost daily intercourse a source of so great pleasure, and of the friendly encouragement to which I shall ever feel indebted for whatever progress I may hereafter make in ornithology."[7]

The new sandpiper was of course just one of hundreds of new animal species then being discovered in North America. Baird well understood the need for zoological exploration so that the full species spectrum could be thoroughly worked out. He therefore urged that all expeditions make large and representative collections each time they entered new terrain. Since it was usually too expensive for the Smithsonian to fund private individuals it was obvious that the best way forward was to place already competent naturalists on as many of the Government sponsored expeditions as possible. Often with the help of General Churchill, he succeeded in placing surgeons on important surveys in a joint medical and scientific capacity: Newberry joined Lieutenant Williamson in California and Oregon; Stimpson was attached to the North Pacific (Rodgers and Ringgold) Exploring Expedition; Coues went to a military outpost in Arizona and afterwards worked with the Northern Boundary Commission and the U.S. Geological Survey in Colorado; and Cooper was with Whitney's Geological Survey of California. When it was necessary to arrange special privileges for fieldworkers in the army, Baird asked the General to send a letter to the commanding officer and in this way, for instance, made sure that John Xántus could collect in the vicinity of Fort Tejon despite official reluctance to let him do so.

Besides the more famous naturalists attached to well-known surveys Baird was associated with more than a hundred other collectors, from a variety of backgrounds. Sergeant Feilner collected birds in California and the Dakotas; Lieutenant Trowbridge collected on the Pacific Coast from California to Washington; Lieutenant Couch collected in northern Mexico, and Colonel Grayson corresponded from western Mexico. Dr Thomas Henry, James Hepburn and William Hutton all sent birds from the western states. Susan Fenimore Cooper, daughter of the famous novelist, sent him fish from New York State.

Baird personally made sure that all the fieldworkers were as well instructed and equipped as possible. Above all he encouraged them with letters, reference books, novels, newspapers and fresh supplies for collecting. After a while the Smithsonian issued a small pamphlet of *Directions for Collecting, Preserving, and Transporting Specimens of Natural History*. It was prepared under Baird's supervision with the co-operation of several unnamed naturalists (though John G. Bell is credited for providing a preservative recipe). In later editions Thomas Brewer supplied more detailed guidelines for collecting nests and eggs, and

Alfred Newton sent instructions from England on how to blow eggs and on the vital importance of verifying the species that laid them. Though the latter section contained sound practical advice, Newton's recommendations were denounced by Xántus as being only suitable for British parks and villages and completely unsuited to the inhospitable wilds of Baja California. Altogether it was a useful manual, listing and describing all the possible items necessary for field collectors, describing how best to skin and dry birds, mammals, reptiles and fish, how to preserve items in liquids, how to prepare skeletons, press plants, pack nests and eggs and even how to gather microscopic organisms from the dust that accumulated upon the sails and decks of ships.

"Collect the commonest species" was Baird's constant plea. Men entering new areas tended to ignore the soon familiar species around them because of their obvious abundance. Yet the common species were often peculiar to that particular habitat and were not found elsewhere. Specimens prized for their rarity by the finder often turned out to be well known from another locality.

One of many recurring problems for the traveller was the transportation of alcohol, because of its weight and its appeal to the unscientifically minded. The *Directions* advised them all clearly:

> "It is sometimes necessary to guard against the theft of the spirit employed by individuals who will not be deterred from drinking it by the presence of reptiles, &c. This may be done by adding a small quantity of tartar emetic, ipecacuanha, quassia, or some other disagreeable substance. The addition of a little arsenic will add to the preservative power of the spirit. A small quantity of soap is said to have a remarkable effect in preserving the color; a little saltpetre appears to have also the same effect."[8]

To protect unscrupulous tipplers from this unwholesome concoction Kennicott further requested that Baird arrange for his alcohol containers to be smeared in creosote!

Although overseeing all the various collectors took up a great deal of Baird's time he was a tremendous organizer with a formidable memory and he was a great letter writer, said to have penned over three thousand in one year on Smithsonian business. Each letter to the field men had a personal touch and almost always contained some form of encouragement, some statement intended to convey the importance of the work, buoying up the collector who might well be striving under the most adverse conditions. If their trophies arrived damaged he offered further advice on packing and transporting, if they had rotted away he tactfully explained better methods of preservation, if they were few in number he praised their rarity or quality. But even allowing for the passage of time between letters it is obvious that Baird had devised a formula for extracting the best results. Often he praised their locations: in 1858 he told Dr William Anderson that Fort Burgwyn, New Mexico, was "certainly the greatest centre of rare birds in the United States";[9] in about 1855 he wrote to Mrs Gardiner at Fort Tejon, California, saying "There is no locality from which [specimens] would be more acceptable";[10] in 1860 he convinced John Xántus that "The Cape [San Lucas] is certainly the most remarkable zoological mine I know of anywhere";[11] and in 1867 he encouraged Robert Ridgway to join a trans Rocky Mountain expedition with the words "It offers more promise of novelty in Natural History than any other part of North America."[12] Above all he praised and flattered the collector's achievements and capabilities. To John H. Clark on the Mexican border he claimed "No American Naturalist has ever added so many species of reptiles and fishes to our fauna in a period even of years, as you have."[13] To Alexander Winchell in Alabama he wrote "I consider the entire despatch from you as one of the most important contributions the Institution has ever received."[14] To Captain George B. McClellan in Mexico he confessed "a little astonishment at the amount and value of the collection".[15] To Robert Kennicott in the Mackenzie River District "...you far exceeded my expectations".[16] To Colonel Grayson on the Pacific Coast "You could readily become known in the scientific world as the Audubon of the West."[17] To Xántus at Fort Tejon "I have seen many collections of specimens but nothing within my experience comes up to the perfection of finish of your skins in every respect."[18] And again to Xántus at Cape San Lucas "I have, I believe, exhausted my stock of superlatives of astonishment and admiration on previous occasions, and can say nothing more, sir, in this way beyond repeating that I never heard of your equal in your line."[19]

All this was transparently designed to get the best out of each of them but there seems to have been nothing cunning, dishonest or devious in Baird's flattering prose. It simply arose from his child-like enthusiasm for acquiring more and more specimens and his academic determination to unravel the secrets of the nation's zoology. Some of his disciples took it all with a pinch of salt; they understood the system and worked it to good effect themselves. Kennicott, passing through Nicaragua on his way to Alaska in 1865, met a Mr Henry Holland and urged Baird to act at once: "He has but recently learned to make skins, but makes very fair ones—He is well equipped for bird collecting, and is keen, quick, and *ambitious* ... I believe you'll find him worthy of careful cultivation—He will be, I think, susceptible to a moderate amount of praise and flattering."[20] Likewise Coues, writing to Baird from North Carolina in 1869, wanted to encourage his hospital steward: "On general principles he is a numbskull, but with streaks of intelligence here and there. After a dozen trials he learned to skin first rate, as you will see ... If you were to write him a letter of his own, with S.I. stamp on it in full vig[nette], saying they are very good, and bestowing a little of that which you know so well how to bestow, it would set him up wonderfully."[21]

Specimens began to arrive at the Smithsonian from all over North America (and also from the North Pacific, South America and Europe). It was the heyday of Government exploration, of Boundary Surveys, Geological Surveys and Pacific Railroad Surveys, the latter series intended to find "the most practicable and economical route for a railroad from the Mississippi River to the Pacific Ocean". The Civil War of course brought many naturalists back east. A few in far-flung places were scarcely affected by it at all, though collecting itself could hold more than enough in the way of anxiety and danger. Xántus, in his usual poor English, informed Baird of yet another calamity at Fort Tejon: "I have great difficulty in preserving the alcoholic specimens, the several severe Earthquaque shocks broke almost everything here; what we can call a mug, jar or botle."[22] Coues wrote to Baird from Arizona: "Last night we shivered under blankets, and blew our numb fingers this morning. By ten o'clock it was hot; at eleven hotter;

twelve, it was as hot as—it could be. The cold nights stiffen our bones, and the hot days blister our noses, crack our lips and bring our eyeballs to a stand-still."[23] Coues survived the climate and the intermittent Apache attacks but Feilner was killed by the Sioux because in his eagerness to collect he failed to take the necessary precautions. And Kennerly and Kennicott both died on 'active service' as it were, each from medical disorders. Baird was under no illusions about all the hardships but in a letter to his special friend, Congressman George Marsh (who was doing some collecting for him in Turkey) he made light of the responsibility he bore:

> "I fear me I have much to answer for in the way of deluding unsupecting young (and even old) men to possible destruction from bite of snake, scorpion or centipede, engulfing in caverns while in search of fossil bones, embrace of Krakens when catching starfish on the seas; or some other undescribed species of calamity, the genus, even, of which is not yet known."[24]

Baird quietly sat out the Civil War in Washington, on one occasion sending Lucy with the ageing General Churchill to Carlisle when the situation looked particularly threatening. Baird had masses of work to do and simply got on with it. Besides looking after his collectors he set up a system of international exchange for Smithsonian Institution publications, personally bundling up hundreds of packages for export to learned societies, receiving in return loads of papers and journals. There was also much written work to be done, especially the zoological reports for the various Government surveys. For these he usually took on an editorial role, writing up only a few branches of zoology and passing the rest on to specialists. For Captain Stansbury's report on the Corps of Engineers' expedition to the Great Salt Lake he prepared the notes on the quadrupeds and birds and co-wrote the reptile section with Charles Girard, and for Major Emory's U.S. and Mexican Boundary Survey he completed the mammals, birds and reptiles.

So too for the Pacific Railroad Surveys: for Lieutenant Beckwith's report on the surveys along the 38th, 39th and 41st parallels he completed the mammals, birds and reptiles; for Lieutenant Whipple's 35th parallel survey only the reptiles; for Lieutenant Williamson's report from California only the mammals; and for Lieutenant Abbot's report from Oregon and Washington only the reptiles again. The Pacific Railroad Reports finally ran to twelve volumes, published between 1855 and 1859. Baird's general report on the mammals of the various surveys, in volume eight, issued in 1857, was re-issued as the *Mammals of North America* (1859). Similarly, volume nine on the birds (in which he was assisted by John Cassin and George Lawrence) first appeared in 1858 and was re-issued as the *Birds of North America* (1860). Baird wrote most of this ornithological classic in sittings averaging about fifteen minutes because he was so busy that he could only fit in sessions now and again! Coues later described its impact:

> "It represents the most important single step ever taken in the progress of American ornithology in all that relates to the technicalities. The nomenclature is entirely remodelled from that of the immediately preceding Audubonian period ... The synonymy of the work is more extensive and elaborate and more reliable than any before presented; the compilation was almost entirely original, very few citations having been made at second-hand ... The appearance of so great a work, from the

hands of a most methodical, learned, and sagacious naturalist, aided by two of the leading ornithologists of America, exerted an influence perhaps stronger and more widely felt than that of any of its predecessors, Audubon's and Wilson's not excepted, and marked an epoch in the history of American ornithology."[25]

In 1870 Baird edited James G. Cooper's manuscript and notes on the land birds of California, which were published as part of the Geological Survey (see Whitney). Four years later, with the assistance of Brewer, Ridgway and Coues, he completed the three-volume *A History of North American Birds* (1874) on the land birds. *The Water Birds of North America* by Baird, Brewer and Ridgway, did not appear until 1884, published as a continuation of the Geological Survey of California. Another important title was Baird's *Review of American Birds in the Museum of the Smithsonian Institution* (1864–72), never completed as it seems to have been superseded by the aforementioned works on the land and water birds.

Despite all these authoritative works an examination of the A.O.U. *Check-list of North American Birds* for 1983 reveals only fourteen new birds described by Baird for species occurring regularly in America north of Mexico. This total is less than those of Wilson, Audubon, Swainson, Lesson and Bonaparte. One reason for this is that very often Baird did not have the time to work through new collections and tended to send the material to Cassin or Lawrence. But the measure of a great ornithologist has never had much to do with the number of new birds described. Baird's importance lies in his reviews of American ornithology and the impetus and direction he gave it during the middle part of the nineteenth century. He is not usually remembered for his work on individual species but since he described so few new birds it is worth listing them: Aleutian Tern, Gila Woodpecker, Yellow-bellied, Least, Gray and Pacific-slope Flycatchers, Couch's Kingbird, Northwestern Crow, Black-capped Gnatcatcher, Virginia's, Grace's and Kirtland's Warblers, Abert's Towhee and Cassin's Finch. It is notable that half these birds are named after people (the scientific name of the Gray Flycatcher being *Empidonax wrightii*). Baird used bird names as another means of encouraging fieldworkers, though Cassin thought he went a little too far and was sometimes reluctant to follow Baird's advice when choosing new names: he was cautious about naming Hutton's Vireo and in another instance refused to name a bird after a non-naturalist army officer.

As time wore on Baird devoted less time to birds in favour of fish. In 1851 he had called again on Kirtland (and been given a new warbler by him) and in 1853 he went with Kirtland and Philo R. Hoy on a fish collecting trip in Wisconsin—one of the most enjoyable field excursions he ever experienced. Baird had also hoped to complete a book on fishes with Louis Agassiz but the project was abandoned because of other commitments. In June 1863 Baird went to Wood's Hole on the coast of Massachusetts for the first time, taking with him his wife and daughter for a vacation since all three of them were in poor health. When the U.S. Fish Commission was created by Congress in 1871 Baird was appointed Commissioner and he chose Wood's Hole as the site for the Commission's headquarters. He directed marine research, sought to improve the east coast fish stocks and on the death of Joseph Henry in 1878 had the added responsibility of becoming the Secretary of the Smithsonian. The extra work load helped cause a further decline in health and though he at last tried to take life easier he died at Wood's Hole on 19 August 1887, at the age of sixty-four.

Obituaries and posthumous tributes tend to dwell on the positive side of a man's career but it is still hard, even after a thorough search of the literature, to find much criticism of Professor Baird. Only Elliott Coues seems to have felt ill-disposed towards him, because Baird had consistently failed to find a place for Coues at the Smithsonian. In a private letter to J.A. Allen, Coues wrote that it was Baird's "settled policy never to have one of his peers or betters about him—so that this establishment [the Smithsonian] is simply a hatching house of henchmen who make an honest living by doing what they are told to do ...".[26] In fact Baird knew Coues inside out, and though he knew him to be an excellent naturalist he was also aware of his tactless, strong-minded and opinionated behaviour which was guaranteed to cause more trouble than he was worth. Baird had a long-standing aversion to controversy or argument of any kind and the consensus of opinion was that he was a generous and gentle man, loved by his staff and colleagues and greatly respected by the international scientific community.

Immediately following his death, tributes poured in from around the world. Amongst those in avian nomenclature was Leonhard Stejneger's Kauai Creeper *Oreomystis bairdi*, 1887. Stejneger had been sent by Baird to Bering Island in 1882 but by 1887 he was back in Washington and came across the new species when working through the Hawaiian collection of his fellow Norwegian, Valdemar Knudsen. Ridgway, working at the Smithsonian as Curator of Birds, had already named some Central American species after the professor: Peg-billed Finch *Acanthidops bairdii* in 1882, Baird's Junco *Junco bairdi* in 1883 (now a race of Yellow-eyed Junco), and Cozumel Vireo *Vireo bairdi* in 1885. Baird's old friend Lawrence had named Baird's Trogon *Trogon bairdii* from Costa Rica and Panama back in 1868. And Gambel had named Baird's Flycatcher *[Myiodynastes] bairdii* from Ecuador and Peru as early as 1847.

Baird's Beaked Whale *Berardius bairdii* was named by Stejneger in 1883 and admirably commemorates Baird's marine interests.

Kauai Creeper

Pierre Etienne Théodore Ballieu† (1828-1885)

PALILA *Loxioides bailleui* Oustalet

Loxioïdes Bailleui Oustalet, 1877. *Bulletin de la Société Philomathique de Paris*, Ser. 7, 1, p. 100: Hawaii

The Palila is an endangered species restricted to the Mamane and Mamane-Naio forests on the upper slopes of Mauna Kea on the island of Hawaii, its numbers having fluctuated in recent years between 2000 and 6000 individuals. Palilas feed almost exclusively on green Mamane seed pods which they tear apart with their powerful hooked bills.

In 1877 Dr Émile Oustalet brought the species to the attention of naturalists in a paper entitled 'Description de quelques espèces nouvelles de la collection ornithologique du Muséum d'Histoire Naturelle [de Paris]'. His description of this large grey and yellow finch was just one of the many contributions he made to French scientific journals during his time at that museum between 1875 and 1905. The bird was so distinctive that he created a new genus especially for it and gave it a specific name commemorating Théodore Ballieu, the collector who had sent the two first known specimens from Hawaii during the previous year. Unfortunately Oustalet (or the printers) made a mistake in the spelling of the bird name which appeared in the *Bulletin de la Société Philomathique* as *Loxioïdes Bailleui*. The collector's name also appears wrongly spelled in a list of correspondents of the Muséum d'Histoire Naturelle[1] and in S.B. Wilson and A.H. Evans's *Aves Hawaiienses* (1890–99). In this last work we learn a little about Ballieu from a footnote to their account of the Palila:

> "The late M.[= Monsieur] Ballieu was an enthusiastic naturalist, and spent some months at Dr. Trousseau's mountain-cottage in the district of Kona on Hawaii, engaged in forming a collection of birds which he forwarded to the Museum of the Jardin des Plantes, with a second collection consisting of fishes."[2]

†No portrait traced

Théodore Ballieu is known to have been the French Consul at Honolulu, on Oahu, from 1869 to 1878 and it may have been on the above-mentioned trip to the island of Hawaii that he found the Palila. Concerning the rest of his collecting activities, Christian Jouanin informs us that Ballieu "sent specimens to the Laboratoire des Mammifères et des Oiseaux du Muséum de Paris from 1872 to 1878, all originating from the Hawaiian Islands, except two from California— probably collected during the travels to and from his office in Hawaii. The Paris Museum is indebted to him for a large part of its collection of Drepanidinae [Hawaiian Honeycreepers and Finches]."[3]

Enquiries at the Ministère des Affaires Étrangères in Paris have revealed a few basic career details about this little-known natural history collector who was born in Paris on 20 December 1828. From 1848 to 1852 he studied law at the Académie de Rennes, and entered the French consular service on 1 February 1852 at the age of twenty-three, remaining in Paris until 2 February 1869. For part of this time, from 1853 to 1855, he was also working there as a lawyer. From 3 February 1869 to 18 February 1878 he spent nine years in Honolulu as French Consul (Second Class) for the Hawaiian Islands. Afterwards he was transferred to Sydney, Australia, for three years, during which he was promoted to Consul (First Class). In March 1881 he was appointed Consul at Port Louis on Mauritius and he remained there until his retirement on 12 December 1882, the date perhaps having been brought forward because of ill health. He died in Paris on 25 August 1885, aged fifty-six. He was married and had four sons.

Much more biographical information exists in the archives of the Ministère des Affaires Étrangères but there is nothing about Ballieu's natural history interests and no portrait of him.

Rather surprisingly, the records at the Paris Museum show that Ballieu did not send back any specimens from Australia or Mauritius.

Palila

John Barrow (1764-1848)

BARROW'S GOLDENEYE *Bucephela islandica* (Gmelin)

Anas islandica J.F.Gmelin, 1789. *Systema Naturae*, Vol. 1, Pt. 2, p. 541: Iceland

Point Barrow, Barrow Sound and Barrow Straits in the Arctic, and Cape Barrow in the Antarctic, are perhaps surprising memorials to a man who, for half his life, held a desk job in London. Sir John Barrow was Second Secretary to the Admiralty for over forty years, and was responsible for the civil administration of the Royal Navy over a period that included the latter part of the Napoleonic Wars. When peace came in 1815, Barrow was able to suggest to Lord Melville, First Sea-Lord, a plan for resuming the search for a sea route north of mainland America to the Far East: the North-west Passage. This quest virtually dominated Barrow's thoughts for the rest of his life; apart from those who actively took part in the search, he contributed more to Arctic exploration than any other man.

In 1818 John Ross sailed to Baffin Bay to search for a route westwards but in Lancaster Sound, believing his way to be blocked by a range of mountains, he retreated and sailed back to England. This unfortunate decision resulted in a long-standing and bitter feud between Ross and Barrow, who was convinced that a more thorough investigation should have been carried out. Barrow was all the more vehement because he had already stated publicly that he believed in the existence of such a 'Passage'. His claims were supported in the following year when William Parry sailed right through Lancaster Sound into Melville Sound and called the narrow opening between, Barrow Straits. Nevertheless, almost thirty years later, sufficient ill-feeling remained for Barrow to devote two chapters in his *Voyages of Discovery and Research within the Arctic Regions* (1846) to a virulent attack upon Ross. Once aroused he had become inexplicably vindictive and Ross's misfortune served as a warning to all other explorers on the many succeeding Arctic expeditions.

Ross's nephew, James Clark Ross, continued to support Barrow, and so did all the other officers involved in Arctic exploration at that time. They had no alternative if they wanted to further their naval careers. Years later, when the younger Ross sighted the Antarctic continent he named some of the first land

BARROW. *SECOND SECRETARY TO THE ADMIRALTY, 1807 TO 1845. HE STRONGLY SUPPORTED THE BRITISH SEARCH FOR THE NORTH-WEST PASSAGE. IN HIS YOUTH HE TRAVELLED IN CHINA AND SOUTHERN AFRICA.*

he saw after Edward Sabine and John Barrow; the latter Ross described, in his narrative of the voyage, as "the father of modern arctic discovery, by whose energy, zeal and talent our geographical knowledge of those regions has been so greatly increased; and we may hope, by God's guidance and blessing attending the exertions of the [Franklin] expedition that has so recently left our shores, he may live to see the great object of his heart, the discovery of a North West passage through Barrow Straits to the Pacific Ocean, accomplished."[1]

To John Barrow the existence and discovery of the 'Passage' was a matter of both personal conviction and national pride. In the *Journal of the Royal Geographical Society* he wrote: "If we should allow the completion to be snatched from us by any other power, we shall sustain a humiliating defeat." Although technically completed by dying members of Franklin's last expedition just a few months before Barrow's death, it was a Norwegian, Roald Amundsen, who first sailed through, but not until 1903–06.

By exerting his influence as a powerful naval bureaucrat Barrow was largely responsible for a great many Arctic expeditions and the explorers whom he had a hand in sending northwards included James Clark Ross, Edward Sabine, John Franklin and Dr John Richardson. Barrow therefore indirectly instigated the study of North American Arctic biology of which Richardson's *Fauna Boreali-Americana* was the most important early work. In the second volume, on the birds, which was issued in 1832 and co-written by William Swainson, a species of duck which they understood to be new was described under the name *Clangula Barrovi*. The "specific appelation [was] intended as a tribute to Mr. Barrow's varied talents, and his unwearied exertions for the promotion of science", though Swainson admitted that "the name of Barrow ... will not be solely indebted to us for its imperishable record".[2] Their specimen was a male shot in the Rocky Mountains by Thomas Drummond; they gave the bird the vernacular name Rocky Mountain Garrot unaware that there were earlier names and descriptions by J.F. Gmelin and Latham. It has since become known as Barrow's Goldeneye in both Britain and North America.

There is therefore no direct connection between the bird and the man, but an examination of Barrow's earlier life reveals that he was not uninterested in natural history. Although his later years were mainly sedentary they contrasted sharply with the first half of his life in which he travelled extensively. He was born of humble parentage near the village of Dragley Beck, on the northern shores of Morecambe Bay in north-west England, on 19 June 1764. His parents' small thatched cottage, with three or four fields attached, faced seawards. Educated mainly at Ulverston, Barrow excelled at mathematics and taught the subject to a one-armed midshipman in return for lessons in navigation. While still at school he spent two months carrying out a survey of nearby Conishead Priory, which proved to be the first of many detailed surveys which he undertook. His parents wanted him to enter the Church in the belief that salvation for the family was best secured by having a son in the clergy. Instead, he became a time-keeper at a Liverpool iron-foundry for three years.

By chance he was introduced to a man who ran an academy at Greenwich, London, and Barrow was invited to teach mathematics there. In his spare time he taught the wife of Sir George Beaumont and the son of Sir George Staunton, and naturally became acquainted with the boy's father. Barrow confided later that he was indebted for all the good fortune in his life to Staunton, for it was he who introduced him to Lord McCartney: Barrow soon found himself employed by McCartney as 'Comptroller of the Household' amongst a party of seventeen on their way to visit the Emperor of China!

Sailing in HMS *Lion* they eventually arrived at Tientsin and then proceeded to Peking. McCartney travelled on to the Royal Palace at Gehol while Barrow stayed behind to supervise a display of gifts for the Emperor. When the Emperor came to Peking to view them Barrow described the visit in his *Travels in China* (1804) as a long and thorough examination. McCartney's party then travelled overland for a thousand miles to Canton, the young Barrow, not yet thirty years old, taking every opportunity to travel independently of the main group.

Back in London Barrow stayed with Sir George Staunton and assisted him in his literary work. When time allowed he botanized at Kew but his studies were cut short when he went away again with Lord McCartney, this time to South Africa as his private secretary. As soon as they arrived McCartney sent him on a mission to reconcile the Boers and Kaffirs, and to obtain more accurate topographical knowledge of the area; at that time there was no map that covered even a tenth of the Cape Colony. After seven years it appears that Barrow was preparing to settle in the Colony, for he married a Miss Trüter and bought a house looking toward Table Mountain. However, the Colony was ceded to the Dutch by the Treaty of Amiens of 1802, and so Barrow, his wife and young daughter moved to England. Barrow then wrote an *Account of Travels into the Interior of Southern Africa* which was published in two volumes in 1801 and 1804. It was the best written account of the Cape Colony, its topography, native inhabitants, mineralogy and natural history, containing frequent references to the birds he had encountered. His map was superior to any that had yet been compiled and the whole work, together with his political dealings with the Boers and Kaffirs, were sufficient recommendation for Lord Melville, who had known Barrow in Cape Town, to appoint him Second Secretary to the Admiralty.

When there was a change of Sea-Lords in 1806 Barrow was removed from office with a £1000 a year pension, but the following year he was reinstated. In the intervening period he wrote his 'Life of Lord McCartney', became a member

Barrow's Goldeneye

of the London Literary Society and was elected a Fellow of the Royal Society through the influence of Sir Joseph Banks. Banks and Barrow often cooperated in the selection and placement of civilian naturalists on Royal Navy expeditions that went all over the world. In time Barrow became one of the main founders of the [Royal] Geographical Society and was Chairman at its first public meeting in 1830. His son John was also an active member of the society.

Barrow rarely wrote for the *Geographical Journal* but he contributed 159 articles to the *Quarterly Review* in which he examined accounts of various voyages, descriptions of foreign countries, recent discoveries in natural history and the arts, naval improvements and a host of other subjects, the majority of which were of a geographical nature. He also found time to compile a 'Life of Peter the Great' (1832), biographies of Lord Howe (1838) and Lord Anson (1839), and *The Eventful History of the Mutiny and Piratical Seizure of HMS Bounty* (1831). The latter remained in print far longer than any of his other works and is one of the better accounts of the mutiny, giving the first extracts from Captain Bligh's journal and the court martial proceedings to which Barrow was able to gain access.

During all this literary work and behind the scenes activity he continued to work at the Admiralty. From 1807 to 1845 he remained Second Secretary under thirteen naval administrations, often under far less competent First Secretaries, and in effect, Barrow was all powerful. He never became First Secretary because that position could only be filled by a member of the House of Commons. An incident in 1815 demonstrating his influence and high-ranking contacts is contained in a letter from Lord Liverpool, the Prime Minister, to Lord Castlereagh, his Foreign Secretary, concerning the recently defeated Napoleon: "Since I wrote to you last, Lord Melville and I have conversed with Mr. Barrow on the subject, and he decidedly recommended St. Helena as the place in the world best calculated for the confinement of such a person ... At such a distance and in such a place, all intrigue would be impossible ..."[3] Barrow knew because he had been there!

When the Duke of Clarence, who had been First Sea-Lord for a short period, became King William IV he made Barrow a baronet. And when Barrow retired, at the age of eighty-one, he received the thanks of the Prime Minister, but the service of plate presented to him by the officers engaged in Arctic discovery is said to have given him the greater pleasure. The accompanying letter was signed by Franklin, James Clark Ross, Parry and George Back, while others who had contributed included Francis Crozier, Sabine and Richardson.

In November 1848, at the age of eighty-four, Barrow died suddenly while writing at his home. He was buried at Pratt Street, Camden Town, where a marble obelisk was set up. In 1850 a large lighthouse-like monument was erected to his memory on the Hill of Hoad, above Ulverston.

BARTRAM. *PHILADELPHIAN ARTIST–NATURALIST WHO ASSISTED MANY OTHER NATURALISTS, ESPECIALLY ALEXANDER WILSON. THE PORTRAIT IS BY CHARLES WILLSON PEALE.*

William Bartram
(1739-1823)

UPLAND SANDPIPER *Bartramia longicauda* (Bechstein)

Bartramia Lesson, 1831. *Traité d'Ornithologie*, Pt. 7, p. 553. Type, by
monotypy, *Bartramia laticauda* Lesson = *Tringa longicauda* Bechstein
Tringa longicauda Bechstein, 1812. In J.Latham, *Allgemeine Uebersicht der
Vögel*, Vol. 4, Pt. 2, p. 453: North America

There are many well-known ornithologists who owe their success to the early
influence of another: for example, Robert Ridgway was a protégé of Spencer F.
Baird, John Xántus was taught the art of taxidermy by William Hammond,
William Gambel was tutored by Thomas Nuttall, and Alexander Wilson was
deeply indebted to the friendship, encouragement, teaching and practical help
given him by William Bartram. Elliott Coues nicely described the latter
relationship when he stated "If Wilson be called the 'father of American
ornithology,' Bartram then may be styled its god-father."

Bartram was a native of Philadelphia and throughout his long life he dwelt
on the west bank of the Schuylkill River at Kingsessing, in the house where he
was born on 20 April 1739. As the son of the celebrated botanist John Bartram
(1699–1777) and as a learned botanist and artist–naturalist in his own right, he
enjoyed an international reputation. In his youth he sent drawings, preserved
birds and observations to George Edwards, which were used in *Gleanings of
Natural History* (1758–64). Several English patrons of science commissioned his
curious and animated paintings and many American plants were introduced to
European gardens from seeds which he sent to his correspondents. His two or
three hundred acres of farmland and woodland, which included about five acres
of garden devoted to the cultivation of native and exotic plants, attracted many
foreign and American travellers. Bartram and Wilson first met in 1802 when
Bartram was in his early sixties and Wilson was in his mid-thirties. The details
of their first encounter have gone unrecorded, but the Scot must often have
come across his friend at work, as described by an earlier visitor:

"Arrived at the Botanist's Garden, we approached an old man who, with a rake
in his hand, was breaking the clods of earth in a tulip bed. His hat was old and
flapped over his face, his coarse shirt was seen near his neck, as he wore no cravat

or kerchief; his waistcoat and breeches were both of leather, and his shoes were tied with leather strings. We approached and accosted him. He ceased his work, and entered into conversation with the ease and politeness of nature's noblemen. His countenance was expressive of benignity and happiness. This was the botanist, traveller, and philosopher we have come to see."[1]

Wilson arrived at Bartram's Gardens as a novice birdwatcher. Four years later, when he resigned as teacher from the nearby Gray's Ferry School, he had launched into the preparatory collections of skins and drawings for his monumental and seminal *American Ornithology* (1808–14). The old gardener had willingly taught him the rudiments of ornithology and together they had found Eastern Bluebirds building their nests while snow still lay on the ground, listened to the first Whip-poor-will of the summer, watched Ospreys plunging into the nearby river and enjoyed the thrill of hearing great flights of honking Canada Geese pass overhead. On rainy days, Wilson sat by Bartram's fireside studying the works of Mark Catesby and George Edwards, books which he could not have afforded to buy for himself. He also scrutinized Bartram's own bird migration records, typically interspersed with comments about plant growth and weather.

Bartram's ornithological knowledge extended far beyond the boundaries of his own acres. He was able to tell Wilson of the birds he had seen on his journeys with his father to the Catskill Mountains of New York State when he was just fourteen, and to Florida when he was twenty-six. Early in 1773 he had set off alone on his longest journey and travelled in the south again for nearly four years, collecting plants and seeds for Dr John Fothergill. The tale of his wanderings was published in 1791 in *Travels Through North and South Carolina, Georgia, East and West Florida, the Cherokee Country, the Extensive Territories of the Muscogulges, or Creek Confederacy, and the Country of the Chactaws, Containing an Account of the Soil and Natural Productions of Those Regions, Together with Observations on the Manners of the Indians*. In this classic volume, an odd blend of accurate scientific reporting and protracted poetic prose, he described in relentless detail the vegetation of the swamps, forests and savannahs but made much less of the indigenous birdlife. Nevertheless, he included a list of 215 species categorized as residents or summer or winter visitors in either Pennsylvania or the Carolinas and Florida.

From his studies, Wilson knew that Bartram's list was the most comprehensive to have been published, but they both realized that it was far from complete. As Wilson considered the daunting challenge of producing a book containing pictures and descriptions of all the North American birds he naturally looked to the old sage for advice and wrote to him on 29 April 1807, saying:

"The receipt of yours of the 11th inst., in which you approve of my intended publication of American Ornithology, gave me much satisfaction; and your promise of befriending me in the arduous attempt commands my unfeigned gratitude. From the opportunities I have lately had, of examining into the works of Americans, who have treated of this part of our natural history, I am satisfied that none of them have bestowed such minute attention on the subject as you yourself have done."[2]

During the next few years, as Wilson travelled thousands of miles up and down the eastern United States in his quest for birds, Bartram began to seek *his* advice. On 11 October 1809, Wilson wrote:

"Thanks for your bird, so neatly stuffed, that I was just about to skin it. It is the Rallus virginianus [Virginia Rail] of Turton, and agrees exactly with his description. The one in company was probably the female. Turton mentions 4 species as inhabitants of the United States. I myself have seen 6."[3]

As successive volumes of the *American Ornithology* were published, they bore witness to the importance of Bartram's influence. Frequently, the author mentioned that he had made his observations in or near Bartram's Gardens, or acknowledged that some detail about migration or distribution had been supplied by William Bartram. It is hardly surprising that when Wilson shot the bird now known as the Upland Sandpiper by the Schuylkill River, near Bartram's Gardens, he named it Bartram's Sandpiper *Tringa Bartramia*. Although he could find no reference to it in the literature available to him, it had in fact already been described by J.M. Bechstein, so Wilson's specific epithet has been lost. In 1831 R.P. Lesson created the genus *Bartramia* especially for this peculiar bird.

Bartram outlived Wilson by ten years, remaining interested in every aspect of the passing seasons. On 22 July 1823 he spent the morning in his study and rising to walk into the garden, he suddenly collapsed and died.

Though Bartram's Garden lies within the city of Philadelphia, a remnant of it still survives. The handsome old stone house where such distinguished visitors as Benjamin Franklin, Thomas Jefferson, Alexander Wilson, Thomas Nuttall and André Michaux once found warm hospitality has been preserved, and is open to the public.

Upland Sandpiper

BELL. *NEW YORK TAXIDERMIST AND FIELD COLLECTOR WHO MADE TWO IMPORTANT WESTERN JOURNEYS. THE PHOTOGRAPH WAS TAKEN IN NEW YORK IN JULY 1864.*

John Graham Bell (1812-1889)

BELL'S VIREO *Vireo bellii* Audubon

Vireo bellii Audubon, 1844. *The Birds of America* (octavo edn), Vol. 7, p. 333, Pl. 485: short distance below Black Snake Hills [= near St. Joseph, Missouri]

SAGE SPARROW *Amphispiza belli* (Cassin)

Emberiza Belli Cassin, 1850. *Proceedings of the Academy of Natural Sciences of Philadelphia* 5, p. 104, Pl. 4: California near Sonoma

For half a century John G. Bell was the best and best-known taxidermist in New York. His shop on Broadway at Worth Street was a stimulating and popular meeting place for both local and visiting naturalists who came to inspect and discuss the latest acquisitions. In November 1841 young Spencer F. Baird arrived in the city, immediately visited the shop and met George N. Lawrence who became an intimate friend. Others who frequented the establishment at this time included Titian Peale, Jacob P. Giraud, Audubon and his sons Victor and John, and Major LeConte and his son John.

When Audubon was planning his expedition to the Upper Missouri the Reverend John Bachman urged him to employ Bell as his taxidermist. Bachman and Audubon were collaborating on *The Viviparous Quadrupeds of North America* (1845–54) and the minister needed to examine a good collection of mammal skins and skulls, but realized that their preparation would be a time-consuming and often unpleasant chore. Audubon was now in his late fifties and no longer enjoyed the great stamina of his youth so wanted to be able to devote himself to drawing and painting. The other participants could not be expected to help much with the taxidermy: Edward Harris, though a keen naturalist, had never learnt to skin his specimens; Isaac Sprague was employed as assistant artist and young Lewis Squires was only qualified to hunt. Bell had already travelled as far afield as Florida and was keen to go along and explore new territory. He agreed to a fee of $500, paid equally by Audubon and Harris, who were to share the collection on their return.

During the long journey to the Yellowstone River and back, Bell not only prepared the best of all the specimens taken but also shot or trapped many of them himself, including some new species. One of these was Bell's Vireo, which he shot on Saturday 6 May 1843 while they were on the Missouri. Audubon noted in his journal that it was a particularly cold and windy morning, making it very difficult for the captain to steer the *Omega*. They had therefore taken shelter for a while in the lee of some high hills and the passengers took the opportunity to go ashore. When Bell returned he brought with him two Gray Squirrels and several vireos, including the new species. Later that month Bell killed another new bird, Le Conte's Sparrow, and with Harris he obtained the first known specimens of Baird's Sparrow, Sprague's Pipit and Smith's Longspur.

Bell's Vireo

When the party reached Fort Union, Bell went collecting almost every day for the next two months. At the beginning of July Audubon recorded, with some exasperation, that Bell was not an experienced big game hunter, for he had allowed a deer to escape after seriously wounding it, but by the end of the summer he had killed a large number of Bison, Gray Wolves, Whitetail Deer and Mule Deer. After a long day on the prairies he often worked until late at night removing and salting down the skins while the others relaxed.

Occasionally Bell camped out overnight, and the beginning of a sortie which he made with two trappers from the fort, in search of Bighorn Sheep, was described by Audubon on 7 August:

"When Bell was fixing his traps on his horse this morning, I was amused to see Provost and La Fleur laughing outright at him, as he first put on a Buffalo robe under his saddle, a blanket over it, and over that his mosquito bar and his rain protector. These old hunters could not understand why he needed all these things to be comfortable; then, besides, he took a sack of ship-biscuit. Provost took only an old blanket, a few pounds of dried meat, and his tin cup, and rode off in his shirt and dirty breeches. La Fleur was worse off still, for he took no blanket, and said he could borrow Provost's tin cup; but he, being a most temperate man, carried the bottle of whiskey to mix with the brackish water found in the Mauvaises Terres, among which they have to travel till their return."[1]

Next day Bell and his guides came back without having even seen a Bighorn, bearing only a young Black-billed Magpie and two deer. However, Bell had taken part in a more satisfying hunting trip at the beginning of August when he spent two nights along the Yellowstone River and brought back the skins of Elk and Porcupine. Among the breeding birds which he noted there were

Peregrine Falcon, Sharp-tailed Grouse, Green-winged and Blue-winged Teal and Long-billed Curlew. He was surprised to find a pair of Bald Eagles with young still in the eyrie so directed one of his companions to climb the tree to chase the eaglet out of the nest—it landed safely on the ground and was taken back alive to the fort.

By the autumn, Bell's collection of animal skins included Bison, Gray Wolf, Grizzly Bear, Bighorn Sheep, squirrels, rabbits, hares, rats, mice, prairie dogs and bats. He had also prepared larks, wrens, flycatchers, grosbeaks, swallows, warblers, finches, orioles, geese, herons, pelicans and cranes. After Bell returned to his business in New York he continued to receive interesting specimens from Alexander Culbertson, the post agent at Fort Union, and passed some of them on to Audubon for *The Viviparous Quadrupeds*.

Six years later Bell headed westwards again to hunt in the newly acquired territory of California. He travelled via the Isthmus and collected some Mexican specimens en route, but little is known about his itinerary thereafter. We do know that he returned to New York in April 1850 with a large and exciting collection of bird skins, among them four new species which were described by John Cassin. These were the Sage Sparrow, Lawrence's Goldfinch, Williamson's Sapsucker and the White-headed Woodpecker. He found both woodpeckers on the western slopes of the Sierra Nevada near Georgetown (about forty-five miles north-east of Sacramento) but had difficulty shooting the White-headed as it kept very high up in the tall pine trees. Acorn Woodpeckers were more approachable and he collected many of these delightfully noisy, gregarious characters. He was intrigued to discover that Bell's Vireo, which he had first encountered in Missouri, also occurred in California. In his notes on the Sage Sparrow *Amphispiza belli* he wrote:

> "This bird I shot first near Sonoma, and afterwards at San Diego; its habits much resemble those of the painted bunting, (*S. ciris.*) I observed several of them amongst

Sage Sparrow

the wormwood at the edge of the hills, sitting on the tops of the small bushes, singing, and when disturbed would dart downwards. The song is quite unlike that of any of our finches, rather low and plaintive".[2]

The choicest part of Bell's collection was purchased by the Philadelphia Academy and written up in its *Proceedings* by Cassin during 1850 and 1851. Bell rarely ventured into print, but when he did so his short papers were published in the *Annals of the Lyceum of Natural History of New York*.

Bell continued to run his business for another thirty years and more, passing on his taxidermic skills to many keen young naturalists. His most famous student was Theodore Roosevelt, who from the age of eleven received regular lessons on preserving and mounting mammals and preparing bird skins. An enthusiastic and talented pupil, he had stuffed hundreds of birds by the time he went to Harvard and besides becoming a big-game hunter he remained a keen birdwatcher for the rest of his life.

When Bell retired in 1884 he moved his remaining stock of several thousand birds and animals to his home at Sparkhill, Rockland County. He was still occasionally harassed by former customers who wanted him to preserve their favourite pet or some trophy of which they were particularly proud. Naturalists who came seeking his advice or just the pleasure of his company were more welcome. One youthful visitor was Frank Chapman, who later recalled their first meeting:

"... one Sunday afternoon, a mutual friend took me to see him. Annoyed by the many callers, who seemed to think that his house was a public museum, Bell was not very cordial, but as I had been introduced as a boy who was interested in birds he led me to a large case of mounted specimens and in rapid succession asked, 'What's this?' 'What's that?' The case contained the characteristic Bell assemblage of brightly colored birds from many countries but, fortunately, I chanced to know the name of every bird to which he pointed. This successful examination evidently removed me from the class of undesirable callers and at its conclusion he said, 'You come with me,' and led the way to his workroom on the second floor.

    On the way we paused before another case of birds and, indicating a small Owl, again he said, 'What's that?' Luckily, this time I confessed ignorance, whereupon he added, 'I asked that question once of Professor Baird, when he was here. He looked the bird over and said, "I guess that is some of your handiwork, Mr. Bell."' Then he explained that one rainy day, having nothing else to do, he had made a cork body and covered it with feathers from a Screech Owl that had just been brought to him."[3]

Four or five years later, when Bell was in his mid-seventies, he and Chapman went on an outing together into the hills north of Sparkhill to try and trap some Eastern Woodrats. Bell led the way to a rock slide where forty years before he had captured some, and then scrambled about at such a pace that Chapman had difficulty keeping up with him. Active to the end, Bell died in October 1889 at his home, seventy-seven years after his birth there on 12 July 1812.

Charles Emil Bendire
(1836-1897)

BENDIRE'S THRASHER *Toxostoma bendirei* (Coues)

Harporhynchus Bendirei Coues, 1873. *American Naturalist* 7, p. 330 (footnote): Tucson, Ariz[ona]

> "I cannot improve on Major Bendire's fine description."
> "I cannot do better than to quote Bendire."
> "Major Bendire has given us a fine series of notes."
> "It is even better brought out by Major Bendire."

These phrases, and many others like them, are familiar to students of Arthur Cleveland Bent's *Life Histories of North American Birds* (1919–68). Bent frequently used such words to precede Bendire's descriptions of nests, eggs, diet and song because the Major was so often the best or only source of information. Also, in quoting Bendire, Bent was able to draw on observations made in a wide range of habitats in many parts of the West and Mid West during a military career that had lasted over thirty years.

Born on 27 April 1836 at König im Odenwald in the Grand Duchy of Hesse-Darmstadt, Bendire began life as Karl Emil Bender. The eldest of five children, he was educated at home until the age of twelve and then spent five years at a theological seminary near Paris, but was expelled because of "youthful peccadilloes"—whatever they may have been. After a short time at home, he and his brother Wilhelm emigrated to New York. Wilhelm quickly became disillusioned with the American way of life and decided to return home, but was lost overboard during the voyage. Karl stayed, and changed his name to Charles Bendire when, at the age of eighteen, he enlisted as a private in the First Dragoons. For the next five years he served with the U.S. Army in New Mexico and Arizona. He then tried civilian life for a year but re-enlisted in 1860 into the Fourth Cavalry. He was promoted through the ranks to Hospital Steward, serving in this capacity throughout most of the Civil War with the Army of the Potomac.

After the war Bendire returned to Germany for three months, then resumed his military career as a First Lieutenant in the First Cavalry. During the next eighteen years his postings were San Pedro, California, to April 1868; Fort

BENDIRE. *U.S. ARMY OFFICER. A FIRST CLASS FIELD ORNITHOL-OGIST WHO ESTABLISHED THE BASIC NESTING DATA FOR MANY WESTERN SPECIES. HONORARY CURATOR OF OOLOGY AT THE U.S. NATIONAL MUSEUM FROM 1883 TO 1897. THE PHOTOGRAPH WAS TAKEN IN 1889.*

Lapwai, Idaho, to June 1871; Camp Lowell, Arizona, to January 1873; St Louis, Missouri, to September 1874; Camp Harney, Oregon, to May 1878; Fort Walla Walla, Washington, to May 1882; and Fort Klamath, Oregon, to September 1883, followed by a year's leave after which he completed his service at Fort Custer, Montana.

Bendire's duties frequently involved him in the exploration of unknown territory. In April and May 1867 he led the first military expedition through Death Valley, California, and he also explored the deserts of southern Nevada. When in permanent camp his treatment of his men was unorthodox, but resulted in them accomplishing far more work than was usual; each morning he set everyone certain tasks, on the understanding that on their completion each man would then be free to hunt, fish or do whatever else he wanted for the rest of the day.

Bendire's habit of devoting his off-duty hours to bird study did not begin until his early thirties, when he was based at Fort Lapwai, Idaho. Though he at first concentrated on egg collecting he gradually became more and more inquisitive about every aspect of the breeding habits of birds, influenced and encouraged by correspondence with S.F. Baird, T.M. Brewer and J.A. Allen. Bendire was a stickler for discipline and never permitted his interests to interfere with the performance of his army duty—it was well known that he would pass by even the rarest bird rather than delay his command.

One of Bendire's most profitable postings, from the ornithological point of view, was Camp Lowell in Arizona, where he spent a year and a half between July 1871 and January 1873. The camp was seven miles north-east of Tucson on Rillitto Creek and by the start of the breeding season Bendire was familiar with the surrounding terrain and most of the resident species. When shortage of time prevented lengthy field trips he often rode along the banks of the creek startling little coveys of Gambel's Quail into noisy, evasive action. By April the river bed had dried up leaving only a few stagnant pools, but the groves of cottonwoods and thickets of mesquite and willow provided cover for many species including Hooded Orioles, Phainopeplas and Ladder-backed Wood-peckers, which he watched back to their nests. By the end of June the eggs of Harris's, Swainson's, Gray and Zone-tailed Hawks had also been added to his collection.

Bendire's first Zone-tailed Hawk nest forever remained a vivid memory. He found it on 22 April, a large, bulky nest in a tall cottonwood tree, about forty feet above the ground in a fork close to the main trunk. He took the single, freshly laid bluish-white egg but refrained from shooting either of the adults and returned on 23 May. On this occasion the sitting bird was so reluctant to fly off that he had to rap on the tree with the butt of his shotgun to flush it. Telling the story later, he continued:

"Climbing to the nest I found another egg, and at the same instant saw from my elevated position something else which could not have been observed from the ground, namely, several Apaché Indians crouched down on the side of a little cañon which opened into the creek bed about 80 yards further up. They were evidently watching me, their heads being raised just to a level with the top of the cañon.

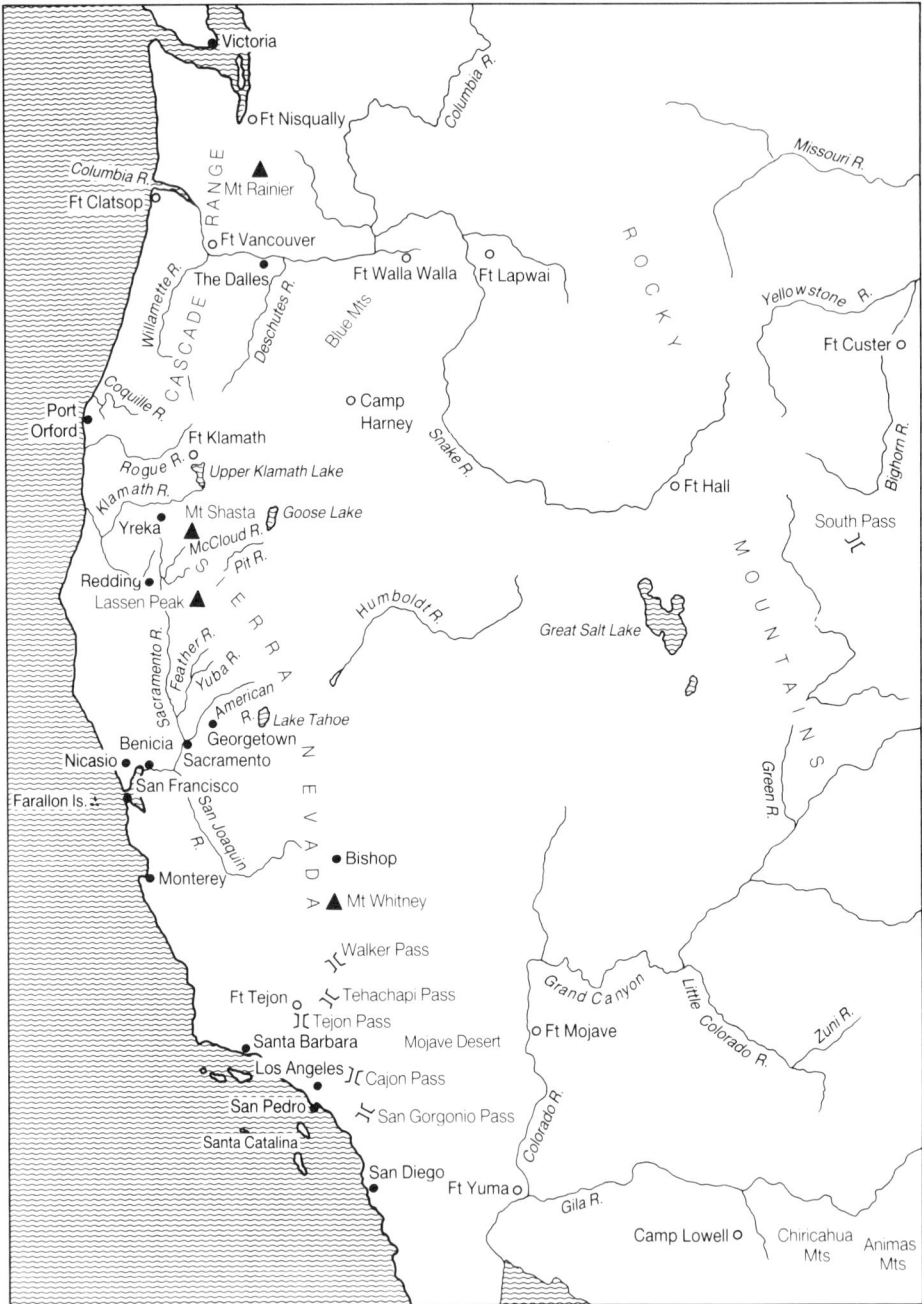

THE WEST: WITH PLACES VISITED BY BENDIRE, HEERMANN, TOLMIE,
WILLIAMSON, ETC.

In those days Apaché Indians were not the most desirable neighbors, especially when one was up a tree and unarmed; I therefore descended as leisurely as possible, knowing that if I showed any especial haste in getting down they would suspect me of having seen them; the egg I had placed in my mouth as the quickest and safest way that I could think of to dispose of it—and rather an uncomfortably large mouthful it was, too—nevertheless I reached the ground safely, and, with my horse and shotgun, lost no time in getting to high and open ground. I returned to the place again within an hour and a half looking for the Indians, but what followed has no bearing upon my subject. I only mention the episode to account for not having secured one of the parents of these eggs. I found it no easy matter to remove the egg from my mouth without injury, but I finally succeeded, though my jaws ached for some time afterward."[1]

By this time Bendire was not just collecting eggs but also studying birds throughout the breeding season and a species that he came to know well was the Gray Hawk. During the second half of April he watched them every day, engaged in noisy courtship chases above the tree tops, but he noticed that they became much quieter once they had paired up. By the end of the month they had begun to build and on several occasions he watched them breaking off living cottonwood twigs, flying very swiftly at the one selected, grasping it in their talons and usually snapping it at the first attempt. Bendire subsequently found four Gray Hawk nests within a ten-mile radius of the camp, all in cottonwood trees and most often in the tallest tree available. His first clutch, of three fresh eggs, was taken from a nest seventy feet above the ground. He described the nest as a rather flimsy affair, constructed from cottonwood and willow twigs, some of them still green and leafy, which he assumed were chosen to provide camouflage. He often watched the hawks hunting lizards, rodents, beetles, grasshoppers and birds and he greatly admired their swift and graceful flight.

While Bendire was at Camp Lowell the owls also attracted his attention and he added the Ferruginous Pygmy-Owl to the avifauna of the United States when he shot a specimen at the end of January 1872. On 17 April he found the first recorded nest of the Spotted Owl and shot the sitting bird after it flew into the branches of a nearby tree. The only other specimen then in any collection was the type, taken by John Xántus at Fort Tejon fourteen years earlier. At first Bendire assumed that it was a Barred Owl (though far outside its range) but when he examined the bird carefully at the camp he realized that it was a species he had never handled before. Elf Owls and Barn Owls also bred in the area, the latter often appearing at the mouths of deserted Badger setts in which they presumably nested. Once, when checking out a dense thicket along Rillitto Creek during the wintertime, Bendire was surprised to come across fifteen Long-eared Owls roosting close together in a small mesquite.

The best known of Bendire's discoveries came on 28 July when he shot a female thrasher and sent it to Elliott Coues. Since Coues had made no special study of the mimic thrushes he sent it on to Robert Ridgway who was working among the skin collections of the Smithsonian Institution. Ridgway concluded that it was a Curve-billed Thrasher of the westernmost form, *Toxostoma curvirostre palmeri*. Unconvinced, Coues asked Bendire for further information and soon received notes comparing the behaviour of the Curve-billed with the other species, as well as a male skin to go with the original female. Bendire also pointed out that the eggs were very different! Coues therefore described the new species as *Harporhynchus Bendirei* and referred to it as Bendire's Thrush, now Bendire's Thrasher.

On leaving Camp Lowell, Bendire was promoted to Captain and sent to a recruiting station in St Louis for about seven months, but his next posting was again in the West, at Camp Harney in Oregon. The camp was situated on the southern edge of the Blue Mountains about 4000 feet above sea level and it provided Bendire with a good variety of habitat. The mountains to the north were well timbered with forests of pine, spruce and fir as well as groves of juniper and aspen; a sparsely vegetated range lay to the south-west, the highest portions permanently under snow; while to the south lay sage brush flats, their monotony alleviated by a chain of shallow, brackish lakes which attracted vast numbers of resident and migratory waterbirds.

Soon after his arrival Bendire spotted a whitish woodpecker and mentioned the sighting in a letter to Thomas Brewer. In his next letter, on 5 December, he had to explain that the supposed woodpecker had on closer examination proved to be a Clark's Nutcracker. He commented, "It is a consolation to me, however, to know that I am not the first one who has made this mistake. [See W. Clark.] It is not to be wondered at that any one who has not seen them before, seeing them for the first time on the wing, should take them for woodpeckers. I am inclined to believe that this bird breeds here in hollow trees, and very early in the season. In an adult female that I shot the other day, some of the eggs in the ovaries were already considerably enlarged."[2] That year he searched for the nests of Clark's Nutcracker without any success, but the

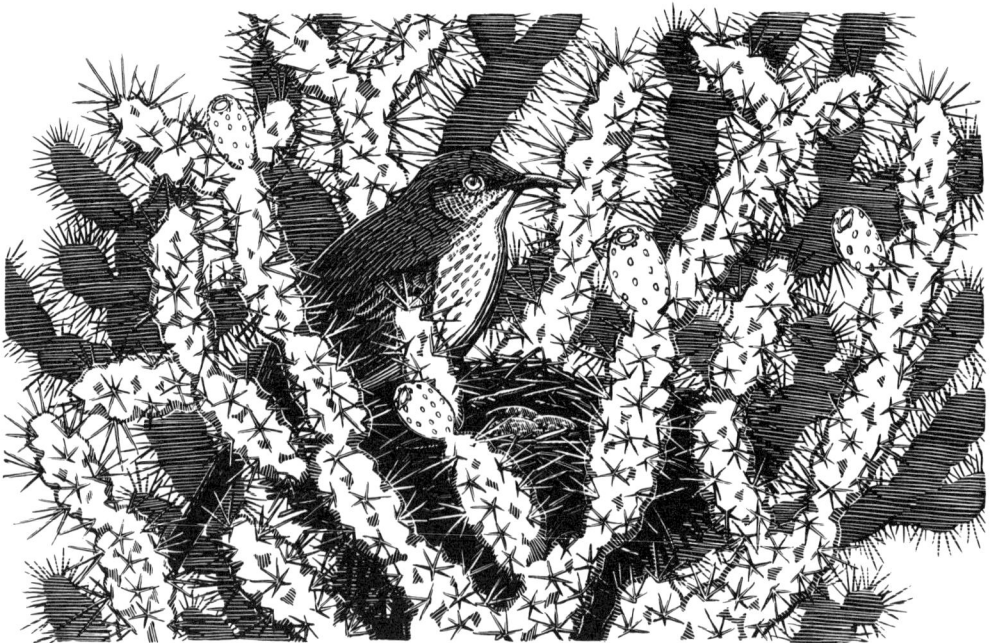

Bendire's Thrasher

following April he at last found one, in a pine tree—though not in a hole. He was much impressed by the devotion of the sitting bird, for it sat tight as he worked his way along the branch and only took flight as he stretched out his hand towards the nest.

Malheur Lake, twenty-five miles south of Camp Harney was a popular hunting ground for Bendire and the other officers. Early in the spring White Pelicans arrived even before the lake was free of ice, later colonizing several of the small islands and laying their first eggs as early as 12 April. Great Blue Herons and Double-crested Cormorants nested among the pelicans but started laying about ten days later; the laying dates, clutch sizes, composition of nests and appearance of the young all being carefully recorded by Bendire on his visits. Other common breeding birds around the lake were Ring-billed and California Gulls, Pied-billed, Eared and Western Grebes, Canada Geese, Mallards, Gadwalls, Black-necked Stilts, American Avocets and huge numbers of American Coots. Short-eared Owls quartered the nearby marshes where Red-winged and Yellow-headed Blackbirds bred abundantly and from time to time a hunting Peregrine Falcon would create widespread panic.

During his first spring at Camp Harney, Bendire visited the lake on 16 April and located two pairs of Ravens; one nesting on a cliff, the other in a large, dead willow. Both nests were difficult to reach, but after several attempts he was successful and returned with the two clutches, each of five freshly laid eggs. In the wintertime the Ravens were gregarious and flocks of up to thirty birds hung around the camp; they picked over the manure piles, loitered knowingly outside the slaughterhouse and stole the scraps thrown to the chickens. Dark-eyed Juncos that had summered higher up in the mountains also appeared around the barracks and for their roost sites appropriated the old nests of Cliff Swallows underneath the eaves. Snow Buntings and Lapland Longspurs were occasional visitors, sometimes a Northern Shrike or some Pine Grosbeaks passed through and one winter a flock of Bohemian Waxwings lingered to fatten themselves on the fruits of the wild rose.

After three and a half years at Camp Harney Bendire was sent to Fort Walla Walla at the other, northern end of the Blue Mountain range, and after four years there he went to Fort Klamath in south-western Oregon. Those three postings gave him an unrivalled knowledge of the birds of the north-western States and as the earliest resident ornithologist in eastern Oregon he provided a long list of first state records. Bendire also collected mammals, fossils and Indian artefacts whenever the opportunity arose and while at Fort Klamath he prepared the skin of a large, dark brown shrew and sent it to C. Hart Merriam who named it *Sorex bendirei*, the Pacific Water Shrew. Other specialists named fossil trees in his honour.

From September 1883 until August 1884 Bendire enjoyed a prolonged period of leave and arrived in New York just in time to take part in the organizational meeting of the American Ornithologists' Union at the American Museum of Natural History. The meetings lasted for three days from 26 to 28 September and among the twenty other founding members present were Elliott Coues, Robert Ridgway, William Brewster, Edgar A. Mearns, Charles B. Cory, C. Hart Merriam, Eugene P. Bicknell and Daniel G. Elliot. With Coues in the chair a provisional constitution was read and adopted, officers were elected and various committees were appointed on classification and nomenclature, migration, avian anatomy, faunal areas, oology and the problem of the European House Sparrow in North America. In October the last issue of the *Bulletin of the Nuttall Ornithological Club* was issued, containing an account of the meetings, and it was replaced in January 1884 by the first issue of *The Auk*. For the rest of his life Bendire eagerly anticipated the annual A.O.U. conferences and rarely missed attending.

After the inaugural meeting Bendire went to Washington D.C. at Baird's request to assume charge, as Honorary Curator, of the Department of Oology at the United States National Museum. The collection had been sadly neglected and its usefulness was limited by the doubtful origins of many clutches. Bendire's enthusiasm and experience ensured that by the end of the year the collection was properly stored, and arranged in such a way that students could readily find what they wanted. The dubious material was thrown out and Bendire added his own cabinet of about 8000 specimens to the collection. As a result of his work, many other ornithologists contributed sets of rare eggs, knowing that they would be safe in Bendire's custody and of real scientific use.

Some years later Bendire summed up his attitude to oology in *Instructions for collecting, preparing, and preserving birds' eggs and nests* (1891) when he explained:

> "Unless the would-be collector intends to make an especial study of oölogy and has a higher aim than the mere desire to take and accumulate as large a number of specimens as possible regardless of their proper identification, he had better not begin at all, but leave the nests and eggs of our birds alone and undisturbed. They already have too many enemies to contend with, without adding the average egg collector to the number. The mere accumulation of specimens is the least important object of the true oölogist. His principal aim should be to make careful observations on the habits, call notes, song, the character of the food, mode and length of incubation, and the actions of the species generally from the beginning of the mating season to the time the young are able to leave the nest. This period comprises the most interesting and instructive part of the life history of our birds.
>
> Do not start in with the idea that because a certain species may be common with you everything must consequently already be known about it, and that your observations would be useless. Rest assured that some new and interesting fact can still be learned by the observant oölogist about even our commonest birds."[3]

Having put this philosophy into practice over a period of many years, Bendire was able to accept another of Baird's propositions—that he write an oology of North American birds. As he gathered material for this huge undertaking the scope of the work increased and the title finally chosen was *Life Histories of North American Birds*. After his retiral from the army early in 1886 Bendire continued as Honorary Curator at the National Museum and laboured on the *Life Histories* for the next eleven years. The first volume, with twelve egg plates, covering the gallinaceous birds, pigeons, doves and birds of prey (including owls) was published in 1892. The second volume, with seven egg plates, appeared in 1895 and included the cuckoos, kingfishers, woodpeckers, nightjars, hummingbirds, tyrant flycatchers, crows, blackbirds and orioles.

The books contained an enormous amount of useful information on the breeding behaviour of birds and their geographical distribution, much of it previously unpublished. They also, of course, contained detailed descriptions of the eggs of each species, obviously written by a man who loved his subject, as is shown in this example:

"The eggs of the [Eastern] Wood Pewee vary in shape from ovate to short or rounded ovate; the shell is close-grained and without gloss. The ground color varies from a pale milky white to a rich cream color, and the markings, which vary considerably in size and number in different sets, are usually disposed in the shape of an irregular wreath around the larger end of the egg, and consist of blotches and minute specks of claret brown, chestnut, vinaceous rufous, heliotrope, purple, and lavender. In some specimens the darker, in others the lighter shades predominate. In very rare instances only are the markings found on the smaller end of the egg."[4]

Unfortunately, only these two volumes were produced, for by 1896 Bendire was suffering from Bright's Disease. Hoping for some improvement in his kidney function he left his Washington home and went to Jacksonville, in north-eastern Florida, but died there just five days later, on 4 February 1897, at the age of sixty. The challenge of revising and continuing the *Life Histories* was taken up by Bent in 1910.

Within two months of Bendire's death, Elliott Coues was boasting in 'Dr Coues' Column' in *The Osprey* that he had "discovered" Bendire. While stationed at Tucson, Bendire had felt slighted by Baird and Brewer, so Coues had engineered Baird's diplomatic intervention with the result that Bendire was thereafter willing to cooperate with both ornithologists. In a manner that must have offended many of his readers, Coues referred to his discovery as a "bumptious and captious German soldier, who was a man to take strong likes and dislikes on very small provocation".[5]

Bendire's genuine friends, who arranged tributes to him in *The Auk*, *The Osprey* and other journals, acknowledged that his reserved manner and honest bluntness had often repelled acquaintances; his erect bearing and strikingly soldierly appearance had also been intimidating. But those few who had won Bendire's respect and had been admitted to his circle of intimates considered themselves privileged indeed. In their company the Major had been a genial companion, loquacious, entertaining, full of bird lore and ever willing to recount anecdotes about his exciting and often dangerous experiences in the West— experiences which could be matched by few other ornithologists.

BEWICK. *ENGLISH ARTIST, BEST KNOWN FOR HIS WOODCUT ILLUS-TRATIONS IN* THE HISTORY OF BRITISH BIRDS. *AUDUBON VISITED HIM AT NEWCASTLE IN 1827, FOUR YEARS AFTER THIS PORTRAIT WAS PAINTED.*

Thomas Bewick (1753-1828)

BEWICK'S WREN *Thryomanes bewickii* (Audubon)

Troglodytes Bewickii Audubon, 1827. *The Birds of America* (folio), Vol. 1, Pl. 18 (1831, *Ornithological Biography,* Vol. 1, p. 96): five miles from St Francisville, Louisiana

"The bird represented under the name of Bewick's Wren I shot on the 19th October, 1821, about five miles from St. Francisville, in the State of Louisiana. It was standing as nearly as can be represented in the position in which you now see it, and upon the prostrate trunk of a tree not far from a fence. My drawing of it was made on the spot ... In the month of November 1829, I had the pleasure of meeting with another of the same species, about fifteen miles from the place above mentioned, and as it was near the house at which I was then on a visit, I refrained from killing it, in order to observe its habits. For several days, during which I occasionally saw it, it moved along the bars of the fences, with its tail generally erect, looking from the bar on which it stood towards the next one above, and caught spiders and other insects ... At other times, it would fly to a peach or apple-tree close to the fence, ascend to its top branches, always with hopping movements, and, as if about to sing, would for an instant raise its head, and lower its tail, but without giving utterance to any musical notes ... I shot the bird, and have it preserved in spirits."[1]

On the next page of his *Ornithological Biography* Audubon went on to say:

"I honoured this species with the name of BEWICK, a person too well known for his admirable talents as an engraver on wood, and for his beautiful work on the Birds of Great Britain, to need any eulogy of mine. I enjoyed the pleasure of a personal acquaintance with him, and found him at all times a most agreeable, kind, and benevolent friend."

As Bewick's life as an engraver unfolds here we shall come in time to the unlikely meeting between the two great bird artists. On the one hand Audubon—flamboyant, ambitious, widely travelled, controversial and French-American, painting birds on a gigantic scale in his own peculiar style. On the other hand Bewick—modest, conscientious, home-loving, conventional and English, using the traditional methods of wood engraving to make illustrations in miniature. They could hardly have been more different.

The lower reaches of the River Tyne, in the very north-east of England close to the Scottish border, was the area in which Thomas Bewick spent almost all his life. He was born in August 1753 at *Cherryburn* farmhouse on the south bank of the river about a dozen miles west of Newcastle. He was the eldest of eight children and as a boy had to help around his father's tenant farm and small coal mine. He showed a natural inclination to draw long before he was fourteen, the age at which he was apprenticed to Ralph Beilby who owned the town's only general engraving business. Here, Bewick learned how to engrave on copper, silver and wood, producing trade cards, labels and billheads as well as inscriptions, scenes and animal illustrations. It happened that Beilby did not especially like working in wood so this type of work tended to be handed to Bewick. As the years went by he improved existing techniques and by his example was able to revive the art of cutting on blocks of wood.

When his seven-year apprenticeship was completed Bewick spent some time at home, made a walking tour in northern England and Scotland and then headed for London to seek his fortune. He readily found work as an engraver but after nine months he was glad to get back home again. He went into a partnership with Beilby that lasted twenty years, during which his best-known works were produced: *General History of Quadrupeds* (1790) and the two-volume *The History of British Birds* (1797–1804). He probably began the preparations for the *Quadrupeds* in 1781 but since much of the work for his books had to be done in the evenings after the general business of the day had been finished it is hardly surprising that it was not completed for nine years. It was an immediate and popular success, fortuitously appearing only a year after Gilbert White's *The Natural History and Antiquities of Selborne*, which was then causing a great awakening of interest in natural history. Seven more editions appeared during his lifetime, to which could be added an 1804 American edition by Alexander Anderson, who copied Bewick's woodcuts and included an appendix of 'Some American Animals not hitherto described.'

In 1779 Bewick's usual line of work was interrupted when he was commissioned by Marmaduke Tunstall to make a woodcut of his herd of wild cattle at Wycliffe on Tees in Yorkshire. Though he never met Tunstall, who was then in London, the wealthy landowner offered him help should he ever wish to embark on a work on the birds of Britain. Tunstall had already published the *Ornithologia Britannica* (1771), a folio pamphlet in Latin, English and French which contains the accepted original description of the Peregrine Falcon, American Pipit and a few other birds. He also possessed a large private natural history museum which he had transferred from London to Wycliffe where it was housed in a specially constructed room. Tunstall died in 1790 but Bewick made arrangements with his nephew for him to view the collection and he accordingly spent two months of the following year drawing the birds. The stiffness and unlikely postures in some of Bewick's woodcuts are due to the ineptness of Tunstall's taxidermist and is particularly evident in some of the rarer species for which Bewick could see no other examples. He soon realized that freshly shot birds were far easier to draw.

The first volume on *Land Birds* appeared in 1797, the second on the *Water Birds* in 1804; the text for the former supposedly by Beilby, the latter by Bewick, in both cases unoriginal and taken from the published works of John Ray, Francis Willughby, and more especially Thomas Pennant, George Montagu and John Latham. It did not matter. It proved to be immensely popular and served to deepen interest in birds throughout the country. Though the engravings are variable in quality, the wealth of detail in some of the backgrounds and in his famous tail-pieces of country scenes gives the work enormous charm and appeal.

Other books illustrated by Bewick include versions of Goldsmith's poems in 1794, Burns's poems in 1808 and Aesop's fables in 1818, besides at least thirty-five children's books in the earlier period up to 1788. As a keen countryman and angler it is hardly surprising that Bewick's next major work was intended to be about British fishes. He had hoped to publish the book in 1826 but he was slowed down by old age, his son Robert's lack of commitment and the difficulty of obtaining good specimens. Only sixteen fish were engraved though Robert had completed over a hundred preliminary drawings. Over forty tail-pieces were prepared and these were utilized in posthumous editions of his *British Birds*, following his death on 8 November 1828 at the age of seventy-five.

Though Bewick had many intimate friends and many well-educated visitors it seems strange that the most vivid (and most often quoted) account of him should be that provided by John James Audubon. The American artist was in Scotland in 1827 hoping to persuade Lizars of Edinburgh to publish *The Birds of America*. On his way south to London he called in at Newcastle (where he

Bewick's Wren

found the Tyne "as dirty and muddy as an alligator hole") and left us with these impressions of Bewick:

"At length we reached the dwelling of the Engraver, and I was at once shewn his workshop. There I met the old man, who, coming towards me, welcomed me with a hearty shake of the hand, and for a moment took off a cotton night-cap, somewhat soiled by the smoke of the place. He was a tall stout man, with a large head, and with eyes placed farther apart than those of any man that I have ever seen: a perfect old Englishman, full of life, although seventy-four years of age, active and prompt in his labours. Presently he proposed shewing me the work he was at, and went on with his tools. It was a small vignette, cut on a block of boxwood not more than three by two inches in surface, and represented a dog frightened at night by what he fancied to be living objects, but which were actually roots and branches of trees, rocks and other objects bearing the semblance of men. This curious piece of art, like all his works, was exquisite ... The old gentleman and I stuck to each other, he talking of my drawings, I of his wood cuts. Now and then he would take off his cap, and draw up his grey worsted stockings to his nether clothes; but whenever our conversation became animated, the replaced cap was left sticking as if by magic to the hind part of his head, the neglected hose resumed their downward journey, his fine eyes sparkled, and he delivered his sentiments with a freedom and vivacity which afforded me great pleasure ... The tea drinking having in due time come to an end, young Bewick, to amuse me, brought a bagpipe of a new construction, called the Durham Pipe, and played some simple Scotch, English and Irish airs, all sweet and pleasing to my taste ... The company dispersed at an early hour, and when I parted from Bewick that night, I parted from a friend ... I revisited him ... on the 16th April, and found the whole family so kind and attentive that I felt quite at home. The good gentleman, after breakfast, soon betook himself to his labours, and began to shew me, as he laughingly said, how easy it was to cut wood; but I soon saw that cutting wood in his style and manner is no joke, although to him it seemed indeed easy. His delicate and beautiful tools were all made by himself, and I may with truth say that his shop was the only artist's 'shop' that I ever found perfectly clean and tidy.
... My opinion of this remarkable man is, that he was purely a son of nature, to whom alone he owed nearly all that characterized him as an artist and a man. Warm in his affections, of deep feeling, and possessed of a vigorous imagination, with correct and penetrating observation, he needed little extraneous aid to make him what he became, the first engraver on wood that England has produced. Look at his tail-pieces, Reader, and say if you ever saw so much life represented before ... As you turn each successive leaf, from beginning to end of his admirable books, scenes calculated to excite your admiration everywhere present themselves. Assuredly you will agree with me in thinking that in his peculiar path none has equalled him. There may be men now, or some may in after years appear, whose works may in some respects rival or even excel his, but not the less must Thomas Bewick of Newcastle-on-Tyne be considered in the art of engraving on wood what Linnaeus will ever be in natural history, though not the founder, yet the enlightened improver and illustrious promoter."[2]

Charles Reed Bishop
(1822-1915)

BISHOP'S OO *Moho bishopi* (Rothschild)

Acrulocercus bishopi Rothschild, 1893. *Bulletin of the British Ornithologists' Club* 1, p. 41: Island of Molokai

For many years the Bishop's Oo was assumed to be extinct. George C. Munro had watched a group of half a dozen on the Hawaiian island of Molokai in 1904, but the species then went unrecorded until 1981, when Stephen Sabo had a brief but convincing sighting of this black and yellow oo in the rain forests of Maui. It seems to have been a rare species even at the time of its discovery in 1892, when Henry C. Palmer wrote: "This bird is rather shy when approached by men, but sometimes it is rather inquisitive, if one keeps quiet, and it generally answers and comes up to the spot when its cry is imitated. The greatest difficulty in procuring specimens is the dense undergrowth."[1]

The naming of Bishop's Oo, in 1893, marked the cooperation between two extremely wealthy men: Charles Reed Bishop and the Honourable (later Lord) Walter Rothschild. Both used their money to found museums of international importance. The Bernice Pauahi Bishop Museum, in Honolulu, was built by Bishop in memory of his wife and was devoted to Hawaiian and Polynesian ethnology and natural history. The Rothschild Museum and Library at Tring, in Hertfordshire, is now a vital extension of the British Museum (Natural History). Rothschild sent out a vast network of collectors, one of whom was Palmer, who spent two and a half years in the Hawaiian Islands, in the course of which he discovered Bishop's Oo on Molokai. When Rothschild described it the following year, he took the opportunity to compliment Bishop, since he had assisted Palmer in various ways and sent photographs, museum publications and duplicate skins to Tring. In return, Rothschild sent him "a box containing specimens of all the species got by my collectors in Laysan" and a copy of his lavish, authoritative work on the Hawaiian birds, *The Avifauna of Laysan and the Neighbouring Islands* (1893–1900), in which he acknowledged Bishop's assistance.

Bishop did not share Rothschild's obsession with natural history. Nor had his riches been inherited. Born at Glen Falls, New York, on 25 January 1822,

BISHOP. *FINANCIER AND PHILANTHROPIST. SEEN HERE WITH HIS WIFE PRINCESS BERNICE PAUAHI IN A SAN FRANCISCO STUDIO IN 1875. HE FOUNDED THE BERNICE P. BISHOP MUSEUM AT HONOLULU.*

Bishop was orphaned in early childhood and raised by his grandparents. At the age of fifteen he left home and worked in stores and on farms until February 1846 when he sailed for Oregon. After a stormy passage round Cape Horn he decided to delay his journey at Honolulu until the spring, but the islands held him for the next fifty years.

At first he took various clerking jobs then became collector of customs at the port. In June 1850 he married the eighteen-year-old Princess Bernice Pauahi despite the opposition of her family who had wanted her to become engaged to either Prince Alexander Liholiho (later King Kamehameha IV) or Prince Lot (later King Kamehameha V). Her parents gradually accepted her decision and presented the couple with a beautiful home, *Haleakala*, which became the centre of social life in Honolulu.

Meanwhile, Bishop organized Hawaii's first bank, which prospered to such an extent that he became financial adviser and privy councillor to the government. On the death of King Kamehameha V he was offered the portfolio of minister of finance and Princess Bernice was offered the throne, but they both declined. As Mrs Bishop, the princess was able to serve as an important link between the Hawaiian and American communities and she enjoyed the role of hostess to visiting naval officers and diplomats as well as her involvement with church and charities. The Bishops enjoyed a very satisfying and successful childless marriage.

After the princess died of cancer in 1884, Bishop frequently visited the United States and in 1894 he moved to San Francisco, becoming president of the Bank of California. He died at Berkeley on 7 June 1915 and after cremation his remains were shipped to Honolulu for interment beside his wife, with royal honours.

Bishop's Oo

THE BLACKBURNE FAMILY IN 1741. ANNA (EXTREME RIGHT, AGED ABOUT FIFTEEN) LATER FORMED A NATURAL HISTORY COLLECTION AT HER HOME NEAR WARRINGTON, ENGLAND. HER BROTHER ASHTON (STANDING BESIDE HER—OR KNEELING THIRD FROM LEFT) COLLECTED BIRDS IN EASTERN NORTH AMERICA.

Anna Blackburne
(1726-1793)

Ashton Blackburne
(c.1730-c.1780)

BLACKBURNIAN WARBLER *Dendroica fusca* (Müller)

Motacilla fusca P.L.S.Müller, 1776. *Des Ritters C. von Linné ... vollständige Natursystem nach der zwölften Latinischen Aufgabe*, Supplement, p. 175: Guyana [= French Guiana]

The Blackburnian Warbler is one of the most beautiful of the North American wood warblers. The bold black head markings, flaming orange throat and streaky flanks all contribute to the striking appearance of the male in breeding plumage. The bird's name is usually said to commemorate Anna Blackburne, an eighteenth century English naturalist, or her brother Ashton who resided at Hempstead, New York, and collected birds for her in eastern North America at the time of the Revolutionary War.

Despite the lady's long interest in New World birds, her many years spent consolidating her natural history museum and her contact with several eminent European naturalists, she is scarcely known at all. Indeed there is little factual material about her life; and even less is known about her brother.

Anna and Ashton grew up in England at Orford Hall, near Warrington, which lies roughly midway between Liverpool and Manchester. Their father, John Blackburne, had a great love for botany and was wealthy enough to devote much of his energy to planting foreign trees and shrubs on his estate as well as growing some of the less hardy exotica under glass. He was so successful with his hot-house plants that they were said to rival those at Kew. On their mother's side, their first cousin was Sir Ashton Lever who was famous as the founder of the Leverian Museum of Leicester Square in London. According to William

Swainson, Lever's "passion for collecting exceeded all the bounds of prudence: every subject of zoology or mineralogy he did not possess was to be purchased, cost what it might."[1] The inclination to collect on a grand scale seems therefore to have come naturally to Anna and Ashton, and to at least two of their brothers whose interests were coins, medals and old prints.

Anna was the fourth of nine children; Ashton being the fifth son.[2] Their mother Katherine died when Anna was fourteen years old and soon, as the rest of the children went their separate ways or died, she was left alone with her father as mistress of Orford Hall. She never married but preferred to be known as 'Mrs Blackburne', a title reflecting her position as female head of the household. Right through her childhood and until her father died in 1787 she remained at Orford, running the hall, helping her father to maintain and enlarge the gardens and building up her own museum which by degrees grew into one of the finest in the area. At the age of sixty-one she moved herself and her collection to Fairfield, a new house built nearby where she spent her last five years.

Anna must have been in frequent touch with Sir Ashton Lever, since their two museums were both expanding at the same time, but no documentary evidence for this has survived and the liaison that we know most about concerns Thomas Pennant: the man who named the Blackburnian Warbler. He lived some forty miles away, across the Welsh border at Downing, near Holywell in Flintshire. He is best known as one of the principal correspondents of the Reverend Gilbert White whose letters to Pennant form a major part of *The Natural History and Antiquities of Selborne* (1789). Pennant had been a friend of the Blackburne family since at least 1768 and he certainly knew them well enough to be their guest for a night in 1772 when he stopped at Orford at the start of his second tour of Scotland. Later, Anna allowed him to examine and describe her North American birds for inclusion in his *Arctic Zoology* (1784–87); her generosity being duly acknowledged in the preface:

> "To the rich museum of *American* Birds, preserved by Mrs. ANNA BLACKBURN, of *Orford*, near *Warrington*, I am indebted for the opportunity of describing almost every one known in the provinces of *Jersey, New York*, and *Connecticut*. They were sent over to that Lady by her brother, the late Mr. *Ashton Blackburn*; who added to the skill and zeal of a sportsman, the most pertinant remarks on the specimens he collected for his worthy and philosophical sister."[3]

From the New York area alone Pennant described just over a hundred species of birds. Sometimes he incorporated Ashton's observations into his text, quoting portions of information that had originally been sent to Anna. There is, for instance, a long section about the Passenger Pigeon which Ashton had seen around Niagara "in vast numbers in all parts, and have been of great service at particular times to our garrisons, in supplying them with fresh meat ...". Numbers soon declined and "everybody was amazed how few there were; and wondered at the reason."[4]

The Blackburnian Warbler described in the *Arctic Zoology* may have been shot by Ashton Blackburne but Pennant gives no details and no dedication to either Ashton or Anna.[5] The name Blackburnian Warbler appears in the margin beside a nine-line physical description of the bird, followed by a final tenth line:

PLACE   Inhabits *New York*.—Bl.Mus.

The bird may therefore be named after Anna (since it was from her museum), or after Ashton (since he probably collected the specimen), or perhaps the name was purposely intended to embrace them both. While other writers have plumped for Anna and a minority have asserted that it was Ashton, it seems to us that Pennant's name was most likely a tribute to them both. The bird could even have been named in a general way after the Blackburne family and the museum.[6]

By the time Pennant's *Arctic Zoology* appeared he had lent his manuscript accounts to Dr John Latham who copied them almost word for word into his *General Synopsis of Birds* (1781–1802). Latham's description of the Blackburnian Warbler appeared in print three years before Pennant's, which has sometimes given rise to the belief that Latham was responsible for naming the warbler. Again, Latham gives no direct dedication to either Anna or Ashton. In 1789 J.F. Gmelin translated the warbler description into Latin for inclusion in the thirteenth edition of Linnaeus's *Systema Naturae*. Gmelin gave Ashton's American birds scientific binomials for the first time, in the process calling the warbler *Motacilla Blackburniae*, using the female genitive suffix to commemorate Anna, but this may or may not have been Pennant's original intention, and it is Pennant's English name that has survived, not Gmelin's scientific binomial. Much later, it was discovered that these early descriptions by Pennant, Latham and Gmelin had all been preceded by another, written by the German naturalist Statius Müller, who had acquired a wintering specimen from French Guiana. Müller's scientific name won priority but Pennant's English name has never been superseded.

Ashton's other birds examined by Pennant and later renamed by Gmelin include specimens from which the original descriptions of thirteen or fourteen North American species were taken. This compares well with Wilson's twenty-five and Audubon's twenty-one species which still remain on the A.O.U. *Check-list of North American Birds* (1983). Professor V.P. Wystrach, who has collated the only reliable information about the two Blackburnes, lists the following

Blackburnian Warbler

species contributed by the Englishman: American Wigeon, Labrador Duck, Red-shouldered Hawk, Yellow Rail, Lesser Yellowlegs, Willet, Short-billed Dowitcher, American Woodcock, Eastern Phoebe, Wood Thrush, Scarlet Tanager, Sharp-tailed Sparrow and Vesper Sparrow. The Dickcissel should perhaps be included. Ashton also supplied specimens for races of the Clapper Rail, Eastern Screech-Owl and White-eyed Vireo.[7]

Besides Pennant, another famous naturalist known well to Anna Blackburne was Johann Reinhold Forster, a Prussian who taught school from 1767 to 1770 only a few miles away from Orford at Warrington Academy. He regularly visited Anna to help her with the classification of insects and later dedicated a genus of plants to Anna and her father. Forster also has a small but notable link with American birds, partly through his association with Pennant and the Blackburne Museum (see Forster).

In addition, Anna had famous correspondents who lived farther afield. She exchanged letters with the German naturalist Peter Simon Pallas during the time that he was based at St Petersburg and she promised him dried plants as well as mineral ores from Derbyshire and Cornwall, receiving in return some of his hard-won birds, plants and seeds from Siberia.

Anna also corresponded with Linnaeus, sending at least two letters to Sweden, mainly about a proposed exchange of specimens. Each letter tells us a little about her expatriate brother. In the first (dated 29 June 1771) she mentions "Having a bro". who lives near new York in north America, who annually enriches my Cabinet with the productions of that Country, if it it [sic] wou'd be agreeable to you I wou'd send you a few Birds & insects, which I believe are not in your Sys. Nat^{ae}. & which he kill'd within 50 miles of that place." In the other (dated 14 October 1771) she boasts that her "Colection of dry'd birds is pritty num[erous]. My cabinet is not destitute of shells, Insects, fish & Fossils, & if my Brother lives will increase fast."[8] Ashton was evidently in poor health and seems to have died a few years after this was written. Although his date and place of death have never been confirmed it seems likely that he died around about the year 1780, aged about fifty, perhaps at Norwalk, Connecticut. If this is correct then Anna's supply of American birds would have ceased before she moved her collection from Orford to Fairfield.

In its new premises, the Blackburne Museum was housed in a special room forty-five feet long that ran the whole width of the second floor. Anna had great plans for laying out a new garden but she died, on 30 December 1793, before her ideas for Fairfield could be realized and she was buried at Winwick a few miles north of Warrington. Afterwards her beloved collection passed to her nephew, John Blackburne M.P., a remnant of it remaining within the family until 1923 when it was loaned to the Warrington Museum. In the 1930s the specimens were to be displayed once again at Orford Hall but they were returned because their condition had seriously deteriorated. A list from that time shows that the ornithological material consisted of one drawer of British birds' eggs, two oak bird cases each with about 200 birds, another case with twenty owls and four corvids, a case of game birds and a case of "Kestrels etc." They were variously described as "Poor and perished", "Many damaged" and "Very dirty". Only one domed glass case with six foreign perching birds was good enough to be retained at Warrington Museum, the remainder were sent to an agent of the family and were very likely just thrown away or burned. Originally there was much more material, all no doubt housed in the finest cabinets available and all carefully labelled but there is no known record of how this portion was dispersed. Her cousin's Leverian Museum fared almost as badly. Sir Ashton Lever's collecting mania eventually caused him severe financial embarrassment and he was forced to dispose of his life's work. To do this he hit on the novel idea of organizing a public lottery: the first and only prize being the whole collection! It was won by Dr Parkinson (the discoverer of Parkinson's Disease) but the specimens became widely scattered when he too eventually parted with it. Some of the surviving portions were acquired by the Vienna Natural History Museum.

The Blackburne family portraits seem to have disappeared. The only picture of Anna and Ashton now available shows them in a family group, reproduced in *Hale Hall, with notes on the family of Ireland Blackburne* (1881).[9] The only other known likeness of Anna is a silhouette cut when she was in her middle years showing her to be a rather matronly figure.

The main source concerning her character comes from an anonymous article in the *Gentleman's Magazine* for 1794 where it is recorded that she was "Sincere and hospitable, of open, candid, and unaffected manners, with a truly good heart, and a clear head, she was highly and justly esteemed, and is now lamented by all who had the pleasure of her acquaintance, and she will be sincerely regretted by the poor, who in her have lost a valuable friend and benefactoress."[10] She was obviously a determined lady with an enthusiastic love for natural history but the extent of her knowledge is impossible to judge as she refrained from publishing anything on the subject. Pennant did not help matters by failing to say anything about her in his autobiography!

In view of the rarity of eighteenth and nineteenth century female naturalists it is remarkable that there should have been a second 'Mrs Blackburn' interested in birds. Jemima Blackburn (no relation, and more often referred to as Mrs Hugh Blackburn) is best known for her striking illustrations in *Birds of Moidart* (1895). She also produced *Birds drawn from nature* (1852) and was the first to depict the young of the Common Cuckoo in the act of ejecting its nest-mates.[11]

BONAPARTE IN 1827. *NEPHEW OF THE EMPEROR NAPO-LEON. SYSTEMATIC ORNITHOLOGIST WHO LIVED IN ITALY FOR MOST OF HIS LIFE BUT WORKED IN THE UNITED STATES FROM 1823 TO 1826.*

Charles Lucien Jules Laurent Bonaparte (1803-1857)

BONAPARTE'S GULL *Larus philadelphia* (Ord)

Sterna Philadelphia Ord, 1815. In W.Guthrie, *Geographical, Historical and Commercial Grammar*, 2nd Am. edn, Vol. 2, p. 319: No locality [= near Philadelphia, Pennsylvania]

In the summertime Bonaparte's Gulls breed in colonies beside ponds and lakes in the coniferous forests of Canada and Alaska. When fall comes the gulls disperse widely, some to the Great Lakes and others to the Pacific and Atlantic seaboards. It was from a wintering individual that Bonaparte's Gull was first described, in 1815, by George Ord of Philadelphia. He called the gull *Sterna Philadelphia*, but in 1831 Richardson and Swainson suggested the name *Larus Bonapartii* in their *Fauna Boreali-Americana*. Though Ord's *Philadelphia* takes priority, the link with Bonaparte has been retained in the vernacular name.

Napoleon Bonaparte had four brothers—Joseph, Lucien, Louis and Jerome. Charles Bonaparte, after whom the gull was named, was the eldest son of Lucien; and Charles's wife, Zénaïde, was the elder daughter of Joseph, therefore she was Charles's cousin. Small and stocky, Charles bore a striking physical resemblence to the Emperor Napoleon and shared his uncle's relentless energy, but his ambitions, although extraordinary, were limited to the world of birds.

When Charles Bonaparte, Prince of Musignano, arrived in the United States at the age of twenty he was well prepared for his ornithological career. Having spent most of his life in Italy he spoke fluent Italian. He could also communicate reasonably well in French and English and was capable of reading and writing Latin with ease, an important requisite for one whose forte was taxonomy and nomenclature. He had already made a thorough perusal of the European ornithological classics, begun a natural history collection of his own and discovered a new species—the Moustached Warbler—in the countryside near Rome. Eager to become familiar with the birds of the New World, he began his study while still crossing the Atlantic by observing the storm-petrels that followed the ship. He soon noticed that there were two species: Leach's, which he saw most frequently during the first part of the voyage, and another, which became commoner as he neared the American continent. He shot a few specimens of each and named the latter *Procellaria Wilsonii*—now Wilson's Storm-Petrel. In his first scientific paper he compared the two storm-petrels with another two and read it before the Academy of Natural Sciences of Philadelphia on 13 January 1824.

New Jersey became the prince's home over the next four and a half years, from the summer of 1823 until the beginning of 1828 (although he was in Europe for most of 1827). He settled beside the Delaware River some twenty miles north-east of Philadelphia at Point Breeze, on the estate belonging to his father-in-law and uncle, Joseph Bonaparte. It was a congenial arrangement for while Charles was in the city or preoccupied with his work, Zénaïde was able to enjoy the company of her father and her sister Charlotte. Their proximity to the cultural capital also pleased Zénaïde, who appreciated the arts, especially poetry and drama.

The prince's expertise and the Bonaparte name assured that he was quickly accepted as a member of Philadelphia's scientific societies. In 1823 he joined the American Philosophical Society, giving his nationality as Italian, an interesting detail since he is usually described as French, because of his Parisian birth (on 24 May 1803). The sudden arrival of this knowledgeable and opinionated young foreigner caused quite a stir, remembered many years later by Miss Malvina Lawson, daughter of Alexander Lawson, the engraver:

> "the advent of the Prince of Munsigno [sic] set the whole Academy [of Natural Sciences of Philadelphia] by the ears. He appeared to make warm friends and equally warm enemies. He would come to father and tell him in high glee of the last *war-whoop* and its effect, laughing heartily. For a time he seemed to take a sort of boyish delight in setting them all by the ears, but he grew tired of the fuss and I think it was one reason for his return to Europe."[1]

Bonaparte's Gull

At the Academy meetings Bonaparte met several important ornithologists, namely George Ord, Thomas Say and Titian Peale. Ord had edited the eighth and ninth volumes of Alexander Wilson's *American Ornithology* (1808–14) after the author's death in 1813. Say and Peale had travelled together to the Rocky Mountains in 1819–20 and returned with a fine collection of new and rare birds.

Ornithology had progressed so much in the decade since Wilson's death that Bonaparte decided to update the *American Ornithology* and began work on the *American Ornithology; or, The Natural History of Birds Inhabiting the United States Not Given by Wilson*. He needed access to the major collections of North American birds for which the friendship of the influential Ord and others paved the way. Since Bonaparte's English was stilted Say offered to edit the manuscript and Lawson, who had engraved Wilson's plates, agreed to do the same for Bonaparte.

Since Wilson was essentially a field ornithologist and Bonaparte was very much a closet naturalist the new work was written from a completely different perspective. For information about the living bird Bonaparte had to rely on friends such as Say, Peale, Audubon, Ord and William Cooper, his own genius lying in his ability to define the specific and generic relationships between the birds. As well as the new species discovered since 1814 he included descriptions of many female and juvenile plumages not previously given. The prince wanted his volumes to be as well illustrated as Wilson's and commissioned the artist–naturalists Alexander Rider and Titian Peale; in 1824 he sent Peale to Florida to collect specimens and make drawings.

In March of that year John James Audubon, then almost unknown, apprehensively approached Philadelphia's scientific clique seeking support for the publication of *The Birds of America*. His friend Dr James Mease took him to Bonaparte, perhaps anticipating that the Italian would be more welcoming than the establishment. Bonaparte was greatly impressed by the bold paintings and took Audubon to a meeting at the Academy so that the other members could view the folio. Alexandre LeSueur, a French naturalist who was then illustrating Say's *American Entomology* was excited by Audubon's originality and eager to encourage him, as were others, but Ord condemned the paintings out of hand. He had no intention of allowing this uneducated, long-haired backwoodsman to usurp Wilson's place as America's leading bird artist. Ord's own prestige was too closely tied up with Wilson's reputation for him to recognize the newcomer's talents and he squashed the nomination for Audubon's membership of the Academy.

Bonaparte also held Wilson in high regard, but he realized that Wilson's pioneering status was too well established to need anyone's protection. He considered using some of Audubon's sketches in his supplement, but first needed his engraver's agreement. Lawson reacted as Ord had done. He took an immediate dislike to Audubon on their first meeting and declared that even if the prince bought the drawings, he would refuse to use them. In the end he relented sufficiently to include just one, of a female Boat-tailed Grackle.

The four volumes of Bonaparte's *American Ornithology* were published between 1825 and 1833. Between 1824 and 1826 Bonaparte also wrote a number of short papers which appeared in the Academy's *Journals*, mostly describing new species from North and South America or revising the nomenclature of species already known. Other important publications from this period were his comparative list of the birds of Rome and Philadelphia, published at Pisa in 1827; his 'Catalogue of the Birds of the United States', which appeared in the first number of the *Contributions* of the Maclurian Lyceum of the Arts and Sciences, also in 1827; and 'The Genera of North American Birds' in *Annals of the Lyceum of Natural History of New York*, 1828.

At the end of 1826 Bonaparte sailed to England and travelled within Europe for a year, visiting numerous museums as well as spending some time with his family in Rome. In London he enjoyed a reunion with Audubon, who was delighted by his friend's unchanged appearance. "His fine head was not altered, his mustachios, his bearded chin, his keen eye, all was the same" wrote the artist on 18 June, 1827. While in London Bonaparte completed the second and third volumes of his *American Ornithology*, taking the manuscript with him when he returned to Philadelphia in the fall. Impatient to move permanently to Italy, he left the supervision of the work in the capable hands of William Cooper, a friend from New York for whom he had named Cooper's Hawk.

At the beginning of 1828 the Bonapartes took their young son to Rome, where they made their home close to Charles's father and other members of the Bonaparte family. The prince now turned to the classification of Italian birds and animals, but from time to time he produced minor publications on other subjects. The last to deal specifically with the North American birds was *A Geographical and Comparative List of the Birds of Europe and North America* (1838). In its preface he compared the works of Audubon and John Gould in such a way as to cause a serious rift in his friendship with the American. Audubon could hardly have appreciated the opinion that: "The merit of M. Audubon's work yields only to the size of the book; while Mr. Gould's work on the Birds of Europe, inferior in size to that of M. Audubon's is the most beautiful work on Ornithology that has ever appeared in this or any other country."[2]

Bonaparte added injury to insult by incorporating into his *List* the names of new unpublished species which he had appropriated from a list of Audubon's. Audubon complained, in a letter to his friend John Bachman, "Bonaparte has treated me shockingly, published the whole of *our* secrets ... after giving me his word of honor that he would not do so."[3] Within a year, in another letter to Bachman, Audubon was referring to his former friend as "that dirty mean fellow, the Bangs of Rome."[4]

Though Bonaparte did not again specialize in North American ornithology, over the years he added many new species to the avifauna of the region, sometimes turning them up as he searched through the vast and often neglected collections in European museums. In all, he named twenty species that occur in Canada and the mainland United States (in some cases only as accidentals). These are Wedge-rumped Storm-Petrel, Black Storm-Petrel, Cooper's Hawk, Swainson's Hawk, Sage Grouse, Semipalmated Plover, Stilt Sandpiper, White-tipped Dove, Key West Quail-Dove, Say's Phoebe, Gray-breasted Jay, Bridled Titmouse, Brown Creeper, Clay-colored Thrush, Fan-tailed Warbler, Pyrrhuloxia, Blue Bunting, Varied Bunting, Yellow-headed Blackbird and Scott's Oriole. Some of these species were not entirely new to science when Bonaparte described them, but had already been inadequately described or given names which were preoccupied.

Bonaparte never returned to America, though for a while he planned to make a great scientific tour there in the company of Louis Agassiz. In the summer of 1842 Bonaparte wrote to him: "You must keep me well advised of your plans, and I, in my turn, will try so to arrange my affairs as to find myself free in the spring of 1844 for a voyage, the chief object of which will be to show my oldest son the country where he was born, and where man may develop free of shackles. The mere anticipation of this journey is delightful to me, since I shall have you at my side, and may thus feel sure that it will make an epoch in science."[5] He was deeply disappointed when the adventure failed to materialize. However he did continue to travel widely within Europe and many naturalists visited him in Rome. In 1836 Audubon's sons, Victor and John, came to stay and were treated to such varied delights as the opera, the Duke of Tuscany's camel farm and the natural history collections of Rome. Bonaparte tried to help them by asking Pope Pius IX to buy *The Birds of America* for the Vatican Library, but his efforts failed after one of the Bonapartes shot a relative of the Pope in a hunting accident!

On the death of his father in 1840 Charles inherited his title (which like his own was a papal title) and for a few years styled himself Charles Lucien Bonaparte, Prince of Canino and Musignano. Until the 1840s the prince was content to be a political bystander, but his strong sympathy for the nationalist movement in northern Italy led to his involvement in the struggles of the Papal States when the agitation spread southwards. In November 1848 the Pope fled from Rome and three months later Bonaparte became vice-president of the legislative council. He called the people of Rome to fight for a united Italy, but the movement was crushed in July by his own cousin, Louis Napoleon, who sent French troops into Italy in order to placate the clergy. Charles Bonaparte escaped and made his way to England and the Netherlands, leaving his family in Rome. Thereafter, he and Zénaïde led separate lives, their marriage having foundered long before Charles had had to flee the country.

For some twenty years Bonaparte had been planning to make a methodical classification of all the birds of the world. Now he began the task in Leyden, despite having had to leave all his own collections, reference books and preparatory notes in Italy because of the speed of his departure. He was encouraged by his congenial friend Hermann Schlegel, the director of the Leyden Museum, who put all the bird skins under his care at Bonaparte's disposal. In May (1850) they travelled together to the Zoological Museum of Berlin and afterwards stayed for a few days with Prince Maximilian who made them welcome at his castle in Neuwied. They also visited the Reverend Baldamus and one day the three ornithologists went out shooting together across a marsh. They emerged to find that a small crowd had gathered to see the foreigners, Schlegel and Baldamus being greatly amused when an old soldier with a wooden leg suddenly spotted the mud-spattered Bonaparte and with tears welling in his eyes shouted "Vive l'empereur! vive l'empereur!" The Emperor's nephew turned round smiling and emptied the contents of his pockets into the old man's hat.

Soon after their return to Leyden Bonaparte learned that Louis Napoleon was willing to let him settle in France if he so desired. Drawn by the attractions of the Museum d'Histoire Naturelle Bonaparte quickly departed for Paris. Two years later J.W. von Müller was able to write that: "In the French metropolis he is now the leader, the supporter, and the protector of all foreign naturalists, whom he receives in his hospitable house." Among Bonaparte's more famous guests were John Gould, Carl J. Sundevall, Philip Sclater and Gustav Hartlaub.

The first volume of *Conspectus Generum Avium* appeared before the end of 1850, but the research for the next volume necessitated visits to museums in Germany, England, Belgium, Holland, Portugal and Spain. In 1854 Prince Bonaparte—as he was known in France—applied for the directorship of the Jardin des Plantes but was unsuccessful.[6]

His last years were marked by frantic activity as he sought to complete the second volume of the *Conspectus*. Progress was threatened by illness, which only served to make him work all the faster in a vain attempt to complete his mammoth project. After his death—on 29 July 1857—Gustav Hartlaub commented on the author's dogged determination:

> "The illness allowed its victim to take no more rest, either mental or physical, during his last years; with dreadful haste, driving him on from one hardly started task to another, it tortured and hounded him, in spite of his tremendous resistance, into the grave. 'The more I have to put up with, the more I work,' said Bonaparte, when I found him, during one of my last visits, writing in his bath."[7]

The great taxonomic survey was never finished, but the second volume was published at the end of 1857, after some editing by Schlegel. As with all works of classification, Bonaparte's met with criticism, one of his harshest judges being Elliott Coues, who wrote in 1880:

> "In my view Bonaparte's services to the science of Ornithology ceased in 1850. The sum total of his subsequent contributions to the subject, until death cut short his schemes, is not only a worthless but a pernicious aggregate. In his later years, Bonaparte simply played chess with birds, with himself as king: *le roi s'amuse!* Scheme followed scheme, tableau tableau, conspectus conspectus, with perpetual changes, incessant coining of new names, often in mere sport—it was nothing but turning a kaleidoscope. It may have been fun for him, but it was death to the subject."[8]

A much more appreciative assessment of Bonaparte's achievements was given in 1951 by Erwin Stresemann when he concluded:

> "There is no denying to Bonaparte the fame of having outstripped all his contemporaries in knowledge of species, especially since he studied in more museums in all sorts of countries than any other scholar; and because his *Conspectus* is essentially not a product of literary diligence but the precipitate of experience, it is often consulted even today by systematists. His arrangement of genera and their relationship to families and orders differs from our 'modern' system less than other contemporary efforts, because Bonaparte, already inspired by the theory of evolution and uninfluenced by the dogmas of natural philosophers and anatomists, could rely for essentials on his eye, which he had been sharpening for years during his careful comparative studies. Certainly this self-confidence not infrequently led him into error, but in such cases only the comparative anatomy of a later period was in a position to confute his views."[9]

Zénaïde Laetitia Julie Bonaparte[1] (1801-1854)

WHITE-WINGED DOVE *Zenaida asiatica* (Linnaeus)

Zenaida Bonaparte, 1838. *A Geographical and Comparative List of the Birds of Europe and North America*, p. 41. Type, by tautonymy, *Zenaida amabilis* Bonaparte = *Columba zenaida* Bonaparte = *Columba aurita* Temminck
Columba asiatica Linnaeus, 1758. *Systema Naturae*, 10th edn, Vol. 1, p. 163. Based on 'The Brown Indian Dove' Edwards, *A Natural History of Birds*, Vol. 2, p. 76, Pl. 76: in Indiis [= Jamaica]

ZENAIDA DOVE *Zenaida aurita* (Temminck)

Columba Aurita Temminck, 1810. In Knip, *Les Pigeons, Les Colombes*, p. 60, Pl. 25: Martinique

MOURNING DOVE *Zenaida macroura* (Linnaeus)

Columba macroura Linnaeus, 1758. *Systema Naturae* 10th edn, Vol. 1, p. 164. Based mainly on 'The Long-tailed Dove' Edwards, *A Natural History of Birds*, Vol. 1, p. 15, Pl. 15: in Canada, error [= Cuba]

Zénaïde Bonaparte was born in Paris on 8 July 1801, the elder daughter of Joseph and Julie Bonaparte. Her uncle, Napoleon I, made her father King of Naples in 1806 and King of Spain two years later. After Waterloo, Joseph Bonaparte left France for the United States but Zénaïde and her sister Charlotte remained with their mother in Europe. At the age of twenty-one Zénaïde accepted an arranged marriage with her nineteen-year-old cousin Charles, a self-taught naturalist who was one of Lucien Bonaparte's sons.

After their wedding in Brussels the young couple spent some months in Italy, then sailed to America in the summer of 1823 to join Zénaïde's father on his estate at Point Breeze, near Bordentown, New Jersey, some twenty miles north-

ZENAÏDE BONAPARTE IN 1820. *NIECE OF THE EMPEROR NAPOLEON AND DAUGHTER OF JOSEPH BONAPARTE, SHE MARRIED HER COUSIN CHARLES LUCIEN BONAPARTE.*

east of Philadelphia. Their first home was a white house with green shutters, linked to Joseph's villa by an underground passage and surrounded by parkland and farms.

Charles Bonaparte was quickly accepted by Philadelphia's scientific societies and various bird collections were made available to him so that he could begin a systematic study of the North American avifauna. In 1824 Zénaïde gave birth to their first child whom they diplomatically named Joseph-Lucien after his two grandfathers.

Early in 1828 Charles took his family to Italy. They spent some months in Florence with Zénaïde's mother and then went on to Rome where they made their home near that of Charles's father. Other members of the Bonaparte clan had already settled in Rome, among them their grandmother Letizia, the redoubtable 'Madame Mère'. In November Lucien Louis was born and Zénaïde bore ten more children: Julie, Charlotte, Marie, Augusta, Napoleon, Mathilde, Albert, Albertine, Alexandrine and Leonie, the last-named four dying in infancy.

When Zénaïde was not fully occupied with the demands of her large family she delighted in poetry and translated the German didactic poems of Friedrich von Schiller. Her husband concentrated on his ornithological publications and the last to deal directly with the American avifauna was *A Geographical and Comparative List of the Birds of Europe and North America* (1838). It was in this list that he created the genus *Zenaida* which includes six species; the White-winged, Zenaida, Eared, Mourning, Socorro and Galapagos Doves. Only the Mourning and White-winged Doves are regular breeders within the United States. The Zenaida Dove is a West Indian species, accidental on the Florida Keys where it bred in the past.

Charles briefly dabbled in politics and became a leader of the radical party in Rome. When French troops arrived in the spring of 1849 he was driven from the city, and carried on his bird work in Great Britain and the Netherlands until he was allowed to settle in France. Zénaïde had no desire to accompany her husband and stayed on in Rome. For the rest of their lives they lived apart and in 1853, after Zénaïde pressed a financial claim against him, Charles applied to Napoleon III and the Pope for a separation. The request was granted the following year, Napoleon decreeing that the younger children should remain

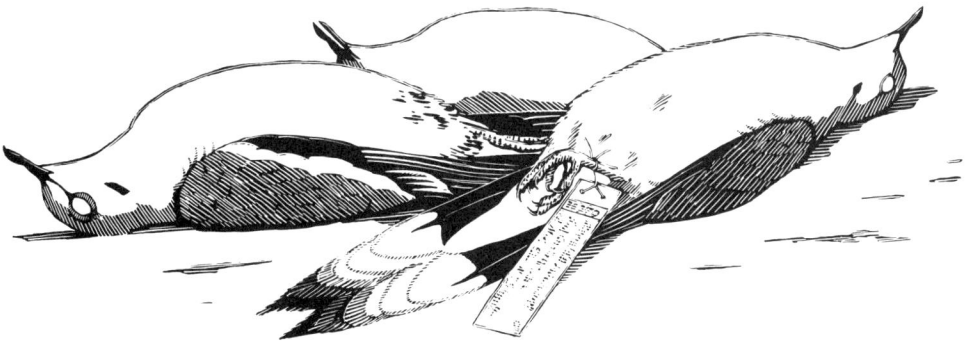

White-winged, Zenaida and Mourning Doves

with their mother. Charles, seriously ill in Paris, nevertheless sent his brother Pierre to fetch his fifteen-year-old son, Napoleon, who because of the circumstances was allowed to join his father.

However it was Zénaïde who died first. Concerned about her physical condition she went to Naples for a holiday, and while there she passed away on 8 August 1854 at the age of fifty-three. Charles, though constantly suffering from weakness and pain, outlived his wife by three years.

Matteo Botteri
(1808-1877)

BOTTERI'S SPARROW *Aimophila botterii* (Sclater)

Zonotrichia botterii P.L.Sclater, 1858. *Proceedings of the Zoological Society of London* (1857), p. 214: vicinity of Orizaba, [Veracruz,] in Southern Mexico

Matteo (or Mateo) Botteri was born on 7 September 1808 in the port of Lesina, on the Dalmatian island of the same name (now Hvar, Yugoslavia). For many centuries Dalmatia had been controlled by the Republic of Venice but in 1797 it was ceded to Austria. By the time of Botteri's birth the country was under Napoleonic rule, but in 1814 Austria regained Dalmatia and Botteri's mother was one of the casualties of the struggle. After his father's death the orphan was taken into the care of his uncle who was a canon in Lesina Cathedral. The boy became imbued with an abiding love for the Roman Catholic Church and its liturgies but his passion for natural history and a desire to travel proved to be more dominant forces.

Little is known about Botteri's early manhood but he certainly travelled to the Greek mainland, the Ionian Islands and Turkey where he made zoological and botanical collections. He took a particular interest in Adriatic fish and marine plants and corresponded on these topics with other European naturalists.

Early in 1853 he agreed to visit Mexico in order to collect and cultivate plants for the Royal Horticultural Society of London. In July he crossed the Adriatic Sea and travelled overland through Germany and France to England. He reached Vera Cruz in July 1854 and immediately headed inland to the ancient city of Orizaba, some four thousand feet above the sea, where he had been instructed to make his base. He was so impressed by the beauty of the surrounding mountains and the friendliness of the local people that he soon decided to stay for the rest of his life. Botteri not only fulfilled his obligations to the Royal Horticultural Society but also worked for the Academy of Sciences of Paris and made collections of every kind, especially of land molluscs and algae. He sent his surplus, uncommissioned material to his European agent, Samuel Stevens, who was a natural history dealer in London.

In 1857, on receiving a collection of birds from around Orizaba, Stevens invited Philip Sclater to make an evaluation. Sclater identified over 120 species

BOTTERI. *PROFESSOR OF NATURAL HISTORY AT ORIZABA COLLEGE, MEXICO. HE SENT SOME OF HIS PLANT AND BIRD COLLECTIONS TO EUROPE.*

and in the *Proceedings of the Zoological Society* published notes on thirty-eight of them, two of which were new to science: the Slaty Vireo and Botteri's Sparrow. After naming the sparrow, Sclater added "I hope M. Botteri will forward better specimens of this interesting species (the examples in the present collection being badly preserved), so as to allow me to make a more accurate investigation of its differential characters."

When the Royal Horticultural Society ran short of funds, Botteri supported himself by tutoring. Pupils were not hard to find, since the inhabitants of Orizaba had come to regard the foreigner as a kind of walking encyclopaedia. As a naturalist his interests embraced geology, mineralogy, botany, conchology, icthyology, entomology and lichenology. As a linguist he had mastered "a dozen languages" which included Latin and classical Greek, modern Greek, French, Spanish, Italian, German and English. In literature his tastes encompassed Virgil, Homer, Ossian and the Italian poets and to those interests could be added history, geography, astronomy, navigation, etymology and ethnology. It seems that Botteri became a popular figure in Orizaba, his great learning, infectious enthusiasm and proverbial generosity winning him many friends. The only personal criticism to have survived the years concerns his preference for honesty rather than tact, though one friend excused him by noting that his frankness could be like medicinal bark: bitter but beneficial.

Botteri returned to Europe only once, during 1863–64, and on his return he became Professor of Languages and Natural History at Orizaba College. He founded a small museum there and continued to collect for himself and others, sending some of his material to the Smithsonian Institution. In 1866 he became a corresponding member of the Entomological Society of Philadelphia. Though he was offered a more prestigious post in Mexico City he chose to remain in Orizaba, where he continued to teach at the college until his sudden death on 3 July 1877, at the age of sixty-eight.

During the early 1870s Botteri's Sparrow was first added to the avifauna of the United States by Henry W. Henshaw who worked the deserts and mountains

Botteri's Sparrow

of south-eastern Arizona with the Wheeler Survey during two field seasons. Sometimes guided by Apaches he hurried from one locality to another and collected fourteen specimens of Botteri's Sparrow around Camp Grant, Camp Crittenden and Cienaga. The sparrow's range also extends into south-western New Mexico and southern Texas.

Johann Friedrich von Brandt (1802-1879)

BRANDT'S CORMORANT *Phalacrocorax penicillatus* (Brandt)

Carbo penicillatus J.F.Brandt, 1838. *Bulletin de l'Académie Impériale des Sciences de St.-Petersbourg* 3, col. 55: no locality [= Vancouver Island]

J.F. Brandt described Brandt's Cormorant in 1838,[1] but there are no skins of this large cormorant at St Petersburg dated earlier than 1841, so the original specimen must have been lost or destroyed. Since the origin of the type specimen is unknown, neither the collector nor the precise location from whence it came can ever be determined.

Brandt's Cormorant is resident on the Pacific coast of North America, mainly between Vancouver Island and Lower California, so the few Russian naturalists who visited the area in the years prior to 1838 and who possibly supplied Brandt with a specimen will be mentioned below. Even if it is not possible to decide who first found the bird the exercise will be worthwhile, since it places these Russian investigations in their proper context with regard to early North American ornithology: most North Pacific birds were described in Europe by Russian, German or British naturalists long before Lewis and Clark had crossed the continent and before either Wilson or Audubon had set foot in America. Brandt's Cormorant happens to be one of the exceptions.

Brandt was born in Jüterbog near Brandenburg, Germany, on 25 May 1802, growing up during the latter part of the Napoleonic War and its aftermath. As a boy he enjoyed finding and identifying plants and was greatly encouraged by an uncle who taught him at the local school and who inspired him still further by presenting him with a copy of Humboldt's 'Travels'. Brandt's father was a doctor and wanted his son to follow in the same profession, so he sent him to the Lyceum at Wittenberg about a hundred miles to the north-west. When he was nineteen years old he moved back closer to home, to Berlin University, where he came under the influence of Hinrich Lichtenstein who was then beginning to make the Berlin Zoological Museum into one of the most commendable in Europe. Indeed Lichtenstein later became a useful and influential contact for him. For a while Brandt greatly enjoyed being sidetracked into

BRANDT. *DIRECTOR OF THE ZOOLOGICAL MUSEUM AT ST PETERSBURG. HE DESCRIBED SOME OF THE NATURAL HISTORY MATERIAL BROUGHT BACK FROM THE RUSSIAN AMERICAN POS-SESSIONS.*

zoology and botany but he soon had to curtail these interests in order to concentrate on his medical exams, which he passed successfully in 1826. A nine-month period as a doctor's assistant discouraged him so much that he started lecturing in botany and pharmacology at the university, writing only the occasional paper of a medical nature. Once beyond the age of twenty-nine he never practised medicine again, because he moved to St Petersburg to become Director of the Zoological Museum at the Academy of Sciences.

He soon found that the museum needed urgent attention. The last part of the eighteenth century and the first part of the nineteenth century had been a frantic period of Russian expansion and scientific exploration but many of the zoological collectors had suffered misfortunes. Steller, for instance, had been forced to leave behind valuable material on Bering Island and specimens of his from Kamchatka and other parts of Siberia that arrived safely at St Petersburg later decayed and disintegrated. Between 1768 and 1774 Pallas, Georgi, Falk, Lepechin, Güldenstädt and S.G. Gmelin had roamed through much of eastern Europe and Siberia contributing enormously to the collections but in 1795 Pallas suddenly moved from the Baltic to the Crimea, taking with him many types and other specimens which were subsequently lost. More encouragingly, new material began to arrive as the fur-bearing mammals of the North Pacific began to be exploited following Bering's second expedition. A good proportion of these exciting new items were from the Russian American possessions, which then included the Aleutian Islands, much of coastal Alaska southwards to Sitka and a few isolated outposts dotted down the coast, almost to San Francisco Bay.

There was a temporary lull in exploration when Brandt arrived in Russia in August 1831 and he was able to initiate a period of intensive study that was matched at the same time by Fischer von Waldheim at Moscow, by Edouard Eversmann at Kazan and Karl Kessler at Kiev. He began to augment the meagre stock of reference books and since most of the indigenous animals of the country were not represented in the museum he set about the long process of acquiring them, either by exchanging or purchasing specimens or by encouraging field collectors.

From the 1840s onwards specimens of all kinds began to filter back from the extremities of the empire in a second phase of expansion. Brandt was on hand to receive some of the discoveries of Severtzov and Prjevalski from Central Asia, and of Middendorff, Maack, Schrenck and Radde from Siberia and the Amur. Conversely, fresh material from Russian America gradually declined, and ceased altogether in 1867 when Alaska was sold to the United States.

Partly for his own benefit, Brandt began his written work at St Petersburg with a review of the zoological publications held at the Academy. This was followed by some 300 titles of which fully two-thirds were mainly about birds and mammals. His first ornithological publication, *Descriptiones et icones animalium rossicorum novorum vel minus rite cognitorum, Aves* (1836), was intended to be the start of a series on the zoology of the empire but only this part on some of the birds was ever issued. He afterwards became interested in avian osteology, publishing several articles on this subject in the *Mémoires* of the Academy that were later bound together and issued as a separate volume in 1837.

In the following year Brandt published a short paper in the Academy's *Bulletin* entitled 'Observations sur plusieurs espèces nouvelles du genre Carbo ou Phalacrocorax'. The skins for this study on cormorants had been found in the museum and included a skin of the bird that has since become known as Brandt's Cormorant. Although he described it as new it may have been stored there for some time. He knew nothing of its habits nor of its origins and said only: *Patria, quod maximopere dolendum, ignota.*

Who then could have collected the cormorant? The Russian exploration of America began with Bering and Chirikov but hardly any of the collections survived until Brandt's day (see Steller). Other expeditions soon followed, many of them having on board naturalists or at least physicians with zoological interests. Thus Dr Carl Merck was with the Joseph Billings expedition and collected in Kamchatka and on Kodiak Island between 1785 and 1794; Baron Langsdorff was on A.J. von Krusenstern's voyage and collections were made at Sitka, the Pribilof Islands and the Aleutians between 1803 and 1806; Johann Friedrich Eschscholtz was with Otto von Kotzebue from 1815 to 1826 and Baron von Kittlitz was with Captain Lütke from 1826 to 1829, both of these latter naturalists collecting at Sitka and several Bering Sea locations. Ferdinand von Wrangel, explorer and one-time governor general of Russian America, either collected his own material or acquired specimens from land-based Russians in the region. Of all these men only Langsdorff and Eschscholtz visited the American coast between Vancouver Island and California and could therefore have collected Brandt's Cormorant. Langsdorff passed Vancouver Island, visited the Columbia River estuary and was at San Francisco Bay from March to May 1806 but he afterwards lost most of his plants and birds in a gale. Eschscholtz was at Fort Ross (just north of San Francisco) in the autumn of 1824 and although 165 species of birds were observed or collected during the course of his whole voyage no detailed list is available. Even so, Eschscholtz seems the most likely Russian to have brought back the cormorant skin.[2]

Although Brandt had large numbers of bird skins under his supervision and gave names to many of them, only two from the North Pacific remain valid,

Brandt's Cormorant

the aforementioned cormorant and the Spectacled Eider *Lampronetta fischeri* which he named in honour of his counterpart at Moscow, Fischer von Waldheim. Brandt did also write the first description of the lovely Red-legged Kittiwake, a species generally restricted to the environs of the Aleutian Chain and the Pribilofs, but he sent his notes to Carl Bruch who included them in a work on the gulls that appeared in the first volume of the *Journal für Ornithologie* (1853). Though Bruch had not even seen the specimen Brandt lost the credit. Some other species described by Brandt have since been relegated to subspecific rank, including the now rare Aleutian Canada Goose.

But the main reason for his names not now being valid is that he had been preceded in the field of scientific nomenclature by Pallas and J.G. Gmelin. The former had described his own birds from the Siberian side as well as some of those collected by Steller and Merck from Kamchatka and Alaska. Gmelin had given scientific names to the birds brought back from Cook's voyage up the Alaskan coast (see Cook). Few species therefore remained to be described from this region when Brandt was active. Some new generic names for the auks that he created in 1837 still survive: *Brachyramphus* (Marbled and Kittlitz's Murrelet), *Synthliboramphus* (includes Xantus's, Craveri's and Ancient Murrelet), and *Ptychoramphus* (Cassin's Auklet).

An interesting aspect of Brandt's career was his proposed ornithological survey of the Russian American possessions. He engaged Wilhelm Pape to depict forty-four species in a series of seven plates from material at the museum. Such Alaskan species as the Red-faced Cormorant, Snow Goose, Emperor Goose, Steller's Eider, Black Oystercatcher and numerous auks were included, and though most were indeed coastal or pelagic species there were some land birds such as Rufous Hummingbird, Steller's Jay, Blue Grouse and Aleutian races of Savannah, Song and Lincoln's Sparrows. Brandt appears to have started the work around the year 1835 but he set it aside and then abandoned it altogether when Alaska was sold. The skins used must have come from the collections of Kittlitz, Wrangel, Langsdorff and Eschscholtz but again it is impossible to assign particular engravings to the birds collected by particular naturalists. The simple but pleasing pictures would have provided the very first illustrations of Brandt's Cormorant, Black Oystercatcher and Red-legged Kittiwake as well as the now extinct Spectacled Cormorant, but they remained unpublished until 1984.

Brandt's unfinished projects suggest that he was overworked. However, he found time to make a trip to the Crimea for mammoth relics and also went to the Caucasus, mainly in search of fishes. He occasionally returned to his home area and travelled in other parts of western Europe, making at least one trip to England. Perhaps his interests and priorities simply changed as more and more material arrived at the museum from various parts of the world and his unfinished works became the inevitable by-product of his long academic career. He died on 3 July 1879 at Merreküll in Estonia at the age of seventy-seven.

As a postscript it may be added that a brother of Brandt's was also well known throughout the museum world.[3] As a successful natural history dealer, based in Hamburg, he had the advantage over his competitors of receiving many duplicate skins from the Russian expeditions and was thus able to supply collectors and museums with rare and valuable material from Central Asia, Siberia and Alaska.

Aglaé Brelay[†] (fl. 1839)

ROSE-THROATED BECARD *Pachyramphus aglaiae* (Lafresnaye)

Platyrhynchus Aglaiae de La Fresnaye, 1839. *Revue Zoologique* [Paris] 2, p. 98: Mexico [= Jalapa, Veracruz]

"In dedicating this species to Madame Brelay, our sole aim has been to render homage to the particular zeal with which she busies herself with ornithology and assists Monsieur Brelay in the formation of his large collection, which already numbers several thousand items. We are far from approving of the habit of giving new birds the names of women, who are often strangers to the love of ornithology ... We do not think that personal names are admissible except when they recall that of some naturalist, writer, traveller, artist or zealous collector who has already rendered or is capable of rendering some service to the science."[1]

Aglaé Brelay must therefore have been quite an enthusiast to win this tribute from Baron Frédéric de Lafresnaye—one of the foremost nineteenth century

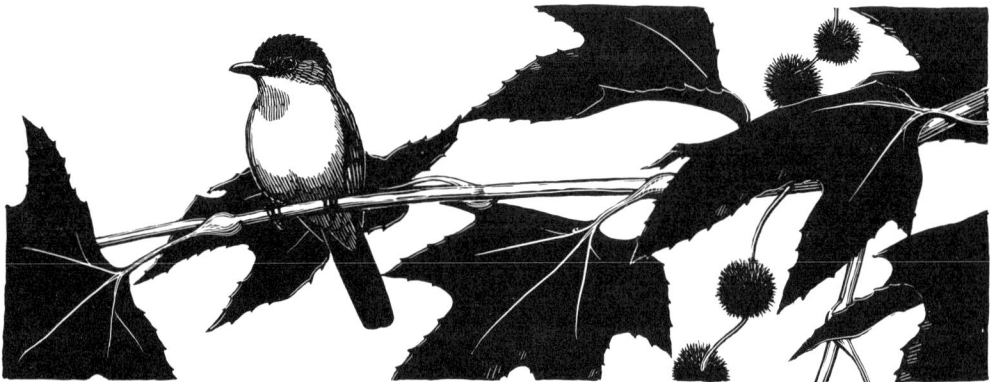

Rose-throated Becard

†No portrait traced

French ornithologists. Unfortunately, nothing is known about her or her husband. Lafresnaye described a few of their birds in the *Revue Zoologique* for April 1839 but there is no introduction to 'QUELQUES OISEAUX NOUVEAUX de la collection de M. Charles BRELAY, à Bordeaux'. There is just a list of eleven birds described in Latin, all from "Mexico" except one from "Rio Grande". Only the Chestnut-capped Brush-Finch, Brown-backed Solitaire, Spotcrowned Woodcreeper and the Rose-throated Becard proved to be new birds. The latter sometimes breeds north of the border in Arizona and near the Rio Grande in Texas.

BREWER IN 1879. *NEWSPAPER EDITOR, PUBLISHER AND PROMINENT OOLOGIST. HE CO-AUTHORED* A HISTORY OF NORTH AMERICAN BIRDS *WITH BAIRD AND RIDGWAY.*

Thomas Mayo Brewer (1814-1880)

BREWER'S SPARROW *Spizella breweri* Cassin

Spizella Breweri Cassin, 1856. *Proceedings of the Academy of Natural Sciences of Philadelphia* 8, p. 40: Western North America, California, and New Mexico [= Black Hills, North Dakota]

BREWER'S BLACKBIRD *Euphagus cyanocephalus* (Wagler)

Psarocolius cyanocephalus Wagler, 1829. *Isis von Oken*, col. 758: Mexico

Thomas Mayo Brewer first came to the notice of ornithologists at large when his observations began to appear in Audubon's *Ornithological Biography* (1831–39). In 1840, when Brewer was twenty-six years old, he produced a new edition of Wilson's *American Ornithology* that cost much less than any previous edition and was therefore accessible to a much wider range of people. Both these works were widely read at home and abroad and Brewer's name became a familiar one in ornithological circles.

His few biographers say nothing of his early years other than the fact that he was born in Boston on 21 November 1814, the son of Colonel James Brewer, a Revolutionary patriot who was involved in the Boston Tea Party. Thomas Brewer graduated from Harvard College in 1835 and from the Medical School three years later. After a few years as a doctor he abandoned his profession, not to become a full-time naturalist but to exercise his inclination for politics and writing. He was a regular contributor to the *Boston Atlas* and later became its editor. Afterwards he was involved in the publishing firm of Swan & Tileston, which changed its name to Brewer & Tileston in 1857 when he became a partner. He continued in publishing and book-dealing until his retirement in 1875. In the following short account of his ornithological work it should therefore be remembered that his achievements in this field were carried out as a sideline to the main business of his life. Yet he was one of the best known ornithologists of his day, the editor, author, or co-author of a number of classic publications. This is not altogether remarkable since there were few full-time professionals offering competition and most American naturalists had to earn their living by one means or another.

In 1835 Brewer joined the Boston Society of Natural History and over the next forty years contributed papers at irregular intervals mainly relating to the birds of New England. His earliest paper appeared in the first volume of the society's *Journal* in 1837 in which he added forty-five extra species to a previous catalogue of the birds of Massachusetts by Professor Edward Hitchcock. Three years later his single-volume octavo edition of Wilson's work was issued. The uncoloured and drastically reduced illustrations seem very unhelpful but the text was immensely useful. He supplemented Wilson's original text with a synonymy and gave a critical commentary on Jardine's British edition, as well as adding a new appendix updating the number of birds then known in the United States and Canada.

By now Brewer had become friendly with Audubon to whom he sent a fund of notes and specimens, particularly of east coast species. His observations most often referred to nests and eggs—Brewer's particular speciality. Each contribution in the *Ornithological Biography* is preceded by an acknowledgement from Audubon, often along the lines of "My young friend Mr. T.M. Brewer says..."

After the *Ornithological Biography* was completed Audubon went with a few friends up the Missouri River to collect more birds and mammals for his current projects. On 20 June 1843, during their sojourn at Fort Union, Edward Harris and John Bell went out to look for more Sprague's Pipits, a new bird which they had found the previous day. They "came back bringing several small birds, among which three or four proved to be a Blackbird nearly allied to the Rusty Grakle [Rusty Blackbird] but with evidently a much shorter and straighter bill."[1] In 1844, in the octavo edition of *The Birds of America*, volume seven, page 345, Audubon named this

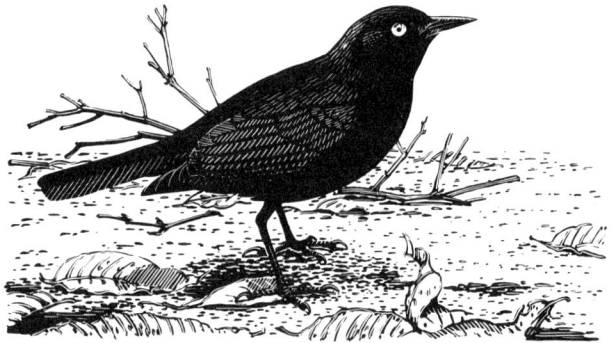

Brewer's Blackbird

new addition to the North American avifauna after his friend Brewer, and the name Brewer's Blackbird has been with us ever since. Audubon's scientific name *Quiscalus brewerii* has not survived because Johannes Wagler had described the species fifteen years earlier from a Mexican specimen.

In 1846 Brewer went to see Audubon and his two sons at *Minnie's Land*. The artist, surrounded by grandchildren, was now sixty-one years old and showing signs of his age. Writing of his visit afterwards Brewer found that "The patriarch had greatly changed since I had last seen him. He wore his hair longer, and it now hung down in locks of snowy whiteness on his shoulders. His once piercing gray eyes, though still bright, had already begun to fail him. He could no longer paint with his wonted accuracy, and had at last, most reluctantly, been forced to surrender to his sons the task of completing the illustrations to the 'Quadrupeds of North America'."[2]

As Audubon's powers declined he was superseded by another generation of ornithologists of which there were four leading lights: Spencer F. Baird, John Cassin, George N. Lawrence and Thomas M. Brewer. Baird and Brewer were great friends: one of their earliest encounters was in May 1844 at Baird's home town of Carlisle, Pennsylvania, when the two young men enjoyed a wonderful few days of egging which Baird reported to his brother William as follows:

> "… we rode over to Wagoner's gap. There we got a lot of Cliff Swallows eggs with some Barn ditto. On Friday—We went down the spring & over to Tempe, getting Rough Wing Swallows eggs, Indigo Bird, & Chipping Sparrows; also on Saturday we went down to Sam Miller's where we got Red head Woodpecker, Dove, Robin, Blue Bird, Field Sparrow & Larks eggs & some imperfect Field plover, the young having just come out. This last lays an immense egg, being nearly allied to the Willit in this respect. From Miller's we went over to Middlesex, where we got several Rough-Wing Swallow's nests, 16 eggs. We had a great deal to talk about, & wished that you were here … we all liked him very much."[3]

Brewer combed all the different east coast habitats for birds and set up a network of American and European correspondents with whom he exchanged or bought skins and eggs. He is said to have accumulated the finest private collection of eggs then known in North America. In 1846 when Baird visited Brewer in Boston it was a pretty fine show; and of course he continued to add to it throughout his life. The English ornithologist Alfred Newton was not so impressed by it. When writing from Boston to Canon Tristram he reported:

> "I have seen Dr. Brewer and his collection, of neither is much to be said; the Dr. is reserved to an astonishing degree for a Yankee, and has evidently never enjoyed anything like those glorious days of last September when you met me at Cambridge [England], the memory of those talks *de omnibus avibus*, etc., has cheered me many times in the last eight months … His collection is extremely moderate considering the scope of it and what must have been great opportunities, the only egg I *coveted* a Hooded Merganser's, of which you have one."[4]

A few years before Newton's visit, in June 1850, Brewer had made one of his more extended field trips to New Brunswick and Nova Scotia, principally to observe the breeding habits of some of the northern birds, claiming success beyond his expectations. Sometimes following Audubon's old trail of 1833, Brewer several times doubted the validity of the old man's observations, querying the real identity of a supposed Harlequin Duck nest and believing him outrightly mistaken in his description of American Tree Sparrow eggs. On the other hand he was able to defend Audubon's much questioned assertion that Herring Gulls on White Head Island bred in high trees, since he had seen the nests there for himself. Amongst the highlights for Brewer were his visits to Leach's Storm-Petrel colonies on the islands around Grand Manan.[5]

Brewer was now a family man. In 1849 he had married Sally Coffin, daughter of Stephen Coffin of Damariscotta, Maine. They had two children, a son who died at an early age, and a daughter. Like the Baird's only child (also called

Lucy) she too had a bird named after her: the Honduran Emerald *Amazilia luciae* (Lawrence, 1867).[6]

All of Brewer's work from the preceding springs and summers, the hours of careful observation, the nest finding and collecting, the correspondence with other enthusiasts, the thorough examination of the sometimes contradictory early literature, the checking and double checking, at last came to fruition for Brewer in 1857. His 'North American Oölogy; being an Account of the Habits and Geographical Distribution of the Birds of North America during their Breeding Season' was published in volume eleven of the *Smithsonian Contributions to Knowledge*. Part one comprised 132 pages supplemented by five plates of seventy-six eggs of birds of prey, swifts, swallows, goatsuckers and kingfishers collected by Audubon, John Bachman, Jean Louis Berlandier, Brewer, John H. Clark, Adolphus Heermann, John Krider, James Trudeau and Gustavus Würdemann. It was Brewer's most important work, bringing him the greatest credit. Unfortunately no more parts were ever issued partly because of the high cost of the plates.

But Brewer's knowledge of the breeding habits of birds was not lost, because his remaining bird biographies found their way into the first part of Baird, Brewer and Ridgway's *A History of North American Birds*. The three-volume 'Land Birds' appeared in 1874, Brewer accounting for two-thirds of the text. The year following its appearance, Brewer retired from his publishing business and was set to devote the remainder of his life to his family and to ornithology, but he lived for only five more years, two of them spent in Europe. After a short illness, he died in Boston on 23 January 1880 at the age of sixty-six. He had just completed his portion of the manuscript for the 'Water Birds', the second part of *A History of North American Birds*. His egg collection of 3500 lots of almost a thousand species he bequeathed to the Harvard Museum of Comparative Zoology.[7]

Brewer should be held in high esteem as a noted American oologist but he is often remembered instead for his part in the infamous 'Sparrow War'. In 1853 the grain-eating House Sparrow, of Asiatic origin, had been introduced into America from Britain where it had been established for centuries. The bird was imported partly for nostalgic reasons and partly to help consume insect pests. Ornithologists such as J.A. Allen, William Brewster, Henry Henshaw and Elliott Coues were strongly against this foolhardy introduction but Brewer sided with the sparrows. Bitter argument raged in the press and in the pages of the *American Naturalist* as each side attacked the other over a period of years, chiefly in the 1860s and 1870s.

After Brewer's death Henshaw was able to recall him without acrimony: "Though possessing a somewhat peppery disposition, he was a most kind and courteous gentleman, and was particularly fond of young people, and ever ready to lend a helping hand or speak a word of cheer to the aspiring young ornithologist"[8] (a tribute at variance with Witmer Stone's assertion that Brewer was "not one to encourage the development of new ornithologists").[9] Henshaw made no mention of House Sparrows and neither did Allen when he wrote an obituary of Brewer in the *Proceedings of the Boston Society of Natural History*. Not so Elliott Coues, who had always been the most vicious in his attacks on

Brewer. In the first volume of *The Osprey* (1897) Coues called him "a narrow-minded, prejudiced, and tactless person" and elsewhere declared that "Brewer had made a fool of himself about the Sparrows for years." Only nine weeks after Brewer's death Coues wrote (in a private letter, thank goodness): "We both know that Brewer was a cantankerous old ass at the time he had the good taste to fall asleep in Jesus."[10]

As is well known, the House Sparrow has since spread into every state and Canadian province inflicting immense economic loss and untold ecological damage through its grain-eating habits. Brewer was wrong, but at least he was on the winning side.

Ironically, Brewer's name is commemorated in the name of a native North American 'sparrow'. In 1856 John Cassin separated the Brewer's Sparrow from the similar Clay-colored Sparrow from specimens in the Philadelphia Academy's collection. Cassin recorded that he had "much pleasure in embracing the present opportunity to dedicate a bird of the United States to my esteemed friend Thomas M. Brewer, M.D., of Boston, one who to the highest abilities and social qualities adds an ardor in devotion to Ornithological science rarely paralleled."

Brewer's Sparrow

BULLER. *NEW ZEALAND LAWYER AND AUTHOR OF* A HISTORY
OF THE BIRDS OF NEW ZEALAND *(1872–73). THIS ENGRAVING
FORMS THE FRONTISPIECE TO ITS* SUPPLEMENT *(1905), SHOWING
HIM IN HIS ROBES AS DOCTOR OF SCIENCE, UNIVERSITY OF
CAMBRIDGE.*

Walter Lawry Buller
(1838-1906)

BULLER'S SHEARWATER *Puffinus bulleri* Salvin

Puffinus bulleri Salvin, 1888. *Ibis*, p. 354: New Zealand

"On the 6th of November, about six miles west of Point Pinos [California], two white-breasted shearwaters dashed up to the boat—one a pink-footed, the other a slender bird without conspicuous mottling on the sides of the head. The first glance revealed that the bird was a stranger. It was only a few yards away and I had to wait a moment for it to pass astern and get within proper range. A successful shot brought it down in perfect condition for a specimen ...

Upon consulting the literature it was found that the specimen agreed with the descriptions of Buller's shearwater, and was the fourth one known to science. The bird had been secured in a region far remote from the supposed habitat of the species, the types and third specimen having come from New Zealand seas. It may confidently be expected that persistent observation off Monterey will add to the list of pelagic wanderers from austral regions."[1]

In this short quote, published in 1900, Leverett M. Loomis not only describes the way in which he secured the first Buller's Shearwater to be taken in North American waters but also shows that at the beginning of this century Monterey Bay was already recognized as part of an outstanding area for "pelagic wanderers". Nowadays, Buller's Shearwater is avidly sought after as a scarce but regular visitor off the California coast from its New Zealand breeding grounds.

The English and scientific names of this shearwater commemorate Walter Lawry Buller, New Zealand's first notable native-born ornithologist. Quite a number of men had worked on the birds before him. The two Forsters collected in the country during Cook's second great voyage and the French zoologist R.P. Lesson discovered a few more species in 1824 and 1827. G.R. Gray added an ornithological appendix of eighty-four species to Ernst Dieffenbach's *Travels in New Zealand* (1843) based on the explorer's skins, then Gray made a more complete list when he wrote up the birds for *The Zoology of the Voyage of H.M.S. Erebus and Terror* (1844–75). Minor discoveries were made by other French and British travellers and in 1848 Titian Peale, ornithologist with Captain Charles Wilkes's U.S. Exploring Expedition, described a dozen birds following their short stay at the Bay of Islands in 1840. But Buller made by far the longest study and was the first to bring all the available information together in a comprehensive way, much of it from his own field experience. He remains the most celebrated early ornithologist of the country and his works are still often consulted, not least because of the drastic changes to the avifauna that have taken place over the past hundred years or so.

Buller's magnificent *A History of the Birds of New Zealand* (1872–73) raised the number of known species to 147 and was immediately praised for its readable, erudite text and its thirty-five exquisite hand-coloured lithographs by J.G. Keulemans. The first edition of 500 copies was soon accounted for and the two-volume second edition of 1000 copies, with some additional species, was a sell out not long after it appeared in 1887–88. One of the new species in the new edition was Buller's Shearwater, which for Buller was a chance find, a gift from the sea:

> "The only example of this fine Petrel I have had an opportunity of examining in the flesh was picked up by myself on the ocean-beach near the mouth of the Waikanae river on the 1st October, 1884, having been blown ashore ...
>
> It is remarkable for its length of neck and tail. Indeed at first sight it looks more like a small Shag than a Petrel, and several of the Maoris at Waikanae to whom I showed it declared that it really was a Kawau till I pointed out to them its tubular nostrils."[2]

Keulemans's drawing of the shearwater was accompanied by very little text since Buller knew nothing about it. Osbert Salvin, a great expert on petrels who had published the first description of the species earlier in the same year, knew nothing more since he had only seen Buller's single specimen and another purchased from Henry Whitely, an English taxidermist and dealer, who could only say that it had originated from "New Zealand".

Buller's interest in birds can be traced back to his earliest years. In the preface to the first edition he confessed that the idea of writing a history of the birds

Buller's Shearwater

of New Zealand had been "one of the day-dreams of my early boyhood". He was so pleased with the reviews of the first edition that he had five pages of them appended to the second edition!

The son of a Cornish missionary, he was born on 9 October 1838 at Newark Wesleyan Mission Station at Pakanae and was educated at Wesley College in Auckland. Moving with his parents to Wellington in 1854 he took a year out because of poor health and devoted himself to scientific, legal and Maori language studies. Around this time he was befriended by William Swainson who was in self-imposed exile from his partially successful career as a zoological writer and illustrator in Britain. Swainson had known almost all the best European naturalists and no doubt inspired Buller with his tales of the great bird collections and the inside details on the production of such famous works as *The Birds of America* and the *Fauna Boreali-Americana*.

At the age of seventeen Buller was appointed Government interpreter at Wellington and in 1859 was made Native Commissioner for the Southern Provinces. In 1862 he married Charlotte Mair and became a justice of the peace. He thereafter held a number of judicial posts and in 1871 went to England as secretary to the Agent-General. He was called to the bar at the Inner Temple in London but three years later he returned to Wellington where he practised law for the next twenty-four years. During this time he was gathering together material for his books as well as publishing over seventy scientific papers on birds. Even his stay in London had been arranged, with the collusion of the New Zealand Government, to allow him to see the first edition of his *History* through the presses. Afterwards, at the request of the government he produced his small *Manual of the Birds of New Zealand* (1882) to provide a cheaper elementary book for popular use.

Many learned societies bestowed honours upon Buller for his zoological work and he was knighted by Queen Victoria for his services as a Commissioner. He moved to London in 1898 and died there on 18 July 1906. A year before his death his two-volume *Supplement to the History of the Birds of New Zealand* brought the state of knowledge up to date with much additional discussion. Keulemans provided some more colour plates and this time there were a number of photographs of birds and bird habitats.

After Buller's death his personal collection of birds went to Canterbury, New Zealand. Three earlier collections had already been disposed of: one to the Dominion Museum, Wellington; a second to Lord Walter Rothschild (who had always been a frequent purchaser of skins from Buller); and the third was sold for a thousand pounds to the Carnegie Museum in Pittsburgh. More of Buller's birds are now also in America because the Rothschild collection was bought by the American Museum of Natural History at New York in 1931.

BULWER. *ENGLISH CLERGYMAN AND AMATEUR NATU-RALIST, MAINLY INTERESTED IN SHELLS. THE PORTRAIT SHOWS HIM OUTSIDE THE CHURCH DOOR AT HUN-WORTH, IN NORFOLK, AT THE AGE OF SIXTY-SIX.*

James Bulwer
(1794-1879)

BULWER'S PETREL *Bulweria bulwerii* (Jardine and Selby)

Bulweria Bonaparte, 1843. *Nuovi Annali di Scienze Naturali, Bologna* 8 (1842), p. 426. Type by monotypy, *Procellaria bulwerii* Jardine and Selby
Procellaria Bulwerii Jardine and Selby, 1828. *Illustrations of Ornithology*, Vol. 2, Pl. 65 and text: Madeira or nearby

"We are indebted to the kindness of Mr. Bulwer, during some years a resident in Madeira, for the subject of this plate which we consider as yet undescribed. It is not to be found in the works of Latham or Shaw, or indeed in any other which we have had an opportunity of consulting; and from its marked characters it is not a species that would be easily overlooked. The length of our specimen is about ten inches; it inhabits Madeira, or the small islands adjacent."

This extract, together with a description of the petrel, appeared in 1828 in *Illustrations of Ornithology* by Sir William Jardine and Prideaux John Selby. They named the species *Procellaria Bulwerii* and gave it its English name. 'Bulwer' now appears in both parts of the scientific name and also appears in the French, Spanish, Dutch and German names, yet the Reverend James Bulwer had few connections with ornithology other than his correspondence with Jardine. He was more interested in conchology and archaeology, and was better known for his association and friendship with the philosopher and traveller Alfred Lyall and the famous landscape artist John Sell Cotman—at his best, one of the finest of all English watercolourists.

James Bulwer was the son of James Bulwer of the Manor House, Aylsham, in Norfolk and his mother Mary Seaman was originally from Norwich. He was born on 21 March 1794, educated at North Walsham School and was later at Jesus College, Cambridge, where he obtained his B.A. in 1818 and his M.A. in 1823. As an undergraduate he took drawing lessons from Cotman and became a Fellow of the Linnean Society on the strength of his knowledge and interest in 'Testaceous Mollusca'; one of his three proposers was William Elford Leach.

In 1818 Bulwer was made a deacon and in 1822 a priest. Between these ordinations, and after moving to Ireland, he married Eliza Redfoord of Dublin by whom he had three sons and one daughter. In 1823 he was appointed perpetual curate of Booterstown, Dublin, but by 1831 he was in Bristol and in

1833 he was curate of St James's Chapel, Piccadilly. Between his time in Ireland and London he spent at least two winters, perhaps four, travelling in Spain, Portugal and Madeira. One of these trips is recorded in *Rambles in Madeira and in Portugal, in the early part of 1826*. Written by his companion Alfred Lyall, but for some reason published anonymously, it gives a general account of their wanderings over the island. It was Lyall's first visit and because he knew little about natural history he benefited greatly from Bulwer's company: "I am very happy in my guide—having spent the last spring here, B. is perfectly at home in the country, of which he seems to know every lane and dingle. He is moreover a very quick observer and an ardent naturalist …".

The appendix is credited to Bulwer and contains a reference to a former visit to Madeira which included a twenty-mile crossing, in a large rowing boat, to a barren rocky island: "A party of us made an excursion to the middle or great Deserta, and passed the night there … We slept in a tent which we had brought with us. All night long there was a singular noise, like that of children crying, which we found proceeded from the mother Carey's chickens [petrels]. During the day these birds hide in the rabbit burrows."

At present, three species of tubenose are known to breed on that island: Cory's Shearwater, Bulwer's Petrel and Band-rumped [Madeiran] Storm-Petrel; and Little Shearwater and Soft-plumaged Petrel breed nearby. The sound that Bulwer described most probably came from the Cory's Shearwaters, nevertheless, it seems likely that Bulwer's Petrel was first collected on this brief expedition in the spring of 1825.

The second part of *Rambles in Madeira* deals with their return journey. In May 1826, after five months on the island they took passage for England but broke their voyage at Lisbon to make excursions to some of the surrounding villages. Lyall commented that "[We] suffered less from our beds than we had feared—they were clean; and I think we commonly had them to ourselves." One presumes that he was referring to fleas or bedbugs. Bulwer is said to have published *Views of Cintra in Portugal* in 1828 but this cannot be traced. In another winter he visited Spain.

When Bulwer sent his sons to King's College School he was able to rekindle his friendship with Cotman who became the school's drawing master in 1834.

Bulwer's Petrel

Two of Bulwer's sketches from Spain inspired Cotman's watercolours, *Alhambra at Granada* and *Toledo*, and another from Madeira forms the basis for *Cliffs on the North-east side of Point Lorenzo*. They later worked together on drawings of scenery and antiquities in the Minehead district of Somerset.

As curate of St James's Chapel, Bulwer was present at the coronation of Queen Victoria in 1838 but a year later, when forty-four years old, he left London to return to Norfolk where he spent the rest of his life in quiet obscurity. At first he was the curate of Blickling with Erpingham, near Norwich, but later he changed curacies and moved to Old Stody Lodge and then the rectory at Hunworth. As a clergyman naturalist, Bulwer cannot be compared with Canon Tristram or the Reverend Gilbert White; he collected only a few birds and wrote only one short paper related to natural history, which concerned some molluscs from the Irish Sea. Nevertheless he gained pleasure from adding to his large collection of shells from time to time and also, between 1847 and 1879, contributed eleven papers to *Norfolk Archaeology*.

In 1945 there were still a few people alive in the parish who could remember Bulwer "as an old man with a very long white beard who was taken about the village in a kind of push chair". He was buried in Hunworth graveyard on the north side of the church, and the cross erected over his grave bears the enigmatic inscription: Sacred to the Memory of / James Bulwer M.A. / of Jesus College / Rector of Hunworth and Stody / from 1848–1879 / died 11 June 1879 / That by any means I preach to others, / I myself might be a castaway. The last two lines are from I Corinthians Ch.9 v.27, and refer to St Paul's hope that after years of preaching the Gospel to others he himself would not have failed to earn his Father's approval and heavenly reward.

After Bulwer's death no obituary appeared in any scientific or archaeological journal, nor in *The Times*. It is also strange that in Jardine's own catalogue of birds there is no mention of the type specimen of Bulwer's Petrel nor of the Reverend James Bulwer. The main collection of over 8500 skins was not sold until several years after Jardine died. The British Museum tried to buy it but their offer was turned down as the sum was considered to be inadequate for so large and so precious a collection and it was eventually dispersed by public auction. As a result of bad organization and limited advertising the sale was poorly attended and all the birds together fetched only £217 2s 6d—less than the museum's earlier offer. To his surprise, the buyer for the Natural History Department of the British Museum purchased the type specimen of Bulwer's Petrel amongst an unspecified lot of mounted birds for only a few shillings![1]

CASSIN. *ONE OF THE WORLD'S LEADING NINE-
TEENTH CENTURY TAXONOMISTS, HE STUDIED THE
HUGE COLLECTION OF AMERICAN AND FOREIGN
BIRDS AT THE ACADEMY OF NATURAL SCIENCES IN
PHILADELPHIA.*

John Cassin
(1813-1869)

CASSIN'S AUKLET *Ptychoramphus aleuticus* (Pallas)

Uria Aleutica Pallas, 1811. *Zoographia Rosso-Asiatica*, Vol. 2, p. 370: Russia ad Oceanum orientalem [= North Pacific Ocean]

CASSIN'S KINGBIRD *Tyrannus vociferans* Swainson

Tyrannus vociferans Swainson, 1826. *Quarterly Journal of Science, Literature, and the Arts of the Royal Institute of Great Britain* 20, p. 273: Temascáltepec, [state of] México

CASSIN'S SPARROW *Aimophila cassinii* (Woodhouse)

Zonotrichia Cassinii Woodhouse, 1852. *Proceedings of the Academy of Natural Sciences of Philadelphia* 6, p. 60: near San Antonio, Texas

CASSIN'S FINCH *Carpodacus cassinii* Baird

Carpodacus cassinii Baird, 1854. *Proceedings of the Academy of Natural Sciences of Philadelphia* 7, p. 119: Camp 104, Pueblo Creek, New Mexico [= 10 miles east of Gemini Peak, Yavapai County, Arizona]

    During the middle years of the nineteenth century John Cassin was one of America's foremost ornithologists and the first American to become equally familiar with the birds of both the Old and the New Worlds. All of his written work was done from one place—the back room of the library at the old Academy of Natural Sciences at the corner of Broad and Sansom Streets in Philadelphia. There, surrounded by books and zoological journals, he arranged, identified, classified and catalogued bird skins from all over the world. An important part of his work was the determination of previously undescribed species, of which he added nearly two hundred to the ornithological literature.

    Of these new species twenty-six are in the 1983 edition of the A.O.U. *Checklist of North American Birds* (which includes Central America). Twelve occur in the United States: Ross's Goose, Heermann's Gull, Williamson's Sapsucker, White-headed Woodpecker, Kauai Oo, Hutton's Vireo, Philadelphia Vireo,

Rufous-crowned Sparrow, Brewer's Sparrow, Black-throated Sparrow, Sage Sparrow and Lawrence's Goldfinch. Cassin named seven of the aforementioned birds after fellow naturalists (in the scientific names), thus demonstrating his approval of the habit of so honouring those who contribute to ornithology. His own thorough systematic research was so greatly respected by his colleagues that he enjoyed the distinction of having four species from the United States named for him by four different naturalists.

The first eponym was given in 1845 by William Gambel during the early years of Cassin's work at the Academy. Gambel had found some auklets along the Californian coast which he named *Mergulus Cassinii*, not realizing that they had already been collected by the Russians. Though Pallas's scientific name takes priority, Gambel's suggestion still appears in the English name of Cassin's Auklet.

The tyrant flycatcher now known as Cassin's Kingbird was first described by William Swainson in 1826 and still retains the scientific name he gave it. Cassin's name was first linked with the bird in 1850 when George Lawrence mistakenly described a Texan specimen as new to science and called it *Tyrannus Cassinii* Cassin's Tyrant Flycatcher.

Cassin's Sparrow was discovered by Dr Samuel Woodhouse, the surgeon and naturalist who travelled from Texas to the Zuni and Colorado Rivers with Lieutenant Sitgreaves in 1851–52. During the course of the expedition Woodhouse was wounded in the leg by a Yavapai arrow and bitten on the hand by a rattlesnake. The pain which he suffered in his arm prevented much collecting during the latter part of the journey, but he had already taken Cassin's Sparrow and another new bird, the Black-capped Vireo. He described both species at a meeting of the Academy of Natural Sciences of Philadelphia on 27 April 1852, and of Cassin's Sparrow he reported that:

> "This interesting bird I shot on the prairie near San Antonio, on the 25th of April, 1851, and, at the time, took it for the Z. savannah (Wilson,) [Savannah Sparrow] which it much resembled in its habits, but upon examination it proved to be totally distinct. I found but one specimen, which is a male.
>
> I have named this in honor of my friend Mr. John Cassin, the Corresponding Secretary of the Society, to whose indefatigable labor in the department of Ornithology we are so much indebted."

The fourth species, Cassin's Finch, was collected in New Mexico during the winter of 1853–54 by Dr Caleb Kennerly and H.B. Möllhausen, naturalists attached to the Pacific Railroad Survey led by Lieutenant A.W. Whipple. The finch was named *Carpodacus cassinii* by Spencer F. Baird of the Smithsonian Institution, who was one of Cassin's greatest friends. A constant flow of correspondence passed between Washington and Philadelphia for they maintained a superficial rivalry which in fact masked a very profitable collaboration. On one occasion when Baird was about to describe several new species, including the White-headed Woodpecker, Cassin took great delight in writing to point out that he had already named it and added "perhaps you had better not describe any of them. Send them this way!". Baird often passed on to Cassin the collections which he received from the Pacific Railroad Surveys, knowing that they would be carefully studied. One lot that he sent for comment included a beautiful brown and crimson finch from Arizona. On writing back to Baird, Cassin suggested that he name it *Cassinii*—which he did!

Cassin's friends greatly valued his advice on ornithological matters and found him to be kind-hearted and loyal, but people who knew him less well were often repelled by his rather brusque manner. In June 1845, when he met Audubon for what was probably the first and the last time, the two great ornithologists had a heated argument as to who had first discovered Harris's Hawk—Harris himself being the mutual friend who had brought them together.

The systematic study of foreign birds, on a very large

Cassin's Kingbird

scale, was made possible for Cassin by the generosity of Dr Thomas B. Wilson, who purchased so many important collections for the Academy that by 1856 it had become the largest bird department in the world. Among the most valuable accessions were the Masséna (Rivoli) collection, John Gould's Australian birds and eggs, Jules Bourcier's parrots and tanagers and Captain W.E. Boyes's Indian collection. Altogether, Dr Wilson acquired some 26,000 birds for the Academy and also provided all the essential reference books. For more than twenty years Cassin monopolized and jealously guarded the collection, it being well understood by all other students at the Academy that he would tolerate no interference. His devotion to the work and the quality of his output ensured that his privilege went unchallenged.

He therefore had every advantage on his side—except time. All his ornithological work had to be done in his leisure hours, as, unlike Thomas Say who had in earlier years worked virtually round the clock at the Academy, Cassin had a wife and family to support. On first moving to Philadelphia at the age of twenty-one he had worked as a merchant, then held a position in the U.S. Customs House. Later he took over the management of a printing and lithography business which demanded much of his energies. He was also hindered, on occasion, by attacks of arsenical poisoning, a common hazard for naturalists at that time.

Of Cassin's many publications the best known are his *Illustrations of the Birds of California, Texas, Oregon, British and Russian America* (1853–56) and *Birds of North America* (1860) by Baird, Cassin and Lawrence. The former was the first book by an American ornithologist to use trinomials and it was extremely useful to those interested in the western and southern birds. Pioneering naturalists such as Heermann, Bell, Couch, Clark, Woodhouse, Kennerley, McCown and

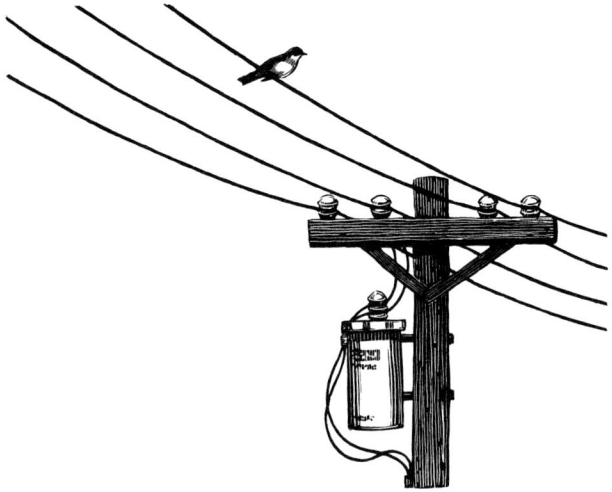

McCall shared their field notes
with Cassin so that he was able
to give much previously unpublished information about the habits of fifty species. Many more
were briefly treated in the synopsis. An example of its immediate
value can be found in the correspondence between John Xántus
and Baird from the time when
the collector was stationed at Fort
Tejon in southern California. He
constantly consulted his copy
and often referred to it in his
letters. In 1858 he found a vireo
near the fort and named it *Vireo
cassinii*. It has long been treated
as a race of Solitary Vireo but
may deserve to be separated as
Cassin's Vireo.[1]

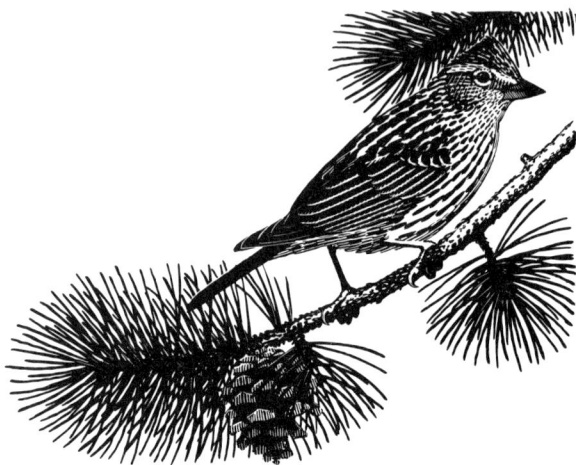

Cassin's Finch

Cassin contributed ornithological sections to a number of U.S. Government
publications. For example, he classified the birds from the U.S. Expedition to
the Dead Sea and River Jordan (1852); he described many Chilean birds taken
during the U.S. Naval Astronomical Expedition to the southern hemisphere
(1855) and he wrote up the results of Perry's expedition to the China Seas and
Japan (1856). After the return of the U.S. Exploring Expedition led by Lieutenant
Charles Wilkes, the mammals
and birds collected on the four
year voyage (1838–42) were
written up by Titian Peale, one
of the expedition naturalists.
Wilkes was so dissatisfied with
Peale's production that he gave
Cassin the responsibility of
rewriting the report, which
appeared as volume eight of
Mammalogy and Ornithology
(1858).

The *Proceedings of the Academy of Natural Sciences of
Philadelphia* was the usual out-
let for Cassin's papers and
among the most notable were
those based on the collections

Cassin's Sparrow

of Paul du Chaillu, an American of French extraction who explored in West
Africa under the sponsorship of the Academy. In 1861 du Chaillu visited London
and gave a series of sensational public lectures at which he exhibited the first

stuffed gorillas ever to be seen in Europe. At a time when Darwin's theories were shocking the nation, du Chaillu's vivid and highly fictionalized accounts of dangerous encounters with fierce and terrifying apes were popular stuff. The title of his book, *Explorations and Adventures in Equatorial Africa*, belied its romanticized contents. Cassin's treatment of the many interesting and novel birds which du Chaillu procured was of a rather different style but just as fascinating for ornithologists.

When the Civil War began in 1861 Cassin laid aside his research for a while and at the age of forty-seven joined the army. Captured by the Confederates, he was incarcerated in the notorious Libby Prison to the detriment of his health, and he died on 10 January 1869, at the age of fifty-five, just four years after his release.

The style of his work was later assessed by Elliott Coues, who wrote:

> "He was patient and laborious in the technic of his art, and full of book-learning in the history of his subject; with the result that the Cassinian period [of American Ornithology], largely by the work of Cassin himself, is marked by its bookishness, by its breadth and scope in ornithology at large, and by the first decided change since Audubon in the aspect of the classification and nomenclature of the birds of our country."[2]

The extent of Cassin's achievements ensured that he was not quickly forgotten. Twenty-one years after his death, the Delaware Valley Ornithological Club founded their journal and named it *Cassinia*. The title was entirely appropriate as John Cassin was connected not just with Philadelphia, but also with Delaware County, Pennsylvania, where he was born on 6 September 1813. His father, who was a Quaker, owned a large frame house on a forty-acre tract of land near Media and it was in the surrounding woods and fields that John first watched and identified birds. In later life, though he worked obsessionally at his self-appointed task, he occasionally admitted to Baird that he yearned to escape from the city and go out birding—especially in the spring. Dr T.M. Brewer, who greatly valued Cassin's field notes, described him as "an ardent lover of nature, and a close observer of living birds, both in their wild wood haunts, and under the open sky". It would therefore be wrong to stereotype

Cassin's Auklet

Cassin as a typical closet naturalist, for he obviously possessed a fair amount of fieldcraft. In September 1842 he discovered a new species while out birdwatching in Bingham's woods near Philadelphia and recorded: "It was in the upper branches of a tree of considerable height, engaged in capturing insects, and attracted my attention by its slow and apparently deliberate movements." In his original description Cassin compared it with the similar but much more widespread Warbling Vireo, emphasizing the distinct supercilium and more vivid coloration of his new species. He named it *Vireosylvia philadelphica*, now *Vireo philadelphicus*—the Philadelphia Vireo.[3] The name is historically interesting but apt to be misleading, as the vireo is only seen around Philadelphia while on migration. It breeds in woodlands across Canada and the most northerly States and winters in Central America.

John Henry Clark (c.1830–c.1885)

CLARK'S GREBE *Aechmophorus clarkii* (Lawrence)

Podiceps clarkii Lawrence, 1858. In S.F.Baird, J.Cassin and G.N.Lawrence *Reports of Explorations and Surveys ... for a Railroad [Route] from the Mississippi River to the Pacific Ocean*, Vol. 9, p. 895: California and New [*sic*] Mexico [=Laguna Santa María, Chihuahua, Mexico]

Professor Baird had many men working in the field, diligently acquiring all kinds of zoological items for the Smithsonian Institution in Washington where he was Assistant Secretary. Often Baird had personally instructed them in how best to preserve and care for their specimens. Some, like Adolphus Heermann, Robert Kennicott and John Xántus contributed huge amounts to the Smithsonian, while others such as Lieutenant Couch, Captain McCown and William Hutton gathered relatively small collections. For a few years John H. Clark was among the former category, being noted both for the wide range of species and the sheer quantity of material amassed.

Though indeed a dedicated and successful collector, well known in his youth as a pioneer naturalist involved in the exploration of the American West, surprisingly little is now known about Clark. His dates of birth and death, and his activities during certain periods remain unrecorded in the usual references available to biohistorians. Said to have been a native of Anne Arundel, Maryland, he is assumed to have been born around 1830, calculated from the fact that he was a student at Dickinson College, Carlisle, Pennsylvania, from 1847 to 1851. At College he was affectionately known as Adam because of his exceptional mathematical capabilities which supposedly likened him to Adam Clark, a famous mathematician. Caleb Kennerly, another great supplier of specimens to the Smithsonian was also a student there. Both received instructions from young Professor Baird who taught at the college from 1846 to 1850 before his museum appointment and both were frequent visitors to the Baird household, being held in high regard by Baird, his wife and their daughter Lucy.

Baird was no doubt delighted when George Lawrence described and named Clark's Grebe from a wintering specimen obtained by Clark in the north Mexican state of Chihuahua. The description appeared in Baird, Cassin and Lawrence's

CLARK, J.H. *MATHEMATICIAN AND SURVEYOR. HE SENT NATURAL HISTORY SPECIMENS TO THE SMITH-SONIAN INSTITUTION WHEN WORKING FOR THE U.S. AND MEXICAN BOUNDARY SURVEY.*

contribution on birds in volume nine of the Pacific Railroad Survey reports published in 1858. The bird was subsequently demoted to a race of the Western Grebe and only recently, in 1988, was it restored once more to full specific rank. It will be many years before all the differences between these two grebes can be satisfactorily worked out, not least because of their similar appearance and partially overlapping breeding ranges.

The type locality indicates Clark's main collecting ground: not Mexico in particular but the area on either side of the border along which he served in the U.S. and Mexican Boundary Survey from 1851 to 1855. The survey started in 1848 at the conclusion of hostilities between the two countries and Baird, ever watchful for an opportunity, secured positions for both Clark and Kennerly. Leaving Dickinson College early in 1851 before graduation, Clark started out with the survey as assistant computer under Colonel J.D. Graham. He may have travelled with the great plant hunter Charles Wright who was with the team in 1849 and from 1851 to 1852. Unfortunately the early years were marred by poor organization and mismanagement which resulted in a number of leadership changes. In August 1854 a fourth Chief Commissioner was appointed, this time Major William Emory who had gained field experience on a recent expedition from Kansas to California. Emory managed to pull things together and increase the general morale so that the overall quality of the scientific work improved. Besides Charles Wright, other men attached to the survey included Lieutenant Couch, Lieutenant Michler, Major Rich and Captain Van Vliet and civilians Arthur Schott, John Potts and Charles C. Parry. Kennerly was physician and naturalist and Clark took over as astronomer when Emory was promoted.

Though he was perhaps at first unsure of himself in the desert environment, Clark's enthusiasm was unrestrained. He never chose to specialize in any particular field, though circumstances sometimes forced him to do so, as indicated

Clark's Grebe

in part of a letter written to Baird from Frontera, near El Paso, on 4 July 1851 as he made his way westwards:

"We reached El Paso on the 25th June, having made the trip in 45 days; of all the barren waterless regions on the face of the earth I want to see no worse than I experienced on this route. There are stretches of from 50 to 100 miles without living water, without grass, and without wood enough to boil a pot of coffee.

There is nothing in abundance save the lizards which make up in number and variety for the scarcity of everything else in objects of Natural History. I flatter myself that I have made a good collection as far as the lizards are concerned."[1]

Despite the obvious difficulties Baird usually managed to keep his young friend supplied with the necessary equipment for collecting. The preparation of bird skins was less of a problem than the preservation of the reptiles and fishes for which Baird had to send copper containers of alcohol which weighed many pounds and were a nuisance during rough travelling. It is probable that Clark worked in a very similar manner to Henry Henshaw who has wonderfully described the paraphernalia of the itinerant field collector of the period:

"My equipment as naturalist was simple enough. A pair of roomy saddle-bags enabled me to carry a few bottles for the reception of small specimens, especially insects, and a supply of cartridges, cotton, matches, and other trifles of like nature. I also carried an insect net attached somewhere about my person, while a good double-barreled shot-gun, slung on the horn of the saddle, completed my everyday outfit ...

Two stout boxes, one for supplies such as powder, shot, arsenic, cotton, and the like, and the other fitted with trays in which to dry and carry bird and mammal skins, a copper tank of alcohol, enclosed for prudential reasons in a stout locked box, and a plant press, were also part of the naturalist's impedimenta. My skinning table was improvised by placing one collecting box on top of the other, and a folding stool enabled me to sit down and to skin birds with reasonable comfort, although several hours work usually developed a number of different sorts of backache.

As we seldom slept two nights in the same place, and as the bird and mammal skins had to be dried while being transported on mule back, I used to place them in paper cones, which I tacked securely to the bottom of the trays. Provided all went well, the skins dried in very good shape ..."[2]

In his usual manner, Baird encouraged Clark letter by letter: "My Dear Adam ... Don't neglect the birds, but collect all you can. And by all means get nests and eggs ... Collect everything you can find, especially the very COMMONEST species."[3] Two years later he was still urging him on: "Get everything, and look particularly for crabs of all kinds, especially the minute ones. Worms, shells, star fish, &c. secure with all diligence. Look sharp for the Hydragiras, the little banded fish ..."[4] Clark also found time for some of the larger mammals procuring at Santa Rita, New Mexico, the type of the 'Copper Mine Grizzly'. He even sent back a squirrel as a pet for Lucy Baird who was only four years old.

In 1853 Professor Agassiz, the famous Swiss-born scientist working at Harvard, sent a letter to the Smithsonian wanting to know why Baird had been fighting to gain control of the Mexican Boundary Survey collections sent back by Major Emory. Baird's reply explained the reasons for his actions:

"I assure you that I have not the slightest desire to monopolize Major Emory's things nor any other and shall always be willing to share such mutual desiderata ... You must note, however, the grounds on which I objected to your superintendence of the Major's things. It was not a desire to have them for the Smithsonian Institution but on account of the fact that they were all made by a protégé of my own, brought up and trained by me, between whom and myself there exists the warmest affection: for whom I got the appointment of computer, and who was in no way obliged to make collections, this not being his office, but did make them for the love of me."[5]

Fortunately Clark's collections continued to come to Baird who, in order to mollify Agassiz, promised to send him some fish since he was expecting another despatch from Clark that fall.

The survey was finished in 1855 and the task of compiling the scientific reports began. The first volume of the *Report on the U.S. and Mexican Boundary Survey* was published in 1857 and was divided into a 'Narrative of the Survey' and geological reports. Another volume appeared two years later divided between the botany and zoology; the former section was mainly by John Torrey, with George Engelmann, as usual, specializing in the cacti. Almost all the zoology was by Baird but he used additional notes on the mammals by Schott, Kennerly and Clark, notes on the birds by Couch, Kennerly and Clark, and notes on the reptiles by most of the above-named naturalists with further help on this last group provided by Charles Girard and Robert Kennicott at Washington. The entire fish section was by Girard who was working as Baird's assistant.

Other publications relating to the Boundary Survey appeared in a number of society transactions. Two important papers by Asa Gray concentrated on Charles Wright's plant collections. Six more were by John L. LeConte who worked through the insects collected by Clark and the others, presenting the results in the *Proceedings* and the *Journal of the Academy of Natural Sciences of Philadelphia*. Though Clark spent time on the east coast after the survey and helped in the preparation of these various publications he seems never to have published anything in his own name on natural history. Some of his bird notes were included in John Cassin's *Illustrations of the Birds of California, Texas, Oregon, British and Russian America* (1853–56) where Cassin asserts that the information gained about Audubon's Oriole "by so accurate and careful an observer as Mr. Clark, deserves especial attention".

In 1857 Clark was in Nebraska with Colonel Johnston surveying its southern boundary with Kansas. One of Clark's personal priorities had been to see the great herds of Buffalo which still roamed the plains. Writing to Baird from the Quapaw on 28 October he had to admit that he had been successful:

"I never believed half that was said about the great number of buffalo and the few met as [we] went west confirmed my conviction. As we returned, for fifteen days there was not one [day that] we did not see thousands, ten thousands, hundred thousands of Buffalo. The meat of the old bulls is scarcely eatable. A fat cow or calf does tolerably well, but there are a great many things to eat in this world that I like better. I got completely surfeited–disgusted in fact with buffalo. I ate buffalo, I drank buffalo, I smelt buffalo, there was nothing but buffalo in sight. They eat up all the grass, they saturate all the water, and perfume the very air ... I was very anxious to see the buffalo and am satisfied ..."[6]

In the following year Clark became one of the United States Commissioners of the U.S and Mexican Boundary Commission and held this office until 1862. Skins of Bell's Vireo, Red-eyed Vireo and Northern Mockingbird dated May 1860 from Fort Cobb, Arkansas, are mentioned in Baird's *Revue of American Birds* (1872) and Bendire notes a clutch of Black-shouldered Kite's eggs collected by Clark near Fort Arbuckle on 9 May 1861 which indicate that Clark still kept his hand in.[7] At the close of each field season Clark returned to Washington where, according to Lucy Baird, he was "usually a member of our family". Clark is next heard of in 1874 as astronomical observer with the Wheeler expedition. From July 7 to November 7 he was in charge of the field surveying station beside Great Salt Lake at Ogden, Utah, and seems at other times to have been in Wyoming and Oregon. But Clark was not mentioned by Henshaw in his recollections of those exciting times and he did not list him when he named C.G. Newberry, Henry Yarrow, Charles Aiken and Ferdinand Bischoff as collectors. Not a single bird seems to have been contributed by Clark since none is mentioned in Henshaw's ornithological appendix to Wheeler's *Annual Report upon the Geographical Explorations west of the One Hundredth Meridian* (1875).

Clark's last known communication with Baird is dated April 1882 and contains a surprise:

> "When I was at the Dalles for Wheeler I got a cold which has run into a bad catarrh—then two years ago I broke my left ankle which is not yet well & not likely ever to be. I was confined to the house from last November till nearly spring from nervous prostration—heart & stomach sympathizing. This is enough to ail one man isnt it? I didn't think I would but I reckon I had just as well—I am married nowadays—have a boy nearly two years old and a girl as many months. What do you think of that?—don't lose your breath. You can tell Mrs. Baird there is no left handed business about it—the only possible wrong in the matter being the extreme youth of the wife and mother. Still I believe under eighteen years ...
>
> My Dear Prof I am often gloomy and despondent, and when in that mood reflecting know many of those I have known & been associated with in life have already made the final journey. I sometimes feel that it can't be long till I join that 'innumerable caravan that marches one by one' to everlasting rest."[8]

Though the letter was headed Noble, Illinois, it gives us no clue as to where Clark had settled down, as he said that he was away from home. There is no further biographical information at Dickinson College, the Smithsonian or the Maryland State Archives to throw any further light on Clark's last years.[9] He was now over fifty years old and from the weary tone of the letter it seems unlikely that he lived for many more years. It would certainly be interesting to know what became of John H. Clark since he did so much for zoology in the mid part of the nineteenth century and was praised so highly by Baird for his work along the Mexican border:

> "One of these days, when the results of the expedition are published, people will be astonished to find how much one man can do under difficulties. It all depends on training, don't it? No American Naturalist has ever added so many species of reptiles and fishes to our fauna in a period even of years, as you have."[10]

William Clark
(1770-1838)

CLARK'S NUTCRACKER *Nucifraga columbiana* (Wilson)

Corvus columbianus Wilson, 1811. *American Ornithology*, Vol. 3, pp. xv, 29, Pl. 20, Fig. 2: shores of the Columbia [= Clearwater River, about two miles north of Kamiah, Idaho]

Clark's Nutcracker is one of the hardiest inhabitants of the high western coniferous forests of North America. It builds its treetop nest early in the season and begins to lay its pale green, lightly spotted eggs in February or March, often incubating through heavy blizzards of hail, sleet and snow. When the young hatch they are fed on regurgitated pine seeds, cached in the fall when the seeds were ripe. A pair of nutcrackers will hide many thousands of seeds, burying just a few at a time, often on cliff ledges and south-facing slopes. Many other corvids also store food, but Clark's Nutcracker does so to a remarkable degree, memorizing the landmarks so well that it can often relocate hundreds of sites even when they are covered by snow. Opportunistic feeders, they also scavenge at campsites, prey on eggs and nestlings and hunt for beetles, ants, grasshoppers and other insects.

The first observation on their diet was made by their discoverer, Captain William Clark, who thought that he had found a new species of woodpecker because of the bird's general shape and bounding flight. In his diary for 22 August 1805 he wrote:

> "I saw to day [a] Bird of the woodpecker kind which fed on Pine burs its Bill and tale white the wings black every other part of a light brown, and about the Size of a robin".[1]

Captain Clark was on his way from the Missouri River to the Pacific Ocean on the historic expedition that he led with Captain Meriwether Lewis. This first encounter with the nutcrackers was by the Lehmi River while Clark was making a reconnaissance with eleven men of the party, prior to choosing the next stage of their route across the continental divide. On their return journey a year later they killed a nutcracker and prepared the skin while lingering by the Clearwater River, waiting for spring to reach the Rocky Mountains. At the conclusion of the expedition skins of Clark's Nutcracker, Lewis's Woodpecker and Western

CLARK, W. *CO-LEADER OF THE CELEBRATED LEWIS AND CLARK*
EXPEDITION OF 1804–1806.

Tanager were given to Alexander Wilson, who described these new species in the third volume of his *American Ornithology* (1808–14). It seems wholly appropriate that Clark should have such a distinctive, hardy, resourceful and endearing species named in his honour.

He was born on 1 August 1770 at Charlottesville, Virginia, the ninth child of John and Ann Clark. When William was fourteen years old the family left their plantation and moved West, taking with them their slaves, horses, stock and furniture. William's brother, General George Rogers Clark, was already living at Louisville, beside the Ohio, and there they built their new home, *Mulberry Hill*. Red-haired William developed into a sturdy six-footer and by the age of nineteen (and possibly much earlier) he was helping to defend Kentucky settlements against Indian attack. In March 1792 he joined the U.S. Army as a lieutenant of infantry and during the next four years, while serving under General Wayne, he gained experience in handling squads of men under military discipline, practised negotiation and diplomacy in his contacts with the Spanish, continued to fight against the Indians and learned engineering skills. On at least one campaign he served with Meriwether Lewis, who was four years younger and a more junior officer. In the summer of 1796 Clark resigned his commission and returned to *Mulberry Hill*, which he inherited three years later on the death of his parents.

Two weeks before his thirty-third birthday, Clark received an unexpected letter from his old friend Lewis offering him joint leadership of an expedition across the continent, seeking a river route to the Pacific. He accepted without hesitation and in May of the following year, 1804, they set off up the Missouri in charge of three boats and about forty-five men, one of them being Clark's negro slave, York. The events of the next twenty-eight months constitute one of the best known and least tarnished parts of American history, but those unfamiliar with Lewis and Clark's exploits can find an account of them in the chapter on Lewis (and a map in Townsend, p. 450).

The success of the expedition, from the scientific as well as the geopolitical viewpoint, was largely due to the compatible qualities of the two leaders, who worked in complete harmony and exploited each other's strengths and talents to the full. Clark was a good judge of men, warm and companionable. Quick-

Clark's Nutcracker

witted and intuitive, his swift actions saved the company from disaster on more than one occasion, and being able to empathize with the Indians he quickly formed good relationships with them and earned their respect by his frankness and his skill in treating their ailments. Experienced in surveying and adept at reading the terrain, he took on the role of cartographer and for many years his maps of the regions traversed were greatly valued for their reliability. Lewis had the benefit of a more academic education and some scientific training. A bit of a dreamer, he was inclined to be reserved, anxious and generally humourless but he cared well for his men. He had a talent for speechmaking and on the initial meetings with various tribes his eloquence was of the utmost importance.

Five months after his return to civilization, Clark resigned from the army (in February 1807) and became superintendent of Indian affairs for the Louisiana Territory, serving also as a brigadier-general of militia. He married Julia Hancock of Fincastle, Virginia, and in January 1808 brought her to St Louis where they made their home. Indians became increasingly frequent visitors to the town after Clark became superintendent; many had business with 'Redhead' as they called him and no other white man inspired so much confidence among the natives. Clark will always be remembered for his part in the early exploration of America, but his later role as a mediator between the white man and the redskin was of equal importance for the expansion of the United States.

Meriwether Lewis was made governor of Missouri Territory. On his sudden death in 1809 Clark was asked to succeed him, but he felt obliged to decline, although he accepted the job in 1813 and continued until Missouri achieved statehood in 1821, when he again held the post of superintendent of Indian affairs until the end of his life. During these latter years he appointed agents and factors, received delegations of chiefs at his home and occasionally escorted some of them to Washington. He constantly tried to persuade the government to treat the Indians in a humane and honourable fashion.

In 1820 his wife died, leaving him with four children to raise. The next year he married again, to a widowed cousin of his first wife, née Harriet Kennerly. They had one son, Jefferson Kennerly Clark, who in 1843 joined Sir William Drummond Stewart's bizarre expedition to the Rockies, along with many other young men—including Dolly Heermann.

William Clark died in St Louis on 1 September 1838, at the age of sixty-eight, while staying at the home of his eldest son, Meriwether Lewis Clark. His second wife had predeceased him.

James Cook
(1728-1779)

COOK'S PETREL *Pterodroma cookii* (Gray)

Procellaria Cookii G.R.Gray, 1843. In E.Dieffenbach *Travels in New Zealand*, Vol. 2, p. 199: New Zealand

Captain Cook made three long voyages of discovery—voyages on which he charted immense areas of ocean and coastline. In this chapter we shall only concentrate on the period most relevant to the ornithological history of Canada and the United States when he explored the north-west coast and discovered the Hawaiian Islands. Though a few Russian naturalists had collected before them in Alaska, Cook's men were the first to do so in Hawaii.

Cook's talent for seamanship and surveying allowed him to rise through the ranks of the Royal Navy despite his humble beginnings near Whitby, in Yorkshire. For the first of his world voyages he was given command of the *Endeavour* and from 1768 to 1771 explored much of the southern Pacific, visiting New Zealand, the east coast of Australia and a number of islands including Tahiti; the discoveries of his naturalists Joseph Banks and Daniel Solander astonished the world. Cook sailed again in 1772, re-visited New Zealand and circumnavigated Antarctica, taking with him the German naturalist Johann Reinhold Forster and his son George.

Returning to England in 1775, it was not long before Cook set out again, this time to investigate the possible existence of the North-west Passage by an opening in the North Pacific. Sailing to New Zealand once more, with the *Resolution* and the *Discovery*, he went from there to the Hawaiian Islands by way of Tonga, the Society Islands and Line Islands. After a brief call at Kauai they moved eastwards towards the Oregon coast, making their first anchorage off Vancouver Island in Nootka Sound at the end of March 1778. More collecting was done a few weeks later in Alaska at Prince William Sound. (For a map of Alaska see p. 318.) After a frustrating interlude in Cook Inlet and Turnagain Arm where no eastward passage could be found they made a stop in the Aleutians at Unalaska. Cook then pushed on past St Matthew Island and through the Bering Strait until ice prevented any further northward progress. If he had

COOK. *ENGLISH NAVAL OFFICER AND EXPLORER. ON HIS THIRD GREAT VOYAGE, BIRDS WERE COLLECTED IN THE HAWAIIAN ISLANDS AND ON THE PACIFIC COAST OF NORTH AMERICA FROM VANCOUVER NORTHWARDS TO THE BERING STRAIT. THE PORTRAIT IS BY JOHN WEBBER, THE EXPEDITION ARTIST.*

managed to round the northern cape of Alaska he would certainly have perished as he was not adequately equipped for a winter so far north. Instead he turned back to Hawaii only to be ignominiously killed there in a petty squabble between his crew and the islanders.

After Cook's death on 14 February 1779 Captain Clerke took command and tried again for the North-west Passage in the following season but he made no extra progress. When Clerke died, Lieutenant King brought the two ships safely back to England, arriving home in October 1780.

During the voyage at least 120 bird species were collected. Those from the Pacific north-west were mainly taken by the surgeon-naturalists William Anderson and William Ellis and by the artist John Webber. The two latter men produced all the artwork and undertook almost all the collecting after Anderson's death from tuberculosis in August 1778, shortly after they entered the Bering Strait for the first time. Anderson did not therefore play any part in the expedition into the interior of Hawaii in February 1779 when most of the Hawaiian specimens were obtained.

Initially, the birds were displayed in London in the private museums of Sir Ashton Lever and Sir Joseph Banks, where they were examined by Dr John Latham. With the exception of the Iiwi, which was part of a small group of birds taken by one of Cook's sailors to Germany and described by George Forster, the novelties were all described in Latham's *General Synopsis of Birds* (1781–85) and again, in some instances, in Thomas Pennant's *Arctic Zoology* (1784–85). Latham and Pennant only gave the birds English names and it was left to J.F. Gmelin, a German naturalist, to reap most of the credit for naming and describing these species since he was the first to give them scientific binomials. Gmelin published his descriptions in the 13th edition of Linnaeus's *Systema Naturae* (1788–89). This explains why Gmelin's name is associated with so many birds from the Pacific islands, New Zealand and North America. About seventy of his scientific names of birds from Cook's third voyage are still valid, the remainder being disallowed either because his names were already in use for some other species or because the birds had already been described elsewhere.

Eventually the specimens became widely dispersed, many being lost or destroyed—a sad fate for these historic items. Banks's portion of the birds and the drawings of Ellis and Webber were deposited at the British Museum (Natural History) but the skins have not survived. Those in Lever's museum were disposed of by a lottery won by Dr James Parkinson, who displayed them in London for about twenty years. In 1806 this collection was auctioned off; William Bullock, the Earl of Derby and an agent for the Imperial Museum of Vienna being among the principal purchasers. While some of Cook's specimens still remain in Vienna, those in the Bullock museum were resold and became scattered amongst various European museums. The Earl of Derby's collection went to Liverpool Museum. In many cases the localities and dates on the Ellis and Webber drawings are more reliable than those attached to the birds.

Among the American type specimens described by Latham and Gmelin are the Rufous Hummingbird, Varied Thrush and Steller's Jay from Vancouver Island; the Surfbird, Marbled Murrelet and Golden-crowned Sparrow from

Prince William Sound; the Savannah Sparrow from Unalaska; and the Fork-tailed Storm-Petrel, Ancient Murrelet and Whiskered Auklet from various Bering Sea localities.

From Hawaii the following types were secured: Hawaiian Rail, Elepaio, Hawaiian Thrush, Hawaii Oo, Ou, Common Amakihi, Hawaiian Akialoa, Common Akepa, Iiwi, Hawaii Mamo and Apapane. Mention should also be made of the types of the Wandering Tattler and Bristle-thighed Curlew, which were found by Cook's men while wintering in the Society Island group.

In the expedition narratives by Cook and King casual mention is made of a number of shearwaters and petrels. Though there is nothing to suggest that they ever saw Cook's Petrel it is possible that they did so, unknowingly, since this bird breeds in New Zealand and is a rare but regular wanderer in the northern summer to the coastal waters between California and the Aleutian Chain, and off the Hawaiian Islands. However, Cook's Petrel was not named until some sixty years after Cook's death when G.R. Gray wrote an ornithological appendix to Ernst Dieffenbach's *Travels in New Zealand* (1843). Though Gray reported that the type specimen came from New Zealand he gave no special reason for naming the bird after Cook,[1] but it seems suitably named since its range covers the same area as much of Cook's own oceanic wanderings.

Cook's Petrel

William Cooper
(c.1798-1864)

COOPER'S HAWK *Accipiter cooperii* (Bonaparte)

Falco Cooperii Bonaparte, 1828. *American Ornithology*, Vol. 2, p. 1, Pl. 10,
Fig. 1: near Bordentown, New Jersey

William Cooper's part in American ornithology was a minor one and he
would be almost forgotten if Charles Bonaparte had not given him lasting fame
by naming Cooper's Hawk in his honour. Cooper was born in New York in
or around 1798, but the exact date of his birth and many other salient facts
about his life are unrecorded. With the advantage of inherited wealth he was
able to dabble in natural history at leisure and in 1821 he went to Europe for
nearly three years to study zoology.

On his return to New York he became a leading member of the Lyceum of
Natural History (now the New York Academy of Sciences) of which he had
been a founder. With John Torrey and James De Kay he controlled and
organized the affairs of the society and assured the success of the *Annals*,
launched in 1823. He took great pride in the spacious new purpose-built halls
on Broadway which were completed in 1836, for they added greatly to the
prestige of the Lyceum. Cooper contributed to the society's natural history
collection but for a man who "devoted his life to the study of Nature" his
publications were few. For the *Annals* he wrote a few papers on fossils, ferns,
bats, and in 1825, on a new species of bird, the Evening Grosbeak, sent to him
from northern Michigan by Henry Schoolcraft. During the 1820s Cooper began
to write 'The Birds of Long Island' but his manuscript was never published.

Cooper gave assistance to Charles Bonaparte, the Italian systematist who
concentrated on the American avifauna for a few years. When Bonaparte returned
to Rome in 1828 Cooper took on the responsibility of editing the second and
third volumes of the *American Ornithology; or, The Natural History of Birds
Inhabiting the United States Not Given by Wilson*. Bonaparte rewarded his
colleague by naming Cooper's Hawk *Falco Cooperii* in the second volume. In
his lengthy dissertation on the new accipiter Bonaparte mentioned that he had
handled seven or eight specimens. One of these was a male he had shot himself
while it was feeding on a Ruffed Grouse. He added that "Mr. Cooper, the

COOPER. *ONE OF THE FOUNDERS OF THE NEW YORK LYCEUM OF NATURAL HISTORY AND FATHER OF THE ORNITHOLOGIST DR JAMES G. COOPER.*

friend to whom we have dedicated this species, has recently favoured us with an accurate description of a specimen of a somewhat larger size, shot in the early part of November, on the eastern part of Long Island."

The naming of the hawk caused considerable enmity between Cooper and Audubon. The artist cared little for scientific priority and again and again he named previously described species after his own friends and patrons. Lord Stanley, Earl of Derby, had won Audubon's favour by subscribing to *The Birds of America* and in 1828 the Cooper's Hawk appeared in it as Plate 36 under the name of *Falco Stanleii* (later *Falco Stanleyii* in *Ornithological Biography*). That November, Cooper wrote to Bonaparte angrily asserting that he would not permit Lord Stanley to "carry off honors" meant for himself. When Audubon visited the Lyceum in September 1831 Cooper took perverse pleasure in showing him bird skins which he longed to paint, but denied him the permission to do so.

In the spring of 1837 Cooper took his family to live on a farm at Guttenberg on the New Jersey side of the Hudson River. Consequently he became only a corresponding member of the Lyceum until 1853, when he took up residence in Hoboken and served again as vice-president until his death on 20 April 1864. Latterly he had taken a particular interest in conchology and made numerous dredging excursions to as far north as Nova Scotia and south to the Bahamas. His last publication was a report on western molluscs (in volumes one and twelve of the Pacific Railroad Survey Reports) collected by his eldest son, James Graham Cooper (1830–1902). It could be said that William Cooper's greatest achievement was the inspiration and training of this son who became the better-known ornithologist, famed for his studies of West coast birds and honoured in the title of the Cooper Ornithological Club.

In recent years a strange discovery of William Cooper's has attracted considerable interest. On 24 May 1833 Cooper shot a moulting shorebird at

Cooper's Hawk chasing Passenger Pigeons

Raynor South on Long Island and the unique skin was preserved at the Smithsonian Institution. Twenty-five years later Baird described it as a new species, *Calidris cooperi*, Cooper's Sandpiper. Its taxonomic affinities have long been debated and it is currently placed in Appendix C of the 1983 A.O.U. *Check-list of North American Birds*. Its many similarities to the perplexing Cox's Sandpiper (described in 1982 from two South Australian specimens) have been pointed out by John B. Cox who believes that both Cooper's and Cox's Sandpipers are hybrids.[1]

Charles Barney Cory
(1857-1921)

CORY'S SHEARWATER *Calonectris diomedea* (Scopoli)

Procellaria diomedia Scopoli, 1769. *Annus 1 Historico-Naturalis*, p. 74: No locality [= Tremiti Islands, Adriatic]

Dear Sir:

You are cordially invited to attend a Convention of *American Ornithologists*, to be held in New York City, beginning on September 26, 1883, for the purpose of founding an AMERICAN ORNITHOLOGISTS' UNION, upon a basis similar to that of the "British Ornithologists' Union." The place of meeting will be announced hereafter.

The object of the UNION will be the promotion of social and scientific intercourse between American ornithologists, and their co-operation in whatever may tend to the advancement of Ornithology in North America ...

Those who attend the first meeting will be considered *ipso facto* Founders of the AMERICAN ORNITHOLOGISTS' UNION. Active and Corresponding Members may be elected in due course after organization of the UNION, under such rules as may be established for increase of membership. Details of organization will be considered at the first meeting.

Should you favor this proposition, and propose to attend the first meeting, please so signify to any one of the undersigned.

J.A. ALLEN, Cambridge, Mass.
Editor of the Nuttall Bulletin
ELLIOTT COUES, Washington, D.C.
Assoc. Editor of the Nuttall Bulletin
WILLIAM BREWSTER, Cambridge, Mass.
President of the Nuttall Club[1]

Charles Cory was very pleased to be one of the forty-eight ornithologists invited to become Founders of the A.O.U. He had a reputation for being an amateur and a playboy so he would have seen it as recognition of his talents and an acknowledgement that he was one of the more active ornithologists of the period. On 26 September he joined twenty other men from diverse parts of Canada and the United States at the American Museum of Natural History at New York for the organizational meeting. Allen and Professor Baird could not attend but they were nevertheless elected as Founders. During the three-day

CORY. *HERE SHOWN AMONG THE FOUNDERS OF THE AMERICAN ORNITHOL-OGISTS' UNION, 1883. 1: N.C. BROWN. 2: E.A. MEARNS. 3: T. McILWRAITH. 4: J.B. HOLDER. 5: D.G. ELLIOT. 6: C. ALDRICH. 7: J.M. WHEATON 8: A.K. FISHER. 9: R.W. SHUFELDT. 10: R. RIDGWAY. 11: H.W. HENSHAW. 12: S.F. BAIRD. 13: G.N. LAWRENCE. 14: E. COUES. 15: C.F. BATCHELDER. 16: C.B. CORY. 17: C.H. MERRIAM. 18: J.A. ALLEN. 19: W. BREWSTER. 20: M. CHAMBERLAIN. 21: D.W. PRENTISS. 22: H.A. PURDIE. 23: C.E. BENDIRE. 24: E.P. BICKNELL. 25: H.B. BAILEY.*

meeting officers were chosen (Allen as President, Coues and Ridgway as Vice Presidents), committees were formed and the constitution was adopted. A further twenty-four active members were elected, this second batch counteracting the rather eastern bias caused by the absence of several western naturalists who had been unable to make the journey.

At a meeting of the A.O.U. Council at Washington in December a title was chosen for the new journal, the choice lying between *The American Ornithologist* and *The Auk*. The latter name was adopted after an impassioned plea from Baird who wished to follow the precedent set by the B.O.U. who had named their journal *The Ibis*. *The Auk* was duly published as a continuation of the *Bulletin of the Nuttall Ornithological Club*, appearing from 1884 onwards. The original Great Auk motif for the title page was taken from Cory's limited edition double elephant folio *Beautiful and Curious Birds of the World* (1880–83)—though a new auk design was later produced by Fuertes. Cory contributed the first paper in the new journal, on some birds from Santo Domingo. In 1903 he was elected President of the A.O.U. and served for two years.

Charles Cory (born into a very wealthy Boston family on 31 January 1857) had developed his interest in birds during childhood. He became a member of the Nuttall Club at the age of nineteen and remained a member for the rest of his life. One paper that he wrote for the *Bulletin* had interesting consequences. His 'Description of a New Species of the Family Procellaridae' (1881) described how he had obtained a shearwater on 11 October 1880 at Chatham, Cape Cod, and after showing it to some fishermen they had returned in the afternoon with several more that had flown close to their boat. These birds were not a new species but were wandering individuals of the Atlantic race of the Mediterranean Shearwater, which breeds on the Canaries. The first use of the name Cory's Shearwater for them dates back to Herbert K. Job who suggested it in the last volume of the *Bulletin*, before it became *The Auk*. Much later the name was adopted as a general term to encompass the Mediterranean and Cape Verde races as they are indistinguishable in the field.

Encouraged by the activities of other club members and an abundant supply of money, Cory was often absent from his studies at the Lawrence Scientific School at Harvard. In 1877 he made a trip to Florida and the next year escaped again, to the Gulf of St Lawrence, quickly producing *A Naturalist in the Magdalen Islands* (1878). After a few months at the Boston Law School he quit for good and thereafter devoted himself to a life of pleasure: travelling, collecting and indulging in his many sporting and literary interests. Frank Chapman once remarked that he had "never met a man so gifted as Charles Cory. He had the inherent potentialities as well as the means to win marked success in a surprising number of widely different fields."[2] Cory excelled at billiards, pistol shooting and golf, often winning championships but never allowing sporting competitions to take over from ornithology. He made the birds of the Caribbean his own special study, so much so that he formed the best and most complete bird collection from the area and produced several standard texts.

Despite the comparatively late date of Cory's activities there, little had been done to bring together the information gathered by earlier collectors. The English naturalist Philip Gosse had written about the birds of Jamaica (1847), Alcide d'Orbigny had worked through La Sagra's collections from Cuba (1839), Felipe Poey had produced a classic account of Cuba's animal life (1851–58) and there were scattered observations in earlier general treatises or isolated papers in journals. Recently Henry Bryant (1850s) and Frederick Ober (1870s) had collected in the Caribbean and Dr Gundlach was still active on Cuba. Cory sailed to many of the islands and paid collectors to go to others so that he could at last bring the data together. He was very thorough, even including two isolated islands then known as Old Providence and St Andrew which lie well away from the main groups close to the coast of Nicaragua. One of his earliest successes was the discovery of the winter home of the Kirtland's Warbler on Andros when even its breeding grounds were still unknown.

In 1880 Cory's *Birds of the Bahama Islands* was published, detailing a number of new records for the islands and some undescribed winter plumages. In 1885 *The Birds of Haiti and San Domingo* was hailed as a "meritorious work" and "a most useful volume" deserving "great commendation"—high praise from the editors of *The Ibis* who had poured scorn on his *Beautiful and Curious Birds of the World* because it was too costly for most people. In 1886 Cory brought out *A List of the Birds of the West Indies, including the Bahama Islands, the Greater and Lesser Antilles, excepting the islands of Tobago and Trinadad* which was expanded upon with plates and corrections in *The Birds of the West Indies* (1889)—again excluding Tobago and Trinidad because the avifauna there is more closely allied with mainland South America.

Cory formulated the first good bird lists for many of the islands but few species remained to be discovered partly because of centuries of sporadic collecting by Europeans. According to the 1983 A.O.U. *Check-list of North American Birds* Cory is credited with the original descriptions of only seven species from the region: Red-fronted Parrotlet from the east coast of Costa Rica; Ridgway's Hawk, Green-tailed Ground Warbler and Chat Tanager from San Domingo; St Lucia Black Finch from St Lucia; and the Vitelline Warbler and Grand Cayman Thrush from Grand Cayman, the latter species now extinct.

By the early 1890s Cory's collection numbered about 19,000 specimens. It had been housed at his home in Boston, then with the Boston Society of Natural History but when the Field Museum opened in 1893 Cory transferred the collection to Chicago where it formed the nucleus of the ornithology department. Cory was made honorary curator—which meant absentee curator—allowing him to continue his usual extravagant pursuits. His marriage to Harriet Petersen, in

Cory's Shearwater

1883, had not cramped his style either. His trips to Europe, hunting in the Rockies, collecting in the West Indies and sporting events all continued apace, Harriet often joining him on field trips, especially to Florida, where they spent part of every winter.

He now began to concentrate on a series of highly original field guides illustrated with small woodcuts of heads, feet or other unmistakable parts of each species. The *Key to the Water Birds of Florida* (1896), *How to Know the Ducks, Geese and Swans of North America* (1897), and a similar title on the shorebirds (1897) were followed by his two volume *The Birds of Eastern North America* (1899). They were said at the time to be "about as near foolproof as anything that has been devised for the identification of birds."[3] Most were bought by duck hunters.

Cory himself, of course, belonged to the so-called shotgun school of ornithology but he was living at a time when great changes were taking place in the attitudes of leading ornithologists. Audubon Societies were springing up, formed chiefly to counteract the depredations of the millinery trade. In 1902, as Cory was emerging from an A.O.U. meeting, an earnest young man, Harry C. Oberholser, handed him a flyer with details of an Audubon Society meeting for the following night. "What are Audubon Societies?" asked Cory. When told that they protected birds Cory replied "I am not interested. I do not protect birds. I kill them."[4] Though it spoils a good story, this was not altogether true: for several years Cory had been safeguarding non-game birds on his thousand acre estate at Hyannis on Nantucket Sound, one of the first bird sanctuaries in the United States.

In 1906 Cory suddenly lost his entire fortune as a result of some disastrous investments. He renounced most of his former associates, changed the habits of a lifetime and at the age of fifty turned to earning his living. A salaried position was found for him as Curator of Zoology at the Field Museum, so he and his family went to live in a small house in Chicago. For the rest of his days he was virtually deskbound, even museum-sponsored collecting expeditions being carried out by his assistants. He began at once on *The Birds of Illinois and Wisconsin* (1909) and followed it up with *Mammals of Illinois and Wisconsin* (1912)—each of them winning critical acclaim. In 1909 he wrote the *Birds of the Leeward Islands*, another Field Museum publication. Until his death he worked away on the *Catalogue of the Birds of the Americas* (1918–30), a monumental unillustrated museum publication listing the type descriptions and synonyms of all the New World birds. He died at Ashland, Wisconsin, on 31 July 1921, after some months of partial paralysis, and the catalogue had to be completed by Carl Edward Hellmayr. It remained a most important reference for taxonomists for many years, ranking alongside Ridgway's *A Manual of North American Birds* (1887) and *Birds of North and Middle America* (1901–19).

The longest and most detailed memorial to Cory was written by Wilfred H. Osgood, a Field Museum colleague, who contributed an affectionate fifteen-page biographical sketch to the thirty-ninth volume of *The Auk*.

COSTA. *SAVOYARD SOLDIER, POLITICIAN AND NATU-
RALIST WHO FORMED A VALUABLE COLLECTION OF
HUMMINGBIRDS.*

Louis Marie Pantaléon Costa, Marquis de Beauregard (1806-1864)

COSTA'S HUMMINGBIRD *Calypte costae* (Bourcier)

Ornismya Costae Bourcier, 1839. *Revue Zoologique* [Paris] 2, p. 294: la Californie [= Magdalena Bay, Baja California]

Costa's Hummingbird was first collected by the French naval surgeon Adolphe Simon Néboux at Magdalena Bay, on the west coast of Baja California, when his ship *La Vénus* anchored there between 25 November and 6 December 1837. After Néboux's return to Paris his new discovery came into the hands of Jules Bourcier who named it Oiseau-Mouche de Costa *Ornismya Costae* for "M. le *marquis de Costa*, de Chambéry, qui possède une très-belle collection d'oiseaux".[1]

Then in his early thirties, Costa was a dedicated collector of hummingbirds, having been interested in ornithology, and mineralogy, since about the age of fifteen. He was serving with the Royal Sardinian Piedmontese Army, being a native of Isère (now a department of south-eastern France) which was in Savoy, a part of the Kingdom of Sardinia. Costa began his military career at the age of fifteen when he became an equerry to Prince Charles Albert. Over the years Costa rose to become Captain of the Cavalry and Head Equerry to Charles Albert, who succeeded to the throne in 1832. In 1848 Costa took part in the unsuccessful campaign against Austria, then left the army to serve in the Sardinian Parliament as a senator for Savoy. After Savoy was annexed by France in 1860 he was offered a seat in the French Senate but turned it down. Instead, he settled at *La Motte*, his chateau near Chambéry, studied archaeology and became chairman of the Savoy Academy. He died at his home on 19 September 1864, on his fifty-eighth birthday.

In 1878 the Costa collection of hummingbirds was purchased by Adolphe Boucard. He considered it to be one of the best he had ever seen, there being many rare species including a male and female Costa's Hummingbird that he believed to be the type specimens. Hummingbirds held a special fascination for Boucard as they did for many other European naturalists; Jules Bourcier, John Gould, Etienne Mulsant and the Duke of Rivoli being some of the best known.

Boucard was attracted not just by their beauty and collectable size, but by their commercial value, for he was one of the last of the French natural history dealers who operated on a large scale during the nineteenth century.

Boucard travelled extensively throughout America, at one time collecting in Mexico for Philip Sclater. During 1851 and 1852 he stayed in San Francisco and at his home on the edge of Chinatown he kept up to sixty caged hummingbirds—mostly Anna's and Rufous—feeding them on fresh flowers and insects. He attempted to ship some of them alive to Europe, but even the most robust died during the voyage. Dealers annually imported hundreds of thousands of bird skins for the fashion trade; as public opinion began to turn against such uncontrolled destruction Boucard defended his profession in an impassioned diatribe in *The Hummingbird* (1891):

> "I must say a few words to the general public, and especially to the fair sex of both worlds, to explain that it will make very little difference to the wingy tribes, if Ladies condemn themselves in not wearing as adorns to their perfections the most brilliant jewels of Creation ... I am a Naturalist of forty years' standing, and have travelled all over America from Cape Horn to California. I have explored thoroughly the United States, Mexico, Central America, part of South America, and what I can warrant is this. In the southern parts of the United States, Mexico, Central America, and in Nicaragua, I have seen thousands and thousands of specimens of various species of Herons, Spoonbills, Ducks, Geese, Tanagers, Sparrows, Swallows, Humming Birds, etc. ... What are about one million or two millions of birds sent annually to Europe; chiefly from Brazil, Trinidad, Colombia, South America, and from India, against such number of birds as Nature can boast of.
>
> Even supposing that the fashion would continue for ever, it is my opinion that certain species of Birds are so common that it would take hundreds of years before exhausting them."[2]

He finished by arguing that the birds would die in any case, whether or not women wore them, since if left in the wild they would be eaten by birds of prey or other animals!

Costa's Hummingbird

The year before this was written a small group of influential ladies had founded the Royal Society for the Protection of Birds in order to stop the slaughter for the millinery trade. Though Boucard was by then living in London, we can safely assume that he did not become a member of Britain's foremost conservation organization. Six years later, in Boston, the Massachusetts Audubon Society was likewise established for the protection of wild birds. The collectors' heyday was over.

COUCH. *U.S. ARMY OFFICER AND UNION GENERAL. SLIGHT AND RATHER FRAIL, HE WAS A GOOD PROFESSIONAL SOLDIER BUT COULD BE IMPATIENT, BLUNT AND CAUSTIC. IN 1853 HE MADE A LOW-BUDGET COLLECTING EXPEDITION INTO NORTH-EASTERN MEXICO.*

Darius Nash Couch
(1822-1897)

COUCH'S KINGBIRD *Tyrannus couchii* Baird

Tyrannus couchii Baird, 1858. In S.F.Baird, J.Cassin and G.N.Lawrence *Reports of Explorations and Surveys ... for a Railroad [Route] from the Mississippi River to the Pacific Ocean*, Vol.9, pp. xxx, 170, 175: New Leon and San Diego, [Nuevo León,] Mexico

Surgeon-naturalists were sufficiently numerous in the latter part of the nineteenth century for E.E. Hume to write a full volume on *Ornithologists of the U.S. Army Medical Corps* (1842). Soldiers without medical training who collected birds, such as Darius N. Couch, George B. McClellan, John P. McCown and George A. McCall, were much less common. In retrospect, the ornithological achievements of the latter group seem slight, but their discoveries and observations played a key part in determining the range and distribution of many species. Sometimes only the army could enter the more inhospitable regions where the bird life was then so little known. The observations of some of the above-named officers were of value, for instance, to John Cassin, who used their records in his *Illustrations of the Birds of California, Mexico, Oregon, British and Russian America* (1853–56). One section of such notes, provided by Lieutenant Couch, offers a little insight into the depth of feeling sometimes induced in naturalists when collecting birds. Some collectors seem to have been completely heartless, capable of silencing thousands of songsters without remorse, but Couch was less sure of his part-time vocation:

"Early one morning, an old man, who had daily called on me, with his wife and six nude pickaninnies, presented himself, and wished as usual to take me to a spot where great numbers of rare birds were to be found. Gladly assenting, we were out of sight of the rancho at sunrise of one of the magnificent mornings only known in tropical latitudes. It was the day after a severe *norther*, and the whole feathered kingdom was in motion. My guide soon called my attention to two *calandrias*, as these birds [Audubon's Orioles] are called by the Mexicans, which were quietly but actively seeking their breakfast. The male having been brought down by my gun, the female flew to a neighboring tree, apparently not having observed his fall; soon, however, she became aware of her loss, and endeavored to recall him to her side with a simple *pout pou-it*, uttered in a strain of such exquisite sadness, that I could scarcely believe such notes to be produced by a bird, and so greatly did they excite my sympathy, that I felt almost resolved to desist from making further collections in natural history ..."[1]

Despite this, Couch went on to complete his most important bird work on this trip, exploring parts of north-eastern Mexico during a period of leave of absence whilst loosely attached to the U.S and Mexican Boundary Survey. A few years later he married and left the army to settle down for civilian life in Taunton, Massachusetts, little realizing that the major part of his military career, as a Union General, was yet to come. In 1861 he could not ignore the call to arms and volunteered his services at once. He was given command of a division of volunteers in the Peninsular campaign, distinguishing himself at the battle of Fair Oaks. Illness first contracted in Mexico afterwards induced him to offer his resignation but McClellan greatly valued his expertise and instead he was appointed major-general of volunteers. Couch fought at Antietam, Fredericksburg and Chancellorsville. He then took command in Pennsylvania where he assisted at the battle of Gettysburg. In 1864, as the Civil War dragged on, he was assigned to a division of the 23 Corps and arrived in time to lead it at the battle of Nashville, when his horse was shot from beneath him.

Couch's early training for these cataclysmic years had been received at West Point and in the Mexican War. Born on 23 July 1822 on a farm near the town of South East, Putnam County, New York, he entered the Military Academy in 1842 at the age of twenty. He graduated four years later in the same class as future generals McClellan, 'Stonewall' Jackson, Grant, Burnside, Franklin and several other distinguished commanders. Jesse Lee Reno was his room mate for three years.

Couch, like John P. McCown, was commissioned into the 4th artillery and was sent to Mexico. He joined his battalion at Montclara shortly before the long drawn out, bloody and inconclusive battle of Buena Vista on 22 and 23 February 1847. In May he was sent to Point Isabel but became so seriously ill that he was sent home in August on sick leave. He next served in Pennsylvania and North Carolina and around this time may perhaps have first made the acquaintance of John Cassin at Philadelphia and Professor Baird at Washington. At the beginning of 1853 he started his aforementioned year's leave of absence in order to make a zoological expedition into northern Mexico—his success in this venture being the reason why his name is known at all to present-day American birders.

By 15 February he had been in Brownsville for two weeks and wrote to Baird outlining his achievements and immediate plans:

"Some 25 or thirty birds badly put up, a few snakes, quadrupeds, and insects is all I can show—and yet I have worked; the weather, however, must bear part of the blame, having had many rainy days ...

The season for collecting insects is not favorable; too early. As the warm weather approaches, the field will be richer.

I have hired two trusty servants–and hope to secure my mules to-morrow–there is not a little to do in completing my necessities."[2]

The following month Baird replied encouragingly, noting that the previous evening had been spent with General Scott and Colonel Cooper who had "promised everything". Baird had no doubt been negotiating or devising schemes to improve the Smithsonian Institution's natural history acquisitions from Texas and Mexico. In concluding the letter Baird warned that Couch's collection from New Orleans had not yet arrived.

During his time at Brownsville Couch was involved in some curious negotiations of his own. He made several trips across the border to Matamoros to see the widow of the French botanist Jean Louis Berlandier. This man had been sent to Mexico by Swiss naturalists who had underrated the difficulties faced by their collector and had poured scorn on the 55,000 specimens he had managed to send back! Berlandier decided never to return to Europe and settled at Matamoros where he had become, in the words of Couch, "universally beloved from his kind amiable manner, and regard for the sick poor of that city, being always ready to give advice and medicine to such without pay".[3] Berlandier had drowned while crossing a river in 1851 and Couch was now able to buy his whole collection, accumulated over twenty-four years, for $500, to save it from obscurity and ruin. Couch outlined its extent for Baird: "There are about 150 bottles of diff. sizes, diff. species of vertebrata, mostly snakes, lizards, etc.;—a few birds; several cubic feet of minerals; a box of plants; some twelve square feet of insects nicely preserved in glass cases; paintings of all the different indian tribes in Old Mexico; Sketches of Mex. scenery, meteorological reports, observations with piles of manuscripts relating to his labors. It's very valuable, probably been abused somewhat."[4]

After a month at Brownsville Couch was able to send off Berlandier's material with his own meagre contributions, overjoyed at being able to leave the Gulf Coast at last to commence the most exciting part of his expedition into north-eastern Mexico. (For a map see p. 498.) Entering the State of Nuevo León it

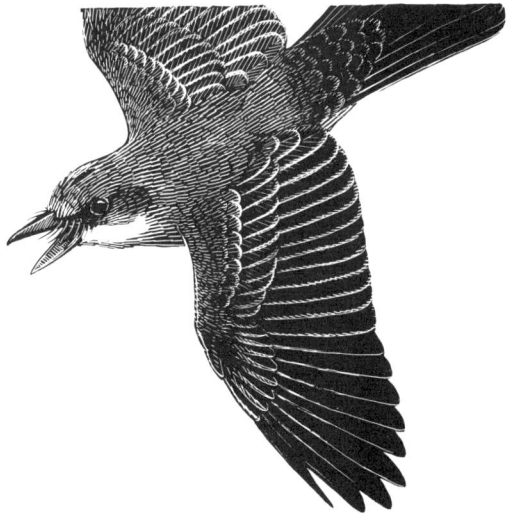

Couch's Kingbird

was not long before he had added the bird now known as Couch's Kingbird to his collection: in March and April 1853 he shot specimens at New Leon and at nearby San Diego. After making his way to Monterrey he pushed on to Agua Nueva collecting all the while, eventually covering considerable portions of the States of Tamaulipas, Nuevo León and Coahuila. His observations on birds such as the Greater Roadrunner, Black-throated Sparrow and Audubon's Oriole found their way into Cassin's *Illustrations*, adding in a small way to its appeal.

After crossing the border back into the United States Couch headed for Philadelphia and the Smithsonian, taking with him those items he had not sent back already. In order to secure priority for himself he wrote up some of his new bird discoveries as soon as possible in the April 1854 number of the *Proceedings of the Academy of Natural Sciences of Philadelphia*. His 'Descriptions of new Birds of Northern Mexico' appears to be his only ornithological publication, and though spread across two pages it could have been fitted into just one. Three species are described: a junco, Scott's Oriole and the Chihuahuan Raven. Only the latter proved to be completely new but the name Scott's Oriole is derived from Couch's proposed scientific name *Icterus Scottii* which was first suggested in this paper.[5]

By March 1854 Baird had completed the "enumeration and determination" of Couch's collection but the skins of the kingbirds from New Leon and San Diego were not described by Baird until 1858 when he decided to give Couch's Kingbird full specific status in the ninth volume of the Pacific Railroad Reports. When Baird came to write up the zoology for the U.S. and Mexican Boundary Survey, published in 1859, there was of course a section on the ornithology. Sometimes bound up separately and known as 'The Birds of the Boundary' it ran to thirty-three pages with contributions by Couch, Caleb Kennerly and John H. Clark. Couch's notes are usually the shortest and those for Couch's Kingbird are typically disappointing: "Very common and noisy; found among the large trees of gardens and luxuriant river bottoms."[6] To make up for this, plate XI of the forty-five hand-coloured plates shows an Empidonax flycatcher, the tail only of a Tropical Kingbird and a glorious illustration of Couch's Kingbird.

A few months after Couch's return from Mexico in August 1854 he married Mary Caroline Crocker. Eight months later, while serving at Fort Leavenworth, Kansas, he resigned from the army and took employment at the Taunton Copper Company run by the Crocker family. After six peaceful years Couch was caught up in the Civil War and served with distinction, as related earlier, from 15 June 1861 until 26 May 1865. After the war he stood as Democratic candidate for the Governorship of Massachusetts but was defeated. He then made a number of career changes, being a customs official at the port of Boston, president of a Virginia mining and manufacturing company and, after moving to Connecticut in 1870, he became quartermaster-general of the state for two years and adjutant-general for another two. He died at Norwalk on 12 February 1897 at the age of seventy-four and was buried at Taunton. There is no evidence that he continued his interest in birds in the post-war years. He is said to have written an account of his natural history expedition into Mexico but it was never published.

Grace Darling Coues†
(1847-1925)

GRACE'S WARBLER *Dendroica graciae* Baird

Dendroica graciae (Coues MS) Baird, 1865. *Revue of American Birds*, Vol. 1, p. 210: Fort Whipple, near Prescott, Arizona

The Coues family name is Norman-French, originally pronounced 'Coo-ès' but Anglicized to 'Cows' following the emigration of one Peter Coues from the Channel Island of Jersey to the United States in 1735. His grandson, Samuel Elliott Coues, became a successful shipping merchant at Portsmouth, New Hampshire. When his first wife died he married again and the three surviving offspring of this union were Elliott Coues (born 1842), Louis Dwight Coues (born 1845) and Grace Darling Coues (born 4 September 1847). The children spent their earliest years in New Hampshire but in 1854, before Grace was seven years old, her father accepted a position at the U.S. Patent Office in Washington. For the next twelve years she lived at 397 12th Street where she was joined by an adopted sister, Lucy.

While little is known about Grace's early years much has been written about Elliott, her soon-to-be-famous brother. When he was very young he showed a strong passion for wildlife and since he lived in Washington it was almost inevitable that he should meet Professor Baird, who had been appointed Second Secretary to the newly formed Smithsonian Institution in 1850. He showed such promise that when he was seventeen years old Baird fixed him up with a place on an expedition to Labrador. Afterwards he graduated from medical school in Washington and gained valuable surgical experience treating Civil War wounded, but in 1864 Baird stepped in once again to secure a position for him at an army post in the West.

When Assistant Surgeon Coues reported at the army headquarters in Santa Fe he learned that he was being posted to Fort Whipple, Arizona. To complete the final leg of his journey he rode south to Albuquerque to catch up with a huge supply train of some ninety-two wagons headed for the fort. For the next five hundred miles "Doctor Coues never ceased ... making excursions along the

†No portrait traced

flanks of the column and arriving in camp with many specimens. Clad in a corduroy suit of many pockets and having numerous sacks and pouches attached to his saddle, he regularly rode out of column every morning astride of his buckskin-colored mule ... Rarely did we see him again until we had been some hours in the following camp but we heard the discharge of his double barreled shotgun far off the line of march ... When he sat upon the ground and proceeded to skin, stuff and label his specimens, he was never without an interested group of officers and men about him."[1]

At the end of June they made a three-day halt at Fort Wingate before attempting the crossing of the Zuni Mountains. At Whipple Pass on 2 July Coues hurriedly obtained a single specimen of a new *Dendroica* warbler.

Four weeks later the party at last arrived at Fort Whipple, which was set in a valley at 7500 feet; to the south and west the country was mountainous and forested; to the north and east it was much more open and bare. Coues emphasized two distinct disadvantages of the area when he wrote to Baird: "*No water whatever*; and ornithologizing will require much wind and muscle. The Apaches are so hostile and daring that considerable *caution* will have to tinge my collecting enthusiasm, if I want to save my scalp."[2]

Even so, by November of the following year Coues had accumulated over 600 bird skins as well as some mammals, reptiles, amphibians, insects and plants. Amongst the birds were nine more examples of the new *Dendroica* warbler collected around the fort between 11 August and 29 October. He found the yellow and black inhabitant of the pines to be "the most abundant bird of its kind, excepting Audubon's Warbler [Yellow-rumped Warbler]".[3] As he wrote up his notes Coues said that he wanted to name the new species after his sister Grace in the hope "that my affection and respect keep pace with my appreciation of true loveliness of character". When Baird came to publish the description of the bird he followed Coues's proposal and chose for the type specimen one of the nine skins from the fort, rather than the first one taken on the outward journey through New Mexico.

By 4 March 1865 Coues had begun to write to Baird pleading with him to

Grace's Warbler

use his influence to get him back East. Desert life had quickly lost its appeal, new material was now much harder to come by and he still had no reference books to work from. In early September he managed to make a month-long excursion westwards to Fort Mojave and down the Colorado Valley to Fort Yuma, enjoying the change of scene and securing Abert's Towhee, Wood Stork and other species that he had never seen before including the second recorded specimen of Le Conte's Thrasher. In mid-October he was at last recalled and headed West from Fort Whipple at the first opportunity, pleased to be able to spend a few days in California with the ornithologist James G. Cooper before taking the steamer home.

'Ornithology of a Prairie-Journey, and Notes on the Birds of Arizona' appeared in 1865 in *The Ibis*, the journal of the British Ornithologists' Union, and his 'List of the Birds of Fort Whipple, Arizona' was published in 1866 in the *Proceedings of the Academy of Natural Sciences of Philadelphia*. Both publications served to bring international acclaim to Coues who was just twenty-three years old when he returned to Washington. Grace's Warbler and the Gray Vireo were his only new bird species but he had discovered new subspecies of the Western Wood Pewee, Bell's Vireo, Solitary Vireo and Black-chinned Sparrow. In the succeeding years, as he served on various government expeditions, Coues developed a unique grasp of the North American avifauna and became one of America's greatest ornithologists.

Meanwhile, Grace Coues was delighted with the safe return of her elder brother. Her other brother, Louis, had died suddenly on 19 March 1864 only three and a half months before the discovery of Grace's Warbler. Gracie, as she was called, spent the winter of 1866–67 with Elliott while he was serving in Columbia, South Carolina, and she was no doubt present at his wedding in the city on 3 May. In Portsmouth, New Hampshire, on 15 October, Gracie married an old friend of Elliott's, Charles Albert Page, a former war correspondent who had joined the Consular Service. While living in Zurich they had two children: Louis in 1869 and Charles in January 1870; followed by two more while based in London: Charlotte in December 1870 and George in 1872. When her husband died in London in 1873 she returned to the United States with the children and moved in with her mother in Washington. They were all joined for a while in 1881 by Elliott who had separated from his wife.

On 10 November 1884 Gracie again married a friend of her brother's, this time Dana Estes who had just returned from a short trip to Europe with Elliott. The wedding was held at her brother's home at 1726 N Street, Washington. Though they had no children of their own Estes had three by his first wife who had died the previous year. Dana Estes was a director of Estes & Lauriat and dealt with the wholesale and manufacturing side of the business while his partner, Charles E. Lauriat, was in charge of retail. The firm published Elliott Coues's well-known *Key to North American Birds* (1872), *Check List of North American Birds* (1873–74), *Birds of the Northwest* (1874) and *Fur-bearing Animals* (1877). Though natural history was not their speciality they also produced Charles Cory's *The Birds of Haiti and San Domingo* (1885), Isaac Sprague's *Flowers of Field and Forest* (1886) and were among the first to publish the work of Louis Agassiz Fuertes.

Gracie's wealthy husband took up archaeology and foreign travel and treated her to visits to London and tours in Italy and Egypt. In 1904 he was the first American to journey up the Nile to Uganda and the Congo and in 1908 he sailed down the east coast of Africa, landed at Mombasa and travelled 2500 miles overland to Cape Town. This last trip weakened his health so severely that he died on 16 June of the following year at their home in Brookline, Boston.

Two of Gracie's sons were trained at Estes & Lauriat, and with their brother formed L.C. Page & Co. Louis and George remained in Boston but Charles later left for Australia. Their sister Charlotte married J. Warren Stearns and lived in San Diego, California. Gracie went out to San Diego for the last few years of her life and died there on 27 December 1925, aged seventy-eight.

Federico Craveri
(1815-1890)

CRAVERI'S MURRELET *Synthliboramphus craveri* (Salvadori)

Uria Craveri Salvadori, 1865. *Atti della Società Italiana di Scienze Naturali, e del Museo Civile di Storia Naturale. Milano* 8, p. 387: Golfo della California, Lat. 27°10′12″ [N.], Long. 110°10′45″ [W.], [= Raza Island, Gulf of California]

In 1865 Count Tommaso Salvadori noticed an undescribed murrelet among a small collection of Mexican bird skins in the Zoological Museum of Turin University. He named the new species *Uria Craveri* in honour of its collector, who was by then a well-known scientist teaching nearby at Bra.

Federico Craveri was born at Turin on 29 July 1815, the elder son of Angelo Craveri, a lawyer and politician. At the age of twenty-five Federico left Italy and for the next fifteen years lived in Mexico City. After graduating in chemistry and pharmacy he continued to work at the university as an assistant chemist and during his vacations climbed some of the many volcanoes in the vicinity, including Popocatepetl (17,881 feet). At the beginning of 1847 he was joined by his brother, Ettore, who stayed in Mexico for two years. Having the same scientific interests they enjoyed various field trips together until Ettore had to return to Italy, because of the death of their father, to look after his museum at Bra.

During the next few years Craveri's work took him to many mining districts, including the picturesque city of Guanajuato, at the centre of Mexico's silver mining industry. In October 1855 he left Mexico City and went to the port of Mazatlán because the Mexican government had asked him to assess the potential for a guano industry in the Gulf of California. Guano, which is formed in dry climates from the excrement of seabirds and other marine animals, had been used by Peruvians to fertilize their crops for centuries, but regular trade in guano between the New and the Old Worlds did not develop until the 1840s. The chemical composition and manure value of different sites varied greatly. Craveri therefore had to visit many islands up and down the gulf to assess quantities and to take samples for chemical analysis.

He began the five and a half month cruise at the end of January 1856 and on

CRAVERI. *ITALIAN CHEMIST, METEOROLOGIST AND NATU-*
RALIST WHO WORKED FOR TWENTY YEARS IN MEXICO. IN
1856 HE VISITED MANY OF THE BIRD COLONIES IN THE GULF
OF CALIFORNIA.

21 April landed on Raza Island, where he took the type specimen of Craveri's Murrelet. Though he wrote at length about the appearance of the low, flat-topped rock, its volcanic origins and the composition of the guano, he made no mention of the seabirds. Presumably he failed to realize that his murrelet was a species unknown to science, since it lay undescribed for another nine years— but then Craveri was a mineralogist and chemist, not a great ornithologist. Nineteen years later, in April 1875, the birds of Raza Island were more carefully observed by Dr Thomas H. Streets. In the midst of a huge colony of Heermann's Gulls he found Craveri's Murrelets incubating their eggs wherever there was cover under the rocks. Most of the murrelets had a full clutch of two eggs and Streets was the first to describe their diverse ground colours and markings.

Craveri must have been impressed by the marine life of the gulf but he made only a small, rather varied zoological collection, perhaps because the demands of his job gave him little opportunity to do otherwise.

During the late summer and autumn Craveri was employed in examining the silver mines around Mazatlán, then in December he returned to surveying the Mexican islands for another seven months, this time along the Pacific coast of Baja California. In January he discovered several uncharted islands, one of which he named Isla Elide—a memorial to two unhappy love affairs, each with a girl of the same name. He took official possession of the islands in May after visiting San Francisco to establish Mexican ownership.

In July Craveri returned to Mazatlán, then travelled south to visit the mines of the Sierra Madre, his last Mexican project. In May 1858 he left the country for a twelve-month tour of parts of Canada and the United States. With his professional interest in mining and precious metals, it was not surprising that Craveri headed for Vancouver in order to spend six weeks on the Fraser River where thousands of gold prospectors had tried their luck during the previous two years. He then sailed south to San Francisco

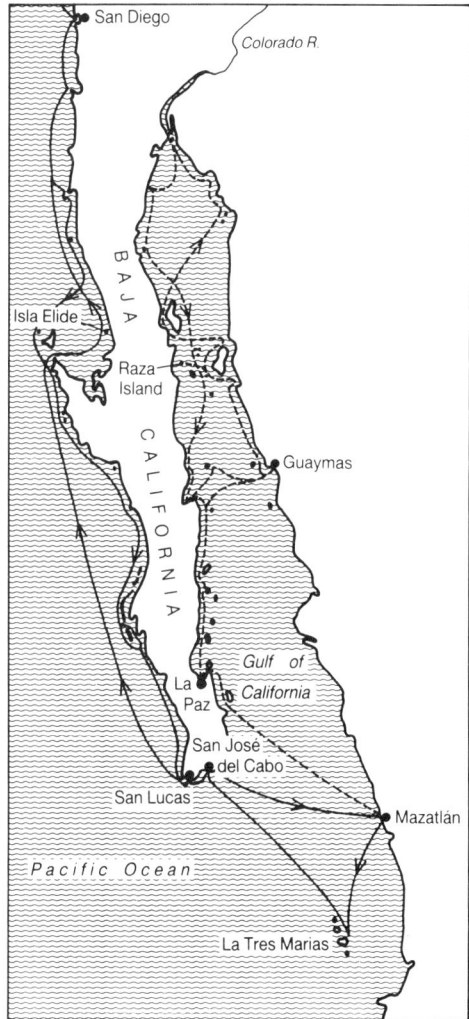

CRAVERI'S VOYAGES
------- 1856 ———— 1856–57

and from the end of September until the middle of February he travelled in the Sierran foothills between Downieville on the Yuba River and Mariposa, to the east of Merced. From San Francisco he sailed to Panama, crossed the Isthmus, and took ship for Havana where he spent several weeks before continuing on his way to New Orleans. His journey then took him to Louisville, across to St Louis, up the Mississippi to St Paul, on to Chicago and through the Great Lakes. After viewing the Niagara Falls he enjoyed sight-seeing in the major eastern cities and finally embarked for Europe from Gloucester, Massachusetts, at the end of July. Even then he was reluctant to return home and spent time in England and France, before eventually reaching Bra in September 1859 after an absence of nineteen years.

Now aged forty-four, he settled at Bra. His mineralogical and natural history collections were added to the museum and with his brother he strove to increase the usefulness of the establishment by developing a chemical laboratory and other research facilities of use to local industry and agriculture. The Craveris became immersed in meteorological studies, the practical difficulties of which drove Federico to modify and redesign various instruments to measure wind strength and light intensity. Many of his observations and innovative theories were publicized in a stream of papers, written in French and Spanish as well as Italian. From 1861 he taught at the Technical School in Bra, becoming Professor of Physics, Chemistry and Natural History. In 1876 he was awarded the title of 'Cavaliere della Corona d'Italia'.

On 30 June 1866 Federico married Elisa Tarabiono whom he had met while visiting friends in Ivrea. Well educated and widely read, she was the author of many poems, some of which she had published. She was only half the age of Federico, but was deeply devoted to him and supportive in every way until she died giving birth to a stillborn son, on 20 September 1870.

In 1866, the year of their marriage, a daughter, Cornelia, was born to Craveri, but Elisa was not the mother. Craveri did not officially adopt Cornelia until 1884, although she was by then staying with him at Bra. His only child, she later became Cornelia Salvadori and died in 1952.

Federico Craveri pursued his scientific interests to the end of his life, his last publications appearing in 1887. An active member of the Italian Alpine Club,

Craveri's Murrelet

he also remained physically fit and his death on 14 April 1890, at the age of seventy-four, was sudden and unexpected. Just two days before, he had given the Craveri Museum to the town of Bra; the Museo Civico Craveri di Storia Naturale is now an important centre for the study of meteorology and ornithology.

DOLE. *JUDGE, DIPLOMAT AND STATESMAN WHO BECAME THE FIRST GOVERNOR OF THE TERRITORY OF HAWAII. SEEN HERE WITH HIS WIFE OUTSIDE THEIR RESIDENCE IN HONOLULU.*

Sanford Ballard Dole
(1844-1926)

CRESTED HONEYCREEPER *Palmeria dolei* (Wilson)

Himatione dolei S.B.Wilson, 1891. *Proceedings of the Zoological Society of London*, p. 166: Maui

The type specimen of the Crested Honeycreeper—or Akohekohe—was a juvenile, shot by Scott B. Wilson in July 1888 while he was exploring the district of Kula, on Maui. Wilson was an English zoologist, then aged twenty-four, who had been collecting birds in the Hawaiian Islands for over a year. When he subsequently published the original description of the Crested Honeycreeper as *Himatione dolei* he wrote: "I have named it after my friend Mr. Dole, whose name is so well known amongst those ornithologists that have studied the Hawaiian avifauna."

Dole's ornithological publications were the most comprehensive sources of information about the Hawaiian birds then available to Wilson as he began his study of the endemic species. Sanford B. Dole, the son of American missionaries, had been born at Honolulu on 23 April 1844 and spent some of his early years on the outlying island of Kauai. Though he had left the islands in 1866 in order to study law in Massachusetts, he had returned to practise on Oahu and his first ornithological paper appeared soon afterwards, in 1869. Entitled 'Synopsis of Birds hitherto described from the Hawaiian Islands' in the *Proceedings of the Boston Society of Natural History* it enumerated forty-eight species. According to Walter Rothschild, who reviewed all the Hawaiian ornithological literature preceding his *Avifauna of Laysan* (1893–1900), Dole's paper was "Most valuable, though full of errors".

In 1876 Dole studied the collection of stuffed birds belonging to J.D. Mills of Hilo, Hawaii, and found an undescribed red, black and grey drepanid, now extinct, the Ula-ai-hawane. Dole wrote that it was "a bird of remarkable beauty, its peculiar combination of colours producing a most harmonious and elegant effect". He gave it the scientific name *anna* as a tribute to his wife, Anna Prentice Cate (1841–1918), a native of Maine, who had become Mrs Dole in 1873. *Ciridops anna* was first described in Thrum's *Hawaiian Almanac and Annual for 1879* (published 1878) as part of Dole's revised list of fifty-three species. He made no further contributions to the ornithological literature, though he was a lifelong corresponding member of the A.O.U. from the time of its organization in 1883. As president of the board of trustees of the Bishop Museum, it is likely that his appreciation of Mills's irreplaceable collection led to its purchase by Charles R. Bishop soon after the museum was founded.

In 1884 Dole was elected to the Hawaiian legislature and became prominent in the reform movement that forced King Kalakaua to accept a new constitution in 1887. Dole then served as a judge of the Supreme Court until 1893 when Queen Liluokalani lost her throne and he was elected President of the Provisional Government. He pursued a policy of annexation to the United States but President Cleveland opposed such a move, favouring instead the restoration of the monarchy. Since it was evident that annexation could not be achieved while Cleveland remained in office, the Republic of Hawaii was declared on 4 July 1894 with Dole as President. His administration was beset with difficulties but relations with the United States improved after the inauguration of President McKinley and early in 1898 Dole visited Washington to negotiate an annexation treaty. He was made the first Governor of the Territory of Hawaii in 1900 and held office until 1903. He resigned to become the U.S. District Judge and remained on the bench until 1915.

After retirement "The Grand Old Man of Hawaii" remained an influential and outspoken figure, still active in various societies and organizations until shortly before his death in Honolulu on 9 June 1926, at the age of eighty-two.

Crested Honeycreeper

Johann Gotthelf Fischer von Waldheim (1771-1853)

SPECTACLED EIDER *Somateria fischeri* (Brandt)

Fuligula (Lampronetta) Fischeri J.F. Brandt, 1847. *Novam Avium Rossicarum Speciem*, pp. 1–18, Pl. 1: St Michael, Alaska

The type specimen of the Spectacled Eider was taken by Ilya Vosnesensky who collected for the St Petersburg Academy of Sciences. In 1839, at the age of twenty-three, he sailed from the Baltic port of Kronshtadt and for the next ten years was based at various Russian settlements along the coasts of Alaska and eastern Siberia. He called at Bering Island, the Kuril Islands, the Aleutians and the Pribilofs, and on Kodiak Island took sixty species of birds. Periodically he dispatched his collections to Johann Friedrich Brandt, the Director of the Zoological Museum at the Academy. The new eider arrived there with material from St Michael, an outpost on the southern shore of Norton Sound, near the Yukon Delta.

After Vosnesensky returned to St Petersburg, Brandt took advantage of his wide experience of Russian-American birds and they worked together on various publications. When Brandt described the Spectacled Eider in 1847 he did not name it after the collector but after his distinguished colleague at the Moscow Academy, Dr Fischer von Waldheim, who was then in his seventies. The old man had been one of the dominant Russian naturalists for over forty years, but he was a native of Saxony, born in Waldheim on 13 October 1771. As the son of a linen weaver his education would have been very limited but for various benefactors who rewarded his early scholastic achievements by sending him to grammar school in Freiberg where he became friendly with Alexander von Humboldt, the two youths sharing a fascination with natural history. At Leipzig Fischer studied medicine and gained his doctorate, then travelled to Vienna and Paris with Humboldt.

In 1798 Fischer became Professor of Natural History at Mainz, on the Rhine, and there married Catharina Renard in 1801, who mothered three daughters and a son. In 1804 he was called to be Professor of Natural History and Director

FISCHER. *ZOOLOGIST AND PALAEONTOLOGIST AT THE*
MOSCOW ACADEMY, HE TOOK A MORE ACTIVE INTER-
EST IN BIRDS DURING THE LAST YEARS OF HIS LIFE.

of the Natural History Museum at the Moscow Academy, and in August of the following year he founded the Société Impériale des Naturalistes de Moscou. He continued to study zoology and during this period he was primarily engaged in the classification of invertebrates. Between 1820 and 1851 the fruits of his entomological studies were published in the five-volume *Entomographia Imperii Rossici*. He was also engaged in studying the fossils from the wooded countryside around Moscow, his zoological and palaeontological investigations earning him the title of 'Russia's Cuvier'. Many fossils were called after Fischer but his wide-ranging interests also included mineralogy, and his labours in that branch of science were honoured in an aluminium phosphate known as Fischerite.

In September 1812 Moscow was occupied by the French until the Russians started a great fire that forced Napoleon to withdraw but destroyed many of the city's finest buildings, including the Natural History Museum and Fischer's own library. After the war Fischer and his fellow scientists immediately planned a new museum and the acquisition of collections for it.

Fischer's mark on Russian ornithology was slight in comparison with his other areas of research because he only turned to the subject during the latter part of his life. He is chiefly known by birdwatchers because of his link with two species; the Spectacled Eider which breeds on the Arctic tundra and Pander's Ground Jay, an inhabitant of the great sandy Russian deserts, which he described and named in 1821.

In 1835 Fischer was elevated to the nobility, adopting the title of von Waldheim because of his great and lasting affection for his native town. That same year he also became a privy councillor and throughout his long life many honours were heaped on him by Russian and European scientific societies. He died in Moscow in October 1853, aged eighty-two, three years after the death of his wife.

There were a number of other nineteenth century naturalists with the name of Fischer, so the specific name of *fischeri* does not always refer to Fischer von Waldheim. Fischer's Lovebird is a tribute to Dr Gustav Fischer who explored in eastern Africa from 1876 to 1886, and Fischer's Fruit Dove was named for Dr Georg Fischer, an army surgeon in the Dutch East Indies who collected for the Darmstadt Museum.

Spectacled Eider

FORSTER. *GERMAN NATURALIST, AUTHOR AND TRANS-LATOR. HE WORKED THROUGH SOME OF THE EARLIEST NORTH AMERICAN SPECIMENS TO BE SENT TO EUROPE BUT IS BEST KNOWN AS THE NATURALIST ON COOK'S SECOND PACIFIC VOYAGE. HE IS DEPICTED HERE ON TAHITI WITH HIS SON GEORGE.*

Johann Reinhold Forster
(1729-1798)

FORSTER'S TERN *Sterna forsteri* Nuttall

Sterna Forsteri Nuttall, 1834. *Manual of the Ornithology of the United States and of Canada*, Vol. 2, p. 274. New name for *Sterna hirundo* (*not* Linnaeus) Richardson, 1832. In W. Swainson and J. Richardson, *Fauna Boreali-Americana*, Vol. 2 (1831), p. 412: on the banks of the Saskatchewan [River] [= about 10–50 miles west of Cumberland House, Saskatchewan]

Johann Reinhold Forster was born at Dirschau (Tczew), near Danzig and was educated locally, then at Berlin and at Halle University. He spent twelve years as a minister in a country parish, occupying his leisure with intensive scientific study before leaving in 1765 to carry out a commission for the Empress Catherine to inspect new colonies being established east of the Volga. After falling out with the Russian authorities he emigrated to England and secured a position at Warrington Academy, east of Liverpool. For three years he taught natural history and modern languages then, in 1772, he and his son George were chosen as naturalists for a three-year voyage with Captain Cook. On their return the Forsters endeavoured to pay off their debts by writing accounts of the voyage and by translation work. Both Forsters later returned to the Continent, the elder becoming Professor of Natural History at Halle University for nineteen years until his death on 9 December 1798.

While with Cook he studied South African raptors, woodland birds from obscure tropical islands and oceanic and Antarctic seabirds. He recognized five new species of penguin and wrote at length about three kinds of albatross. His major zoological treatise, the *Descriptiones animalium*, was close to completion shortly after his return but was set aside and published posthumously in 1844, edited by Hinrich Lichtenstein. Most species were from the southern Pacific but one new bird that reaches northern waters is the Mottled Petrel, a rare but regular visitor to Hawaiian and southern Alaskan coasts.

Before the voyage, while still at Warrington, Forster began his brief association with Thomas Pennant, the noted author of the *British Zoology* (1766) and *Arctic Zoology* (1784–85). Pennant lived nearby in North Wales and took pity on the impoverished naturalist, who had a wife and seven children to support. He helped Forster a little in financial ways, introduced him to other naturalists and initiated his involvement with North American birds. In 1771 Forster published a *Catalogue of the Animals of North America ... To which are added Short Directions for Collecting, Preserving, and Transporting all Kinds of Natural History Curiosities*. It seems that Pennant had begun the list of species but it was Forster who completed it and added the *Directions*. Most of the specimens were collected by Joseph Banks in Newfoundland in 1766, or by Ashton Blackburne in the eastern United States for his sister Anna's museum, which was located a few miles from Warrington. The Common Nighthawk was described properly for the first time in the catalogue, based on Mark Catesby's illustration of a 'Whip-poor Will' in his *Natural History of Carolina* (1754).

In 1772 Forster wrote 'An account of the birds sent from Hudson's Bay' for the *Philosophical Transactions of the Royal Society*. This time the skins were supplied by the Hudson's Bay Company, who had recently asked its employees to send more specimens to England following the success of James Isham's collections that had helped to illustrate George Edwards's *Natural History of Uncommon Birds* (1750). Andrew Graham sent sixty-four skins from Severn River, Humphrey Marten sent seventeen skins from Albany, Moses Norton sent seventeen skins from Churchill and there were two skins from Ferdinand Jacobs and Thomas Hutchins at York Factory. Forster was allowed to examine them all and he described five new species: Great Gray Owl, Boreal Chickadee, Blackpoll Warbler and White-crowned Sparrow from the Severn River, and the Eskimo Curlew from Albany.[1]

Forster considered one of the other birds to be a variety of the 'greater Tern' *Sterna Hirundo* and though the number tag was missing he thought that it perhaps came from the Churchill River. He wrote a fair description of the bird now known as Forster's Tern: "The feet are black; the tail is shorter and much less forked than that described and drawn in the Br. Zool. The outermost tail-feather likewise wants the black, which that in the British Zoology has. In other respects it is the same."[2] In 1834 Thomas Nuttall woke up to the fact that there was another North American tern and coined the name *Sterna Forsteri* for it in honour of "the eminent naturalist and voyager who first suggested these distinctions [from the Common Tern]".

Forster's Tern

John Franklin (1786-1847)

FRANKLIN'S GULL *Larus pipixcan* Wagler

Larus Pipixcan Wagler, 1831. *Isis von Oken* 24, No. 5 (May), col. 515: Mexico

By any standards, John Franklin led an exciting and adventurous life. He was born at Spilsby in Lincolnshire, England, on 16 April 1786, the youngest son of a family of twelve children. Six weeks before his fourteenth birthday he joined the Royal Navy and a year later was serving as midshipman on H.M.S. *Polyphemus* during the battle of Copenhagen. He then joined the *Investigator* for a circumnavigation of Australia with Captain Matthew Flinders—an uncle by marriage. During the latter part of this remarkable voyage the discovery-ship became unseaworthy and, still suffering from scurvy, the sailors transferred to the *Porpoise* and the *Cato*, both of which were wrecked on an uncharted coral reef. Flinders set off to cross the 800 miles back to Sydney in a six-oared cutter and returned three weeks later to rescue the crews. Young Franklin then made his way home via Canton with a trading fleet that successfully repulsed an attacking French squadron. Franklin was still not twenty years old when he took part in the battle of Trafalgar; forty of the forty-seven men with him on the quarter deck of the *Bellerophon* were killed and his hearing was permanently impaired by the thundering broadsides. He spent the next six years on the *Bedford* in European and American waters and was wounded during hand-to-hand fighting as he led a boarding party onto an enemy gunboat at New Orleans.

At the end of the Napoleonic War the Admiralty, and Second Secretary John Barrow in particular, wanted to re-open the quest for the North-west Passage. In 1818 John Ross was sent to Baffin Bay to seek a route westwards, while Captain David Buchan and Lieutenant Franklin went further north by way of Spitzbergen in command of the *Dorothea* and the *Trent* in an unsuccessful attempt to reach the North Pole, another long-sought goal. When Franklin returned he was considered the best choice for leading an overland expedition into Arctic Canada to shed further light on a possible north-west route.

The Canadian expedition got underway shortly after Franklin arrived at York Factory on Hudson Bay on 30 August 1819. Amongst those with him were midshipmen George Back and Robert Hood and surgeon-naturalist Dr John

FRANKLIN. *ROYAL NAVY OFFICER AND ARCTIC EXPLORER. HE MADE TWO OVERLAND CANADIAN EXPEDITIONS WITH THE ZOOLOGIST DR JOHN RICHARDSON.*

Richardson. Franklin and Back passed the winter at Fort Chipewyan, while Richardson and Hood remained at Cumberland House beside the Saskatchewan River. The second winter was spent together beyond the Great Slave Lake at Fort Enterprise, poised for the main push northwards as soon as conditions allowed. In June 1821 they left the fort for the journey down the Coppermine River to the Arctic Ocean, which was first seen in mid-July. Some of the men now returned, but Franklin, the three other Britons and the rest of the hired hands went eastwards in canoes, charting the coastline as far as Point Turnagain above Bathurst Inlet.

By the time that Franklin decided to head south the season was well advanced and as winter set in the journey across the Barren Grounds turned into a nightmare of fatigue, hunger, murder, cannibalism and starvation. Those who arrived at Fort Enterprise found it deserted and they had to subsist on a few discarded caribou skins until Back succeeded in bringing food and help from further south. Only nine of the twenty-man group survived.

Not surprisingly, nothing at all in the way of natural history research was accomplished in the latter stages and Richardson's cherished rock samples, bird skins and plants of the summer of 1821 were all discarded. Earlier, Richardson and Hood had collected birds and mammals at Cumberland House, and Richardson had collected at Carlton House during an excursion there in May 1820. These specimens were sent by canoe to Hudson Bay and shipped to England where the Earl of Bathurst gave them to Joseph Sabine, the Secretary of the Horticultural Society. On 15 January 1822 Sabine presented to the Linnean Society of London a paper entitled 'Account of the Marmots of North America hitherto known, with Notes and Descriptions of Three New Species'. One of these new mammals was the Franklin Ground Squirrel *Arctomys franklinii*.

After Franklin returned to Britain in October 1822 he was promoted to Captain and made a Fellow of the Royal Society. He started to work on his *Narrative of a Journey to the Shores of the Polar Sea in the years 1819, 20, 21 and 22* (1823) and Sabine, having already studied the expedition collections at some length, contributed the appendix on zoology. Richardson, somewhat cheated by this turn of events, wrote only the appendices on the botany and icthyology. He withheld some of his bird notes from Sabine and thereby lost the chance of credit for finding the Stilt Sandpiper, Yellow-billed Loon and Franklin's Gull, which were later described by other naturalists.

In 1825 Franklin led a second overland journey through Canada to the Arctic Ocean and again took Richardson and Back with him. This time he was able to draw on past experiences and used more men from the Royal Navy instead of the previous mixture of French-Canadian voyageurs and Indians who, despite some valuable attributes, were of independent character and unpredictable. The success of the expedition was further aided by thorough preparation, proper help from the newly peaceable fur-companies and the comparatively easy descent of the Mackenzie River. Franklin's party surveyed westwards from the Mackenzie delta to within 160 miles of Captain Beechey's easternmost limit reached that same year from the *Blossom*. Richardson's men travelled eastwards, exploring 863 miles of new coastline as far as the Coppermine River, which was known to him, of course, from the earlier expedition.

*WESTERN CANADA AND ALASKA: WITH PLACES VISITED BY FRANKLIN,
KENNICOTT AND B.R. ROSS*
-------- *FRANKLIN 1819–22*
-·-·-·-·- *FRANKLIN* ⎫
----- *RICHARDSON'S PARTY* ⎬ *1825–27*

This time Richardson shared the natural history honours with his assistant naturalist Thomas Drummond who instead of going north had trekked westwards to botanize in the Rockies. After leaving the Arctic Richardson travelled south to meet him at Carlton House arriving there on 12 February 1827. Though Drummond did not turn up until 5 April they were still in good time to study the spring migration together. In the final stages of the expedition the Scottish plant hunter David Douglas, returning from the Columbia River, happened to meet the Franklin party at Norway House as they all made their way to Hudson Bay for the voyage home.

Too much had been learned about the flora and fauna of Canada for a mere appendix to be added to Franklin's *Narrative of a Second Expedition to the Shores of the Polar Sea in 1825, 1826 and 1827* (1828). Instead, a multi-volume work on the mammals, birds, fishes, insects and plants was commissioned. The texts were based mainly on the specimens acquired by Richardson, Drummond and Douglas, though the bulk of the latter's faunal material had decayed and could not be examined. Shortage of food had forced Drummond to eat some of his skins.

The *Fauna Boreali-Americana; or The Zoology of the Northern Parts of British America* appeared between 1829 and 1837. The second volume, on the birds, was written by Dr Richardson and William Swainson, with Swainson also providing some fifty bird drawings. Richardson wrote the descriptions, supplied or suggested Latin names and was endlessly patient as Swainson struggled with his other publishing commitments and tried to fit the birds into the hopeless quinary system of classification. Their volume did not appear in print until February 1832, although it is dated 1831 on the title page. By then Franklin's Rosy Gull *Larus Franklinii* had lost priority to a May 1831 description by Johann Wagler of a Mexican specimen, probably collected by Ferdinand Deppe.[1] A number of types were procured by the Franklin expeditions including those for the Trumpeter Swan, White-tailed Ptarmigan, Forster's Tern, Olive-sided Flycatcher, Clay-colored Sparrow and Smith's Longspur.[2]

In 1836 Franklin was appointed Lieutenant-Governor of Tasmania. During his time there he hosted James Clark Ross who was en route to Antarctica and he was able to assist John Gould and his wife when they first arrived to start field work for *The Birds of Australia* (1840–48). Seven years later, on his recall to Britain, he pressed the Admiralty for command of the *Erebus* and the *Terror*, then being prepared for yet another attempt on the North-west Passage. Fully provisioned for three years Franklin had great hopes of success as he made his way westwards through Lancaster Sound in 1845. Not until 1859 was it finally confirmed that Franklin had died on board ship on 11 June 1847, and that the full complement of 134 officers and men had all perished as they struggled southwards after abandoning their ice-bound ships. Only recently has it been shown that lead solder had poisoned their tinned foodstuffs.

Since there were no known zoological discoveries made on this last trip, Franklin's connection with natural history is chiefly limited to the two overland expeditions—of which Richardson, and not Franklin, is the naturalists's hero. It was Richardson's contributions to the botany, lichenology, geology, icthyology, mammalogy and ornithology that made the *Fauna Boreali-Americana* such a success. It was so influential that it is the origin of many English names of birds such as Barrow's Goldeneye, Bonaparte's Gull and Franklin's Gull. Franklin's Spruce Grouse has recently been reduced to subspecific level, though in this case the name was first proposed by David Douglas. As Douglas crossed the Rocky Mountains by the Athabasca Pass he noted that his guide Edward Ermatinger had shot a bird on 1 May 1827:

> "Mr. E. killed on the height of land a most beautiful male partridge, a curious species; small; neck and breast jet black; back of a lighter hue; belly and under the tail grey, mottled with pure white; beak black; above the eye bright scarlet, which it raises on each side of the head, screening the few feathers on the crown ... This being the first I have seen, could not resist the temptation of preserving it, although mutilated in the legs ..."[3]

In the *Transactions of the Linnean Society* for 1829 Douglas named the new spruce grouse "in honour of Captain John Franklin, R.N., the amiable and distinguished Commander of the Land Arctic Expedition, to whom the lovers of American research owe so much".[4]

Though Franklin was not a naturalist, Richardson claimed that Franklin had wished to promote science for its own sake and would have helped him collect if time and other duties had allowed. Very early on in his career Franklin began to appreciate the need for scientific research on expeditions, perhaps influenced by the eminent botanist Robert Brown who was a ship-board companion when he sailed around Australia with Flinders.

A number of Indian birds are named after a Captain Franklin. But these birds are named after Captain James Franklin of the Bengal Cavalry, a student of geology who surveyed in the Central Province of India and agreed to collect and draw birds for the Asiatic Society of Bengal. He worked up his two hundred specimens, including many new species and races, for the *Proceedings of the Zoological Society of London* (1831). This Indian Army officer, an elder brother of the Arctic explorer, was the ornithologist of the Franklin family.[5]

Franklin's Gull

William Gambel†
(1823-1849)

GAMBEL'S QUAIL *Callipepla gambelii* (Gambel)

Lophortyx Gambelii "Nutt." Gambel, 1843. *Proceedings of the Academy of Natural Sciences of Philadelphia* 1, p. 260: some distance west [= east] of California [= southern Nevada]

MOUNTAIN CHICKADEE *Parus gambeli* Ridgway

Parus montanus (not Conrad von Baldenstein, 1827) Gambel, 1843. *Proceedings of the Academy of Natural Sciences of Philadelphia* 1, p. 259: about a-day's journey [west] from Santa Fe, in New Mexico
Parus gambeli Ridgway, 1886. A.O.U. *Check-list of North American Birds*, 1st edn, p. 335. New name for *Parus montanus* Gambel, preoccupied

Publications on the ornithology of California for the decade 1840 to 1850 are dominated by the works of William Gambel. In the *Proceedings* and the *Journal of the Academy of Natural Sciences of Philadelphia* are annotated lists of the birds he encountered in the Rocky Mountains and Sierra Nevada and more especially those found nearer to the west coast. Being a pioneer naturalist of the western regions he was privileged to discover, as we shall see, several exciting new bird species.

Surprisingly few biographical details exist for William Gambel. Little is known of his early years, and of his considerable travels there is scant information, since he never published any account of them and his journals have long since been destroyed or lost. He therefore remains a rather mysterious, yet romantic figure in the early annals of American ornithology. This impression is further compounded by the lack of any existing portrait, so we know nothing about his physical appearance; whether tall or short, fair or dark, of delicate or robust constitution.

The following account has been gleaned from many sources, each of them

†No known portrait

throwing a little light upon our subject but none of them telling us much about Gambel's character or personality. It was not until many years after his death, when memories had dimmed, that ornithologists thought of writing biographical sketches of Gambel. Botanists likewise paid only belated tributes to his talents and fellow travellers have recalled few details of the young man who once accompanied them.

William Gambel was born in Philadelphia in June 1823. His father, William Gamble [sic], had emigrated from northern Ireland and had settled in Lancaster County, Pennsylvania. After serving in the war of 1812, and after the death of his first wife, he moved to Philadelphia and in August 1822 married Elizabeth Richardson, formerly of County Tyrone, Ireland. Ten years later he died of pneumonia, leaving his widow to support a son and two daughters, all under nine years of age, our William being the eldest. Despite his poor financial position Gambel could hardly have had a better start in his life as a naturalist. When he was fifteen years old he went with Thomas Nuttall on an eleven-month trip to the south and must have already had a passionate interest in plants and birds for Nuttall to even consider taking the boy along. Leaving in late 1838 and returning in October 1839, he gained expert field tuition in botanical collecting as well as learning much more on other subjects from Nuttall, since the great man was also one of the best ornithologists of his day and a pioneer mineralogist and conchologist. There are no details of this expedition but from the plants collected it has been deduced that they went at least as far as North Carolina, and perhaps also to South Carolina and Georgia.

In 1840 Gambel set out again with Nuttall, who was on his way to New England. At Cambridge, Nuttall gave a series of lectures, frequented some of his old haunts in the surrounding countryside and updated and finished the second edition of his *Manual of the Ornithology of the United States and of Canada* (1840). Nuttall added to it the new information gained on his recent journey across the Rocky Mountains with John Kirk Townsend and supplemented some species accounts with fresh observations from his friends James Elliott Cabot, Thomas Mayo Brewer and William Gambel. His young companion supplied him with notes on the Northern Flicker, Black-and-white Warbler, Nashville Warbler, Northern Waterthrush and Common Yellowthroat. In return Nuttall named a new species from near Fort Walla Walla, Oregon, the Gambel's Sparrow—now only recognized as a race of the White-crowned Sparrow *Zonotrichia leucophrys gambelii*. It was Gambel's first honour in the field of ornithology and, as an acquaintance now noted, he had become "very smart and capable" and was destined "for distinction worthy of a disciple of Nuttall".[1]

Nuttall then took Gambel on a long spring excursion to Maine, to an area he had explored some years previously. They took ship to Portland in April and set off inland on foot to Paris, later making a second foray into Maine to Bath and Kennebec. They were away for four months, instead of the intended two, returning overland to Philadelphia in late August by way of Springfield, Albany and New York.

In March 1841, either from personal inclination inspired by Nuttall's yarns about the far West, or as a result of more direct persuasion, Gambel set off across the continent to collect plants for Nuttall. Armed with a letter of introduction to George Engelmann, the famous botanist of St Louis, he took a more southerly route than Nuttall and Townsend had done to be sure of breaking new ground. From St Louis he proceeded to Independence, the usual starting place for the great westward treks. Writing home to his mother and sisters at about this time he said, "I hope you will ... not be anxious about me for you know I am a carefree fellow and sooner run than fight."[2] In the second week in May he joined a company of about eighty trappers and traders bound for Santa Fe with forty wagons. Following the regular Santa Fe Trail rather than taking the shorter Cimarron Cutoff, they survived a confrontation with five hundred Arapaho warriors in western Kansas by placating them with a portion of their trade goods. The scattered remains of over seventy Pawnees massacred sometime earlier by the Arapahoes convinced them that they had got off lightly. Later, as they approached the New Mexico Rockies, about four hundred Utes attacked them for a whole morning, but they still made it safely to the Raton Pass and arrived exhausted at Santa Fe on 2 July after completing the journey in record time. Gambel collected in the vicinity for the next two months, but a bundle of his plants from the Raton Mountains, sent back east from there, never arrived.

Gambel was soon keen to continue westwards again, so on 1 September, north of Santa Fe at the village of Abiquiu, he joined John Rowland, William Workman and twenty-five other men heading for California along the Old Spanish Trail. They crossed the Green River later that month and were near the Salt Lakes in October, Gambel becoming the first naturalist ever to collect in Utah, though surviving records of his work there are minimal. Gambel may have left the Rowland–Workman party to explore this area but it seems unlikely that he would forsake the comparative safety of the group to travel alone for

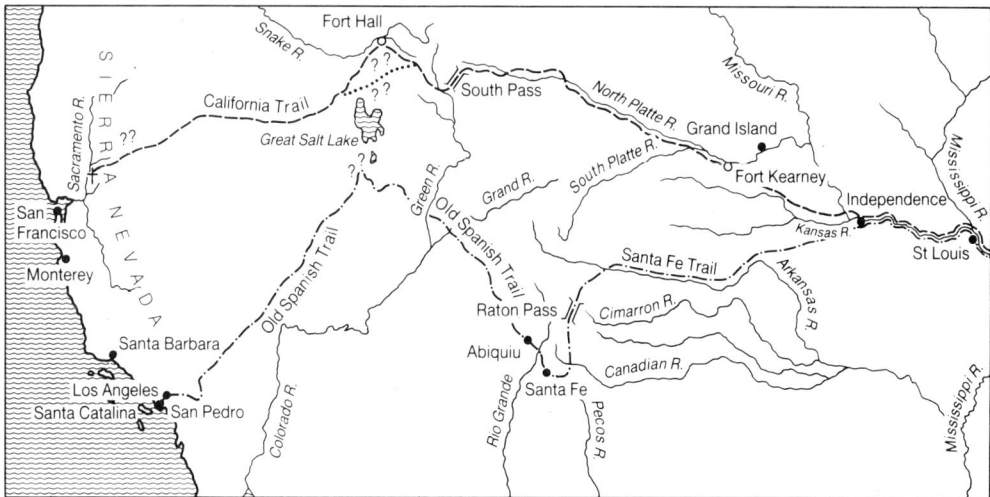

GAMBEL'S ROUTES TO THE WEST
—·—·—·—· *1841*
—————· *1849*

very long. They all seem to have arrived in California in early November and by December Gambel was down on the coast at San Pedro. He planned to collect throughout the coming season and then return home by ship.

About a year was spent in the coastal regions but Gambel evidently found it difficult to collect without being disturbed by Indians. In November, after approaching the U.S. Navy with his troubles, he suddenly found himself offered a position as personal secretary of the commodore of the fleet, Thomas Ap Catesby Jones, who had recently captured Monterey in the mistaken assumption that his country was at war with Mexico. Though soon relieved from duty because of this blunder, Jones helpfully recommended Gambel to his successor and the young naturalist subsequently served on the U.S.S *United States* for Commodore Dallas and later for Captain Armstrong. During the next year or so Gambel visited all the California mission stations up and down the coast to as far north as San Francisco besides various ports in Hawaii, Mexico and South America.

On looking at the localities where Gambel saw his birds, details of some of his wanderings begin to emerge: around the tiny town of Pueblo de los Angeles he delighted in the Anna's Hummingbirds that were in almost every garden; at Santa Barbara the Northern Mockingbirds mimicked Northern Orioles and Western Bluebirds; at Monterey the pines were alive with Pygmy Nuthatches; at the Mission of St John, between Monterey and San Francisco, he shot three Black-shouldered Kites in one day; on Santa Catalina Island he found Ospreys nesting; and at Mazatlán, on the Mexican coast, he thought he saw Lesser Yellow-headed Vultures amongst the large flocks of corvids.

Writing from the west coast he sent a paper to Nuttall for inclusion in the Philadelphia Academy's *Proceedings*, faithfully dedicating to his tutor the first new bird: Nuttall's Woodpecker, which he had found in a willow thicket near the Pueblo de los Angeles. His 'Description of some New and Rare Birds' included nine other species, some of which were indeed very rare. Townsend's Solitaire, for instance, was known only by Townsend's single specimen from Oregon; Gambel secured three more skins, one from beside a small creek in the "Rocky Mountains, between the Rio Colorado and California" and two others from beside a pool on "one of the highest and most desolate mountains". His single specimen of *Fringilla Blandingiana* was long known as Blanding's Finch, named by him after a Philadelphian friend and patron, but it proved to be the Green-tailed Towhee—a Townsend discovery described by Audubon three years previously.

But the two species of most interest to us here are the Gambel's Quail and Mountain Chickadee, both entirely new. When writing about the quail the young naturalist announced: "We met with small flocks of this handsome species some distance west [= east] of California, in the month of November, inhabiting the most barren brushy plains ... Here, where a person would suppose it to be impossible for any animal to subsist, they were seen running about in small flocks of five or six, occasionally uttering a low guttural call of recognition, sometimes of several notes, very different from that of the common species. When flying they utter a loud sharp whistle, and conspicuously display the long crest."[3] Gambel preceded these words with a physical description of the bird but its scientific name is a tribute, or return compliment, given by Nuttall, who received Gambel's manuscript and ensured that it was published by the Philadelphia Academy.

The Mountain Chickadee was first seen by Gambel a day's journey west of Santa Fe and he often saw them later in a variety of woodland types on the way to California. He observed that they were particularly abundant in the Rockies "west of the Great Salt Lake" where they roamed in such large flocks on the high wooded ridges that the trees seemed to be alive with them.[4] His name of *Parus montanus* was already in use for another species so Ridgway later changed it to *Parus gambeli*.

For much of the time that Gambel was away Nuttall fretted

Mountain Chickadee

anxiously, knowing nothing about his movements or his safety. He wrote to Engelmann for news and began to arrange financial aid for Gambel's passage home, unaware that his protégé was earning his own living. Eventually, Nuttall received word that Gambel was intending to return by ship, rounding Cape Horn via the Chilean coast. Unfortunately, Nuttall was in England when Gambel at last arrived back in Philadelphia in August 1845.

"Eureka!" Cassin wrote excitedly to Baird. "Gambel is here with his California birds and others—not very many, but some of the most magnificent specimens I ever saw. He has four new species in addition to those already described: a queer little Parus *crested*, but totally distinct from *bicolor*; another which he calls Parus but is hardly of that genus—more like Setophaga; an extraordinarily large, long-billed bird which he calls Promerops; a new *Mergulus*, like *Alle*, but entirely distinct, with others that need examining. He has also most beautiful specimens of well-known birds, and others not so well known, as *Sitta pygmaea*, *Tyrannula saya*, *T. nigrescens*, *Sialia occidentalis*, etc. Decidedly the gem of his collection is a most superb specimen of *Leptostoma longicauda* Sw., a beautiful cuckoo-like bird which walks on the ground, but I have not time really to tell you about. His *Lophortyx gambelii* Nutt. is splendid, and I can find no description of it in books to which I have access. His description of four new species will be made next Tuesday evening."[5]

Gambel's second ornithological paper appeared in due course, slightly less interesting than the first because he restricted himself to giving only names and physical descriptions. It concerned only five species. One of these was the Greater Roadrunner, which he knew had already been described by Swainson from specimens sent from Mexico by William Bullock. Gambel thought too that it must have been the bird found by Paolo Botta and described by R.P. Lesson but he did not think much of the quality of either of the descriptions. It was still a little-known bird at that time, hence Cassin's obvious excitement. Unfortunately a second species in the paper, *Mergulus Cassinii*, aroused harsh criticism from Cassin:

"Gambel is exceedingly wild about describing, and it is already very difficult to get him to examine birds that he has concluded are new—concluded, I mean in the woods of California without books—with scarcely knowing the names of late ornithologists. The birds that he has described are not examined at all ... and the most doubtful bird, too, probably at least, he has called after me, *Mergulus cassini*."[6]

Cassin, the most experienced ornithologist at the Philadelphia Academy, was correct in this last point since this bird had already been found in the Bering Sea and had been described thirty-four years earlier by Peter Simon Pallas. Ironically, Gambel's suggestion has lingered in the minds of ornithologists to such an extent that the bird is still known as Cassin's Auklet.

The other three species in the paper really were novelties: the California Thrasher, Plain Titmouse and Wrentit, all from the vicinity of Monterey. Gambel tentatively placed the Wrentit in the genus *Parus* but he later created the new genus *Chamaea* especially for it.

The plants collected in the West were shipped across to Nuttall but Gambel declined his friend's invitation to come to England so that his journals could be published and his discoveries sold off to the greatest advantage. Instead he decided to study medicine. He enrolled at the Medical School of the University of Philadelphia and in between times continued his bird work. In April 1846 he published the first of a four-part annotated list of eighty-two species for the Academy's *Proceedings* entitled 'Remarks on the Birds observed in Upper California' (1846–47). This series of papers remains a most important ornithological work since it gave a baseline for all future California studies. It was the fullest account of the birds of the state, all previous reports being scattered through the literature, often in accounts of long sea voyages (such as those of the *Blossom* and *La Vénus*) where little time had been spent ashore, usually only at Monterey or San Francisco Bay.

By quoting just two passages from Gambel's descriptive accounts the reader will see how his observations still maintain much of their original vitality. Here is his entry for the Common Barn Owl:

"This delicate feathered and familiar owl, which hitherto, I believe, has not been known to exist west of the Atlantic coast, I found very abundantly in California, and presenting all the habits ascribed to its European relative. Its favourite resort is in the neighbourhood of the towns and ruined Missions, although it may be found also about farm houses, and occasionally in the prairie valleys, which furnish it with abundance of mice and other small animals for subsistence. It makes its nest under the tiled roofs of the houses of the towns, numbers under one roof, and shows but little fear when approached. I have scarcely ever visited a Mission without disturbing some of these birds, which were roosting about the altar, chandelier, &c., of the chapel, and hearing the ben[e]di[c]tion of the Padre for drinking all the oil out of the lamps. Every where in California, when speaking of it, we are sure to be told of its propensity for drinking the sacred oil; with what truth I cannot say."[7]

The second extract shows that he obviously enjoyed his encounters with the Yellow-billed Magpie, the Californian bird *par exellence*, which he continued to call Nuttall's Magpie, following Audubon's nomenclature:

"I felt great pleasure on arriving at Santa Barbara, in Upper California, in seeing in its native haunts, the distinct and beautiful Magpie, discovered by my friend, the indefatigable naturalist and traveller after whom it is named; among others, a just tribute for the invaluable services he has rendered to natural science, during more than thirty years of his life, spent among us, in untiring investigation of the productions of our country.

In California at least, and where as yet I believe, it has alone been found, this Magpie is exceedingly local, being confined, as far as I have observed, to the immediate neighborhood of Santa Barbara, where among the beautiful evergreen oaks (*Quercus agrifolia*) of the vicinity, it is abundant.

Sprightly and graceful in its movements, it is a favorite with the inhabitants; and when not molested shows considerable confidence, often being seen about the doors of the houses, but becoming remarkably shy and cautious when chased or shot at. During my stay, from frequently shooting at them, although at first they were numerous in small flocks, they at length became so scarce that during the breeding season very few were to be seen, apparently having gone to the ravines of the neighboring mountains, so that I did not find a single recent nest, although the woods were full of those of the last year. The old nests were large, and built loosely of sticks like that of a crow, and situated in the topmost forks of the trees, well concealed by the foliage."[8]

Publication of Gambel's paper firmly established him as a major authority on west coast birds along with Townsend, Nuttall and, to a lesser extent, Tolmie, Drummond, Douglas and Menzies. Gambel superseded them all in respect of the southern species and later consolidated his position by publishing in the Academy's *Journal* (1847–49) an expanded list of 176 species from his western travels. His first series of papers had been incomplete and he now included additional observations of more birds he had seen in California as well as others from the Great Salt Lake, Colorado River and Santa Fe regions. This time there were two interesting plates, the first with his Mountain Chickadee, Plain Titmouse and Wrentit discoveries, the second with a Green-tailed Towhee alongside male and female Nuttall's Woodpeckers. Much of the text is a direct repetition of his earlier papers and the first page once again has a long footnote on the tamed Andean Condors he had seen in the streets of Valparaiso and Calao. Once more, he tells how the residents of Santa Fe provided nest boxes for Mountain Bluebirds and he repeats his story about Nuttall's yellow-headed Anna's Hummingbirds, gleefully pointing out that they did not have any yellow feathers but had been adorned with glutinous pollen whilst feeding from tubular flowers. Amongst the fresh material were interesting details on many California shorebirds such as Black Oystercatcher, Black Turnstone and American Avocet. He also added notes on various land birds seen east of the Sierras such as Ruffed Grouse, Harris's Sparrow and Vesper Sparrow. And he recalled for the first time that when he shot his first Gambel's Quails he thought that he had managed to secure a second new quail species, but whilst he was skinning the Gambel's the other corpses had been snatched away by a couple of hungry Ravens!

On the east coast Gambel was becoming increasingly well known, able to count Adolphus Heermann, Edward Harris and Professor Baird amongst his friends and acquaintances. He had been visited by Audubon on at least one occasion because the artist wanted to compare a thrasher skin from Galveston, Texas, with Gambel's California Thrasher. Spending more and more time at the Philadelphia Academy, Gambel did not now just confine his attention to western birds. In the August 1848 *Proceedings* he described the Singing Quail as new to science from a specimen brought back by a Mr Pease who had been with General Scott in Jalapa, Mexico. Directly below this article he began the descriptions of his two new small mammal discoveries, the Pacific Kangaroo Rat and the California Pocket Mouse. Gambel received some of Heermann's specimens from Florida and made use of some of Heermann's field notes on Snail Kites, Black-whiskered Vireos and egrets in a short paper called 'Contributions to American Ornithology' which was published in December the same year. By studying terns he belatedly concluded in this paper that some of his tern skins from the sand shoals at the mouth of the bay at Mazatlán were of a new species—the Elegant Tern.

As an accomplished botanist his work found its way into the literature in Nuttall's paper 'Descriptions of Plants collected by William Gambel, M.D., in the Rocky Mountains and Upper California' for the *Journal of the Academy of*

Gambel's Quail with two Ravens

Natural Sciences of Philadelphia (1847–50). Gambel's 1841 discovery of the Gambel Oak *Quercus gambelii* on the banks of the upper Rio Grande serves to commemorate this other side of his labours. Three plates drawn by Gambel appeared in Nuttall's *The North American Sylva* (1842–49).

In 1847 Gambel had hopes of becoming the curator at the Philadelphia Academy but this ambition was quashed by Cassin who supported Joseph Leidy's nomination. In March 1848 Gambel qualified as a doctor and in October at St Luke's Church, Philadelphia, he married Catherine M. Towson, a childhood friend. He found it impossible to become established as a doctor in the city but the California gold rush was now underway and he knew he would be able to find work on the west coast. He therefore decided to ship his books and instruments to San Francisco while he made the overland crossing. His bride remained behind, presumably with the intention that she should follow later by ship once William was fully set up and could afford to pay her passage. Cassin now tried to be more friendly towards Gambel since he obviously wanted any fresh discoveries to come his way.

On 5 April 1849 Gambel left Philadelphia with Isaac J. Wistar, a younger man later to become a distinguished Union General and President of the Philadelphia Academy. They travelled by railroad to Cumberland, by stagecoach to Wheeling, by steamboats to Cincinnati and St Louis, and then on to Independence where Wistar had a rendezvous with thirteen youths whom he had agreed to lead on a northern route to the gold fields. Having a doctor with them would be an obvious advantage and a place was found for Gambel with five Virginians who also joined Wistar. After a couple of weeks they all joined Captain D.B. Woods and his Indiana Company. Suffering greatly from rain, mud, stampedes, arduous river crossings and attacks by aggrieved Pawnees, the party still made very good time. On 2 June on the Platte River, about a hundred miles above Grand Island, Nebraska, they overtook a large ox-train with seventy or eighty men led by Captain Boone, from Kentucky. Gambel had found the eager, impatient pace of his travelling companions unsuited to his disposition and unhelpful to his collecting so he readily sought to transfer to Boone's party in a medical capacity—provided he could be excused from chores and guard duty. As Wistar put it: "[Gambel] is an amiable, excellent fellow and very pleasant in conversation ... But he is averse to camp duty and hard work, and fond of taking things easy, and there is no doubt that Boone's large train with plenty of men and animals, and leisurely rate of traveling, will suit him better than our headlong methods, especially as he has formed a warm friendship with Boone ... I never saw Gambel after that separation ..."[9]

The advantages of moving along slowly were soon lost. Boone arrived late in the Sierras. Snow fell early. They waited for the weather to improve but instead it worsened. Too late, they abandoned the wagons and livestock to set off on foot across the mountains guided by friendly Indians. Only Boone, Gambel and a few others survived the crossing and were able to reach the east side of the Sacramento Valley. Descending the Feather River to Rose's Bar on the Yuba River, Gambel contracted typhoid fever while tending some sick miners and because of his weakened condition died almost immediately, on 13 December. He was buried "beneath the towering pines which cluster on a sunny hill-side stretching up from the bright waters of the *Rio de las Plumas*".[10] Another account, by Wistar, suggests that he "was buried on the Bar, which however, as I have understood, has since been entirely removed by hydraulic mining."[11]

The young naturalist was only twenty-six years old when he died. Over the previous eleven years he had achieved a surprising amount and a most promising career had seemed to lie before him. His latest bird and plant collection had been abandoned on the far side of the mountains and even the rest of his belongings were lost when the ship carrying them struck rocks at the entrance of the Golden Gate and sank. During the next few years the remaining new birds of the western regions were discovered by a miscellaneous assortment of field workers of whom John Xántus, John G. Bell and James G. Cooper may be mentioned. Ten years after Gambel's death, his good friend Adolphus Heermann provided the next good study-list of California birds, particularly the southern and desert species.

George Robert Gray
(1808-1872)

CLAY-COLORED THRUSH *Turdus grayi* Bonaparte

Turdus Grayi Bonaparte, 1838. *Proceedings of the Zoological Society of London* (1837), p.118: Guatamala [= Alta Vera Paz, Guatemala]

The Clay-colored Thrush occurs in small numbers in southern Texas where it is a very rare breeder. Its main range is in Central America and Columbia where it favours a variety of open woodland types, plantations and gardens. The first known specimen was collected in Guatemala by Colonel Valasquez de Leon and taken to Europe where it was examined in 1837 by Charles Bonaparte. He announced that the thrush was "A typical species: which I have much pleasure in dedicating to Mr G.R. Gray, a young ornithologist."

George Gray was then twenty-nine years old and virtually unknown because he had not yet published any of the bird catalogues that were to become so vital to nineteenth century ornithologists. He was born in Chelsea, London, in July 1808, the third and youngest son of Samuel Frederick Gray, a botanist and pharmacologist. Educated at the Merchant Taylor's School, at the age of twenty-two he was taken on at the British Museum for twelve months without pay with an additional probationary period. He spent the rest of his life at the museum, being in charge of the bird collection for forty-one years. When Bonaparte's description of the Clay-colored Thrush appeared in the *Proceedings of the Zoological Society* Gray had been at the museum for six years and Bonaparte, who visited all the major museums in Europe to study bird skins, had doubtless met him there.

In 1840 Gray published his first ornithological work, *A List of the Genera of Birds*, a second revised and updated edition with more new names of birds coming out in the following year. In 1843 he wrote up the bird notes for Ernst Dieffenbach's *Travels in New Zealand* and in 1844 began his *List of the Specimens of Birds in the Collections of the British Museum* (1844–48). At the same time he began issuing the first portions of his fifty-part *The Genera of Birds* (1844–49), a mammoth work encompassing all the world's 11,000 known birds, which proved to be of fundamental importance to working ornithologists everywhere. Bound up in three volumes, it was beautifully illustrated by D.W. Mitchell,

GRAY. *THE ARCHETYPAL CLOSET NATURALIST, HE WORKED AT THE BRITISH MUSEUM (NATURAL HISTORY) FOR OVER FORTY YEARS, PRODUCING A NUMBER OF INDISPENSABLE AVIAN CATALOGUES.*

with a few extra plates by Joseph Wolf—"the greatest natural-history artist the world has ever seen", according to Gray's successor Richard Bowdler Sharpe.

In the years to come Gray contributed to various scientific journals and produced more catalogues of birds, either for specific regions such as the tropical islands of the Pacific and New Guinea, or, more often, they were updated lists of all the birds held at the British Museum, the last of which was his *Hand-list of the Genera and Species of Birds* (1869–71). After its completion he had the idea of labelling all the museum's birds according to the nomenclature in his *Hand-list* and had special labels printed with a *Hand-list* number to which he began to transfer all the information from the original labels. The collector's tickets were then stripped off and discarded—the ultimate sacrilege! Gray died on 6 May 1872, having changed the labels for only one genus, thus saving the collection from further injury.

To be fair to Gray, he accomplished a vast amount of useful work. His catalogues were essential for the museum and for the development of ornithology; his texts were always concise and accurate and if they appear desperately dull it is because they were only intended for reference purposes. The slow methodical work of compiling them seems to have suited his character. He had a quiet temperament and was not inclined to strenuous action—Professor Alfred Newton called him "a thoroughly conscientious clerk".[1] Married with no children, Gray led a truly clerk-like existence and though he handled thousands of birds he knew nothing about their habits. After some ribbing from his colleagues about this lack of field experience he once decided to hire a gun and headed off some twenty or thirty miles into the Hertfordshire countryside to shoot some birds—and was immediately arrested for trespassing by a gamekeeper!

George's elder brother John Edward Gray (1800–1875) also served at the British Museum but was much more dynamic and energetic, holding the superior position of Keeper of the Zoology Department from 1840 to 1874. Together they did much to enhance the bird collections, which had increased to over 30,000 items by 1872, but it was in the years immediately following the two Grays that it really expanded. With Albert Günther as Keeper of the Zoology Department and Bowdler Sharpe in charge of the bird col-

Clay-colored Thrush

lections, the number of bird specimens exceeded 300,000 (with an additional 100,000 eggs) by the turn of the century.

GUNDLACH. *CUBA'S GREATEST NATURALIST. HE EXPLORED AND COLLECTED ON THE ISLAND FOR OVER FIFTY YEARS.*

Johannes Christopher Gundlach (1810-1896)

ANTILLEAN NIGHTHAWK *Chordeiles gundlachii* Lawrence

Chordeiles gundlachii Lawrence, 1857. *Annals of the Lyceum of Natural History of New York* 6, p. 165: Cuba

BAHAMA MOCKINGBIRD *Mimus gundlachii* Cabanis

Mimus Gundlachii Cabanis, 1855. *Journal für Ornithologie* 3, p. 470: Cuba

The name of Johannes, Juan or John Gundlach is inextricably linked with the island of Cuba, but Gundlach was born at Marburg in Hessen, on 17 July 1810, and he spent the first twenty-eight years of his life in Germany. By the age of nine his inclinations were evident, for he insisted on learning taxidermy from his eldest brother who had studied the art at Kassel, and all of John's spare moments were spent collecting or preparing specimens. On one occasion while trespassing after birds he spotted an approaching guard and panicked, hiding his gun under his coat with the muzzle upwards. The hammer got caught in his clothing, the cartridge exploded and John's nose and palate were so badly injured that he completely lost all sense of taste and smell. Since all foods were ever afterwards alike to him he ate only to satisfy his hunger—and often when he was engrossed in collecting he even forgot to eat.

At Marburg University he studied Zoology (receiving free tuition because his father had taught there) and graduated in 1837, obtaining his PhD the following year. Determined to explore in some foreign country, he then spent six months studying at the zoological museum in Frankfurt, before sailing for Surinam where a Dutch medical doctor had offered him accommodation and assistance. A fellow student, Charles Booth of Matanzas, invited him to visit Cuba en route and he arrived there at the beginning of January 1839. While staying with

Booth on his coffee estate by the Canimar River, Gundlach heard that his Dutch friend had died. The opportunity to visit Surinam was lost but already he had realized the potential of Cuba. Apart from the occasional trip to Europe, Gundlach spent the rest of his long life investigating the natural history of the island.

Gundlach was fortunate in his friends. During the first few years he became thoroughly acquainted with the local fauna from his base on the Booths' estate, and when they moved to Cárdenas, on the north coast, he moved with them and explored further afield. One of his first discoveries was the endemic Bee Hummingbird which he named *[Mellisuga] helenae* in gratitude to Mrs Booth. With typical unselfishness he passed his notes on to Juan Lembeye, a Havana schoolteacher who obtained credit for the original description when it was published in his *Aves de la Isla de Cuba* (1850). Another new species described in this work was the Cuban Vireo, which Lembeye named *Vireo gundlachii*.

For seventeen years Gundlach limited his collecting territory to the three western provinces (Pinar del Rio, Havana and Matanzas) and the Isle of Pines. Constantly travelling, dressed simply in drill and carrying little in the way of baggage, he became a kenspeckle figure. With his tall, thin, angular frame, awkward movements and kindly blue eyes he was immediately recognized and befriended wherever he went, offered free transport on steamers and trains and

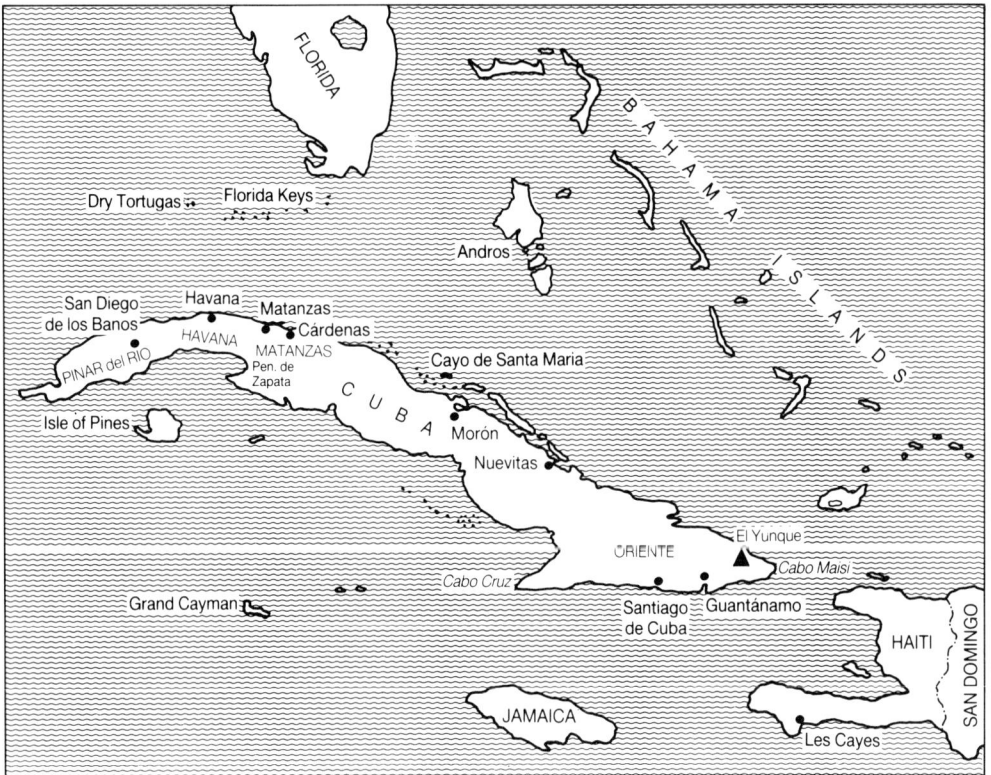

CUBA AND THE BAHAMA ISLANDS

often given food and lodgings as well. On one occasion when his hostess inquired as to his comfort during the night he commented that he always slept well, since he had no worries.

Once Gundlach had got to grips with the commoner species and built up a reference collection of skins he began to publish his observations. He wrote papers in English for the journals of New York and Boston, in Spanish for the annals of Madrid and Havana and in German for the *Journal für Ornithologie*. The latter ran a long series on his work from 1854 to 1861 and from 1871 to 1875.

Gundlach provided the original descriptions for several birds: the Gray-headed Quail-Dove from the lowland forests, the Olive-capped Warbler from the Sierra del Ranges; and La Sagra's Flycatcher, an inhabitant of mangroves and open woodlands, which had previously been inadequately described. Other discoveries were written up for him by fellow scientists, including two species that occur on the Florida Keys: the Bahama Mockingbird and the Antillean Nighthawk. The mockingbird, which he took on the Cayo de Santa Maria off the north coast near Morón, was named *Mimus Gundlachii* by Jean Cabanis in 1855. In 1857, after examining specimens which Gundlach had sent him, George Lawrence named the Antillean Nighthawk *Chordeiles gundlachii* because of "the zeal with which he is investigating the Ornithology of Cuba". Three years later Lawrence received a new bird of prey that occurs only on Cuba and called it *Accipiter gundlachi*—Gundlach's Hawk. Gundlach corresponded with many American ornithologists, but particularly with Lawrence, Brewer and Baird.

In June 1856 Gundlach began his lengthiest exploration of the island, a journey that lasted for three years and took him from Havana to the vast swamps of the Peninsula de Zapata and eastwards along the south coast to Cabo Cruz and Cabo Maisi, with numerous diversions into the mountains of the interior. In the eastern part of the island he found two new birds, the Cuban Gnatcatcher and the Oriente Warbler and in May took their nests and eggs. During the summer of 1858 he collected birds and shells along the immense sand flats of Guantánamo, then headed inland to the Yateras range where the planters vied for the honour of giving him hospitality while he chased after butterflies. Early the next year he left Santiago de Cuba and travelled eastwards to climb El Yunque, then followed the northern coastline to Nuevitas before returning to Havana in August.

Gundlach had collected so many new genera and species that he spent much of the next

Antillean Nighthawk

four years studying the material at his small museum near Cárdenas that he had established back in 1846. Some specimens he sent to specialists in the United States and Europe, but most were retained or shared with the naturalists who had sponsored his journey.

For some years Gundlach had lived with don Simón de Cárdenas and his family on their large sugar estate *Fermina*. In 1864 they persuaded him to move his museum into the upper room of one of their buildings that served as a hospital for the slaves, allowing him to stay and work in one place. This settled period was followed by a trip to Europe at the behest of the municipality of Havana, who commissioned Gundlach to organize the Cuban zoological exhibit at the 1867 Paris Exposition. After the exposition, at which he was awarded a medal and diploma, he went on to Germany to visit family and friends.

The "Ten Years Revolution" began shortly after Gundlach returned to Cuba and on one of his excursions he was arrested by Spanish soldiers who made it clear that he would be foolish to leave the estate again until the troubles were over. Gundlach coped with the frustration by planting around his quarters many carefully chosen food plants for rare butterflies. He was thus able to continue some collecting and studied in detail the life cycles of the butterflies that he had attracted. In time, many of the shyer birds were persuaded to nest in the gardens, and such was Gundlach's gentleness that some would fly in through the open window to the table where he sat working and take bits of tow and cotton to line their nests. One year a pair of Cuban Emeralds even chose to build in an unused chandelier in his room.

In 1873, as the revolution continued, Gundlach left Cuba to visit Puerto Rico for the first time and spent six months collecting there at the invitation of the Jesuit priests in San Juan who wanted a zoological museum for their school. He went there twice more, in 1875–76 and in 1881. These visits led to a number of Spanish and German papers on the birds, mammals, reptiles, amphibians, molluscs, crustaceans and insects of the island.

In 1882 *Fermina* was visited by the Director of the Institute of Havana who hoped to purchase Gundlach's collection for his museum. Gundlach had no intention of parting with his precious collection but agreed to work for the Institute and during the next six years made four excursions to eastern Cuba in search of rare species. He then forsook his travels in order to organize the Institute's existing collections and to set up all his new material. He also concentrated on the compilation of his last large work *Ornitologia Cubana*.

During the winter of 1891–92 Gundlach was visited by Charles Barney Cory, the wealthy Bostonian who had a special interest in Caribbean birds. Cory, who was in his mid-thirties, was amazed at the energy of his friend who though eighty-one years old would enthusiastically tramp for miles. Nor did he mind rising at daybreak to take his guest shopping in the market, where raptors, ibises, pigeons, doves, quails, woodpeckers and meadowlarks were all for sale. For a few days Cory hunted around Havana and studied the bird skins at the museum. Then Gundlach took him to a little town called San Diego de los Banos at the foot of the western mountains where he knew that the bird life was abundant.

In the spring of 1892 Gundlach sold his own huge zoological collection to the Havana Institute Museum for the sum of $8000 in Spanish gold and was appointed its lifelong curator. He had two reasons for this decision. One was the natural desire for his collections to remain intact in Cuba, but the other reason arose from his concern for the Cárdenas family who had shown him great love, friendship and hospitality over the years. Like many other planters, their finances had been ruined by the revolution and they were now in great need since they had been forced to abandon their estate. Gundlach handed over the entire proceeds of the sale to Mrs Cárdenas de Diago, the daughter of his late friend, don Simón de Cárdenas, delighted that he was now in a position to help her.

Gundlach continued to work daily at the museum until the winter, when an attack of influenza developed into broncho-pneumonia. Mrs Cárdenas nursed him tenderly throughout his illness until 17 March 1896, when he slept away peacefully in his own home.

During the last year of his life Gundlach held an eighty-fifth birthday party for his many friends. They realized that he was a happy man, celebrating not just his long survival but the achievement of his ambitions. The collection he had built up over more than fifty years was now completely installed and arranged at the Institute where it was accessible to students and as safe as such perishable things can ever be. Having devoted his life to the study of Cuba's natural history, he had also succeeded in sharing his knowledge with the scientific world through his many publications. His most comprehensive ornithological work was the *Ornitologia Cubana* (1893), which brought together his observations

Bahama Mockingbird

on 263 resident and migratory species. It is still the most comprehensive reference work on the avifauna of the island. When Florentino Garcia's *Las aves de Cuba: Subespecies endemicas* was published in 1987, the author credited Gundlach as the source of most of the information on plumage details, habitat requirements and diet. Specialists in other branches of natural history also continue to hold Gundlach in high esteem. It seems unlikely that his knowledge of Cuban wildlife will ever be surpassed, or even equalled, by another individual.

William Alexander Hammond (1828-1900)

HAMMOND'S FLYCATCHER *Empidonax hammondii* (Xántus de Vesey)

Tyrannula hammondii Xántus de Vesey, 1858. *Proceedings of the Academy of Natural Sciences of Philadelphia* 10, p. 117: Fort Tejon, California

The *Empidonax* flycatchers are drab, difficult to identify birds with grey, olive-green and cream or yellow plumage. *Empidonax hammondii* breeds only in the west, from Alaska south to New Mexico, and it was first discovered in California by John Xántus while he was serving with the U.S. Army at Fort Tejon. In naming the bird after Dr William Hammond, he honoured the man who had first encouraged him to collect for the Smithsonian Institution while they were both based at Fort Riley in Kansas.

Hammond was born on 28 August 1828 at Annapolis, Maryland. When he was five years old the family moved north to Harrisburg and at the age of sixteen he began to study medicine at the University of New York. He obtained his degree in 1848, served his internship at the Pennsylvania Hospital in Philadelphia and then practised for a few months in Maine before deciding to join the Medical Department of the Army. In July 1849 he married and was appointed Assistant Surgeon. The war with Mexico was over and he was immediately sent West, where he served at nine different army posts during the next three years.

At that time Hammond was interested in natural history and like many other medical officers he sent his collections, notes and observations to Professor Baird at the Smithsonian Institution. At Fort Riley he encouraged Xántus to do likewise, and among the twenty-six species that they are known to have acquired from there are Carolina Parakeet, Belted Kingfisher, Eastern Bluebird, Bohemian Waxwing, American Robin, American Goldfinch, Blue Jay and five kinds of woodpecker. Hammond also sent Baird Red-winged Blackbird and Dickcissel from the Republican River, south of Fort Kearney in Nebraska; Vesper Sparrow and Green-tailed Towhee from the Black Hills of South Dakota; Cliff Swallow from Fort Laramie in Wyoming and Mountain Bluebird—then known as the Rocky Mountain Bluebird—from Cooper's Creek. The list is hardly impressive,

HAMMOND. *WHILE SERVING AS A YOUNG ASSISTANT SURGEON IN THE WEST HE COLLECTED BIRDS FOR THE SMITHSONIAN INSTITUTION. HE WAS SURGEON-GENERAL OF THE UNION ARMY FROM 1862 TO 1864.*

but there must have been many others as he also collected around Fort Jones in northern California and elsewhere in the western territories.

Hammond was a big man in every way. At six feet two inches he had a splendid physique, a commanding presence and a forceful personality. He also had enormous energy and vision, but he devoted little of his time to ornithology. His real vocation was medicine. In 1857 he won the American Medical Association prize for an essay on nutrition and two years later he was offered the chair of Anatomy and Physiology at the University of Maryland. This he accepted, but when the Civil War broke out in 1861 he re-entered the Medical Department of the Army and was again commissioned as Assistant Surgeon.

During the war his administrative talents were soon recognized and although still a junior officer he was suddenly appointed head of the Medical Department, with the rank of Brigadier-General. As Surgeon General he introduced many drastic changes to the department, organized the Army Medical Museum, initiated the construction of new military hospitals and founded an army medical school and laboratory. All this was achieved in just two years, for in 1864 he was court-martialled and dismissed from the service after clashing with the Secretary of War, Edwin M. Stanton. Fourteen years later he was fully exonerated but at the time of his dismissal he found himself in financial straits due to the expense of the trial. Loyal friends helped him to resume his career in New York and he later lectured on neurology and psychiatry at several different medical schools and pioneered new methods of treatment. He edited numerous medical journals and published many papers of his own as well as a text-book on the diseases of the nervous system. He also wrote plays and a number of novels, including *The Son of Perdition* (1898).

In 1886 he remarried and two years later moved to Washington D.C. where he established a large sanatorium for the care of patients with nervous and mental illnesses. While working there he developed cardiac disease and though he reduced his responsibilities he died on 5 January 1900, at his home. He was buried at Arlington National Cemetery.

Hammond's Flycatcher

HARRIS. *WEALTHY AMATEUR NATURALIST WHO BEFRIENDED AND ASSISTED AUDUBON. HE TRAVELLED WITH HIM ON THE GULF OF MEXICO AND MISSOURI RIVER EXPEDITIONS.*

Edward Harris
(1799-1863)

HARRIS'S HAWK *Parabuteo unicinctus* (Temminck)

Falco unicinctus Temminck, 1824. *Nouveau Recueil de Planches coloriées d'Oiseaux*, Livr. 53, p. 313: Brésil ... dans les environs du Rio-Grande, près Boa Vista [= Boa Vista, western Minas Gerais, Brazil]

HARRIS'S SPARROW *Zonotrichia querula* (Nuttall)

Fringilla querula Nuttall, 1840. *A Manual of the Ornithology of the United States and of Canada*, 2nd edn, Vol. 1, p. 555: a few miles west of Independence, Missouri

In October 1937 a Philadelphian ornithologist, J. Fletcher Street, was contacted by the Makin family of Moorestown, New Jersey, who asked him if he would like to examine an old bird collection in their barn. Street found twelve bushel baskets of birds stored in old packing cases and a trunk. They had been delivered to the Makins in lieu of cash to pay a plumbing bill and had arrived in a large case of drawers. Mrs Makin had thought that the case would make a fine linen chest so she had demanded the removal of the skins, which were then squashed into the other containers and stored in the barn loft. Later, Mrs Makin's son found that the stuffed birds made handy targets and by the time Street examined the collection, many of the four hundred identifiable skins of two hundred and twenty-eight species were headless, wingless, footless or otherwise mutilated. Many had also lost their labels—but enough intact specimens remained for Street to realize that he was handling the collection assembled a hundred years before by Edward Harris.

Some of the skins had been collected on the historic expedition that Harris and Audubon had made together in 1843 when they travelled up the Missouri to the Yellowstone River. Still bearing labels were an Eastern Meadowlark (Illinois, April, 1843); Western Meadowlark (Missouri River, 1843); four Carolina Parakeets (Kickapou Country, Missouri River, May 1843); Clay-colored Sparrow (Ad. ♀ Fort Union, June 1843); Brown-headed Cowbird (Ad. ♂ Fort Union, June 22, 1843); Black-headed Grosbeak (Fort Union, June 22, 1843); Sprague's Pipit (Ad. ♀ Fort Union, June 24, 1843); Sharp-tailed Grouse (Ad., Fort Union,

June 26, 1843); Western Kingbird (Ad. ♀ Fort Union, June 26, 1843) and Lazuli Bunting (Ad. ♂ Fort Union, June 26, 1843). At least five other species had probably been taken on an earlier trip which Harris and Audubon made to the Gulf of Mexico: American Swallow-tailed Kite, Mississippi Kite, Crested Caracara, Green Kingfisher and Roseate Spoonbill.

Three skins from the Columbia River (two Snowy Egrets and a Gray Jay) bore labels written by John Kirk Townsend and some others were obviously part of Townsend's precious collection, which Harris had bought from the Philadelphia Academy so that Audubon could paint them. Other skins had labels in the youthful handwriting of Spencer F. Baird; most had been taken near Baird's home in Carlisle, Pennsylvania, between 1842 and 1845. Outstanding amongst these was a Passenger Pigeon, shot at Carlisle on 19 October 1844. Many of the better specimens rescued by Street are now safely housed in American museums.

The story of Edward Harris's life and his part in ornithological history is an interesting one, though less well known than it deserves to be, for he is usually remembered in the rather narrow role of friend and wealthy benefactor to Audubon. He began his life at Moorestown, New Jersey, on 7 September 1799, the eldest of three children. His father, Edward Harris Snr, had left England just three years earlier in order to become the American representative of his company, which manufactured hosiery in Leicester. Within two months of arriving in the United States he had married Jane Ustick, the daughter of the pastor of the Baptist Church of Philadelphia. For the next two years they rented a house in the city and then moved across the river to a large farm at Moorestown. Here Mr Harris began the cultivation of raspberries, apples, barley and other crops and exported clover seed to England, all the while shrewdly investing his profits. His wife died in 1810 but he survived until April 1822, by which time his son, Edward, was twenty-two years old and well able to take over the management of the farm and the rest of his considerable inheritance. Young Edward had been prevented by ill health from going to college, but he read avidly on the subjects that interested him—primarily agriculture and game hunting. Astute and innovative, he was also reserved and courteous. He placed a high value on friendship and those who got close to him soon found him to be kind-hearted and generous.

His friendship with Audubon began when the thirty-nine year old artist was in Philadelphia on a tour of the east coast, seeking support for the publication of his bird paintings. Audubon afterwards wrote in his journal:

> "July 12th, 1824. I drew for Mr. Fairman a small grouse to be put on a banknote belonging to the State of New Jersey, this procured me the acquaintance of a young man named Edward Harris, of Moorestown, an ornithologist, who told me he had seen some English Snipe within a few days."
>
> "July 19th. Young Harris, God bless him, looked at the drawings I had for sale and said that he would take them *all* at my prices. I would have kissed him, but that is not the custom in this icy city."[1]

Before leaving Philadelphia Audubon met Harris again and had a hundred-dollar bill pressed into his hand with the explanation that "men like you ought not to want for money." In return, he presented Harris with some drawings of

French birds that he had done while visiting his family at Couëron in 1805. He also promised to send Harris some of the species that were missing from his bird collection.

Soon afterwards, Harris received a letter from his new friend suggesting that they spend a week or two observing the fall migration around Lake Champlain and Lake Ontario but he had to refuse this appealing idea because of illness in his family. Family matters became increasingly dominant during the next few years: in June 1827 he married his cousin, Mary Lang, and honeymooned in England where they spent several months touring the country and visiting relatives. Since her health was delicate they returned to Europe in 1830, hoping that the climate of southern France and Italy would be curative, but she died in Pisa that December. Harris remained a widower for the next fifteen years.

In the autumn of 1836 Harris again came to Audubon's aid. Thomas Nuttall had returned from the Rocky Mountains with Townsend's collection of western birds, many of them little known or completely new. He had deposited them for safe keeping at the Academy of Natural Sciences of Philadelphia until Townsend also returned—but Audubon was desperate to include all the new species in *The Birds of America* without delay. Since Audubon had never been accepted by the ornithological clique at Philadelphia he asked Harris to exert all his influence and his loyal friend offered to purchase the whole collection for $500. This offer was rejected but Audubon petitioned Nuttall, and between them, Harris and Nuttall persuaded the Academy to part with a large selection.

During the winter Harris joined Audubon and his son John in Charleston. Their plan was to explore the coasts of western Florida, Louisiana and Texas but the Seminole War forced them to change their itinerary and in mid-February they started overland to New Orleans, taking a steamboat down the Alabama River to Mobile. Audubon's Gulf Coast expedition journal has been lost, but Harris's has survived and contains a list of the 197 bird species seen between 1 April, when they took the South-west Pass from New Orleans, and 18 May, when they began the return journey from Galveston.

During the first few weeks as they cruised among the islands and bays of Louisiana the great spring migration was at its peak and many wintering species still remained. Around the Grande Terre Islands they saw huge flocks of ducks and shorebirds and amongst those which Harris listed as abundant were Northern Shoveler, Gadwall, Blue-winged and Green-winged Teal, Wilson's Plover, Upland Sandpiper and Willet. At Caillou Island, despite a heavy gale from the west, he went out and was rewarded with the sight of about twenty Least Bitterns which on reaching the island immediately plunged into the marshes and hid so well that neither he nor his dog could flush them. Among the many familiar landbirds that he watched struggling northwards were Northern and Orchard Orioles, Yellow-billed and Black-billed Cuckoos, American Redstarts, Ruby-throated Hummingbirds and many different kinds of vireos, warblers and flycatchers. On 24 April they reached Galveston Bay and the next morning, after another gale, they witnessed a spectacular fall of migrants. Thousands of weary birds were hiding in the grass, flitting around the vessel or struggling, exhausted, in the water.

The naturalists stayed at Galveston for two weeks, then left for Houston, the new capital of Texas, to meet Sam Houston. By the end of May they were back in New Orleans; the Audubons headed directly for Charleston where Maria Bachman was making preparations for her wedding to John, but Harris enjoyed some solitary collecting on the Mississippi before joining them there. In some respects the trip had been a disappointment, for no new birds had been discovered, but they had learned a good deal more about the distribution of certain species and had enjoyed the drama of migration.

The following year Harris was travelling again, this time revelling in the novel experience of crossing the Atlantic on Brunel's steamship, the *Great Western*, which was returning to Bristol after her maiden voyage to New York. Having made the trip many times under sail, Harris was greatly impressed by the reduced motion of the paddle steamer and did not miss a single meal. The purpose of his journey was to purchase European stock for his estate and he spent several months in Scotland visiting farms, markets and agricultural shows, as well as calling on the Audubons, who were then based in Edinburgh. At the beginning of 1839 he visited Normandy in order to buy some of the sturdy Percheron draught-horses he wanted to introduce to the United States. He chose three, but only one reached Moorestown alive and so, within six weeks, he was on his way back to France, this time returning with two stallions and two mares in perfect condition. One of the stallions, named Diligence, was exhibited at shows and fairs all over New York, Pennsylvania and New Jersey for the next five years until Harris decided to give up farming. He then sold his stock by auction and Diligence was acquired by John G. Bell. A year later, when the horse died, Bell wrote to Harris informing him of the sad news, cheerfully adding that the skin was already in a barrel of alcohol in preparation for stuffing!

Harris and Bell knew each other well as they had travelled together to the Upper Missouri in 1843. Harris had initially been worried by Audubon's intention to take Bell along as he feared that the self-employed and successful taxidermist would expect to retain a large part of the anticipated collection. Harris had never mastered the art of skinning specimens and wanted an assistant who would understand that all the proceeds were to be shared between himself and Audubon. Fortunately Bell agreed to work for a fee alone and the cost was shared by the two beneficiaries. Other assistants were required and Audubon tried to persuade young Baird to accompany them but his family intervened, afraid for his health and safety. Instead Audubon took an energetic young neighbour, Lewis Squires, as secretary and general dogsbody. Isaac Sprague joined them as assistant artist. By this time Audubon and Bachman were collaborating on *The Viviparous Quadrupeds of North America* and the primary reason for the trip was to collect mammals and learn as much as possible about their habits.

When the party left Philadelphia on 13 March Harris began a diary that he kept for the next seven months. He noted that his baggage consisted of an inflatable India rubber bed, nine trunks, one box, two guncases and four guns, four bags, one carpet bag, one cloak and his pointer Brag, who proved to be a valuable member of the team. Travelling by coach they crossed the Alleghany Mountains in a snowstorm and at Wheeling boarded the first of a succession of steamers that took them down the Ohio and up the Mississippi to St Louis.

The last part of this journey was a fit initiation for the expected hardships ahead of them. The filthy steamer was crowded with adventurers, the scanty food was greasy and revolting, the washing facilities comprised one cake of soap and a single towel for all on board, and when it rained outside it also rained within. Harris must have remembered the *Great Western* with a certain amount of longing. Their relief on reaching St Louis was marred somewhat by the news that they would have to endure a delay of four weeks until another steamer would be ready to depart for the Upper Missouri.

While Audubon made final preparations in the town, Harris, Bell, Sprague and Squires moved out to Edwardsville, a few miles away on the east side of the river, and hunted nearby. When they at last embarked on 25 April, Harris wrote up his journal in a pessimistic mood:

"At last we are off on our long talked of expedition. The captain had great difficulty in getting the trappers on board, they were nearly all drunk and it was about noon before we got under way, amid the shouts and yellings and firing of guns from our drunken trappers. Yesterday one who had enlisted for the trip was arrested for counterfeiting and murder. How many more of like character we have among us would be difficult to know, all we do know is that they appear to be the very offscouring of the earth, worse than any crew of sailors I ever met with. Our boat makes a very poor attempt at stemming the current of this great river, now more filled with water than it has been for many years—we look for a tedious passage."[2]

In fact, the *Omega* reached Fort Union in the record time of seven weeks (for a map see p. 405). On their way upriver the party went ashore at least once every day when the crew stopped to cut wood for the boilers. On these occasions the naturalists were frequently accompanied by their rowdy fellow-travellers, keen to blast any animals that came within range.

On 4 May, below the Black Snake Hills, Harris shot a small bird with a black crown and throat, ash-coloured cheeks and a pink bill. Audubon was convinced that it was a new species and wrote in his journal "I am truly proud to name it *Fringilla Harrisii*, in honor of one of the best friends I have in this world." He should have realized that it had already been named by Thomas Nuttall, who had discovered the bird—now known as Harris's Sparrow—just fifty miles away near Independence when he was ascending the Missouri with Townsend back in 1834. These first known specimens were found on the eastern edge of their wintering range, the south-central States. In the breeding season Harris's Sparrow frequents the stunted boreal forests of Canada.

During the next week they passed Council Bluffs where Lewis and Clark had experienced their first taste of Indian speech-making forty years before and a few miles upriver they noticed the grave of Sergeant Floyd, the only casualty of that great pioneering expedition. On the 14th Harris found a breeding colony of Great Blue Herons and shot four of them—as well as a Raven that he had watched stealing their eggs during the disturbance. On the 22nd he watched a meadowlark and was amazed by its song, the bubbling flute-like notes being quite different from the clear whistle that he knew so well. Two days later, Harris and Bell went off together and shot several Western Meadowlarks, which Audubon later described, realizing that they were a different species from the Eastern Meadowlark, although almost identical in plumage.

On board the *Omega* Harris took on the role of company physician. When one of the trappers cut his foot with an axe he treated him as best he could and other patients soon sought him out. After arriving at Fort Union on 12 June, Harris's medicine store of pills, emetics, salts and quinine was frequently put to use. Bell, however, was a less than willing patient when he took ill in July and Harris only succeeded in bleeding him after much persuasion. Bell was no fool.

Fort Union was an important frontier post of the American Fur Trading Company, built close to the north bank of the Missouri River about six and a half miles

Harris's Sparrow

above the mouth of the Yellowstone. The principal trading post on the Missouri, it was enclosed by twenty-foot high palisades of hewn cottonwoods. Inside was the company office, the mess room and a number of apartments, one of which was used by Audubon's group for the next two months. An adjoining building contained a retail store, a large warehouse, the fur store, accommodation for clerks, hunters and other employees, an ice-house and a kitchen. Under the balcony were workshops for the cooper and blacksmith, stables for horses, cattle and bison calves, hen-houses and a dairy.

Most days Harris went out hunting, sometimes with Audubon, often with Bell, occasionally accompanied only by Brag. On one of their early excursions Harris and Bell fired together at a small bird that turned out to be a new pipit, which Audubon named after Sprague. During July and August the naturalists took part in Bison hunts and soon obtained a fine skin for their collection. Despite this the killing continued and the pages of both Harris's and Audubon's journals are full of long, detailed accounts of the chase and slaughter. Usually only the tongue was cut out for food, the corpses being left to the Turkey Vultures, Ravens and wolves. Audubon was so carried away by the prevalent passion for big-game hunting that he forgot his original objective. Later, after returning home to *Minnie's Land*, he refused to let Bachman see his journal, knowing what his friend's reaction would be. Eventually the minister got hold of it and took it back to Charleston. He soon wrote angrily to Audubon, "I am afraid the broad shadows of the elk, buffalo and big-horn hid all the little marmot squirrels, jumping mice, rats and shrews. Why man, what poor trappers you proved yourselves to be ... Do you know what I should have done? ... make a brush fence a foot or eighteen inches high ... and I would have had a couple every night."[3]

By mid-August it was time to start the return trip and Bell carefully packed the collections for safe transit in their forty-foot mackinaw boat. On 12 August Harris and Bell enjoyed their last Fort Union Bison hunt together and on the 16th the party embarked with some other passengers, as well as a few live animals in cages—a Swift Fox, a Badger and a small deer. Harris kept only a brief record of the uneventful homeward journey, but continued the geological observations that he had begun on the outward passage, later presenting his conclusions at a meeting of the Philadelphia Academy. On 19 October they reached St Louis and the party divided: Audubon, Bell, Sprague and Squires departed for Cincinnati, leaving Harris to carry on alone to Florida. At Louisville Harris was delayed by a bad attack of asthma but reached New Orleans in mid-November and went on to Tampa where he was grieved by the death of Brag.

As previously mentioned, Harris relinquished his farming interests in 1845 and now turned to the civil and cultural affairs of Moorestown. He also courted another cousin, Mary G. Ustick, and they married on 6 May 1846 when he was forty-six. Harris took his bride on vacation to Europe and while in Germany they called on Prince Maximilian of Wied-Neuwied who showed them an extraordinary array of paintings and Indian artefacts from his own Missouri River expedition. On their return the Harrises settled in Moorestown and four children were born to them there, the third being named Edward after his father.

Audubon died in January 1851 and for long afterwards Harris sent gifts of money to Lucy Audubon. He also kept in touch with Bachman who sent him an interesting letter in March 1852:

"My Dear Sir ...
Do you remember a small animal, a Spermophile, that resembles 'Says' S. Lateralis,' [Golden-mantled Squirrel] that you brought to me. I took it for that animal, although Says' description did not exactly suit it. Since then, I have received Says' species, and on comparing them, I find that yours is a *new* species which I have named *Spermophilus Harrisii*. Now, as you have been flying to immortality on the wings of woodpeckers and other birds, you may be unwilling to submit to the slow process of riding thither on the back of a Marmot Squirrel. But you must endure it, as I was compelled to do, when a shabby fellow in the back country, who had never seen me, walked some miles to show me a dirty little urchin, without shoes and stockings, hat or clean face, whom he had named *John Bachman*. Now what do you know of the history of this little name-sake of yours? Where was it procured, and did it live in communities like the rest of its species? I see it has cheek-pouches.
·Mrs Bachman and Victor join me in kind remembrances to you and to Mrs. Harris."[4]

The "small animal" is now known as the Yuma Antelope Squirrel *Ammospermophilus harrisi* (Bachman), an inhabitant of the Sonoran desert. The woodpecker was the dark western race of the Hairy Woodpecker *Dryobates villosus harrisi* (Audubon) which was long known as Harris's Woodpecker. The other birds were of course Harris's Sparrow, which Audubon named during the Missouri River trip, and Harris's Hawk, which Audubon illustrated in *The Birds of America* (1837). He copied the hawk from a specimen collected "between Bayou Sarah and Natchez" by a Louisiana gentleman whose name was unknown to him. Wrongly assuming that the bird was undescribed, Audubon called it Harris's Buzzard *Buteo Harrisii*, "after my friend EDWARD HARRIS Esq., a gentleman who, independently of the aid which he has on many occasions afforded me, in prosecuting my examination of our birds, merits this compliment as an enthusiastic Ornithologist".

Edward Harris died at Moorestown on 8 June 1863. Since the outbreak of war in 1861 he had been intensely concerned about the state of the Union and aided the cause to the best of his ability. It is said that almost his last words were "Is Vicksburg taken?" His widow was left to rear their young family and it seems that their fortunes dwindled for it was her son, Edward, who eventually gave the bird collection to the Makins in order to settle a bill. However, *his* son, William Ustick Harris, presented a collection of his grandfather's papers, journals and relics to the Alabama State Department of Archives and History. Among them are some letters sent between Harris and other naturalists including James Trudeau, Richard Harlan, Townsend, Nuttall, Bachman and Baird. There are also some very attractive bird drawings by Harris, much in the style of Audubon, including one of a Black-and-white Warbler on a passion flower.

Harris's Hawk

Adolphus Lewis Heermann (c.1827-1865)

HEERMANN'S GULL *Larus heermanni* Cassin

Larus Heermanni Cassin, 1852. *Proceedings of the Academy of Natural Sciences of Philadelphia* 6, p. 187: San Diego, California

Dr Adolphus Heermann was one of that small group of tough, determined, pioneering field ornithologists who first searched the West for new birds and their nests and eggs. Heermann's Gull, his best known discovery, was just one of many unknown or little known birds, mammals, reptiles and plants that he brought back to Philadelphia. Yet all memory of this larger-than-life character was almost lost until 1907, when Witmer Stone, realizing that Heermann deserved a place in ornithological history, wrote a short biography in *Cassinia* after contacting the few naturalists who still remembered him. One of Stone's correspondents was Henry Dresser, a prominent English ornithologist who had known Heermann well. He told Stone:

"When I arrived at San Antonio on the 16th of September, 1863 ... I at once looked him up. I found him a strong, broad, sturdy man of about fifty (perhaps a year or so older), but I never asked his exact age. His hair and beard were tinged with gray, and he must have been a very strong man, but was then rather lame, and stumbled now and then, and it afterwards proved that his lameness was *locomotor ataxia*. He and a younger [older] brother had some house property in San Antonio and a tolerably large rancho on the Rio Medina, a ride outside San Antonio. ... During the whole time I was in Texas my headquarters were San Antonio, and after a time I arranged with A.L. Heermann to take one of the rooms in his bungalow, and lived with him when in the town. This I did at the request of his brother, who thought it would be well if some white man were there to look after him, so that he should not be neglected, and I for my part was glad to be there, so that we could work together at ornithology. Besides I had two thoroughbred horses, which were well trained and used to firing, for I always fired from the saddle when out shooting, so I could and did take Heermann out with me, and being quite sound from the waist upwards, he was all right when once in the saddle. Heermann had a small collection of birds in his rooms at San Antonio, and he and I collected together, and I kept my birds there also.

Heermann's lameness increased quickly, and during the latter part of my stay in Texas it was so bad that he could scarcely walk far, and stumbled terribly, so

HEERMANN. *SURGEON AND NATURALIST WITH THE PACIFIC RAILROAD SURVEY IN CALIFORNIA, ARIZONA AND NEW MEXICO. HE WROTE ONE OF THE BEST EARLY ACCOUNTS OF CALIFORNIA BIRDS.*

at last to avoid a fall we used, when he rode with me, to strap his legs to the saddle. I never dared to take Heermann far out, or where there was any chance of danger, and he generally went out only when I was on a collecting trip within a dozen miles of the town. When I could get away we would go out to the rancho and collect there for a day or two. Heermann never married, and told me he never had any idea of 'committing matrimony.'"[1]

Stone was unable to discover Heermann's date of birth but a later biographer delved into the family history and discovered that Dresser had seriously misjudged his friend's age—he was at least fifteen years younger than he looked and was only about thirty-six when they met, not fifty-one or fifty-two. Perhaps his hard life and syphilis had combined to cause the premature ageing.

Heermann was the son of Dr Louis Heermann (1779–1833) who had been born at Kassel in Germany. Baptized Adolf Ludwig, the elder Heermann changed his Christian name on arrival in the United States and at the age of twenty-three joined the U.S. Navy, in which he served as a surgeon for thirty-one years. In March 1821 he married Eliza Potts of Norfolk, Virginia, and they had five sons of whom Adolphus was the youngest. It therefore seems unlikely that he was born before 1827. At around this time his father was stationed in South Carolina where Heermann is stated to have been born. He certainly cannot have been born much later as he is known to have joined an expedition to the Rocky Mountains in 1843 and if born in 1827, he would then have been fifteen or sixteen years of age. His brother Theodore, who also went along on that trip, was just a year or two older.

The Heermanns' first western adventure was a bizarre experience. The colourful caravan that they joined was totally unlike any of the military surveying teams, emigrant trains or trading outfits that then crossed the prairies. It was a pleasure trip, commanded by the wealthy Scottish laird Sir William Drummond Stewart. Stewart was not averse to hardship; a veteran of the Peninsular Wars he had grown bored with the peacetime routines of the British Army and sought new challenges in North America, travelling with the Rocky Mountain Fur Company traders on several occasions. However this, his last journey to the Rockies, was to be completely different. He took with him wagonloads of luxuries: foodstuffs such as tinned meats, fish and fruits, choice wines and whiskeys, pickles, cheeses, spices, condiments, tea, coffee and sugar; and for his personal comfort a splendid crimson tent, a Persian carpet, fine linens, exotic furs and for the finishing touch, an intricately patterned Turkish incense-burner that rather created the impression that he was in reality some rich eastern potentate.

For company he persuaded his old friend William Sublette to come along as captain of his large and motley band. On one of Stewart's previous journeys he had employed the artist Alfred Jacob Miller with great success. This time he sponsored a writer, Matt Field, who was the assistant editor of the New Orleans *Picayune*. There were also numerous 'young gentlemen' and their servants, four European botanists, a physician, a small group of Jesuit priests bound for the Columbia River and many hunters and hired hands. Adolphus Heermann's exact status within the group is unknown.

The party assembled in St Louis at the end of April 1843 and on meeting John James Audubon there Stewart invited him to join his expedition, but

Audubon had already decided to spend the summer on the Upper Missouri. One or both of the young Heermanns made the acquaintance of Audubon at this time, unaware that their mother had flirted with the artist during the early days of her marriage when he gave her drawing lessons in New Orleans—until Dr Heermann abruptly dismissed him.

At the beginning of May the more privileged members of Stewart's expedition sailed upriver to Westport while the others made their way there by driving the wagons, carts, horses and pack mules. For the next two or three weeks the whole party was forced to camp near the Missouri, impatiently waiting for the prairies to green up sufficiently to nourish their animals. At a nearby campsite Lieutenant John Frémont was making his final preparations for his second surveying trip, this time to the Oregon settlements on the Columbia River.[2] Stewart suggested that they all travel together, but Frémont chose a more southerly route.

At the end of the month the caravan started along the Kansas River, then headed north-westwards towards the Platte, with Sir William and Sublette riding together at the head, reliving old memories. Their part in the expedition, their opinions, decisions and all their doings are well documented—but of young Heermann's impressions and activities we have only a few tantalizing glimpses from the journalistic pen of Matt Field, who mentions that Heermann quickly established himself as the ornithologist of the party and that his name was soon shortened from Adolphus to Dolly.

On 10 June when the travellers were well up the Platte, one of the experienced hunters who had ridden ahead to scout for game suddenly came galloping into the camp yelling that Buffalo were up ahead. The men leapt onto their horses and took chase and some of the beasts were killed. Field tells us that immediately afterwards: "Up rode 'Dolly the Great' on his big mule, with his arms and legs flying like the limbs of a pasteboard man pulled into rapid action by a string, exclaiming 'That's my calf!'"[3] Heermann enjoyed many more hunts, but a month later he lost his fine mare in a Buffalo chase. Next morning he and two other men returned to the camp at dawn, their mules and the remaining mounts loaded down with five fat cows.

On the Fourth of July Heermann was rudely awakened by gunfire, occasioned not by a Sioux war party as some at first assumed, but by a twenty-six gun salute to the States of the Union. In case there was a genuine attack, half of the men were ordered to remain sober throughout the day, but for all there was feasting, patriotic singing and noisy high jinks. Next evening they reached the Laramie River where they halted for several days. The botanists busied themselves searching for interesting species, and Heermann, already a keen collector of birds and eggs, did likewise. Further west, by the Sweetwater River, he spotted a Swainson's Hawk nest in a willow tree. Though the young had already fledged he climbed up and found one addled egg, which he blew. He also shot both adults and one of the young birds.

At the beginning of August they at last reached their destination by the calm waters of Fremont Lake. The tents were pitched in a large green meadow and for the next two weeks everyone enjoyed the carefree delights of hunting and fishing in some of the most beautiful surroundings imaginable. Sir William sent a message to the Snake Indians, who soon arrived bringing their wives and young girls. Carnival time began with running competitions, horse racing, dancing, drinking, feasting, gambling and story-telling as well as the serious business of bartering otter pelts, beaver pelts, buffalo robes and bearskins for blue beads, mirrors, fish hooks and knives.

The celebrations ended on 16 August with the loading of the wagons. Next morning, at sunrise, the men were roused and discipline was restored as they began to retrace their steps back to St Louis. Many of the men, eager to get back home and recount their adventures, would never sleep under canvas again, but for Heermann it had been an extraordinary initiation into the rough and risky business of life on the frontier.

At some time after his return Heermann must have moved to Philadelphia, for in April 1845, when he was accepted as a member of the Philadelphia Academy of Natural Sciences, he was recorded as a resident of the city. When William Gambel returned from California that same year, Heermann was inspired by his discoveries and his accounts of collecting there. They became close friends and Heermann began to consider the possibility of visiting the far West. In the meantime, he obtained his M.D. from the University of Maryland in 1846 and made a useful contact that April when he met Spencer F. Baird who was on a visit to the Academy. During the next few years it is likely that Heermann earned his living by working as a physician, but in March and April of 1848 he was collecting birds with John Krider on the south-west coast of Florida around Charlotte Harbor and on Key West. He must also have visited Europe at some unknown date (but prior to 1849) for in his ornithological report for the tenth volume of the Pacific Railroad Reports he mentions having several times picked up Swifts in the streets of Geneva after the birds had fallen to the ground during aerial squabbles.[4]

In 1849 he at last set off on his first expedition to California, apparently travelling by the Isthmus route, as he is known to have collected birds at Mazatlán on the western coast of Mexico. In California he made Sacramento his base and from there made excursions to the gold mines on the American

Heermann's Gull

River and north to the sources of Cottonwood Creek in the forested coastal mountains. He also investigated the Calaveras and Consumnes Rivers and spent time in the south of the territory. At numerous locations along the coast he found Heermann's Gulls which had until then escaped the attention of naturalists. They were particularly common around San Diego where he studied their feeding habits and watched them as they ducked after small fish in the company of Western Gulls, or fought for the scraps thrown overboard from the vessels entering the harbour. They also scavenged along the beach and hunted among the kelp beds in the deep offshore waters. When the spring came Heermann sailed out to the Coronados Islands, some twenty miles south of San Diego, and found his new gulls breeding there.

From San Francisco he visited the Farallon Islands and hunted sea lions as well as seabirds. Though he never hesitated to take an interesting clutch of eggs he was shocked by the massive harvesting of murre eggs that took place five days a week throughout the breeding season. Heermann watched in fascination as the men raced into the crowded colonies while flocks of Western Gulls flew just ahead of them, predating the exposed eggs as the panic-stricken seabirds took flight. Heermann devoted some time to watching a pair of Black Oystercatchers that he was sure had a nest, but he failed to find it. He did succeed in adding a Tufted Puffin egg to his collection—after first wrapping his hand in a thick handkerchief, having learnt by painful experience that the birds had sharp and powerful bills.

In the fall he built up his bird skin collection by shopping at San Francisco market, where amongst many other species there were large numbers of Tricolored Blackbirds. One spring he found a place near Shasta where many of the blackbirds nested:

"They had chosen a space of several acres, covered with thickets of alder and willow bushes, in the immediate vicinity of a stream of water. The nests were placed so closely to each other that I could often, without advancing, put my hand in six or eight nests. They were built of mud and straw, and lined with fine grasses. The eggs, four in number, were of a light-bluish cast, marked with spots and lines of dark umber, and a few light purple dashes. When I discovered this breeding place the young were nearly all hatched, and nightly the wolves and foxes came to devour those which had fallen from their nests during the day. I camped alone one night, in the vicinity of this place, hoping to obtain some of these depredators, but, although they would not permit me to sleep, on account of their howlings, they were cunning enough to keep out of my sight."[5]

Heermann spent three years in California and eventually returned to Philadelphia with the largest collection of nests and eggs then to have been made in the region. He must have been determinedly single-minded to have concentrated on birds when hordes of miners were heading into the foothills. But then, to Heermann, a well-marked clutch of eggs was just as precious as gold. On 28 August 1852 John Cassin wrote from Philadelphia to his friend Baird:

"DEAR PROFESSOR ...
    Heermann has arrived from California with a collection of about 1200 Bird skins—I have not seen them all, but expect to tomorrow—I have a portion of them, brought in his trunk, amongst which are a Humming bird, *T. Alexandrii*

Bourcier, new to our Fauna—a Wren *T. mexicanus*—and an undescribed Finch—
also a Squirrel which Le Conte says is new—not very strongly marked however,
being like the common Gray Squirrel, but I think quite distinct—also the greatest
kind of a lot of nests and eggs …
Ever yours,

JOHN CASSIN."[6]

The "undescribed Finch" which Cassin described as a new species—*Emberiza
rostrata*—has since proved to be only a subspecies of the Savannah Sparrow,
now *Passerculus sandwichensis rostratus*. He also described Heermann's Gull.
The Black-chinned Hummingbird (now *Archilochus alexandri*) that Heermann
collected at Sacramento had not previously been reported north of Mexico.

During the winter Heermann discussed one hundred species from his bird
collection in 'Notes on the Birds of California, observed during a residence of
three years in that country' which was published in the *Journal of the Academy
of Natural Sciences of Philadelphia* in January 1853. The water birds were not
included, since they had been delayed in transit. That winter Heermann also
prepared a complete catalogue of the egg collection at the Academy, but overall
he contributed little to the literature, since he considered writing to be a terrible
bore.

Within two months of returning to the east coast Heermann applied to join
a proposed expedition to Japan and Baird supported the plan because of
Heermann's proven collecting abilities and his fluency in French, which would
have been useful in contacts with the Russians. Cassin wanted Heermann to
collect in Texas the following spring but neither venture materialized. Instead,
he returned to California.

This time he served as surgeon and naturalist on a Pacific Railroad Survey led
by Lieutenant R.S. Williamson. Their task was to explore southern California
with the object of finding the best passes through the mountains by which the
routes along the 32nd and 35th parallels might reach the coast. In June 1853
the team disembarked at San Francisco and spent several weeks nearby at Bernicia
before starting south. On 21 August they reached Tejon Pass and camped in
the grassy valley where Fort Tejon was to be built the following year. After
checking out numerous other passes they explored part of the Mojave Desert
and by December were at Fort Yuma, beside the Colorado River. Later that
month, in San Diego, it was decided that Heermann would return East via
Tucson, El Paso and San Antonio with Williamson's second-in-command,
Lieutenant John G. Parke, in order to survey between the Gila River and the
Rio Grande.

Once back in Philadelphia Heermann handed over his collection of reptiles
to Dr Edward Hallowell, but prepared his own reports on the birds, published
in 1859 as part of volume ten of the Pacific Railroad Reports. He listed and
discussed 142 species taken between San Francisco and Fort Yuma and included
a number of birds that had previously gone unrecorded in California, including
the Vermilion Flycatcher and the Verdin.

He was disappointed that he had failed to collect a California Condor or its
egg, though since he had seen the birds on both visits he was able to include
some observations:

"Whilst unsuccessfully hunting in the Tejon Valley, we have often passed several hours without a single one of this species being in sight, but on bringing down any large game, ere the body had grown cold, these birds might be seen rising above the horizon, and slowly sweeping towards us, intent upon their share of the prey. Nor, in the absence of the hunter will his game be exempt from their ravenous appetite, though it be carefully hidden, and covered by shrubbery and heavy branches; as I have known these maurauders to drag forth from its concealment and devour a deer within an hour. Any article of clothing thrown over a carcass will shield it from the vultures, though not from the grizzly bear, who little respects such flimsy protection. The California vulture joins to his rapacity an immense muscular strength; as a sample of which it will suffice to state that I have known four of them, jointly, to drag off, over a space of two hundred yards, the body of a young grizzly bear, weighing upwards of a hundred pounds ...

At Santa Cruz I saw three or four pairs of vultures constantly, from February to October. At almost all times they could be seen sailing far overhead; but I did not, after much watching, trace them to their nests."[7]

In his report on the birds collected between Fort Yuma and San Antonio he listed just twenty-five species, but included many little-known birds such as Abert's Towhee, Pyrrhuloxia, Curve-billed Thrasher, McCown's Longspur, Gila Woodpecker, Mountain Plover and Aplomado Falcon. Heermann was unable to learn much about the falcon from the available literature, but noted: "I saw this bird twice on the vast plains of New Mexico, near the United States boundary line, before procuring it; flying over the prairies in search of small birds and mice, and at times hovering, as is the wont of our common sparrow hawk [American Kestrel]."

Heermann's activities during the next seven years are largely unknown, except that he passed the winters of 1854–55 and 1855–56 in San Antonio. He must have continued to collect reptiles since Dr Hallowell described sixty-nine of Heermann's specimens in the *Proceedings of the Academy of Natural Sciences of Philadelphia* for 1856 and thirty species in 1857, all of them from Texas.

In 1862 Heermann was in Philadelphia. By then he was walking with difficulty and so he spent a good deal of time in the library of the Academy, laboriously copying pictures of birds from the books there. He then placed his small paintings at the back of the trays of eggs in his cabinet. It would seem that he had only recently aspired to becoming an artist for Dr Nolan, the librarian, astonished and delighted him by demonstrating how the desired shade could be obtained by mixing a few primary colours. This trick had never occurred to Heermann, who had been trying to match each shade of plumage to the nearest colour in his box of paints. Nolan later recollected that Heermann once dropped a full tray of valuable eggs and immediatcly gave vent to one of the displays of artistic profanity for which he was famous.

Soon afterwards Heermann returned to San Antonio where he spent the last few years of his life. In September 1863 Henry Dresser arrived in the town and later commented: "I had, before leaving England, heard of his death, and therefore was both surprised and gratified to find him alive and comparatively speaking, in good health." Dresser added that after a short while he arranged to stay in the same house and "This was a great advantage to me, as often, when obliged to leave birds half-skinned to attend to business, he would kindly finish them for me; and besides that, he had a thorough knowledge of the adjacent country."[8]

They celebrated Christmas together but when the breeding season came around Dresser's business (supplying lumber to the Confederate States) kept him in Houston for weeks at a time. Heermann was still able to enjoy some nesting around San Antonio and he encouraged wrens to build on his ranch by putting up cigar boxes with a hole cut out of the front; Carolina, Canyon and Bewick's Wrens all took advantage of these new sites. Nest boxes were a novel idea to Dresser, who was not slow to note their ingenious practicality: "we had only to lift up the lid in order to see the progress they had made ... and when we wished to take one we took box and all, thus having the nest ready packed!"[9]

At the end of July Dresser returned to England, taking with him many Texan bird skins and eggs. Towards the end of the next year he received a letter from Theodore Heermann, informing him that on 2 September Adolphus had gone out alone to collect birds. Apparently he had stumbled and fallen and his gun, going off, had killed him.

HENSLOW. *ENGLISH NATURALIST AND CLERGYMAN. PROFESSOR OF BOTANY AT CAMBRIDGE UNIVERSITY, HE BEFRIENDED AUDUBON IN 1828.*

John Stevens Henslow
(1796-1861)

HENSLOW'S SPARROW *Ammodramus henslowii* (Audubon)

Emberiza Henslowii Audubon, 1829. *The Birds of America* (folio), Vol. 1, Pl. 70 (1831, *Ornithological Biography*, Vol. 1, p. 360): opposite Cincinnati, in state of Kentucky

Henslow's Sparrow is an uncommon, secretive bird that breeds in wet shrubby fields and rank meadows across the north-eastern states and winters in pine woods from South Carolina to Texas. It was first discovered in Kentucky by John James Audubon who afterwards became more familiar with it in its winter habitat. When he wrote about its haunts in his *Ornithological Biography* he explained that: "In naming it after the Rev. Professor HENSLOW of Cambridge, a gentleman so well known to the scientific world, my object has been to manifest my gratitude for the many kind attentions which he has shewn towards me."

Henslow was just one of many Englishmen whose kindnesses to Audubon were rewarded by eponyms. Most of these people had no connection with American birds and, fortunately, most of Audubon's bird names have since become obsolete. Henslow was at least an important naturalist—though not an ornithologist.

Born at Rochester in Kent, on 6 February 1796, he inherited his passion for unravelling the secrets of nature from both his parents. By the time he went to Cambridge University in 1814 he was already an able entomologist and conchologist. After graduating in 1818 he spent some years geologizing, teaching, and studying for the ministry, being ordained in 1824; for the rest of his life he successfully served both science and the Church. After eight years as a curate in Cambridge he was made a vicar at Cholsey in Berkshire where he lived during the long university vacations. In 1837 he was given a parish at Hitcham in Suffolk, to which he moved two years later with his wife and children. He devoted much time and energy to improving the lot of his parishioners by establishing schools and athletic clubs and encouraging horticultural progress.

His long career as Professor of Botany at Cambridge began in 1827. In the following year Audubon visited the city in search of subscribers for *The Birds of America*, arriving from London on 3 March. He quickly delivered his letters of introduction and on the 5th showed his portfolio to many of the students and staff. Henslow invited him to dinner on the 7th and two days later the visitor recorded his impressions of the city:

"Sunday, March 9. Cambridge on a Sunday is a place where I would suppose the basest mind must relax, for the time being, from the error of denying the existence of a Supreme Being; all is calm—silent—solemn—almost sublime. The beautiful bells fill the air with melody, and the heart with a wish for prayer. I went to church with Mr Whewell at Great St. Mary's, and heard an impressive sermon on Hope from Mr Henslow. After that I went to admire Nature, as the day was beautifully inviting."[1]

It is not surprising that Audubon found Henslow to be a sympathetic and encouraging confidante, for the Professor had a great knack for nurturing talent. Though his students were sometimes awestruck by the depth and breadth of his knowledge, he was adept at making them feel at ease and succeeded in enthusing and inspiring many. The most famous of his pupils later wrote that: "He was free from every tinge of vanity or other petty feeling; and I never saw a man who thought so little about himself or his own concerns. His temper was imperturbably good, with the most winning and courteous manners; yet, as I have seen, he could be roused by any bad action to the warmest indignation and prompt action."[2]

The student was Charles Darwin. Together they made many excursions into the flat Cambridgeshire countryside and Darwin later stated that no other man had so profoundly influenced his life. In 1831, when Henslow was asked to suggest a suitable naturalist for the *Beagle*, he recommended Darwin. Many years later, after the publication of *On the Origin of Species* (1859), it was Henslow who chaired the famous meeting of the British Association for the Advancement of Science, in June 1860, at which the Bishop of Oxford crossed swords with Thomas Huxley.

Henslow enjoyed excellent health for most of his life, but he was eventually smitten with respiratory and cardiac disease, which caused him to collapse at the end of March 1861. Death came peacefully on 16 May, ending a long life of fruitful service. He was survived by two sons and three daughters. The eldest daughter, Frances, married Joseph Hooker, who had sailed to the Antarctic with James Clark Ross and later botanized in the Himalayas.

Henslow's Sparrow

Jens Wilken Hornemann
(1770-1841)

ARCTIC REDPOLL *Carduelis hornemanni* (Holböll)

L[inota] Hornemanni Holböll, 1843. *Naturhistorisk Tidsskrift* 4, p. 398: [Amera-likfjord], Greenland

The Arctic Redpoll has the distinction of often being the most northerly wintering passerine. The differences between this species and other redpolls were first noticed by Carl Peter Holböll, a Danish zoologist whose long association with Greenland lasted from the 1820s until his death in 1856. In 1840 he wrote (in Danish) 'Ornithological Contributions to the Fauna of Greenland' which appeared in the *Naturhistorisk Tidsskrift* (1842–43). It contained his description of the Arctic Redpoll, which he named *L[inota] Hornemanni* after J.W. Hornemann, the Director of the Botanic Gardens at Copenhagen, where Holböll's father was the head gardener.

Jens Wilken Hornemann was born on 6 March 1770 in Marstal on the small Danish island of Aerø, which lies quite close to the Baltic coast of Germany. He was educated by his father until he was thirteen years old, then sent to live on the much larger island of Fyn with an uncle, Pastor Morten Bredsdorff, from whom he evidently derived his interest in natural history. After three years he was sent from the beautiful countryside that surrounded the manse to Copenhagen, to study under Jens Bindesböll. Hornemann matriculated at the University in 1788 but by then his father had died and he had to supplement his income by teaching.

When a natural history society was founded in the city in 1789, Hornemann attended the lectures and became closely associated with Martin Vahl, the principal instructor, with whom he developed a lifelong friendship. There seemed to be little hope of earning a living from his interests and so Hornemann decided to study medicine; he changed his mind in 1793 when he won an essay competition with *Forsøg til en Dansk økonomisk Plantelaere* which concerned the economic importance of Danish plants. It was a youthful work, incomplete and unoriginal, but was useful as a travelling flora and so gained a wide circulation; many of today's Danish plant names stem from its first publication

HORNEMANN. *DANISH BOTANIST. AUTHOR OF THE* FLORA DANICA *AND FRIEND OF THE HOLBÖLL FAMILY. HE WAS SIXTY YEARS OLD WHEN THIS PORTRAIT WAS PAINTED.*

in 1796. It went into a second edition, and a third that was completely revised and expanded to cover the whole Danish Kingdom and so included plants from Norway, Iceland and Greenland.

In 1798, aided by public and private funds, Hornemann visited Germany, the Pyrenees, the south of France and Paris, and in 1800 he spent five months studying the various herbaria in London. On his return to Denmark, through the influence of Vahl who was then Professor of Botany at Copenhagen University, Hornemann was appointed Lecturer at the Botanic Gardens. Not long after this, in October 1801, he married Marie Judithe, the daughter of Professor Claus Hornemann. Vahl died two years later and Hornemann not only took over his lectures but also continued with Vahl's *Flora Danica* and in 1808 became Professor of Botany. In later years many of his students admitted that he had not been an original or inspiring teacher, but his friendly and obliging nature made him a popular figure at the University among both students and colleagues and on his sixtieth birthday they paid him the unusual compliment of organizing a torch-light procession. His wife died only two months later and Hornemann himself was in poor health for the last five or so years of his life. Eventually he had to abandon his duties entirely and died on 30 July 1841, at Copenhagen.

For thirty-five years he had been the leading Danish botanist. His most important contribution, for which he had collected most of the specimens, was the *Flora Danica, or Differences of the Vegetation in the Danish Provinces* (1824). In eighteen parts, with 1100 plates, it was especially good for the vascular plants but the section on the non-flowering plants was considered to be poor. It was later translated into German and English.

Hornemann had few connections with ornithology but he was deeply interested in the exploration of Greenland. He supplied a list of the plants of that island

Arctic Redpoll

to W.A. Graah, one-time travelling companion of C.P. Holböll, and this was added to Graah's 'Narrative of an Expedition to the east coast of Greenland', published, in Danish, in 1837. Hornemann also took a keen interest in the exploits of Holböll for he knew the young man and his family well. F.L. Holböll, his head gardener, had worked with him for many years and together they had brought the Botanic Gardens up to a very high standard.

William Hutton[†]
(fl.1851)

HUTTON'S VIREO *Vireo huttoni* Cassin

Vireo Huttoni Cassin, 1851. *Proceedings of the Academy of Natural Sciences of Philadelphia* 5, p. 150; *ibid* 6, Pl. 10, Fig. 1: Monterey and Georgetown, California [= Monterey, California]

"Calling that Vireo after your friend Hutton is one of the severest things. I don't want to do it—when he gets better known I will call something after him. This kind of thing is bad enough at best, but to name a bird after a person utterly unknown is worse than that. I do not doubt his entire capability but I don't like to thrust honors upon him."[1]

But Professor Baird's wishes prevailed and soon John Cassin was writing to him again requesting further details:

"Please give me the name in full of Mr. Hutton, his name, style, address, business, locality, age whether young or adult, present pursuits, occupations and whereabouts to the best of your knowledge and belief and I must have it early as the paper goes to the printer about Wednesday or Thursday next."[2]

If we knew the answer to all these questions we would know a lot more about Hutton! After reading Cassin's description of Hutton's Vireo we are left little the wiser. Cassin merely announced that the vireo was "presented with many other birds, by Mr. William Hutton, a zealous and talented young naturalist now resident at San Diego, in California, in honor of whom I have taken the liberty of naming it."[3]

Elsewhere, Hutton's name appears a few times in Baird, Cassin and Lawrence's *Birds of North America* (1860) where the sources of certain specimens held at the Smithsonian Institution are scattered through the text with dates and locations. From this we learn that he collected around Washington, D.C., in March 1844, and April and May 1845. He must then have headed out to the West, just before the California gold rush, since he was at Monterey from May 1847 to November 1848. According to the label on the type specimen he found *Vireo huttoni* there in June 1847, but Joseph Grinnell thought from the state of

†No portrait traced

the plumage that it was more likely to have been obtained in September.[4] Hutton was in San Diego in 1851 and also collected at La Paz in Mexico.

In May 1857 Hutton was again collecting near Washington, D.C., but after that he is never heard of again. Perhaps he died during the Civil War. Whatever his fate he must have been a competent field ornithologist to have acquired the shy and retiring Hutton's Vireo which is so fond of disappearing into the densest thickets, the males rarely showing themselves even to sing.

The history of the discovery of Hutton's Shearwater does not help in tracing further information on William Hutton. The shearwater is a New Zealand species and is named for Frederick Wollaston Hutton (1836–1905), author of a *Catalogue of the Birds of New Zealand* (1871) and other works relating to that country.[5] Indian birds with the name of Hutton commemorate Captain T. Hutton (1807–74) who served with the British Army during the Afghan War of 1839–40 and collected birds around that time.

Hutton's Vireo

Robert Kennicott
(1835-1866)

WESTERN SCREECH-OWL *Otus kennicottii* (Elliot)

Scops Kennicottii Elliot, 1867. *Proceedings of the Academy of Natural Sciences of Philadelphia* 19, p. 99: Sitka, Alaska

"The general hue of this curious little owl is a reddish-brown, mottled and blotched with black …

The specimen was procured at Sitka by the expedition engaged in laying the telegraph which is to connect the two great continents of America and Asia, and is one of the most interesting additions which the indefatigable naturalists attached to this band of zealous workers have made to the Avifauna of North America.

The example from which my description is taken is unique, and belongs to the Academy of Natural Sciences of Chicago, by the kindness of whose officers I am enabled to bring this bird to the notice of ornithologists …

In bestowing on this owl the [scientific] name which I trust it is ever destined to bear, I simply express the desire which I am sure is felt by all ornithologists, to render honor to him who, combining the intrepidity of the explorer with the enthusiasm of the naturalist, twice penetrated the forbidding cheerless districts of the far north, in order to extend the knowledge of his favorite science; and who perished in his early manhood, in the full tide of his usefulness, on the banks of the Yukon.

Ornithology has met no greater loss, in these later days, than in the death of Robert Kennicott."[1]

Writing these words in 1867, Daniel Giraud Elliot went on to promise a life-size illustration of the new 'Kennicott's Screech-Owl' in his *New and heretofore unfigured species of the Birds of North America* published in fifteen parts from 1866–69 and bound together in two volumes. Volume one contains a delightfully atmospheric plate by Joseph Wolf, showing the owl with glistening orange-yellow eyes, perched on a mossy pine limb with the moon breaking out from behind thick grey clouds. It was a beautiful addition to Elliot's written tribute to one of the legendary bird men of mid-nineteenth century America: a man who discovered no new birds, published little about them but was nevertheless one of the most enthusiastic and energetic of all the zoological field collectors.

KENNICOTT. ZOOLOGICAL COLLECTOR FOR THE SMITH-
SONIAN INSTITUTION AND CHICAGO ACADEMY OF SCI-
ENCES. FAMOUS FOR HIS NORTHERN TRAVELS – THREE
PRODUCTIVE YEARS ON THE GREAT SLAVE LAKE, MACKEN-
ZIE AND YUKON RIVERS, AND A FUTILE EXPEDITION TO
WESTERN ALASKA.

His wide-ranging interests are shown by the remarkable diversity of material gathered by him in the far north woods of the Yukon, North West Territories and interior Alaska. He capitalized on his brief time there by encouraging many Hudson's Bay Company officials to continue the work, with amazing results.

Although born at New Orleans on 13 November 1835, Kennicott was taken as an infant to Chicago and grew up at *The Grove*, at Northfield, eighteen miles north-west of the city. His early attraction to natural history can probably be attributed to his rural surroundings and the support he received from his father, a leading Illinois horticulturalist. Young Robert suffered from poor health but always seemed to improve when he was out and about. He would often return home bubbling over with excitement, carrying some new fish or snake.

At the age of sixteen he went to Cleveland, Ohio, to study for the winter of 1852–53 under a friend of his father's, Dr Jared P. Kirtland. Kirtland taught him about reptiles, fish, birds and insects and later, from Dr Philo R. Hoy at Racine, Kennicott learned still more about birds, particularly passerines. Both men encouraged the lad to get in touch with their good friend Spencer F. Baird at the Smithsonian Institution and from November 1853 Kennicott's life began to be very much tied up with the activities and aspirations of that museum. He also became involved with the Chicago Academy of Natural Sciences, which was established in 1856; from 1857 to 1864 Kennicott was its devoted secretary, striving to promote its general development and generously enhancing its zoological collections in particular.

Kennicott became busy on a variety of projects. In October 1856 he received a commission, through Baird, to write up the 'Quadrupeds of Illinois, injurous or beneficial to the Farmer' for the *Agricultural Report of the Patent Office* (1856–57). In the winter he also studied medicine, but when his health broke down this was discontinued so that he could go north in the spring to collect for the Northwestern University of Evanston. By July he was on the Red River near Winnipeg, developing a taste for the Canadian forests, lakes and rivers. From mid-December until April 1858 Kennicott was at the Smithsonian working through the reptiles brought back by the various Railroad Surveys, his results appearing in the reports of Williamson, Cooper, Beckwith and Whipple. He also examined the reptile collections of John H. Clark made under Major Emory on the United States and Mexican Boundary Survey. From May to July Kennicott was back at Evanston but in August he was off again, pursuing insects in northern Wisconsin with Henry Ulke. By October he was back at the Smithsonian where he spent some of the winter collating Lieutenant William P. Trowbridge's collections from California and Oregon.

These times at Washington seem to have been among the happiest for Kennicott. He became an intimate friend of Professor and Mrs Baird and their young child Lucy, also counting among his friends William Stimpson who was sorting through the collections of the North Pacific Exploring Expeditions of Ringgold and Rogers. Stimpson was sufficiently well off to hire a cottage for himself, Kennicott, August Schönborn and other high-spirited young men. They all contributed to the housekeeping and called themselves the Megatherium Club. Contacts made during this period included James G. Cooper, John S. Newberry and Ferdinand Hayden, each of whom afterwards travelled with early surveying expeditions in the West. Cooper thought that Kennicott was rather wild.

Early April 1859 saw the start of Kennicott's first and most important major expedition in the way of natural history research. Little had been done in the boreal forests of Canada, or in the Arctic tundra beyond, since the days of Dr John Richardson's two overland journeys with John Franklin, which though important zoologically had been devised mainly for geographical exploration. One of the main aims of Kennicott's venture was to find the unknown breeding grounds of many familiar birds that migrated northwards each spring. Large numbers of nests and eggs of Arctic species would not only upgrade the Smithsonian collections but would be of value as exchange material. (For a map of western Canada see p. 190.)

On his way north Kennicott called on Dr Kirtland at Cleveland, and spent a few days with his parents and sisters at Chicago before taking a boat to Collingwood on Georgian Bay, Lake Huron. From there he made a short overland journey to Toronto to see Richardson's friend Dr John Rae, from whom he gained much valuable advice on travel in the north. Taking a boat to Lake Superior he arrived at Fort William, opposite Isle Royale, on 10 May, and then began the more adventurous part of the journey in the cheerful company of Canadian voyageurs making for Lake Winnipeg. He saw thousands upon thousands of ducks on the lakes, while the forest edges bustled with Cedar Waxwings, Black-backed Woodpeckers, Red-breasted Nuthatches, American Robins and Yellow Warblers, and he once found a fresh Pine Siskin nest. Though he managed to shoot Spruce Grouse, Ruffed Grouse and Swainson's Thrushes his overall success was poor. His journal explains his exciting mode of travel and the reasons why there was so little time left for collecting:

"The three canoes were each about thirty-six feet long, four in width, and two and a half deep in the middle; the bow and stern taper to a compressed point and are suddenly elevated a foot or two. The outer shell is formed entirely of birch bark, placed with the fibres transversely in the same position as it grew upon the tree, but with the inside outward …

One of these canoes, after some use, weighs, when empty, about three hundred pounds, and it will carry nearly three thousand pounds, besides the crew …

The proper crew of such a canoe is eight men, a *bowsman, steersman*, and six *middlemen* …

All, except the steersman, keep perfect stroke in paddling, and when the three canoes kept abreast, as they sometimes did, all singing and keeping stroke together, the efect was very exhilarating. The Iroquois paddle with great rapidity, making about forty strokes per minute, on the average, dipping the paddle a foot or eighteen inches into the water and pulling with very considerable force. When it is considered that this is kept up, exclusive of several short resting spells of ten or fifteen minutes each, and of the stops for breakfast and dinner, from twelve to fifteen hours per day, some idea may be formed of the extreme powers of endurance possessed by these voyageurs …

Our manner of travelling was to me very novel and interesting. At three or four o'clock in the morning we were aroused by lusty shouts of *levé! levé!* from the guides, when all hands were quickly engaged in launching the canoes and loading them, and often we were paddling briskly off within ten or fifteen minutes of the first awakening. No stop was made for cooking before starting ... We stopped for about an hour to cook breakfast at from eight to ten o'clock. Dinner was generally eaten without stopping to cook ... We camped at night at about eight or nine o'clock, when the first care of the voyageurs was to unload the canoe and take it ashore, when it was turned up on edge in such a position as to shelter the crew who slept under it ...

No time was ever lost on portages, and, unless it was a very long one, none of the men ever stopped a moment to rest."[2]

On 10 June Kennicott had a meeting at Norway House with Governor Simpson who promised him every assistance by the Hudson's Bay Company personnel who controlled the area. On 24 July, on the notorious Methy Portage, he fell in with chief trader Bernard R. Ross and travelled with him to Fort Simpson on the Mackenzie River. The fort was on a small island just below the mouth of the Liard River and like other Company posts it was arranged on three sides of a square with an open front. Constructed of spruce logs, some of them weather-boarded, the main house was positioned centrally at the back with the stores and servants' quarters at the sides. Kennicott spent much of the winter here with Ross, who knew the area well, knew the other Company men and had himself already collected items for British museums and the Smithsonian. Together they outlined their plans for collecting birds and mammals and whiled away the long hours of darkness playing whist or singing.

Very early on Kennicott realized that comparatively little would be achieved by his solo efforts and he decided to elicit the help of factors and traders by personal contact and encouragement. Since these men were primarily concerned with getting the pelts of fur-bearing mammals brought to them in the winter and early spring they had comparatively little to do in the summer months. Kennicott therefore spent his first autumn travelling hundreds of miles between isolated outposts in the vicinity seeing Alexander McKenzie at Fort Liard, John Reid at Big Island and Laurence Clarke at Fort Rae. He loved travelling by dogsled and convincingly acted the part of the rugged sourdough as he went from post to post. As spring approached he decided to spend the collecting season with William Hardisty at Fort Resolution on the south shore of Great Slave Lake. Extracts from Kennicott's journal reveal how each new day brought fresh excitement for him:

"*May 7th.*—Warm and pleasant. A few geese seen, but no large bands; all *Bernicla canadensis*, except a single white-fronted goose. A solitary swan seen, and a pair of mallard ducks ... The flowers of the alder beginning to show ...

May 10th.—A few white-fronted and Canada Geese arriving. Frogs croaking in the woods. A good many mallard ducks seen, all in pairs. Saw a golden-eye duck, and a blue-winged teal ... I see no Canada jays now; they are evidently going back into the woods to breed.

May 12th, Sunday.—Two inches of snow last night. The open ground about the fort is clear of snow, but it is still deep in the woods, and there is no getting about without wading through deep water ... Saw *Hirundo bicolor* [Tree Swallow] ...

May 15th.—Warm ... Saw a good many stickle-backs trying to ascend from the lake into the small runs from swamps. Secured one specimen of *Plectrophanes lapponicus* [Lapland Longspur], a male ...

May 17th.—A large flock of snow-geese passed this morning, the first I have seen here. A flock of *Plectrophanes nivalis* [Snow Bunting] about the fort this morning, nearly all, if not quite all, females; the males appear to have left ...

May 18th.—... Shot a pair of green-winged teal, and *Podiceps cornutus* [Horned Grebe], male. Alders in full flower.

May 19th.—A good deal of snow remains in places in the woods, though it has generally melted off. Ice still solid on the lake ... Shot five species of Tringidae ...

May 21st.—Large flocks of *Plectrophanes lapponicus*, associated with which are a few *P. nivalis*, mostly females ... To-day I obtained two specimens of *P. pictus* [Smith's Longspur], both males ...

May 23rd.—... Golden plover arriving in large flocks, and in full plumage, with all the under parts black. Got off the Indians several specimens of *Bernicla Hutchinsii* [Canada Goose] ...

May 25th.—An Indian brought nine eggs of *Anas boschas* [Mallard]. Shot a male *Setophaga ruticilla* [American Redstart] ...

May 26th.—Saw a number of *Dendroica aestiva* [Yellow Warbler], and an *Empidonax*. Shot *Numenius borealis* [Eskimo Curlew]. The base of the bill is not yellowish in life, but brownish flesh color.

May 27th.—Large bands of white geese passing. Very few other kinds seen.

May 28th.—Went with Mr. Hardisty and all the men to the mouth of the Slave River to hunt geese, and spent three days there. The geese were abundant, but they were wild and flew high. I, however, spent little time trying to shoot geese, and only bagged two. The *Anser Gambellii* [White-fronted Goose] was the most abundant ... A small snow-goose, of which I have seen fine specimens, and which the hunters say is common, is apparently *A. albatus* [Snow Goose], or *Rossii* [Ross's Goose]."[3]

At the end of the season he sent many rare skins and eggs southwards but had no intention of spending any more time in that area. In August he went further north, down the Mackenzie River beyond Fort Good Hope almost to the Beaufort Sea. He then made his way westward over the Richardson Mountains to La Pierre's House, a small outpost on the headwaters of the Porcupine River. After a day there he was off down the Porcupine to Fort Yukon, where he arrived on 28 September to stay with James Lockhart for ten months.

In deciding to work the Yukon Flats Kennicott had made an excellent choice. Ducks, geese and shorebirds were everywhere but he knew, as all nesters know, that the season is a short one, especially in Arctic Circle latitudes where no time at all should be lost:

"From the last of May till now (June 24th) Lockhart and I have been at work generally about eighteen hours out of every twenty-four. As it is light all night (indeed for a week we see the sun at midnight, by refraction, I suppose), we pay little attention to the time of day, but just work as long as we can keep awake. We start off from the fort with several Indians and canoes, and go through a series of lakes, making portages between these and the various small rivers (both lakes and rivers are very numerous), thus making a turn of fifty to one hundred miles in two or three days. We always go with at least two canoes and a party of four, and when we enter a lake one of the occupants of one canoe hunt in it through the grass at the edge of the lake where the loons, grebes, and canvas-back ducks nest, while his companion wades in the shallow water among the grass, near shore, where we get *Fulix marila* [Greater Scaup] and *F. affinis* [Lesser Scaup] eggs, and sometimes a nest of *Dafila acuta* [Pintail], that is near the water, or a canvas-back duck in shoal water. The nests are found by seeing the female rise from them. For widgeon's eggs we hunt through the bushes, and for pin-tail ducks, too, generally. When we find spots that seem to promise good breeding ground ashore, we leave the canoes and hunt through the woods and open, dry spots. We camp during the middle of the day at some good point for collecting, nominally to sleep, but, in fact, we sleep very little. I was at one time out three days, in which time I slept only once, and then scarce six hours, when I had already been forty-eight hours without sleep. I am making up for last winter's hybernation."[4]

After this orgy of collecting Kennicott and Lockhart spent days writing notes and labels, blowing eggs and preparing skins, often referring to a recent letter from Baird which outlined better ways of preserving their trophies:

"I wrote to all the gentlemen suggested by you by the winter's packet, enclosing egg drill and pipe to each. Will try and write again this spring. I hope they will get the drills in time for this year's season. Many of last year's eggs were greatly broken up, partly by being cracked in opening, partly in the carriage. I don't think moss is good for the purpose. Paper twisted round the big eggs, like the globular sugar plums, is next to enveloping in cotton, the best packing. The eggs should not be in large lots, but divided up into smaller. I shall try to make up a package of books for each contributor of specimens ...

It was very well to take out stuffing from large birds before forwarding. If paper can be had, better wrap up each specimen, so as to keep them from becoming greasy. Don't pack small specimens inside of larger."[5]

August once more meant the end of the season and Kennicott went back across the mountains to Peel's River Fort (Fort McPherson), staying there until December. Then he retreated into the mountains again to stay with James Flett at La Pierre's House where he remained for much of the winter checking his trap round or brooding on his isolation and loneliness. Occasionally, at least, he thought of home and the lovely woods around *The Grove*. "I used to like evergreens best, but I fear the too constant view of spruces will give me a distaste for them. And oh! how I long to see even the bare branches and shaggy bark of the hickory." Christmas morning saw Kennicott travelling with his dogs in the mountains, pausing only to smoke his last cigar to the health of his family and the 'Megatheria'.

Towards the end of January, still based at La Pierre's House, he wrote a rough draft for a paper on twenty-three kinds of fish that he had found in the Yukon and Mackenzie Rivers and other northern waters. But mostly he looked forward to the spring and summer, which he planned to spend on the Anderson River, further north than he had ever been and right in the heart of some of the most interesting shorebird breeding grounds in the world. In the meantime, to break the monotony of winter, he moved down to Peel's River Fort run by C.P. Gaudet, where he worked through some of his collections and continued his trapping. In the early spring he went south, upriver on the ice to Fort Simpson to see friends and make practical arrangements for the forthcoming invasion of summer migrants. Here, news from home put an end to his plans for glory. His father was critically ill and he heard about the Civil War for the first time. The whole of the field season was wasted as he made the journey south to Chicago where he arrived on 17 October 1862. On the way, at the Methy Portage, he lost his much-prized barrel of northern fishes.

At home he was dissuaded from enlisting and instead went to Washington to work at the Smithsonian again. During the winter there he wrote a short monograph on ground squirrels and sorted through some of the hundreds of birds, mammals and other items from the north, which ranged from butterflies and bumble bee lice to Eskimo pipes and Indian costumes. But he had already been away too long, for in April he wrote to his friend Roderick MacFarlane bemoaning his circumstances:

"Don't be in a hurry to leave the North, McFarlane. I assure you there is little comfort in the outside world unless you love work more than I do. My best consolation while working so constantly is the thought that the sooner I get thro the matter in hand, so much the sooner I'll start North again. In fact I hope to tie my garters and belt and start from St. Paul for Fort Churchill on the Bay, a year from next January ...

It's all very well to *talk* of the delights of the civilized world, but give me the comfortable North where a man can have some fun, see good days, and smoke his pipe unmolested. D--n civilization."[6]

In the spring Kennicott went back to *The Grove* and was present when his father died on 4 June. The Smithsonian had agreed to allow Kennicott's share of the Canadian collections to be set up at Chicago and generously promised to augment these with their duplicates from every department of natural history. He was elected curator of the Chicago Academy and the whole of 1864 was spent co-ordinating the packing, transportation and new arrangement of these specimens (the bulk of which were lost in the great Chicago fire of 1871).

In 1865 Kennicott's plans to work up the birds and mammals for the Smithsonian's proposed 'Report on Arctic Zoology' (to make a worthy successor to Richardson's *Fauna Boreali-Americana*) had to be postponed and were soon disrupted entirely. The Western Union Telegraph Company was planning to lay an overland telegraph line across Alaska to Siberia with the intention of connecting the homes and business centres of North America with those of Europe. Kennicott was persuaded that his experience would make him a valuable member of the surveying team and he was promised a group of scientists to help collect more Arctic material. Accordingly, on 19 March he arrived at New York with a party of naturalists amongst whom were William Healey Dall, Henry Martyn Bannister and Ferdinand Bischoff, a German taxidermist. Charles Pease, a grandson of Jared Kirtland and a boyhood friend of Kennicott's, also went along.

When they arrived at San Francisco in April disillusionment quickly set in. There were endless delays and disputes between the expedition leaders, surveyors, wire stringers and scientists. Kennicott found himself confronted by men hostile to all his suggestions despite his greater experience. He was soon consumed with anxiety about the success of the whole venture. Eventually two surveying teams were organized, one of which set off on a futile quest up the Fraser River from Vancouver. Shortly after their departure Kennicott went back to his room "and while sitting on the edge of the bed, talking to one of his companions, the color suddenly left his cheek, and he fell back pulseless for several minutes on the bed. The immediate production and use of the strongest brandy by his friend brought him through this attack, which, if he had been alone, might have proved fatal; and he was confined to his room for several days."[7]

On 9 August Kennicott and his team were all assembled at Sitka where they spent a further two weeks. Bischoff was posted in the town and some useful collecting was done on the island since it was here that the type specimen of the Western Screech-Owl *Otus kennicottii* was secured.

In the Aleutian Islands Kennicott spent a most enjoyable single day collecting, one of his last good field days. An unnamed bitter opponent and rival, seeing him "full of life, fun, and irrepressible energy" found it impossible not to admire him and stated later that this day totally changed his opinion of Kennicott and made him wish that they could have got to know each other in this way sooner, believing that they could then have been friends instead.

At last they arrived at St Michael on the shores of Norton Sound where provisions and equipment were off-loaded even though it was now mid-September. (For a map of Alaska see p. 318.) While Dall and some of the others crossed over to work on the Siberian side before returning to San Francisco for the winter, Kennicott tried to make good use of the end of the season. There was little or no time left for collecting as all hands helped to take the gear overland to Kaltag and up the tree-lined west bank of the Yukon to the Russian fort at Nulato. Unable to travel far until the spring thaw, the party had to wait

Western Screech-Owl

there for the ice to break up. Kennicott, troubled by the animosity of his associates and stressed by his unaccustomed irksome responsibilities, became deeply unhappy. To console himself he explored the mountains west of the fort trying to map the immediate vicinity, but most of the time he worried and fretted, sometimes wondering if he was a failure. On 12 May he burst into action, paddling out into the Yukon opposite Nulato to save the life of a Russian whose canoe had been damaged by giant ice floes. That night Kennicott could not sleep. Fearing another attack similar to the previous one he drew up instructions in case he should die and at about four or five o'clock in the morning went for a walk beside the river.

Later that day he was found slumped on the beach beside maps scratched in the sand. He had died of a heart attack, or, as William Dall put it, "struck down by disease of the heart, aggravated by exposure, privation and anxiety."[8] Three days later the thaw was almost complete and they could soon have set off upriver. Instead, Kennicott's body was placed in a double wooden coffin, sealed with pitch, taken downriver by Pease and Bannister and returned to *The Grove*. Dall arrived back at St Michael in July and took charge of the scientific work. Ten months later at Nulato, on the anniversary of Kennicott's death, he erected a tablet and a wooden cross hewn out by one of the Russians:

TO THE MEMORY
of
ROBERT KENNICOTT,
NATURALIST,
who died near this place,
May 13th, 1866,
aged thirty.

The expedition was abandoned when it was learned that an undersea cable had been laid across the Atlantic. The only good news for the explorers was that Alaska had been bought from the Russians and was now a United States Territory. Dall decided to stay on to continue with the work he had planned, spending many years in Alaska and becoming its foremost naturalist. Bischoff collected successfully at Sitka, Kenai and Kodiak Island but years later when in New Mexico he went on a collecting trip and was never seen again.

A bird long known as Kennicott's Willow Warbler was found in 1866 at St Michael by the Telegraph Expedition. Now classified as a race of the Arctic Warbler, its scientific name *Phylloscopus borealis kennicotti* (Baird, 1869) still commemorates the young naturalist who had attempted so much. Writing in 1915, Dall considered that Kennicott had directly and indirectly added more to the Smithsonian collections than any other man. Even after his death the fruits of his friendships with the men of the Hudson's Bay Company were still being harvested. Reid, Clarke, Hardisty, Lockhart, Gaudet and many others continued to send in all manner of specimens. Roderick MacFarlane outshone them all by collecting thousands of items from the Fort Good Hope and the Anderson River regions. On the tundra between the Anderson River and Franklin Bay he collected thirty clutches of Eskimo Curlew eggs and found some of the first known eggs of the Hudsonian Godwit, Sanderling, Baird's Sandpiper, Buff-breasted Sandpiper and Stilt Sandpiper.[9] Kennicott would have loved to have been with him.

Jared Potter Kirtland
(1793-1877)

KIRTLAND'S WARBLER *Dendroica kirtlandii* (Baird)

Sylvicola kirtlandii Baird, 1852. *Annals of the Lyceum of Natural History of New York* 5, p. 217, Pl. 6: near Cleveland, Ohio

On 13 May 1851 Dr Kirtland was presented with a dead warbler by his son-in-law, Charles Pease. Spring migration was then at its height and Pease had shot the bird that day, probably on Kirtland's own land which covered eighty-three acres along the shore of Lake Erie, just to the west of Cleveland. Kirtland examined the bird's yellow underparts, the blue-grey crown and the streaky plumage of the back, and on comparing it with other warblers in his collection found that none was of the same species. Professor Baird, who stopped by for a day after attending a meeting in Cincinnati, also examined the warbler, which on dissection proved to be a male. Kirtland gave it to him and Baird published the first description of the migrant, dedicating it "to Dr. Jared P. Kirtland, of Cleveland, a gentleman to whom, more than any one living, we are indebted for a knowledge of the Natural History of the Mississippi Valley".

Kirtland never learnt anything more about Kirtland's Warbler. Its wintering grounds in the Bahamas were not discovered until a year after his death when Charles Barney Cory visited Andros Island in January 1879. The breeding grounds remained a mystery for even longer and were not found until June 1903 when the ornithologist E.H. Frothingham went trout fishing on the Au Sable River in Oscoda County, on the lower peninsula of Michigan. He heard an unfamiliar song coming from the Jack Pine plains north of the river, went to investigate and one of his companions shot the songster. It was identified at the University of Michigan by Norman A. Wood who immediately set out for the Au Sable River to look for a nest.

On 2 July Wood heard his first singing male and during the next five days located four more and one female, but found it much easier to hear the birds than to see them. On 8 July he tried a new area some distance from the river, a tract of several hundred acres that had been swept by fire about six years before and was now covered with Jack Pines between three and ten feet tall. Here he heard a singing bird and spotted its mate fluttering anxiously nearby.

KIRTLAND. *PHYSICIAN, TEACHER, HORTICULTURALIST AND NATURALIST WHO LIVED AT CLEVELAND, OHIO.*

In his pocket notebook he recorded: "I watched her and she seemed very uneasy (having just been flushed from the nest). I began looking carefully on the ground, as I had made up my mind it would be found there. Suddenly I saw the nest! ... In the nest were two young birds a few days old and, as luck would have it, one beautiful egg ... Pinkish white, thinly sprinkled with chocolate brown spots gathered in a wreath at the larger end."[1]

Since the 1950s Kirtland's Warbler has been intensively studied in the Jack Pine forests of Michigan. Apart from a few pairs in Wisconsin it breeds nowhere else and is an endangered species, its numbers often fluctuating considerably from year to year. Its specialized breeding habitat is now managed by controlled planting and burning. The story of its conservation would have fascinated Jared Potter Kirtland, who was essentially a practical naturalist with a great concern for man's relationship with the world around him.

In everything Kirtland studied, whether botanical or zoological, he added to scientific knowledge and much of his research brought human benefit, particularly in the realms of horticulture and apiary. He cared greatly about the quality of the environment, which he said "exerted an immense influence over character", and he constantly encouraged farmers to consider the aesthetic value of their land as well as its productivity. Throughout his lifetime the forests of Ohio were being felled for agricultural expansion and he was deeply concerned about the effect on wildlife. In a letter to Baird, written in January 1851, Kirtland complained about the rapid decrease in number of land and freshwater molluscs and attributed the decline to the pollution of the rivers from farmyards and factories, as well as to the overgrazing of woods by pigs. In a newspaper article written the previous year, Kirtland lamented that Golden Eagles were rarely seen along the shores of Lake Erie, though they had once been commonplace; and that Bald Eagles, which fifty years before had visited almost every farmstead during cold weather, were now much reduced in number.

Few people were as aware of the relentless habitat destruction as Dr Kirtland. His own father had instigated many of the changes, since he was a stockholder in the Connecticut Land Company which bought from the state most of the lands in its Western Reserve in the north-east corner of Ohio. Turhand Kirtland had set out for the new territory with a party of surveyors and settlers in 1798 and for the next thirty-six years was Chief Agent for the company on the frontier. Jared, the youngest of four children (born 10 November 1793) was left behind in Wallingford, Connecticut, to be reared by his grandfather, a physician and horticulturalist.

In 1810, at the age of seventeen, Jared Kirtland saw Ohio for the first time when he went to visit his family at Poland, close to the Pennsylvania border. He immediately began to investigate the trees, shrubs and other plants of the region, his expertise soon being recognized by the neighbouring farmers who came to learn the techniques of budding and grafting that he had experimented with since childhood. When Kirtland's grandfather died the next year, bequeathing to him his medical library and sufficient funds for his further education, he returned East, joined the first medical class at Yale, studied also at Pennsylvania University and obtained his M.D. in 1815. That same year he married Caroline Atwater of Wallingford and established a practice there. Two and a half years later they moved to the nearby town of Durham but after the death of Caroline and his younger daughter Kirtland decided to go West.

With his only surviving child, Mary Elizabeth (who later married Charles Pease) he moved to Poland, Ohio, in 1823. Two years later he married Hannah Tousey and spent the rest of his life in the state. He practised medicine at Poland for the next fourteen years (serving also as a representative to the Ohio Legislature for six years), then took up the Professorship of the Theory and Practice of Medicine at Cincinnati Medical College until 1842, when he moved to Willoughby Medical School near Cleveland. The next year he became a founder

Kirtland's Warbler

of the Cleveland Medical College and taught there until his retiral in 1864. Despite his varied medical and public service responsibilities he always found time for horticulture and zoology.

During 1837 and 1838 Kirtland took part in the Geological Survey of Ohio with responsibility for the zoological section. When he turned his attention to the birds he realized that he would have to acquire some skill in preparing skins and in time became extremely accomplished in the art, with the result that many young people came to him for training. He kept records of the arrival and departure dates for migrant birds and was soon expressing the opinion that ornithology was "not less fascinating than conchology"! His check list of the birds of Ohio, in the *Second Annual Report of the Geological Survey of Ohio* (1838) was the first ever published and listed 222 species. It was remarkably complete and accurate and served as a model for other state lists.

At the end of 1840 Kirtland bought a tract of land at Rockport, five miles west of Cleveland, where he built a stone cottage and created one of the show farms of Ohio. In 1852 he was visited by the Reverend John Bachman who returned south with glowing descriptions of the extensive orchards, the arboretum, the greenhouses, the apiary with its hundreds of hives and the brilliantly gay flower gardens. Kirtland was so generous with grafts from his fruit trees that he sometimes had to remind importuning visitors that he was not running a nursery. He produced over thirty new varieties of cherry by hybridization and was dubbed the 'Cherry King' by his contemporaries.

Kirtland and several friends founded a local newspaper and, as editor, Kirtland's purpose was to awaken a general interest in natural history, particularly amongst farmers. Another vehicle through which he influenced the community was the Cleveland Academy of Natural Sciences to which he was unanimously elected the first (and only) president, in 1845. Private meetings were held in his room at the college and a natural history museum was established there, open to all. The Academy also ran public lectures and in December 1846 a large crowd gathered to hear Kirtland speak on 'The Nature and Habits of the Fish of Ohio'. In 1869 the Academy changed its name to the Kirtland Society of Natural Sciences, with Kirtland remaining president until 1875. Towards the end of the century the society became inactive, but was later merged with the Cleveland Museum of Natural History, which now houses Kirtland's bird collection.

Kirtland's attachment to Ohio and the local opportunities for research prevented him from much travelling further afield. In 1853 he joined Baird on a short trip to Wisconsin; in 1857 he visited Indiana and occasionally he returned to New England to visit friends, attend scientific meetings and establish personal contact with other naturalists. On one such trip, in 1854, he wrote to young Robert Kennicott that he had been buying old books "by cartloads" and that some of them were about Arctic voyages—armchair travelling recently having become one of his favourite hobbies. In 1869, when Kirtland made his longest journey, he was seventy-five years old, but he had decided to see the wonders of Florida. He took the opportunity to visit Bachman who, though four years older, was still preaching in Charleston, South Carolina. During the Civil War Kirtland had been concerned about his friend and had written to many of his former students, serving as surgeons in the Union Army, asking them to assist Bachman if by any chance they found him in need. Like so many of Bachman's guests, Kirtland became a friend of the family and continued to correspond with one of Bachman's daughters after the pastor's death.

Kirtland outlived his second wife by several years but enjoyed the company of his grandchildren and his daughter, who took good care of him during his final months when he was confined to bed. He died peacefully on 10 December 1877, at the age of eighty-four.

The obituaries that followed not only dwelt at length on the great usefulness of his life, but also drew attention to the attractiveness of his personality, particularly his cheerfulness, lack of pretension and lively sense of humour. One anecdote that illustrates these aspects of his character was recounted by Harold F. Mayfield:

"A prominent horticulturalist had traveled from the East to consult with Dr. Kirtland. He stopped his carriage at the Kirtland home and, seeing an old man in torn straw hat and overalls hoeing among the flower beds, asked, 'Is this the home of Professor Kirtland?' The man replied, 'It is.' 'Is Professor Kirtland at home?' 'He is.' 'Well, my man, come and hold my horse.' The old man advanced and held the stranger's horse while the visitor smoothed his coat and went to the door. His knock was answered by Dr. Kirtland's daughter. The visitor inquired, 'Where can I find the renowned Professor Kirtland?' The daughter answered, 'He's out there in the street holding some man's horse.' "[2]

KITTLITZ. *GERMAN ARTIST AND NATURALIST. HE VISITED ALASKA DURING THE ROUND THE WORLD VOYAGE OF THE* SENJAWIN.

Friedrich Heinrich von Kittlitz (1799-1874)

KITTLITZ'S MURRELET *Brachyramphus brevirostris* (Vigors)

Uria brevirostris Vigors, 1829. *Zoological Journal* 4 (1828), p. 357: San Blas [Mexico], error [= North Pacific]

"Eight years ago when I [Frank Kleinschmidt] shot my first Kittlitz murrelet in the ice pack of Bering Sea, an Eskimo, looking at the bird, said, 'Him lay egg way up in snow on mountain.' I ridiculed the idea then of this bird laying its egg in the snow far from the sea on the mountain side, but, keeping a constant lookout, expected to find its breeding place on the rocky islands of Alaska or Siberia, perhaps in company with the auks and murres. Now, however, I found the Eskimo's words corroborated and the murrelet's solitary egg laid in just such a strange place as he described ...

During my recent expedition I spent the time between the first and middle of May cruising in Chatham Strait, Icy Strait, and Glacier Bay. Among other specimens we collected quite a few marbled murrelets and also several Kittlitz murrelets. It was the height of the breeding season of these two species, for we found in every specimen fully or partly formed eggs, most of which, however, were broken in the collecting. However, I preserved, of the Kittlitz murrelet, one fully formed and colored egg, besides several broken ones.

On June 5, while lying at anchor off Pavloff Bay, Alaska Peninsula, a trapper and miner came on board, who saw me preparing skins of the Kittlitz and marbled murrelets. He recognised the Kittlitz immediately, and said it was strange that a water bird should lay its egg far inland, high on the mountain sides, in the snow. Upon closer questioning he said he meant that the egg was laid, not on the snow, but far above timber line on the mountain, in bare spots, amid the snow ...

On June 6 I was hunting brown bear for the Carnegie Museum, in company with this man, and while crossing a high divide, a Kittlitz murrelet flew past us. 'There is your bird,' called the trapper immediately; 'it has a nest here somewhere.' On June 10 I saw with my glasses a she-bear and two cubs far up in the snow of Mount Pavloff. To reach them I had to climb several miles inside the snow line, with only here and there a few bare spots to give me a much-desired walking ground, when close to my feet rose a Kittlitz murrelet. There on the bare lava, without even the pretension of a hollow, lay a single egg."[1]

This, the first published account of a Kittlitz's Murrelet nest, was written in 1913 while Kleinschmidt was working for John E. Thayer. It was an exciting discovery for the bird had long been something of an enigma; the first skin for an American museum had not been taken until 1877 when E.W. Nelson visited Unalaska Harbor, and the bird was considered to be extremely rare until large flocks were found at Glacier Bay in 1907 on the Alexander Alaska Expedition. One reason for the very slow gathering of information about the murrelet's range, status and habits was its limited distribution in the far north: in summer confined to the coastline of Alaska and north-eastern Siberia and in winter moving only as far as the ice-free waters of the more southerly parts of its breeding range.

For the history of the earliest discovery of the bird we must go back to the first half of the nineteenth century when all of this territory was Russian. J.F. Brandt, Director of the Zoological Museum in St Petersburg, found the murrelet skin amongst others that had been collected by Kittlitz on the voyage of the Russian corvette *Senjawin* and consequently, while working on a five-page review of the auks in 1837, he named the bird *Brachyramphus kittlitzi*. This name persisted for many years (appearing in Ridgway's *Manual of North American Birds* in 1887) but it was eventually shown that Brandt's description had been preceded. The real type specimen, described by Nicholas Vigors, was collected during Captain Beechey's famous voyage to the North Pacific on H.M.S. *Blossom* between 1825 and 1828—at around the same time as the *Senjawin* expedition. The given type locality (off San Blas in Mexico) is now considered to be erroneous as it is well outside the recorded range of the species. Through traditional use, Kittlitz is retained in the vernacular name, although Vigors' scientific name has priority. The following account of Kittlitz's life concentrates on his activities in the North Pacific since this is the area in which he found the murrelet, but he is most often mentioned in connection with the South Pacific islands where he was the only ornithologist ever to observe some of the endemic species that are now extinct.

Baron Heinrich von Kittlitz was born on 16 February 1799 in Breslau (now Woctaw, Poland), the son of a Prussian army officer. By the age of fourteen he too had become a soldier but while stationed near Frankfurt he became friendly with Eduard Rüppell who encouraged his interest in natural history. He was also influenced by Maximilian, Prince of Wied-Neuwied, of Missouri River fame.

In 1821 Rüppell left on his first lengthy expedition to Egypt and sent back zoological collections to Professor Cretzschmar of the Senckenberg Natural History Society. Cretzschmar asked Kittlitz to illustrate many of the specimens

Kittlitz's Murrelet

for the *Atlas zu der Reise im nördlichen Afrika* [Atlas of Travels in northern Africa] (1826) and this project clearly demonstrated Kittlitz's artistic talents. Although he had risen to the rank of Captain, Kittlitz left the army at the age of twenty-six determined to pursue a career as a naturalist. Using his mother's family connections he gained a place on the scientific staff of a Russian expedition to Alaska and Siberia, under the command of Captain Friedrich von Lütke of the corvette *Senjawin*. Among the personnel were two other artists: the mineralogist Alexandre Postels and the physician–naturalist Carl H. Mertens, both of whom helped to widen Kittlitz's scientific expertise during the next three years.

The *Senjawin* sailed from Kronshtadt on 20 August 1826 carrying cargo for Sitka and Petropavlovsk with instructions to survey parts of the Bering Sea and the Caroline Archipelago. After a mid-winter crossing of the Atlantic they rounded Cape Horn and sailed up the coast to Valparaiso in Chile where Kittlitz appreciated several weeks ashore. The *Senjawin* then headed out into the Pacific and for more than two months the monotony was broken only by occasional sightings of seabirds and cetaceans. By the time they reached the small Russian port of Sitka, Kittlitz was determined to spend his time in Alaska as productively as possible. (For a map of the route of the *Senjawin* in the North Pacific see Steller p. 411.)

For the next five weeks, from 23 June to 1 August, the three naturalists lived in a wooden house near the fort. So that they could explore the coast the governor provided them with three-seated *baidarkas*—sealskin craft with a frail wooden frame—and native oarsmen. Kittlitz hunted daily and soon American Robin, Varied Thrush, Ruby-crowned Kinglet, Orange-crowned Warbler, Wilson's Warbler, Rufous Hummingbird and Hairy Woodpecker were added to his bag, along with Steller's Jays which were common around the edges of the woods, and Ravens that hunted along the beaches and about the buildings. He devoted many hours to pursuing Black Oystercatchers and eventually succeeded in hitting one. Other shorebirds taken included Killdeer, Lesser Yellowlegs and Wandering Tattler. Around the wooded islands he stalked Harlequin Ducks, Great Blue Herons and Belted Kingfishers, often stopping to watch Bald Eagles as they hunted for fish or perched immobile on the branches of dead trees. With Mertens he studied botany and made two drawings that later appeared in *Twenty-four Views of the Vegetation of the Coasts and Islands of the Pacific*. One was a detailed scene of a clearing in the forest and the other was a more dramatic view of a long, narrow inland lake separated from the sea by a ridge of mountains.

At the beginning of August the *Senjawin* set sail for the Aleutian Islands and though Kittlitz saw petrels almost every day he had no opportunity to collect them. They spent ten days in the sheltered harbour at Unalaska where a small group of Russians lived with the Aleuts. Sitting on the beach, Kittlitz drew some small birds feeding on bare patches of gravel between tangles of blue lupine and yellow daisies, with the hillside rising behind to high bare peaks where snow still glistened in the deep gullies. He climbed the treeless slopes to hunt for ptarmigan but they were far too wary for him, and so their exact identity was unresolved.

Their next destination was the Pribilofs but the naturalists were unable to land on either St George or St Paul because of rough seas. At St Matthew Island their experience was the same, even though they remained offshore for a week so that Lütke could make a detailed survey of the coastline. The cliffs were still thronged with auks, Walrus were hauled out on some of the beaches and a few Polar Bears were wandering about, but Kittlitz could only fume with frustration, much as Steller had done on his even more tantalizing voyage to Alaska nearly ninety years earlier. When they anchored off Bering Island on 20 September some of the hunters who were based there came out to the vessel but again Kittlitz had to remain on board, unable to explore the island where Bering and Steller had been shipwrecked.

Towards the end of the month the *Senjawin* reached Petropavlovsk, on the east Kamchatkan coast, and anchored there for five weeks. Loons had gathered on the calm surface of Avatcha Bay, Red-faced Cormorants loitered on the rocks with outstretched wings and after a few days flocks of Red-necked Phalaropes turned up. There were also various gulls in the harbour but Kittlitz had great difficulty distinguishing the different species.

On 1 November the expedition departed for the South Pacific where they overwintered and here Kittlitz made some of his most important ornithological observations. To the east of the main Caroline group he discovered the Kusaie Island Crake, a small black rail that inhabited the damp and shady parts of the forest. He was the only naturalist ever to see the bird alive and prepared two skins that are now in the Leningrad Museum. Also extinct is the Kusaie Island Starling that he collected in the mountainous forests where it was probably already rare. Returning northwards they visited the Bonin group, some seven hundred miles south of Tokyo, and Kittlitz was able to land on Peel Island, which had a good harbour. He later described two new species from here, the Bonin Island or Kittlitz's Thrush and the Bonin Woodpigeon, both of which were exterminated by the end of the century.

By early June the *Senjawin* was again at anchor in Avatcha Bay. By this time Kittlitz was tired of being at sea and decided that he would achieve far more by remaining on Kamchatka for the summer than by returning to the Bering Sea with the others. During the next four months he explored the southern tip of the peninsula, sometimes accompanied by a Russian friend and several assistants. One of their excursions was to Starichov Ostrov, an island off Avatcha Bay densely populated by auks and gulls. One of the common breeding birds. was the Tufted Puffin and at the beginning of July they were incubating their eggs. Kittlitz made a drawing of his two barefooted companions gathering scabird eggs into buckets while auks and cormorants wheeled around the steep grassy cliffs. Using hooks they dragged the incubating puffins out of their burrows and then strung the corpses on belts around their waists, taking a good many in this way for food. Before leaving the island, Kittlitz lost his gun when it slid down a rocky slope and fell into the sea, but he doubtless sought out a replacement at Petropavlovsk before he set out on his next trip. Somewhere along the coast he collected his Kittlitz's Murrelet, but he gave no precise location for the skin, which was simply labelled 'Kamchatka.'

Kittlitz made long journeys upriver into the interior, attracted by the volcanic snow-capped peaks of the central mountain chain, and he frequently halted to sketch the vast pine forests, deciduous woods or grassy plains through which he travelled. He worked his way around to Bolsheretsk on the west coast, then returned by horse and canoe to Avatcha Bay where he arrived on 15 October and was reunited with his colleagues on the *Senjawin*.

They then continued the round-the-world voyage back to Europe, via the Caroline Islands, the Philippines and the Cape of Good Hope. On 12 July 1829 Kittlitz disembarked at Le Havre, thinking that he would travel faster overland to St Petersburg, but he was delayed by illness and missed the reception at which Tsar Nicholas bestowed bonuses on the other members of the expedition. After failing to obtain a position at the Academy of Sciences where his collections of 300 bird species and 200 drawings were deposited, Kittlitz returned to Germany but found himself handicapped by frequent bouts of illness that had first afflicted him in the Pacific. He devoted all his limited energy to describing his new birds, completing his accounts of the journey and contributing both notes and illustrations to Lütke's three-volume narrative, *Voyage autour du monde ... sur le corvette Le Séniavine* (1835–36).

Early in 1831 Kittlitz travelled to Egypt, determined to explore north-eastern Africa with Rüppell, but on the journey up the Nile he became so nauseated, weak and depressed that he had to stay in his bunk. Rüppell insisted that he return to Germany, where he did improve, but recurrent attacks prevented him from joining any other expeditions.

He produced several books that were based partly, or wholly, on the *Senjawin* voyage and among these were *Über einige Vögel von Chili* [On some Birds from Chile] (1830) which contained a number of descriptions of new species that he had discovered around Valparaiso; *Über einige noch unbeschriebene Vögel von der Insel Luzon, den Carolinen und den Marianen* [On some undescribed Birds from the Islands of Luzon, the Carolines and the Marianas] (1831) containing coloured portraits of ten species new to science; and *Kupfertafeln zur Naturgeschichte der Vögel* [Engravings Illustrating the Natural History of Birds] (1832–33). This last book was a great disappointment for Kittlitz, as owing to a lack of money, only a few copies were ever produced, even though some of the eighty birds were being shown for the first time, such as the Bonin Woodpigeon and the Kusaie Island Starling. It also contained a painting of Kittlitz's Plover, one of the few birds he had been able to collect in Egypt. The aforementioned *24 Vegetationsansichten* [*Twenty-four Views*] was published in Germany in 1844 and in London in 1861, his only book to have been translated into English during his lifetime. Although Kittlitz made the original sketches during the voyage, he did not engrave them until many years later, by which time he considered that he had become sufficiently skilled at the process of etching. An extract from Kittlitz's *Denkwürdigkeiten einer Reise nach dem russischen Amerika* (1858), covering the voyage from Valparaiso to Petropavlovsk, is now also available in English.[2]

Kittlitz married at the age of forty-five and became the father of two sons and a daughter. In 1865 his wife died and Kittlitz passed away nine years later, on 10 April 1874, at Mainz, aged seventy-five.

LAWRENCE. *NEW YORK BUSINESSMAN AND ORNITHOL-OGIST.* CO-*AUTHOR* OF BIRDS OF NORTH AMERICA *WITH BAIRD AND CASSIN, HE AFTERWARDS SPECIALIZED IN NEOTROPICAL ORNITHOLOGY.*

George Newbold Lawrence (1806-1895)

LAWRENCE'S GOLDFINCH *Carduelis lawrencei* Cassin

Carduelis Lawrencei Cassin, 1852. *Proceedings of the Academy of Natural Sciences of Philadelphia* 5 (1850), p. 105: Sonoma and San Diego, California [= Sonoma, California]

In 1635 John and William Lawrence, aged seventeen and twelve respectively, arrived with their sister and mother at Plymouth Colony from St Albans, Hertfordshire, England. Nine years later they moved to Long Island, settling successively at Hempstead and Flushing, John Lawrence becoming mayor of New York in 1672. William continued to live at Flushing and it is from this branch of the family that the ornithologist George Newbold Lawrence was descended. He was born on 20 October 1806 in New York City and died there eighty-eight years later on 17 January 1895, his entire life having been spent in the area.

When Lawrence was a boy there was much more incentive to study birds there than now. In the summer months he went with his parents to their country house *Forest Hill*, which was about eight miles from City Hall overlooking Manhattan and the Hudson River, the primeval forest then stretching unbroken all the way to Fort Washington Point. In September Lawrence noted that Passenger Pigeons would sometimes appear in huge numbers above the woods, and one day more than a hundred were shot near the site of General Grant's tomb. During migration American Robins were gunned down in the trees where 3rd Avenue and 20th Street now lie and Eastern Meadowlarks would flock to the pastures now overlain by Broadway and 40th Street. And Lawrence later recalled seeing six drake Labrador Ducks going rotten at Fulton Market because nobody wanted to buy them![1]

At the age of fourteen young George was given his first gun and began to shoot and collect enthusiastically, attempting to preserve the birds without skinning them since he knew nothing about the proper preparation of birds, not yet having made contact with John G. Bell, the great New York taxidermist. In 1820 he started work as a clerk in his father's wholesale druggists company, becoming a partner at the age of twenty and director in 1835; until his retirement at the age of fifty-six he could only devote some of his time to birds. In the meantime he received some inspiration from a brief acquaintance with Audubon who moved into a property near *Forest Hill* with his family. The artist–naturalist was already ailing when Lawrence met him but he got to know his two sons Victor and John rather better. It was not until Lawrence was thirty-five years old that he received the stimulus to devote his life to birds in a scientific way. In 1841 Bell invited his young friend to his taxidermy shop to hear Spencer F. Baird expounding his ideas on muscle attachment and bird flight. Almost immediately Lawrence's aims and ideals changed and in 1846 he produced his first paper describing the Lawrence's Black Brant. With Baird in Washington directing operations, Lawrence in New York, Cassin in Philadelphia and Brewer in Boston the scene was set for the so-called Bairdian Epoch of American ornithology, an exciting time for both field and closet ornithologists. Almost any expedition could expect to return from foreign parts with novelties, and the majority of their discoveries were worked through by one or other of the foursome.

In 1850 Cassin named a new goldfinch *Carduelis Lawrencei* in the *Proceedings of the Academy of Natural Sciences of Philadelphia*. Bell had just returned from a trip to California with a choice collection of birds. His notes on Lawrence's Goldfinch, a bird that breeds solely in California and Baja California, were published in full by Cassin:

> "This bird I first observed at Sonoma. In habits it much resembled our common [American] Goldfinch (*C. tristis.*) The flock, out of which I shot these two, was feeding in company with the small black headed species (*C. psaltria,*) [Lesser Goldfinch] on the seeds of plants growing near the ground, and when disturbed alighted upon the nearest bushes. When flying, they keep up a constant chattering or calling like our common species. I also saw this bird at San Diego, feeding as above in company with the crimson fronted Bullfinch, (*Erythrospiza frontalis,*) [House Finch] in the open prairie. I never saw it in the mountains."[2]

Having already named a bird after Bell on the previous page (the Sage Sparrow *Amphispiza belli*), Cassin completed the account of the goldfinch with a long dedication:

> "I have named this bird in honor of Mr. George N. Lawrence, of the city of New York, a gentleman whose acquirements, especially in American Ornithology, entitle him to a high rank amongst naturalists, and for whom I have a particular respect, because, like myself, in the limited leisure allowed by the vexations and discouragements of commercial life, he is devoted to the more grateful pursuits of natural history."[3]

As the years went by, more and more species were named after Lawrence by friends and correspondents, including such European naturalists as Hans von Berlepsch and Philip Sclater and fellow Americans J.A. Allen, Charles B. Cory and Robert Ridgway. By 1891 some twenty-two species had been named in his honour, mostly foreign birds. Lawrence's Warbler, named from a specimen taken near Chatham, New Jersey, has since proved to be a hybrid Blue-winged and Golden-winged Warbler; and several others are not now recognized as full species.

Lawrence's Goldfinches, with Lesser Goldfinch at right

The year 1858 saw the culmination of a vast amount of ornithological research when volume nine of the Pacific Railroad Reports was published. It was not, as one might expect, limited to the results of the various recently completed surveys but dealt with all the known birds found north of Mexico, making it the most complete and detailed work up to that date. Most of the 1000 page text was by Lawrence's friend Baird. Cassin wrote pages 4–64, 689–753 and 900–918, while Lawrence contributed pages 820–900 dealing with some of the groups of water birds. It was reissued two years later with some minor alterations and with additional plates as *Birds of North America* (1860); and another edition emerged in 1870. Two contemporary ornithologists who appreciated its qualities are best qualified to comment upon its immediate impact. Daniel G. Elliot considered that it "created a revolution in the technicalities and methods of American ornithology, sweeping away all the old land-marks, and introducing a new era, a new system, and practically a new science."[4] Elliott Coues, in much the same vein, thought that it "exerted an influence perhaps stronger and more widely felt than any of its predecessors, Audubon's and Wilson's not excepted, and marked an epoch in the history of American ornithology."[5]

With the initial publication of this work, Lawrence virtually abandoned the study of northern species to concentrate on neotropical ornithology. For the rest of his career he specialized in the birds of the West Indies and Central and South America, never developing any inclination to write about their distribution, relative abundance or habits but always pursuing that which he most enjoyed:

the slow, methodical and painstaking elucidation of new species. Nothing pleased him more than to receive a collection from some hitherto unexplored region which he could catalogue, and from which he could determine new forms by careful reference to his own skins and those at the neighbouring museums of his friends. Among those he worked through were some of the birds collected by Johannes Gundlach in Cuba, by Francis Sumichrast in south-western Mexico, by Colonel Grayson on the islands of Tres Marias and Socorro, and by James McLeannan and Frederick Hicks in Colombia.

The A.O.U. *Check-list of North American Birds* (1983), which includes the West Indies and Central America, lists well over seventy species first described by Lawrence. Since those occurring *north* of Mexico are so few it is worth enumerating them. In the *Annals of the Lyceum of Natural History of New York* for 1851 he worked through the small collections made by Captain McCown in Texas, giving scientific names to the Ash-throated Flycatcher, Olive Sparrow and McCown's Longspur. In the same journal for 1857 he added to his list of new species the Black-tailed Gnatcatcher from Texas and the Antillean Nighthawk from Cuba (which breeds in the Florida Keys). Western species described by him are Le Conte's Thrasher and the California Gull in the *Annals* for 1851 and 1854 respectively; and in the Pacific Railroad Reports, volume nine (1858), are the Western Grebe, Clark's Grebe and Pacific Loon, the latter two only recently accepted again as full species. Lawrence's Black Brant, already mentioned, is now considered to be a (probably extinct) race of the Brant.

Though Lawrence was still contributing to the literature well into his eighties (the last to *The Auk* in 1891) he had sold his collection of 8000 specimens and over 300 types to the American Museum of Natural History at New York in 1887. J.A. Allen had the task of assessing its importance and recataloguing it, eagerly assisted by Frank Chapman, then a little-known twenty-three year old museum worker who many years later looked back fondly on those early days:

> "... I can think of no task that could have brought me more pleasure and profit. The Lawrence collection was exceedingly rich in species, most of which were new to me. Every few minutes I found some specimen of such exquisite beauty that I could not resist the temptation of showing it to Dr. Allen as though I had made an actual discovery, and he, in his quiet way, shared my pleasure though he had doubtless seen it before."[6]

Ten months after Lawrence's death the 13th A.O.U. Congress was held at the U.S. National Museum in Washington. Elliott Coues arranged for a special memorial evening to be held for two Honorary Members who had died in the past year. Coues delivered the eulogy for Thomas Huxley; Daniel Elliot that for George Lawrence, from which the following quote is derived:

> "I cannot recollect the time when I did not know George N. Lawrence ... Courteous, gentle, simple in his tastes and habits, almost child-like in his deference to the opinions of others in whom he reposed confidence, asserting his own opinions with a modesty that was remarkable, because so rare, Lawrence was a conspicuous example of that personage to whom we all turn with mingled feelings of admiration and respect—a gentleman of the Old School, of the days of our ancestors, when knee breeches and brocaded silks were parts of the ordinary costume, and the manners of the age were characterized by dignity and a respectful

demeanor ... Lawrence never grew old ... he was as eager for news of ornithology and ornithologists as he ever displayed in the days of his activity ... The end was peaceful, and he passed away only a few days after the death of his wife, to whom during the period of her long illness, he had ever exhibited a touching, affectionate devotion rarely witnessed."[7]

William Elford Leach†
(1790-1836)

LEACH'S STORM-PETREL *Oceanodroma leucorhoa* (Vieillot)

Procellaria leucorhoa Vieillot, 1817. *Nouveau Dictionaire d'Histoire Naturelle, Nouvelle édition*, Vol. 25, p. 422: coast of Picardy

"Lot 78—an undescribed Petrel with a forked tail, taken at St Kilda in 1818; the only one known (with egg)."

In the summer of 1819 dealers from all over Europe flocked to an important auction in London. Franco Bonelli, Hinrich Lichtenstein and Coenraad Temminck were there but the petrel was knocked down for £5 15s to Dr Leach of the British Museum. He also bought a Great Auk and an egg for just over £16.[1]

The owner of the collection, and the man who had obtained the petrel, was William Bullock. He described himself as "Silversmith, Jeweller, Toyman and Statue Figure Manufacturer" but he was also a naturalist, collector, traveller, antiquary and showman. Towards the end of the eighteenth century he established a museum at Sheffield and, encouraged by public interest, moved it to Liverpool and in 1809 transferred it to London, establishing himself at 22 Piccadilly and later at Egyptian Hall nearby. The museum was a truly miscellaneous assortment of items which included works of art, armoury, curiosities from the South Seas, Napoleonic relics and various birds.

In those days Bullock's museum was a well-known and popular attraction in central London, and his bird specimens were so well mounted that his competitors were forced to improve their standards. Even though the museum was still attracting a large number of people Bullock decided to sell his collection under the hammer, acting as his own auctioneer. It comprised 32,000 curiosities and over 3000 birds and in May 1819, after the sale had been underway for five days, William Leach made his famous purchase.

Not long after, whilst on a visit to the British Museum, Temminck asked to examine the petrel and in 1820 in his *Manuel d'Ornithologie* he named it

†No portrait traced

Procellaria Leachii.[2] The almost inevitable English name soon followed, along with a protracted debate (which is still resurrected from time to time) over Leach's eligibility for such an honour. To counteract Temminck the Reverend John Fleming proposed that the petrel be called *Procellaria Bullockii*, "in order to do an act of common justice to the individual who had energy to undertake a voyage of enquiry, and sagacity to distinguish the bird in question as an undescribed species."[3] Bullock had already published *A Concise and Easy Method of Preserving Objects of Natural History* (1817); he was a Fellow of the Linnean and Wernerian Societies and would therefore seem to have been capable of writing his own scientific description of the bird. Had he done so it would have been of little consequence as far as the species' present *scientific* name is concerned because, despite the claims of the auction catalogue, there were already three known examples of the petrel in Europe. One was at the Paris Museum, a second was in the possession of Meiffren Laugier and a third, which had been taken in Picardy, was owned by L.A.F. Baillon; the last being the type specimen which Vieillot described in 1817—the year before Bullock obtained the first British specimen.

Leach's Storm-Petrel

An examination of Leach's career shows him to be as well qualified to be commemorated as many another naturalist. He probably achieved more than Bullock in advancing British ornithology and more than most in promoting the study of marine biology. William Elford Leach was born at Plymouth in the family town-house near the Hoe Gate, the youngest of four children. His father, George Leach, was a prosperous solicitor and his mother, Jane Elford, was from a well-known Devon family. From the age of twelve William went to a school in Exeter which was attached to the Devonshire and Exeter Hospital where he first started his studies in anatomy and chemistry. At seventeen Leach began to study medicine at St Bartholomew's, London, but three years later moved to Edinburgh University and then, in 1812, he was given permission to take his M.D. at St Andrew's University. He then returned to London but gave up his new profession in favour of his zoological interests. In 1813 he edited the Reverend George Low's *Fauna Orcadensis* and in the same year was employed initially as an assistant librarian in the Zoology Department at the British Museum, which was then at Montague House. Leach found the zoological collections shockingly neglected. Many of the specimens were unlabelled and unarranged and others had been left to decay in the basements. Sometimes helped by his friend William Swainson, Leach managed to salvage some of the items but was forced on occasions to have evil-smelling bonfires to dispose of those which were considered beyond redemption.

Leach lived in two rooms at the museum, one of which had various skulls placed around it and another, which happened to be decorated with some stuffed bats—the 'skullery and battery' as Leach liked to call them. He was extremely agile and active, almost always running up the stairs in leaps and bounds. One of his exercises involved "leaping over the back of a stuffed zebra which was placed in the centre of a large room ... over which we [Swainson] have seen him vault with the lightness of a harlequin". There is no known portrait of Leach but Swainson considered him to be: "Of a slight form and delicate habit ... with a naturally nervous and irritable temperament."[4] He was also warm, frank and generous and appears to have been well respected and well liked in Britain and abroad.

In addition to sorting through the collections Leach involved himself in arranging the displays and also carried out research, exchanged and purchased specimens or collected them in the field himself. Somewhat surprisingly, in view of his achievements, he was only employed at the museum for about eight years and even then was not always present: in July 1816 he obtained leave for two months but did not return until after the New Year. Nevertheless, he always worked hard, perhaps too hard. His written works for his period at the British Museum include: a *Zoological Miscellany, being descriptions of new and interesting Animals* published in three volumes (1814–17); a *Monograph on the British Crabs, Lobsters, Prawns and other Crustacea with pedunculated eyes* (1815–17); a *Systematic catalogue of the Specimens of the Indigenous Mammalia and Birds that are preserved at the British Museum* (1816); a *Synopsis of the Mollusca of Great Britain* (circulated in type in 1820, but not published until 1852); and the zoological appendices for a number of Arctic and African expeditions of the period. Other contributions by Leach include articles in the *Encyclopaedia Britannica* and numerous papers in the *Philosophical Transactions of the Royal Society* and the *Zoological Journal*. Apart from some crucial entomological work his main accomplishments were his studies of Crustaceans and Molluscs; after the death of George Montagu in 1815, few people in Britain could have known more about the latter two subjects than Leach.

He began to work at night to catch up on the task he had set himself; his main scientific endeavour now being the creation of his natural system of arrangement and classification, especially of Crustaceans. The work was always laborious and often tiring and eventually his health gave way. In 1821 he became paler and thinner, suffered a complete breakdown, and formally resigned his position as Assistant Keeper on 9 March 1822.

His elder sister took him away to the Continent for a prolonged period of rest. They stayed at various villas in the south of France and near Lake Como in northern Italy and they may also have travelled more widely, to Greece and other parts of the eastern Mediterranean. On 26 August 1836, while back in Italy, Leach died suddenly of cholera at the Palazzo San Sebastiano, near Tortona, north of Genoa.

John Lawrence LeConte (1825-1883)

LE CONTE'S THRASHER *Toxostoma lecontei* Lawrence

Toxostoma Le Contei Lawrence, 1851. *Annals of the Lyceum of Natural History of New York* 5, p. 121: California near the junction of the Gila and Colorado Rivers [= Fort Yuma, California]

LE CONTE'S SPARROW *Ammodramus leconteii* (Audubon)

Emberiza le conteii Audubon, 1844. *The Birds of America* (octavo edn), Vol. 7, p. 338, Pl. 488: wet portions of prairies of upper Missouri [= Fort Union, North Dakota]

Le Conte's Sparrow was described by Audubon after he returned from his Missouri River expedition. He called it Le Conte's Sharp-tailed Bunting, noting that the type specimen had been shot on 24 May 1843 by John G. Bell, the well-known taxidermist from New York who had accompanied him. Audubon made no mention of this new discovery in his journal, his most vivid memory of that day being an accident that befell him when he borrowed a small double-barrelled gun from his friend Edward Harris: "the cone blew off, and passed so near my ear that I was stunned, and fell down as if shot, and afterwards I was obliged to lie down for several minutes."[1] It may be that this event made him forget the sparrow when he wrote up his journal, but it seems much more likely that it was some time before he or Bell realized that the small brown bird was a new species.

Audubon wrote the original description of the sparrow at *Minnie's Land*, New York, in 1844, while working on volume seven of the octavo edition of *The Birds of America*. Unfortunately he did not include any Christian names in his tribute:

"I have named this interesting species after my young friend Doctor LE CONTE, son of Major LE CONTE, so well known among naturalists, and who is, like his father, much attached to the study of natural history."

LECONTE. *THE LEADING NORTH AMERICAN ENTOMOLOGIST OF HIS DAY. HE WAS THE SON OF MAJOR JOHN LECONTE OF NEW YORK AND A COUSIN OF JOHN AND JOSEPH LECONTE.*

There has been some confusion about which LeConte Audubon was referring to since the family at that time contained five naturalists. Louis LeConte was a keen botanist who managed a plantation in Georgia and he had two sons, John (1818–1891) and Joseph (1823–1901), both of whom became famous in their respective fields of physics and geology. E.S. Gruson, in *Words for Birds* (1972) linked the sparrow with Dr John LeConte, the son of Louis. This John LeConte had studied in New York from 1838 until 1841 so he may have known Audubon well, but his father was not a major. Elliott Coues in his *Key to North American Birds* (1872) wrote that the sparrow honoured Major LeConte, but it is difficult to understand why he should have thought so. Major John Eatton LeConte, an enthusiastic botanist and zoologist, lived in New York and had one son, John Lawrence.

John Lawrence LeConte, the subject of this chapter, is surely the one. In 1844 he was already a promising naturalist, although only nineteen years old. Doubt could be cast on his eligibility by a modern understanding of the title 'Doctor', as he did not graduate from medical school until 1846; but at that time one could work as a doctor without any training at all. Even at the medical schools, lecture courses were short and students learned primarily by serving a three-year apprenticeship with a practising physician. He would, therefore, have been regarded as a doctor by this time.

LeConte's career as a naturalist owed much to the influence of his father. Within a few weeks of his birth in New York, on 13 May 1825, his mother had died, and the Major was thereafter devoted to his only child and eager that he should share his interests, which although wide, were centred on insects, particularly beetles. The Major sent his son to St Mary's College in Emmitsburg, Maryland, and when the boy's tutors complained that young John was far too interested in birds, bugs, shells and stones the Major shocked them all by responding to this terrible news with great approval, insisting that his son should be given every encouragement!

In 1842 LeConte left school and returned to New York where he entered the

Le Conte's Sparrow

College of Physicians and Surgeons. The following summer he visited his distant relative, Spencer F. Baird, at Carlisle, Pennsylvania, bringing him various items from Philadelphia. On 22 June 1843 Baird wrote to his brother William telling him about the birds:

> "A fine White Ibis white plumage, with tips of Black. These two are very good Skins. *Ardea Herodia*; Short eared Owl & Marsh Hawk not so good. Also Head & tail of Fork tailed hawk, and head of great Crow Blackbird. He says he will get all the rest I want next winter, he is going there again then. He is here yet, but leaves to morrow morning. Dont forget about getting bugs for him."[2]

In the fall LeConte's cousin, Joseph LeConte, also joined the college and together they spent the season of 1844 travelling to the headwaters of the Mississippi via Buffalo, Detroit and the Great Lakes. When Audubon named Le Conte's Sparrow in that same year it was already obvious that LeConte was going to follow in his father's footsteps. His first paper, on the *Carabidae*, was also published in 1844 and on receiving his degree in 1846 he made entomology his vocation and avoided the practice of medicine until the outbreak of the Civil War. His father's considerable wealth meant that LeConte was unhampered by the need to earn his living. He could afford to travel and he made several visits to Lake Superior, once in the company of Dr Louis Agassiz.

Towards the end of 1850 he began a much more adventurous journey, sailing via Panama to California. He landed at San Francisco, collected around San Jose, then headed southwards to San Diego and from there travelled through the Sonoran desert to the eastern border of Arizona. On the way, near the junction of the Gila and Colorado Rivers, he shot a pale, greyish thrasher with tawny undertail coverts, and later gave the skin to George N. Lawrence in New York. Lawrence brought the bird to the attention of the members of the Natural History Lyceum on 8 September 1851, calling it Le Conte's Mocking Bird— now Le Conte's Thrasher. LeConte published his entomological findings in the *Annals* of the Lyceum but in 1852 he and his father moved to Philadelphia and thereafter they both published many of their studies in the periodicals of the Academy of Natural Sciences of Philadelphia.

The LeContes frequently had to refer to the pioneering entomological works of Thomas Say but these were widely scattered and some were in almost inaccessible publications. Twenty-five years after Say's death John L. LeConte collated Say's papers, publishing them in two volumes as *The Complete Writings of Thomas Say on the Entomology of North America* (1859). They included a memoir by George Ord who had been a close friend of Say's and in his youth had travelled with him to Georgia and Florida.

The next few years were eventful ones for LeConte. In 1860 his father died, in 1861 he got married and in 1862 he joined the Medical Corps of the Union Army as a Surgeon. Promoted to Lieutenant-Colonel and Medical Inspector he served until the end of the war with his talents for organization and administration being exercised in new directions.

In the fall of 1869 he sailed to Europe with his wife Helen and two young sons, and spent the next three years visiting many private and public museums to examine their insect collections, travelling as far afield as Algeria and Egypt.

Shortly before the Civil War LeConte had begun a *Classification of the Coleoptera of North America* to be issued in parts, but it was never completed. As time passed it became clear that an entirely new edition was needed but LeConte's health had begun to fail and after 1878, when he accepted the position of Chief Clerk at the U.S. Mint, he could no longer devote himself single-mindedly to entomology. Nevertheless, with help from George Horn, the work was accomplished and LeConte lived to see it well received. Soon afterwards he died at his home on 15 November 1883, aged fifty-eight.

Through his publications and his generous assistance to others LeConte had stimulated a widespread interest in insects that had led to the establishment of societies and periodicals devoted entirely to entomology, including the American Entomological Society of which he was a founding member. About half of all the American insects then known had been named by him, and his cabinet, which went to Harvard, contained about 6000 type specimens collected by amateur enthusiasts and the various United States government surveys. He had been the pre-eminent American entomologist of his generation.

Le Conte's Thrasher

CLÉMENCE LESSON. *FRENCH NATURAL HISTORY ARTIST. WIFE OF RENÉ PRIMEVÈRE LESSON WHO GAVE THE ORIGINAL DESCRIPTIONS OF MANY NORTH AMERICAN BIRDS DISCOVERED BY FRENCH COLLECTORS.*

Marie Clémence Lesson (1796 or 1797-1834)[1]

BLUE-THROATED HUMMINGBIRD *Lampornis clemenciae* (Lesson)

Ornismya Clemenciae Lesson, 1829. *Histoire Naturelle des Oiseaux-Mouches*, p. xlv; 1830, p. 216, Pl. 80: le Mexique [= Mexico]

René Primevère Lesson, the first describer of the Blue-throated Hummingbird, was born during the French Revolution on 20 March 1794, at the port of Rochefort on the Atlantic coast. The son of a low-grade naval clerk, he entered the Naval Medical School at Rochefort in 1810 and the next year was sent to sea as an assistant surgeon. During the following three years of fierce fighting against the British he gained much practical experience but when peace came in 1814 he was discharged and remained ashore for the next eight years. He continued his studies in pharmacy and after a spell at Rochefort Hospital he was appointed a Civil Health officer, then took charge of the Rochefort Botanical Gardens.

In August 1822 he began a round-the-world voyage to South America, the South Pacific and Australasia on *La Coquille*, serving as surgeon and naturalist in the company of Dumont d'Urville, the official botanist. After the ship returned to Marseilles in March 1825 Lesson was sent to Paris to study the expedition collections, which were deposited at the Museum d'Histoire Naturelle. For his part in the success of the mission, Lesson was promoted to Pharmacist First Class and awarded the Legion of Honour. For the rest of his life he kept up a steady flow of publications, among them being *Distribution geographique de quelques Oiseaux marins* (1825); two volumes on vertebrate zoology in Captain Duperrey's *Voyage autour du monde sur la corvette la Coquille* (1826–30); *Manuel de Mammalogie* (1827); *Manuel d'Ornithologie* (1828); *Voyage Médical autour du Monde* (1829) and *Traité d'Ornithologie* (1831). Scattered through his books and papers are the original descriptions of forty species listed in the A.O.U. *Check-list of North American Birds* (1983).

In Paris Lesson met the ornithologist Charles Henry Dumont de Sainte Croix who was to become his father-in-law. Dumont had been imprisoned during the reign of terror but survived to marry Marie Rey, a daughter of Rey de Neuvié, secretary to Jean de Cambacérès. After the restoration of the monarchy Dumont had become a legal adviser to the court and in his spare time had written ornithological articles for *Dictionnaire des Sciences naturelles* (1804–30). Lesson also wrote some sections and collaborated with Dumont on others.

In February 1827 Lesson married Dumont's second daughter, Marie Clémence. Clémence, as she was known, was then twenty-nine or thirty years old, an attractive and well-educated lady who had trained in Paris as a natural history artist, studying for a time under Gerard van Spaendonck, a Dutch master of botanical illustration at the Jardin du Roi. Clémence helped to illustrate at least two of her husband's ornithological works: she is listed among the artists in *Complément des Oeuvres de Buffon* (1828–30) and her name appears on Plate 72 'L'Arlequin' in *Histoire Naturelle des Oiseaux-Mouches* (1830). Her contributions seem sparse, considering Lesson's output, but it may be that motherhood deflected her energies, for marriage brought her a step-daughter, the seven year old Cécile Lesson, whose mother (Zoë) had died three days after her birth in November 1819. On 27 November 1827 Clémence gave birth to a daughter of her own—Anaïs.

Two years after their marriage, Clémence's husband proudly named a new hummingbird "L'Oiseau-Mouche de Clémence" *Ornismya Clemenciae*. The type specimen was owned by the Duke of Rivoli who had acquired it, along with other Mexican birds, in 1829.

In 1832 the Lessons moved to Rochefort because Clémence was consumptive and they hoped that the sea air would restore her health, but she died of cholera on 5 August 1834 during an epidemic that swept through western France. Now twice widowed, Lesson doted on Anaïs, but she died of typhoid fever on 3 November 1838, a few weeks before her eleventh birthday. Lesson expressed his anguish in poetry, and in paintings and sculptures of his cherished daughter, but he was unable to assuage his grief. The Museum Anaïs and *Mino anais*[2] (the Golden-breasted Myna of New Guinea) were among his monuments to her memory. He died, aged fifty-five, at Rochefort, on 28 April 1849.

Blue-throated Hummingbird

Meriwether Lewis (1774-1809)

LEWIS'S WOODPECKER *Melanerpes lewis* (Gray)

Picus torquatus (not Boddaert, 1783) Wilson, 1811. *American Ornithology*, Vol. 3, p. 31, Pl. 20, Fig. 3: No locality [= Montana, about lat. 46°N.]
Picus Lewis "Drap[iez]." G.R.Gray, 1849. *The Genera of Birds*, Vol. 3, App., p. 22. New name for *Picus torquatus* Wilson, preoccupied

On 27 May 1806, during the eastward journey back from the Pacific coast, Captain Lewis noted in his diary that his companions had at last succeeded in shooting a few specimens of a new woodpecker that they had frequently seen in the Rockies and further west. After describing the structure, size and plumage, he added a few observations which indicated that he had studied the living bird and not just the corpse:

> "this bird in it's actions when flying resembles the small redheaded woodpecke[r] common to the Atlantic states; it's note also somewhat resembles that bird. the pointed tail seems to assist it in seting with more eas or retaining it its resting position against the perpendicular side of a tree ... it feeds on bugs worms and a variety of insects."[1]

After their return, as the leaders of the first expedition to cross North America, Lewis and Clark handed over their bird skins to Alexander Wilson who was then working on his *American Ornithology* (1808–14). Wilson named the new woodpecker Lewis's Woodpecker—*Picus torquatus*. Having himself travelled thousands of miles through the wild parts of America, Wilson was a great admirer of the two explorers and reminded his readers that the three new species discovered by them (Lewis's Woodpecker, Clark's Nutcracker and Western Tanager) were "but a small part of the valuable collection of new subjects in natural history, discovered and preserved, amidst a thousand dangers and difficulties by these two enterprizing travellers, whose intrepidity was only equalled by their discretion, and by their active and laborious pursuit of whatever might tend to render their journey useful to science and to their country."[2]

Meriwether Lewis first sought the opportunity to explore the trans-Mississippi country when he was an adventurous nineteen year old, but his friend Thomas Jefferson (then George Washington's Secretary of State) considered him to be

LEWIS. *CO-LEADER OF THE CELEBRATED LEWIS AND CLARK EXPEDITION OF 1804–06.*

too inexperienced. Born near Charlottesville in Virginia, on 18 August 1774, Lewis had been a neighbour of the Jeffersons until the age of ten, when the Lewis family moved to upper Georgia. For three years Lewis spent a good deal of time out hunting and the expertise he acquired was later to prove as vital a part of his education as the next, more academic phase when he studied Latin, mathematics and science. When Lewis was eighteen his step-father died and so he took over the management of the family estate. After two years he joined the local militia during the Whiskey Rebellion and took to soldiering with such enthusiasm that in May 1795 he enlisted in the U.S. Army and served for six years.

In 1801 when Thomas Jefferson was elected president he immediately wrote to Lewis asking him to be his private secretary and Lewis accepted. During the next two years they frequently discussed the possibility of exploring a land route across to the Pacific and in January 1803 Jefferson sought the approval of Congress. Lewis was sent to Philadelphia for a crash course in the use of sextant and chronometer and in June he wrote to William Clark asking him to share the leadership of the expedition. By the end of the year the small force had mustered at a site eighteen miles above St Louis where three boats were built and discipline was established. On the afternoon of 14 May 1804 the party of about forty-five men set out on their voyage up the great Missouri—the first part of their epic journey to the Pacific. (For a map of their route see Townsend p. 450.)

In late fall the explorers reached the Mandan villages in North Dakota and there they built a fort for their use during the six months of winter. When the ice broke up in early April, Lewis sent some of the men back downriver in the largest boat, loaded up with zoological collections, Indian wares and preliminary reports for President Jefferson. The Corps of Discovery then resumed the ascent of the Missouri in the two smaller vessels and six new dugout canoes. They took with them a new member of the team: Sacagawea, a sixteen year old Shoshoni girl who was accompanied by her husband and their two month old baby son.

The navigation of the unpredictable currents of the Upper Missouri was a tedious, frustrating and exhausting business. Progress became even slower in mid-June when they reached the Great Falls of Montana and for the next two weeks they had to portage the boats and all the equipment. Their route along native trails then took them southwards and into contact with the Shoshoni, who sold them horses. At the end of August the explorers hid their dugouts and rode over Lemhi Pass, then turned north towards the land of the Nez Percés. In October they felled five large Ponderosa Pines and in their new canoes followed the Clearwater, Snake and Columbia rivers to the sea; their longed-for arrival turned out to be a miserable anti-climax during a week of incessant rain and high winds.

The Corps of Discovery overwintered on the coast until the end of March when they set out to recross the continent. They travelled together as far as Lolo Pass, where Lewis took a more northerly route to reconnoitre the Marias River, while Clark descended the Yellowstone. On his side trip up the Marias with only three companions, Lewis got caught up in the most serious skirmish of the whole journey when a small band of Blackfeet attempted to steal their horses and two of the Indians were killed. Two weeks later Lewis was shot in the thigh by one of his own men (blind in one eye and myopic in the other) when he mistook his buckskin clad leader for an Elk. The wound was painful and disabling but fortunately the two parties were reunited next day and Clark assumed all the leadership responsibilities for the next few weeks. When they reached the Mandan villages the explorers parted from Sacagawea and her family; then the Corps descended the Missouri as quickly as possible and arrived at St Louis on 23 September 1806, their twenty-eight month journey having been a resounding success.

Despite the irregularities of the expedition—shared leadership and the presence of a woman with an unweaned infant—the objectives had been achieved, only one member of the team had died (probably of appendicitis) and only one encounter with the natives had been violent. The Corps had brought back a vast amount of new geographical and scientific information in maps and journals, strengthened United States claims to territory in the north-west and provided a clear incentive for their people to push the frontier westward.

The natural history results were particularly impressive since despite all the routine chores and ever-varying stresses of the trek both Lewis and Clark found the energy to write about the wildlife, earning themselves a foremost place among American naturalists. They were the first White Americans to mention many of the western species and made small collections of skins and dried plants, but their discoveries were presented piecemeal by various writers with the result

Lewis's Woodpecker

that Lewis and Clark received less credit than their due. Clark's narrative was not published until 1814 and their joint journals were not issued until 1904–5. One of their discoveries was the Pronghorn, the swiftest of all American quadrupeds, first bagged by Clark in September on the initial stage of the journey. It was stuffed and sent off to Washington the following spring as part of a large consignment from Fort Mandan that included six live animals: four Black-billed Magpies, a Sharp-tailed Grouse and a Black-tailed Prairie Dog. The explorers wrote at length about the huge prairie dog towns, and the Grizzly Bears which frightened them on several occasions. The Pronghorn, Black-tailed Prairie Dog and Grizzly were all described by George Ord in 1815 along with the Tundra Swan—or Whistling Swan, as the discoverers christened it. Since they had brought the swan from below the great narrows of the Columbia River Ord gave it the scientific name of *columbianus*. Many of the other birds which Lewis and Clark referred to in their journals, such as Sage Grouse, Northwestern Crow, Pinyon Jay, Long-billed Curlew, Mountain Quail and Western Meadowlark were not properly described until later years.

One of their most surprising ornithological discoveries was made on 17 October 1804 near the junction of the Missouri and Cannonball Rivers. Lewis wrote: "This day took a small bird alive of the order of the [blank space in Ms.] or goat suckers. it appeared to be passing into the dormant state ... the mercury was at 30 a.o. the bird could scarcely move. I run my penknife into it's body under the wing and completely distroyed it's lungs and heart yet it lived upwards of two hours this fanominon I could not account for unless it proceeded from the want of circulation of the bl[o]od".[3] It was to be another forty years before Audubon gave the original description of *Phalaenoptilus nuttallii*, now known as the Common Poorwill, and more than 140 years before the pages of a scientific journal posed the question "Does the Poor-will 'Hibernate?'".[4] It is now known that the Common Poorwill can indeed survive prolonged periods of dormancy in cold weather, an adaption common in mammals but demonstrated in birds only by the nightjars, although swifts and hummingbirds can enter into states of less profound torpidity.

The botanical collections that Lewis made on the outward journey were lost in the Columbia River when one of the canoes was wrecked, but he returned with about 150 plant species, most of which were figured in Fredrick Pursh's *Flora Americae Septentrionalis* (1814). Pursh created the genera *Lewisia* and *Clarkia* and named several species after Lewis; Lewis's Monkey Flower *Mimulus lewisii*, Lewis's Wild Flax *Linum lewisii* and Lewis's Syringa *Philadelphus lewisii*.

On Lewis's return to Washington, President Jefferson appointed him governor of Louisiana Territory, based at St Louis. It was a key position that fully exercised his talents for leadership, diplomacy and pacification but his service was short-lived. While travelling through Tennessee on a trip to Washington, Lewis died on the night of 11 October 1809 at Grinder's Stand, an out of the way inn on the Natchez Trace. He may have been murdered or he may have died by his own hand; the matter has never been resolved.

Félix Louis L'Herminier†
(1779-1833)

Ferdinand Joseph L'Herminier†
(1802-1866)

AUDUBON'S SHEARWATER *Puffinus lherminieri*
Lesson

Pufflnus [sic] *Lherminieri* Lesson, 1839. *Revue Zoologique* [Paris] 2, p.102: ad ripas Antillarum [= Straits of Florida]

The scientific name of Audubon's Shearwater presents a puzzle. It was given in 1839 by René Primevère Lesson, a French naturalist working in Paris. In the *Revue Zoologique* he published a short paper on thirteen birds, from various museums, that he considered to be new and his entire entry for the eleventh (complete with misprint) tells us remarkably little:

> "11. *Pufflnus Lherminieri*, Less.—Corpore suprà nigro, infrà albo; rostro et pedibus nigris.—Long.: 12 poll.—Hab. ad ripas Antillarum.—Mus. Rupifortensis."

Lesson gives no indication as to who L'Herminier was nor why he so chose to name the bird. However, the only well-known naturalists with this name working in the first half of the nineteenth century were Félix and Ferdinand L'Herminier, father and son, both of whom made a special study of the natural history of the West Indian island of Guadeloupe.

Félix Louis L'Herminier was born in Paris on 18 May 1779. When he was sixteen he served as a cabin boy in the French navy but was discharged after

†No portraits traced

fifteen months so that he could study surgery, botany and chemistry at the garrison hospital at Lille for two years. In 1798 he was commissioned as 2nd class chemist and sent to Guadeloupe. After a year there he took part in an expedition to Cayenne, French Guyana, collecting mammals, birds, reptiles and insects as well as live plants and seeds. He then settled down to his duties on Guadeloupe, becoming chief chemist at the military hospital on Marie-Galante and master of the mint at Guadeloupe. In between times, over a period of thirty years, he thoroughly investigated the colony's wildlife.

In 1809 he received permission to go to France but his ship was captured by the British and taken to the United States. He somehow got back to Guadeloupe later that year with more botanical material. Early in the following year the colony was taken by the British, an event which proved to be only a minor interference in his activities, since the island was temporarily made over to Sweden in March 1813. Napoleon returned to power in 1815 and L'Herminier offered to supply gunpowder and medicines to keep the island in French hands but in September it was retaken by the British. He was imprisoned for rebellious activities and then exiled with his pregnant wife and seven children. He moved up through the Antilles, collecting at Antigua, St Barthélemy, St Eustatius, Saba and St Thomas, finally settling at Charleston, South Carolina. In order to survive he sold his "valuable and extensive collection of specimens, the fruit of twenty years' industry"[1] to the Charleston Museum. Thomas Say called it a "handsome collection" when he saw it on his way to Florida in 1818.[2]

L'Herminier was made superintendent of the museum, being its first salaried officer, but he lost the money received for his specimens in an unsuccessful attempt to set up a chemical and pharmaceutical laboratory in the city. Completely ruined, he was employed for a time at an apothecary shop and by 1819 had managed to save enough to return to Guadeloupe, which was now run by the French again. His former titles of King's Naturalist and Director of the Botanic Gardens were restored to him and he received some financial compensation for his recent hardships, but only part of the entitlement because he antagonized two important island officials. He was forced to go to France in 1829 in order to put his case and died in Paris on 25 October 1833, at the age of fifty-four. The natural history material that he collected is now scattered between Guadeloupe, Paris, Geneva and Charleston.

Four years after he had first settled in Guadeloupe L'Herminier's son Ferdinand was born, at Basse Terre, on 20 June 1802. He was sent to study medicine at Paris where he became a pupil of the distinguished naturalist Henri de Blainville. Ferdinand was reunited with his father in 1826 when he returned to Guadeloupe and they enjoyed three years together before his father went on his mission to Paris. Ferdinand became greatly interested in classifying birds by means of their bone structure. In 1827 he published his first paper on avian osteology in the *Annales de la Société Linnéenne de Paris* and followed this with others in *Comptes Rendus* and the *Annals des Sciences Naturelles*. His ideas were sufficiently revolutionary to gain him international recognition and Alfred Newton, in the ninth edition of the *Encyclopaedia Britannica*, devotes over two pages to young L'Herminier's contribution to the development of ornithology.[3]

Ferdinand stayed on at Guadeloupe for the rest of his life, serving as a doctor there for forty years. He became chief medical officer at the civil hospital at Point à Pître and the hospice of Saint Elisabeth, receiving the Legion of Honour for his work following the severe earthquake of 1843. He had been planning a work on the birds of Guadeloupe but the earthquake ruined all his mounted birds, all those preserved in spirit, all his notes and all his books. Undaunted he continued his studies with renewed vigour and published much of his work in the *Magasin de Zoologie* instead. He died at the age of sixty-four, on 11 December 1866, after a two-month illness contracted towards the end of a devastating cholera epidemic that swept through the island.

Which L'Herminier was Lesson thinking of when he named the shearwater? The evidence is inconclusive. Félix had been dead for six years but his name was well known in French natural history circles because of his Guadeloupe specimens. He could have met Lesson in Paris when he was there between 1829 and 1833, sufficiently recently for Lesson to think of him when describing the bird. On the other hand, his son Ferdinand was better known in Paris since he had studied there as a doctor and naturalist. Even though Ferdinand had left the country thirteen years previously his series of papers on osteology were appearing in French journals between 1827 and 1837 almost up to the time when *Puffinus lherminieri* first appeared in print.

It is interesting to check some of the other birds that commemorate the name of L'Herminier. Earliest of these is the Guadeloupe Woodpecker *Melanerpes herminieri*, named in 1830—but once more Lesson fails to specify which L'Herminier. The Forest Thrush *Cichlherminia lherminieri* of the Lesser Antilles was named by Baron Lafresnaye in 1844, apparently after the younger L'Herminier since he explains that Ferdinand had just lost his collection in the earthquake but had given him permission to publish three of his new species.[4] Charles Bonaparte created the genus *Cichlherminia* especially for the Forest Thrush but he made no dedication to one or the other.

It is reported in Urban's *Symbolae Antillanae* (1898–1913) that the labels with the L'Herminier plant collections do not indicate whether the father or the son was the collector. If this is true of the birds then it may be that Lesson himself did not know which L'Herminier he wanted to honour.

Audubon's Shearwater

Thomas Lincoln
(1812-1883)

LINCOLN'S SPARROW *Melospiza lincolnii* (Audubon)

Fringilla Lincolnii Audubon, 1834. *The Birds of America* (folio), Vol. 2, Pl. 193: Labrador [= near mouth of Natashquan River, Quebec]

Lincoln's Sparrow was discovered on Audubon's 'Labrador Expedition' in 1833 and he named the new bird after his young companion, Thomas Lincoln, whom he had known only since the previous year. In his *Ornithological Biography* Audubon gave an account of their first meeting in his chapter on the Spruce Grouse:

"In August 1832, I reached the delightful little village of Dennisville, about eighteen miles distant from Eastport [Maine]. There I had the good fortune of becoming an inmate of the kind and most hospitable family of Judge LINCOLN, who has resided there for nearly half a century, and who is blessed with a family of sons equal to any with whom I am acquainted, for talents, perseverance and industry. Each of these had his own peculiar avocation, and I naturally attached myself more particularly to one who ever since his childhood has manifested a decided preference for ornithological pursuits. This young gentleman, THOMAS LINCOLN, offered to lead me to those retired woods where the Spruce Partridges were to be found. We accordingly set out on the 27th of August, my two sons accompanying us. THOMAS, being a perfect woodsman, advanced at our head, and I can assure you, reader, that to follow him through the dense and tangled woods of his native country, or over the deep mosses of Labrador, where, you know, he accompanied me afterwards, would be an undertaking not easily accomplished with credit. The weather was warm, and the musquitoes and moose flies did their best to render us uncomfortable. We however managed to follow our guide the whole day, over fallen trees, among tangled brushwood, and through miry ponds; yet not a single Grous did we find, even in places where he had before seen them, and great was my mortification, when, on our return towards sunset, as we were crossing a meadow belonging to his father, not more than a quarter of a mile from the village, the people employed in making hay informed us that about half an hour after our departure they had seen a fine covey. We were too much fatigued to go in search of them, and therefore made for home."[1]

LINCOLN. *AT THE AGE OF TWENTY-ONE HE WENT ON*
AUDUBON'S LABRADOR EXPEDITION AND RETURNED TO
RUN THE FAMILY ESTATE AT DENNYSVILLE, MAINE.

Thomas, who was then aged twenty, made such a favourable impression on Audubon that he was invited to join him the following summer on an expedition. Since the artist was then forty-eight years old and beginning to complain that he could no longer travel and work as tirelessly as in his youth, he took along four additional young men to hunt for him; his son John Audubon, George Shattuck, Joseph Coolidge and William Ingalls. At that time the southern coast of Quebec was known as Labrador and Audubon always called this trip his Labrador expedition, even though he barely reached the present border of the province.

On 6 June 1833 the little group sailed from Eastport on the chartered schooner *Ripley*, skippered by Henry Emery, "a small spry Yankee" who had been a schoolmate of Lincoln's. By the next day the landlubbers were all shockingly seasick in the Bay of Fundy, but not too ill to notice the gulls, murres and petrels that flew past. By 12 June they were off the Magdalen Islands and were eager to go ashore and shoot some birds. Despite the bitter cold they all rose at four the next morning and after comforting themselves with mugs of hot coffee they landed on one of the islands between two great bluffs. As the day wore on the sunshine became pleasantly warm and they all enjoyed the ramble, discovering the bird life to be more varied than they had expected. Winter Wrens, Fox Sparrows, Pine Siskins, American Robins and Hermit Thrushes sang from the woods or flitted across their path and on the shore they shot Piping Plovers and Common Terns.

The next evening Lincoln wrote in his journal:

> "June 14th. This morning we set sail for the east end of Anticosti, about noon came in sight of the 'gannet rocks,' [Bird Rocks] which presented an appearance that astonished us. We could scarcely trust our own senses. The rocks, the larger a quarter of a mile broad, looked as if covered with a bank of snow; the surface was literally covered with gannets … As we neared the island they sailed round us by the thousands. Four or five of us took a boat and rowed toward it but found to our great disappointment that the surf ran so high on the S.E. side (which is the only point whence the cliff can be ascended and that with extreme difficulty) that we could not land there."[2]

In stormy weather, with the passengers again shivering and seasick, they continued north-westwards to the Quebec coast and anchored in a small bay five miles west of the Natashquan River. Audubon was appalled by his first view of the bleak landscape:

> "What a country! When we landed and passed the beach, we sank nearly up to our knees in mosses of various sorts, producing as we moved through them a curious sensation. These mosses, which at a distance look like hard rocks, are, under foot, like a velvet cushion. We scrambled about, and with anxiety stretched our necks and looked over the country far and near, but not a square foot of *earth* could we see. A poor, rugged, miserable country; the trees like so many mops of wiry composition, and where the soil is not rocky it is boggy up to a man's waist."[3]

He was further disappointed when he discovered that the nearby islands had been raided by eggers from Halifax just a few days before their arrival. Each day, for the next ten days, the young men went out hunting and the nests most frequently found were those of Common Eider and Great Black-backed Gull. Lincoln and John always worked together, being the strongest and most energetic of the team, as well as being the most experienced shots. In this wild country they were in their element and delighted in exploring as far afield as possible. One day they went inland to investigate some shallow pools and returned with a male Black Scoter, but although they had seen several others they were unable to find any of their nests.

Audubon was keen to sail further east as soon as possible but foul winds delayed them, so on 27 June they all went ashore instead and searched a small valley. They got two Gray Jays and a Ruby-crowned Kinglet, then Audubon heard a song that was new to him and shouted to the others to come over.

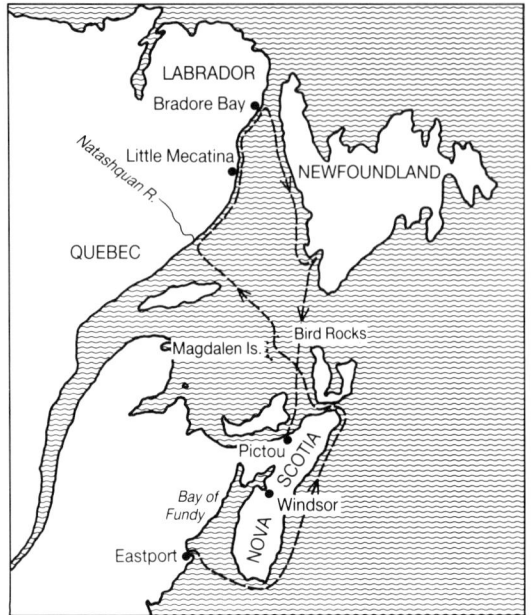

THE VOYAGE OF THE RIPLEY

"They came, and we all followed the songster as it flitted from one bush to another to evade our pursuit. No sooner would it alight than it renewed its song; but we found more wildness in this species than in any other inhabiting the same country, and it was with difficulty that we at last procured it. Chance placed my young companion, THOMAS LINCOLN, in a situation where he saw it alight within shot, and with his usual unerring aim, he cut short its career. On seizing it, I found it to be a species which I had not previously seen; and, supposing it to be new, I named it *Tom's Finch*, in honour of our friend LINCOLN, who was a great favourite among us. Three cheers were given him, when, proud of the prize, I returned to the vessel to draw it, while my son and his companions continued to search for other specimens. Many were procured during our stay in that country. They became more abundant and less shy the farther north we proceeded, but no longer sang, in consequence of the advance of the season. We did not, however, succeed in finding a nest."[4]

During the afternoon the wind at last swung round to the south-west and they got out to sea. The following day they passed several seabird colonies and landed to collect the eggs of Common Murres and Atlantic Puffins. Some of the boys were so intrigued by the puffins that they captured several alive and kept them on board the boat, their antics affording the crew and passengers much amusement.

On the Fourth of July Tom Lincoln was sent ashore at four in the morning to gather the plants that Audubon needed for his background to the illustration of Lincoln's Sparrow, and he returned with Dwarf Cornel, Cloudberry and Bog Laurel. While Audubon spent the rest of the day finishing the painting Lincoln went off again and noticed that the young of Lincoln's Sparrow were now out of their nests and following their parents.

By the middle of July the naturalists were at Little Mecatina harbour, fishing for cod below high, rugged cliffs shared by Ravens and Peregrine Falcons. For four days dirty weather prevented much exploration and when the wind eventually dropped they were tormented by mosquitos even while on board the schooner. Trips ashore proved so unfruitful that they sailed further along the coast but on 22 July, Audubon wrote that "The Caribou flies have driven the hunters on board; Tom Lincoln, who is especially attacked by them, was actually covered with blood, and looked as if he had had a gouging fight with some rough Kentuckians."[5]

At the end of the month they reached the northern limit of their cruise, at Bradore Bay. By this time Lincoln and Shattuck were both bare-footed so they walked some eighteen miles to Porteau where they hoped to buy some Eskimo moccasins. They returned the next day, hungry, grumbling and still without any footwear, but bearing a Pomarine Jaeger. At the beginning of August hundreds of Eskimo Curlews began to arrive to feed on the ripe Crowberries (then called Curlew Berries). Other waders, including Black-bellied Plovers, were also moving through and on 11 August the *Ripley* began her southward journey down the western coast of Newfoundland, past high mountains where snow still lay in drifts. The young men had little work to do while at sea so Lincoln often occupied himself in playing the flute while John accompanied him on the fiddle. Less than a fortnight later they were put ashore on the mainland of Nova Scotia and after travelling overland from Pictou to Windsor they sailed back to Eastport, arriving there on 31 August, nearly three months after their departure.

Audubon was a little disappointed that they had only discovered one new species, but he had improved his knowledge of the northern birds, having seen ninety-three different kinds during the trip. Moreover, he had accumulated a useful collection of bird skins and plants and added twenty-three illustrations to *The Birds of America*. The voyage had been worthwhile.

Many years later when one of Lincoln's sons asked his father what kind of a man Audubon had been he gave a classic down-east reply, "He was a nice man,

Lincoln's Sparrow

but as Frenchy as thunder."[6] William Ingalls, in *his* old age, was asked by
Ruthven Deane to put some of his memories of the historical cruise down on
paper and he summed up Lincoln as being "quiet, reserved, sensible, practical
and reliable".[7]

As might be expected of such a man, Lincoln returned to the family homestead
in Dennysville after his adventure and never again travelled far afield. Though
he studied medicine for a term at Bowdoin College he did not receive a degree
but with his brother Edmund took up the responsibility of managing the family
estate. Their land, which covered 10,000 acres, produced hay, kiln wood,
sleepers, grain and vegetables which were shipped out of Eastport from their
own wharves. Lincoln's diary, which for a time had included such entries as
"Shot a Redstart, a Tawny Thrush, and a White-bellied Swallow" now listed
barrels of apples and bushels of barley.

In 1852, at the age of forty, he married Emma Johnson who became the
mother of Arthur and Mary. After Emma's death he married Mary Eastman
and she had one son, Edmund. Lincoln's family, his business and his involvement
in the anti-slavery movement filled most of his time but he never lost his interest
in nature, although he rarely hunted during his later years.

When Thomas Lincoln died on 27 March 1883, on his seventy-first birthday,
the house passed into the care of his elder son, Dr Arthur Lincoln. The
ornithologist Charles Wendell Townsend (author of *In Audubon's Labrador*,
1918) visited Dr Lincoln and was shown a number of Thomas's watercolours
of birds and shells, which demonstrated considerable talent. The old clapboard
house where Thomas Lincoln was born, lived and died no longer belongs to his
descendants. It is now the Lincoln House Country Inn, the oldest house in
Dennysville and one of the best examples of colonial architecture in Maine.

Robert McCormick
(1800-1890)

SOUTH POLAR SKUA *Catharacta maccormicki* (Saunders)

Stercorarius maccormicki Saunders, 1893. *Bulletin of the British Ornithologists'*
Club 3, p. 12: Possession Island, Victoria Land, lat. 71° 14' S., long. 171° 15' W

"One morning in the year 1884, several of the officers of the [London] Natural
History Museum were surprised at the sight of a little old man ascending with
quite an elastic step the staircase of the upper floor of the Museum and disappearing
into the Botanical Department. He belonged evidently to a by-gone age. A rather
broad-brimmed hat covered a very evident wig, his neck was encircled by a high
stock, his waistcoat was white and very low, exposing a wide front of flannel shirt
of the hues of a Scotch plaid. His swallow-tail coat was of a dark blue with gilt
buttons, and his trousers were of a pronounced shepherd's plaid. I was telling
some of my colleagues afterwards of the wonderful appearance of the old gentleman
I had seen, when a knock came at my door, and on opening it, I found myself
face to face with the individual in question. On his introducing himself as Dr.
McCormick, I could not repress my astonishment and told him that I thought he
had been dead years ago. 'Yes,' he replied, 'I know I ought to have been, but I
am not. I am eighty-four years of age, and I thought, before I died, I should like
to see some of the animals I shot when I was naturalist to the *Erebus* and *Terror*,
as I am writing my memoirs.'"[1]

Richard Bowdler Sharpe, who wrote these words, immediately interrupted
his work as head of the Bird Department to show the veteran collector around
the galleries. As Sharpe listened to the old man reminiscing about penguin
shooting in the Antarctic a friendship began to develop and was deepened by
repeated visits to McCormick's home at *Hecla Villa*, Wimbledon. Six years later,
shortly after McCormick's death, Sharpe received into the museum 142 birds
and eggs from the Falkland Islands and Antarctic seas—the surviving remnants
of the old explorer's small personal collection of duplicates. Many of the other
birds he had shot had long before been deposited in museums as part of the
official natural history material from his various naval expeditions.

The circumstances of capture of some of the birds are outlined in McCormick's
memoirs, which were published at his own expense in two thick volumes entitled
Voyages of Discovery in the Arctic and Antarctic Seas and Round the World

McCORMICK. *ROYAL NAVY SURGEON AND INVETERATE BIRD SHOOTER. HE MADE VOYAGES TO THE ARCTIC AND ANTARCTIC WITH JAMES CLARK ROSS.*

(1884). Divided into four sections, the first describes his part as surgeon and naturalist on James Clark Ross's four-year Antarctic voyage with the ships *Erebus* and *Terror*; the second outlines an earlier voyage to the Arctic that he made under William Parry; the third concentrates on the small contribution he made to the search for the missing Franklin expedition; and it is concluded by a rather disjointed autobiography detailing some of the other periods of his life. From the text it is evident that his passion for birds started in early childhood and continued up to the time of writing.

He was born during the Napoleonic wars on 22 July 1800 at the village of Runham, in Norfolk. He began his 'ornithological career' by blasting an inoffensive Dunnock out of a hedge with an old ship's pistol taken from a prize ship by his Irish-born father, a naval surgeon. The boy soon graduated to better weaponry and, as far as he could recall, the first bird that he shot on the wing was a Grey Partridge. He claimed that the first eggs he found were those of the Yellowhammer—so beautiful with their delicate scribbled markings that they have inspired generations of young egg collectors.

His father died before peace came in 1815 but the young lad was eager to follow his example and studied at Guy's and St Thomas's Hospitals. He entered the navy in 1823 as an Assistant Surgeon and was sent to Jamaica to fill up one of a number of vacancies caused by death from tropical diseases. He immediately detested the humid unhealthy conditions in the West Indies and after two years managed to get himself invalided back to Britain. In 1827 he volunteered for Arctic service with Captain Parry on H.M.S. *Hecla* in an unsuccessful attempt to reach the North Pole. During the voyage he made friends with James Clark Ross, who was then a young lieutenant of about the same age, and at Spitzbergen he spent as much time as possible ashore hunting reindeer.

On his return, McCormick was promoted to Surgeon and posted to the West Indies again. After only three months he succeeded in getting back to home waters and spent 1828 on blockade duty off the Dutch coast. Much to his disgust his new ship was sent to the West Indies and he found himself there for a third time. As usual he occupied some of his leisure hours by strolling around on shore with his shotgun, by searching for birds' nests and tending to captured birds that he sometimes kept on board. Mostly he lived for the days when he could be sent somewhere cooler and he invalided himself home again. The subsequent three years after May 1831 "were spent in two small miserable craft, and for the most part on my old station the West Indies, where I had already suffered so much from the climate and other depressing influences; which I can only look back to with unavailing regret, as so much time, health and energies utterly wasted."[2]

Although McCormick made a habit of complaining to the naval authorities about his postings and lack of promotion, he was given opportunities that a better scientist could have made better use of. His most obvious chance was his appointment in 1831 as surgeon-naturalist to H.M.S. *Beagle* under Captain FitzRoy, but these two famous names are very noticeably absent from his memoirs. Another name never mentioned is that of Charles Darwin, who joined the *Beagle* partly as a civilian gentleman companion to FitzRoy and partly as an additional naturalist. During a delay before leaving England, Darwin wrote to his friend and tutor Professor John Henslow: "My friend the Doctor is an ass, but we jog on very amicably: at present he is in great tribulation, whether his cabin shall be painted French Grey or a dead white—I hear little excepting this subject from him."[3] McCormick soon found that Darwin was in a very privileged position since he had no naval duties to perform and was allowed to spend the maximum amount of time ashore. They did enjoy collecting together in the Azores, and perhaps also at St Paul's Rocks, but by the time they reached Rio de Janeiro, in April 1832, McCormick had become thoroughly jealous and asked to be sent home. Writing once more to Henslow, Darwin noted that he was enjoying friendly terms with all the officers but "as for the Doctor, he has gone back to England.—as he chose to make himself disagreeable to the Captain & to [1st Lieutenant] Wickham. He was a philosopher of rather an antient date."[4] McCormick is not mentioned at all in Darwin's autobiography or in his account of the voyage and FitzRoy only records his name in the initial list of officers.

If only McCormick had been more tolerant and adaptable he could have achieved great things alongside Darwin at Tierra del Fuego, the Galapagos Islands and various parts of Australia and New Zealand. Even though he was unable to match Darwin intellectually he could have made a superb collection of birds. Instead his "excitable and sensitive character" caused him to fall out with his companions and he was transferred from the *Beagle* to another "small miserable craft" and came home in 1834.

McCormick had previously spent a six-month period at Edinburgh studying geology under Professor Jameson. He now passed four years on half pay travelling 3500 miles on foot through England and Wales indulging his interests in natural history and geology. In between excursions he repeatedly asked John Barrow, the Second Secretary to the Admiralty, if there were any more polar trips in the offing. At last, in September 1839, he was at sea again bound for the Antarctic. He spent four years in the southern oceans with temporary bases in Tasmania, New Zealand and Tierra del Fuego. Each year, for three summers, a fresh expedition was undertaken, Captains Ross and Crozier leading the *Erebus* and the *Terror* further south than any previous navigators, discovering new land and charting large portions of the southern hemisphere. Since there was rarely enough sickness on the ships to fully occupy the surgeons, McCormick often whiled away the tedious hours on board the *Erebus* by watching the seabirds. On the first summer voyage he was greatly impressed by his first Snow Petrels and soon devised a scheme to get one, which he vividly described in his memoirs, in his own inimitable style typical of many another entry:

"Being anxious to secure an early specimen of this rare and beautiful bird for the collection, I seated myself in the galley, on the port-side of the quarter-deck, with my old double-barrel gun in my hand, and during the forenoon watched for an opportunity when the bird was hovering over the mast-head, well to windward, so that when shot dead it might fall on board. After a little practice, by taking into calculation the force of the wind and velocity in the flight of the bird, I became very successful in thus bagging my specimens, the eye and hand soon acting in concert, in estimating the angle at which the bird should be fired at, to secure its falling on board. At the fourth shot I had the satisfaction of examining this lovely bird in my hand, it having fallen dead on the taff-rail to leeward, striking against the mizzen-trysail in its descent."[5]

A few days later the expedition's first penguins were shot, Ross allowing a boat to be lowered to retrieve them; they proved to be Adelie Penguins (named already for the wife of the French explorer Dumont d'Urville). On 12 January two boatloads of men, including McCormick, went ashore on Possession Island (opposite Cape McCormick) to claim Victoria Land for the Queen. For the rest of the voyage McCormick seems to have wandered about permanently armed, repeating his earlier feats of marksmanship from the decks, or ashore, whilst securing specimens or food for the crew.

After their return to England some of his work was incorporated into the *Zoology of the Voyage of H.M.S. Erebus & Terror, during the years 1839 to 1843* (1844–75). The sea-mammals were written up by J.E. Gray, the fish by Dr John Richardson and the birds by G.R. Gray. Gray's twenty-page account entitled *1. Birds of New Zealand* (1846) concerned about a hundred species from New Zealand (including the Chatham and Auckland Islands) most of which had already been described by others. There is no mention of McCormick or much about the work of the expedition. When Bowdler Sharpe updated the bird list in 1875 he managed to stretch a point to include ten discoveries made by the explorers elsewhere in their travels but it only served to emphasize the disappointing results. It was most unfortunate that Ross did not take along a better zoologist so that all the bird work could have been included in one volume by one person. Ross's assistant surgeon, Joseph Hooker, managed it well for the botany by producing the *Flora Antarctica* (1847). And it is interesting to note that Hooker considered Ross and himself the only real scientists aboard: "McCormick and I are exceedingly good friends, and no jealousy exists between us regarding my taking most of his department: indeed he seems to care too little about Natural History altogether to dream of anything of the kind; ... He takes no interest but in bird shooting and rock collecting; as of the former he has hitherto made no collection, I am, *nolens volens*, the Naturalist ..."[6]

Despite his shortcomings, McCormick compiled an acceptable geological appendix for Ross's *Voyage of Discovery and Research in the Southern and Antarctic Regions* (1847); he eventually formed a substantial bird collection; he gave the world the first description of the Wandering Albatross on its breeding grounds; he helped secure the first Emperor Penguins; he gave John Gould notes on the habits of Antarctic birds; and he provided an entertaining alternative account of Ross's great voyage that is not nearly so tedious as non-ornithologists would have us believe. He no doubt knew his limitations and concentrated on what he was best at.

McCormick continued his naval duties with a few uninteresting appointments in home waters until the Franklin expedition disappeared whilst searching for the North-west Passage. McCormick approached the Admiralty with plans for a search and rescue operation, but his ideas were considered too dangerous despite the support of Lady Franklin and his friend Admiral Beaufort. Instead he managed to get a position on the search ship *North Star* from which he was allowed to make a three-week reconnaissance in an open boat; no less successful than all the other more elaborate endeavours. He wrote it all up in his *Narrative of a Boat Expedition up Wellington Channel in the year 1852* (1854).

He now stepped up his earlier efforts to win promotion and in 1859 was reluctantly made Deputy Inspector of Hospitals. He soon tried for the honorary position of Inspector of Hospitals but this was refused and he was placed on the retirement list in 1865. Sir Clements Markham, writing many years later, recalled his impressions of him: "old McCormick I knew well too in his stiff satin stock. The old chap was eaten up with conceit and a sad romancer. What he says must be taken *cum grano*. He assuredly considered himself above 'the ordinary run' and I think he was."[7] In his last lonely years McCormick devoted much time to writing up his memoirs. He doted on a tame white duck and enjoyed the company of a free-living hen House Sparrow which he kept around his home for seven years, the two of them often taking their meals together.

Saved until last are the details of Robert McCormick's connection with the South Polar Skua *Catharacta maccormicki*. The bird breeds in the Antarctic but regularly wanders into the northern Pacific and Atlantic Oceans, usually from May to early November when it is most often seen from ships well out at sea off both the west and east coasts of North America. The type specimen was collected in 1841 by McCormick on the day he scrambled ashore onto Possession Island with Ross, Crozier and two boatloads of men. As the weather steadily worsened, the Union Jack was unfurled, the land was claimed for Great Britain, and a toast was drunk to the health, long life and happiness of Queen Victoria and Prince Albert (sherry for the officers, rum for the men). In between times, during the twenty-five minutes on land, McCormick snatched up a few lumps of black rock, bashed an old penguin on the head with his geological hammer and from above the vast colony of penguins shot the new species of skua and stuffed it into his haversack.

South Polar Skua

John Porter McCown (1815-1879)

McCOWN'S LONGSPUR *Calcarius mccownii* (Lawrence)

Plectrophanes McCownii Lawrence, 1851. *Annals of the Lyceum of Natural History of New York* 5, p. 122: high prairies of Western Texas

John Porter McCown was born near Sevierville in eastern Tennessee on 19 August 1815 and grew up to be much interested in art, poetry and literature. Whether he was interested in birds during his early years is unrecorded but it is quite likely that he was. He chose an army career, entered West Point in 1835, graduated in 1840 and as part of his first military duties assisted in the transportation of Indians to Indian Territory. He then served at Detroit in the Canadian border disturbances and from 1841 to 1842 was at Buffalo, New York. In September 1843 he became a 1st Lieutenant in the 4th Artillery, shortly afterwards serving in the military occupation of Texas and in the subsequent war with Mexico, 1846–47. He saw action in the battles of Palo Alto, Resaca de la Palma, Monterrey, the seige of Vera Cruz, Cerro Gordo and at the seige and capture of Mexico City under General Scott. After the battle of Cerro Gordo he was brevetted Captain for his "gallant and meritorious conduct".

As a soldier he travelled through much of the south-western United States and eastern Mexico, mentally noting the birds he saw but not often able to follow up or write down his sightings. Frontier duty that offered sufficient leisure for serious recording eventually coincided with an area blessed with a superabundance of birds when McCown was posted to the lower Rio Grande in Texas. (For a map see p. 498.) In 1849 he was at the Ringgold Barracks (Rio Grande City) and in the following year moved 100 miles downstream to Fort Brown (Brownsville) at the mouth of the river. Both places are excellent for birds during the spring migration and are also noted for several Mexican species that do not occur elsewhere in the United States. McCown sent many of his skins to George N. Lawrence, who published descriptions and notes on the more interesting ones in the *Annals of the Lyceum of Natural History of New York*. In the first of these papers, read on 28 April 1851, Lawrence correctly described the Olive Sparrow as new to science but his descriptions of the Verdin and Orchard Oriole had been preceded by others. The second part of the paper

McCOWN. *CONFEDERATE GENERAL. BEFORE THE CIVIL WAR HE COLLECTED BIRDS ON THE LOWER RIO GRANDE AND IN WESTERN TEXAS.*

detailed eleven species collected by McCown that were new additions to the United States avifauna, notably the Great-tailed Grackle, Green Jay, Vermilion Flycatcher, Pyrrhuloxia and Black-bellied Whistling-Duck.

In another paper, read on 8 September 1851, Lawrence described three clear-cut novelties: Le Conte's Thrasher, supplied by John L. LeConte from Fort Yuma, California; and two species supplied by McCown from Texas: the Ash-throated Flycatcher and McCown's Longspur. Early in 1851 McCown had been on garrison duty at Fort Hamilton, New York, but later in the year was escorting recruits through San Antonio before returning to New York State. The single flycatcher skin was taken near a small watercourse somewhere between San Antonio and the Rio Grande. The longspur was found by chance when McCown fired at a flock of Horned Larks and was surprised to find the new species amongst those killed. Lawrence was rather vague about where it happened:

> "Two specimens were obtained by Capt. McCown on the high prairies of Western Texas. When killed, they were feeding in company with Shore Larks. Although procured late in the spring, they still appear to be in their winter dress; in summer, I have no doubt they assume the gay and ornamented plumage of their congeners."[1]

The longspurs were wintering in Texas after breeding in the north, perhaps in Montana or north-eastern Colorado. McCown knew nothing about their summering areas, nor did he know much about his Ash-throated Flycatcher or Olive Sparrow since they too were chance finds and not birds that he ever became familiar with. This is not to say that he was not a good observer; on the contrary, he was an accurate and honest recorder and an entertaining writer. However, the only ornithological publication that we can trace under his name appeared in the aforementioned *Annals* for 1853 under the title 'Facts and Observations from Notes taken when in Texas'. Eighteen bird species are mentioned including the Burrowing Owl, Red-billed Pigeon, Hooded Oriole, Plain Chachalaca, Greater Roadrunner, Vermilion Flycatcher and Olive Sparrow. Some accounts, such as the one for the Hooded Oriole, are extremely brief: "Common on the Rio Grande. Shy in the woods, yet seemed familiar when in our camp, where we often saw them." For other species he gave much more

McCown's Longspur, with two Horned Larks

detail, so that for the Greater Roadrunner only the following portion will be quoted:

> "Often in my wanderings through the chaparrals on the Rio Grande, I observed piles of broken snail shells, and always near some hard substance, such as a bone or hard piece of wood, which had evidently been used in breaking the shells. I made many conjectures as to the probable animal. I never suspected a bird, that had left these evidences of their peculiar habits. I heard at times—generally in the morning or evening—a sound very similar to that made by some woodpeckers by a rapid beating of their bill upon an old dry tree. This was also a mystery, as I could find no woodpeckers near the place the sound came from. Upon enquiry of a Mexican, I was told that it was the Paisano breaking the snail shells to get at the snail, which explained at once to me both the noise, as well as the shells. I was afterwards so fortunate as to see a bird so occupied. It took the snail in its bill, and beat it upon the hard substance, striking faster and faster, until the shell broke."[2]

Another peculiar bird known well to McCown was the Plain Chachalaca, which he considered to be abundant all the way from the Rio Grande to the battlefield of Cerro Gordo. He found them "exceedingly noisy, both in the morning and evening", their shrill cry being responsible for their onomatopoeic common name. He added that "There is found under the feathers of the Chiac-ka-laca a fly, about the size of a common house fly, but very flat. This fly is exceedingly annoying, for as soon as you commence handling the bird they leave it, and are just as likely to get into the hair of your head or down your neck as to return to the bird or seek a new abode with some living individual of the chiac-ka-laca tribe. This fly is so hard that I could with difficulty kill it between my thumb and forefinger."[3] These louse flies are common parasitic flies familiar to anyone who handles wild birds yet, with the exception of Alexander Wilson and McCown, very few early collectors ever mentioned them. Living amongst and beneath the feathers they behave just as McCown described, though they are more easily killed by rubbing them with a sideways motion.

When John Cassin wrote his *Illustrations of the Birds of California, Texas, Oregon, British and Russian America* (1853–56) he not only quoted from Lawrence and McCown's short papers but also incorporated other notes personally directed to him by McCown. This time Captain McCown expanded a little more generously about the Hooded Oriole:

> "This beautiful Oriole is quite common on the Rio Grande, where it raises its young. When met with in the woods, and far away from man's abode, it is shy, and seems rather disposed to conceal itself, yet a pair were constant visiters, morning and evening, to the vicinity of my quarters (an unfinished building at Ringgold Barracks, Texas). They became so tame and familiar that they would pass from some ebony trees that stood near by, to the porch, clinging to the shingles and rafters, frequently in an inverted position, prying into the holes and crevices, apparently in search of such insects as could be found there, which, I believe, were principally spiders. They would sometimes desist for a moment from this occupation, to observe my movements, and if I happened to be enjoying a cigar after dinner, seemed to watch the smoke with great curiosity. I often offered them such hospitality as was in my power, but could never induce them to touch any food ...".[4]

After his tour of duty in Texas, McCown was sent to North Carolina and New York and seems to have found less and less time for birds as his responsibilities increased and as national events took precedence. After serving in the Seminole Wars in Florida from 1856 to 1857 he was posted to Fort Leavenworth, Kansas. Since he is reported as being on the Utah Expedition of 1858 he must have been with Colonel Johnston's army when it entered Salt Lake City to quell an alleged Mormon rebellion. Later that year McCown was at Fort Kearney, Nebraska, and from 1859 until 1861 he served at Fort Randall, South Dakota.

When the Civil War began he resigned at once to join the Confederate Army as Lieutenant Colonel of the Tennessee Artillery Corps. In October he was promoted to Brigadier General and in March 1862 became a Major General. After his premature evacuation of New Madrid his military competence was called into question and General Beauregard described his action as "the poorest defense made by any fortified post during the whole course of the war."[5] Despite the criticism McCown was given temporary command of the Army of the West in June. Later, while serving with the Army of Tennessee, General Braxton Bragg (who became a personal enemy) thought him his worst divisional commander but still selected him to begin the battle of Stone's River at the end of December. When things went wrong Bragg ordered McCown to be court-martialled for disobeying orders and he was suspended from rank, pay and other advantages for six months. McCown vented his feelings by proclaiming the Confederacy "a *damned* stinking cotton oligarchy ... gotten up for the benefit of [Tennessee governor] Isham G. Harris and Jeff Davis and their damned corrupt cliques."[6] This was not the sort of comment to endear him to his superiors and he passed much of the remainder of the war in relative obscurity in Mississippi. He had a final moment of glory in the last days of the war when he successfully defended a crossing of the Catawba River near Morgantown, North Carolina, with only a single piece of artillery and three hundred men against a division of Union cavalry. When peace came he returned to his home area to teach at a school near Knoxville.

In 1868 McCown went to the new town of Magnolia in south-west Arkansas to visit his brother, Judge G.W. McCown, and decided to settle there. He bought a few acres just outside the city limits on which he planted a variety of trees and flowers as well as the fruit trees from which he derived some income. He lived there quietly and though he never married he had a wide circle of cultivated friends who often called on him at the farm. One room of the house was reserved for relics from the past which included firearms from different wars, spurs said to have belonged to Santa Anna, a Confederate flag and a collection of shells. There were also many rare books of which an original edition of Audubon's *The Birds of America* was given pride of place.

At the age of sixty-four McCown was suddenly taken ill with pneumonia while attending a Mason's meeting in Scott Street House, Little Rock, and he died a week later, on 22 January 1879. His body was taken back to his brother's house from whence it was borne to the quiet wooded cemetery of Magnolia as the band played 'Dixie'.

Peter McDougall[1][†]
(1777-1814)

ROSEATE TERN *Sterna dougallii* Montagu

Sterna Dougallii Montagu, 1813. *Supplement to the Ornithological Dictionary*, text and plate to "Tern-Roseate" [not paged]: Cumbrae Island, Firth of Clyde, Scotland

The Roseate Tern is chiefly a warm-water species which breeds along the coasts of eastern North America, the Caribbean and north-western and southern Africa; other races frequent the Red Sea, the Indian Ocean, the Pacific and the coasts of Australasia. Because Scotland is at the very northernmost extent of the tern's breeding range, it is not thought of as a characteristically Scottish species and its existence was not even suspected until 1812 when Dr Peter McDougall discovered the tern breeding in the Firth of Clyde.

In 1772, or perhaps a year or two before, Alexander McDougall, a merchant in Kilsyth, married Mary Jeffray. Their first child, Anne, was born in July 1773 and thereafter the McDougalls had six more children of whom Peter was the third child and second son. He was born on 18 January 1777 and grew up in and around Kilsyth, a small town which was then becoming prosperous because of the weaving trade and which was conveniently situated some ten or so miles north-east of the centre of Glasgow. In 1789 Peter matriculated at the University, entered the humanity class and finally graduated as M.D. in 1802. Two years later he was to be found in central Glasgow in practice with another physician, called Barrie. The partnership lasted from 1804 until 1809 when McDougall became one of the surgeons at the newly built Glasgow Royal Infirmary. Because of a disagreement over his licence to practise as a surgeon he was not re-appointed in the following year and he returned to his medical practice, now without his former partner. Once again McDougall lived at various addresses immediately to the south of George Square, right in the heart of the city.

On 24 July 1812 he was with two friends on the island of Great Cumbrae, in the Firth of Clyde. Together they went out to two small islets in Millport Bay, each of which is not much more than a hundred yards across; although

†No known portrait

generally rocky around the edges they are sufficiently well vegetated to suit the requirements of Roseate Terns. These birds no longer breed there yet when McDougall visited the islets he reported that he could scarcely walk about for fear of treading on tern eggs. As the parent birds flew excitedly overhead, one of his companions shot one of them and it landed close to McDougall. He picked it up, noticed it was different from the other terns and asked his friends to shoot some more while he himself began to observe the flying terns more closely:

> "The new species was discerned by the comparative shortness of wing, whiteness of plumage, and by the elegance and comparative slowness of motion; sweeping along, or resting in the air almost immoveable, like some species of Hawk; and from its size being considerably less than that of *Sterna Hirundo* [Common Tern]."[2]

Unknown to McDougall, Arctic Terns were probably also present, but the distinction between the Common and Arctic Tern was not made widely known until 1819, seven years after his visit. Several of the Roseate Terns were added to his collection and he mounted three of them, two males and a female, in a glass case together with a 'White House Swallow'. McDougall's friend Captain Laskey soon informed George Montagu of this new addition to the British avifauna and arranged for a surplus specimen to be sent to Devon. Montagu was thus able to publish the first description of the tern, giving it the scientific name *Sterna Dougallii* in order "to make our public acknowledgements to Doctor M'Dougall, for the very liberal and handsome manner in which the history of this interesting bird was communicated to us, and more particularly for the specimen that accompanied it".[3] It later transpired that the bird which Montagu received was not McDougall's best example as the doctor later admitted that "the figure in the Supplement to Colonel Montagu's Dictionary was taken from a specimen, which was wounded in the neck, and, in consequence, the bird was placed in an unnatural position to hide the defect".[4] Montagu's *Supplement* was issued in 1813; only a year later, at the age of thirty-seven, McDougall was dead.

While carrying out his duties as a doctor he caught a fever, which may have

Roseate Tern

been typhus, and he died on 25 April 1814. The Glasgow papers announced his death with the brief epitaph: "His talents were but beginning to be known—his worth was known to many." He was buried at the cemetery in Ingram Street, which is now much neglected with no obvious memorial to the doctor. McDougall must have died unmarried because on 12 May his entire household contents were offered for sale. They included carpets, rugs, beds, bedding, grates, fire irons, mahogany tables and chairs and kitchen furniture as well as an "electrifying machine" and various natural history curiosities. One such item which was not sold was the "horn of a sea unicorn" (narwhal tooth), which someone, somehow, managed to steal and take away unobserved.

McDougall's fine collection of British and foreign birds was dispersed by auction, two months later, on 15 July at Mr Angus's Academy in Ingram Street. The 101 foreign birds, mounted in twelve glass cases, consisted mainly of colourful species such as toucans, parrots and hummingbirds. Most of these came from the West Indies and South America while other birds were from India and North America, but there is no evidence that McDougall collected any of them himself; he lived in one of the most prosperous parts of Glasgow surrounded by wealthy merchants who sent ships from the Clyde to all parts of the world. However, the 150 British birds which were in twenty-seven cases, along with more than seventy-five other uncased birds, give the indication that they had been collected locally, a good proportion probably by McDougall himself. Most of the birds of prey were represented, including a White-tailed Sea-Eagle, and there were many different species of owls, ducks, seabirds, finches and thrushes as well as his case of Roseate Terns. The only species in the British section not regularly occurring in Scotland was the Hoopoe. One extra case contained both British and West Indian birds.

In addition, amongst the miscellaneous assortment of other items on sale, there were three snakes' skins, two snouts of swordfish, an armadillo, an alligator, a bat from Botany Bay, a chest of drawers full of West and East Indian shells, thirty-two numbers of George Graves's *British Ornithology* and five volumes of John Latham's *General History of Birds*. An assortment of over 200 small glass eyes clearly showed that McDougall had done much of his own taxidermy.

William Bullock may have been one of the natural history dealers who attended the sale since McDougall is mentioned as a 'donor' in one of Bullock's catalogues dated 1816, although of course McDougall may have sent him birds long before the sale took place. Apart from the type specimen Roseate Tern which Montagu described and which is now in the possession of the British Museum, it seems most unlikely that any of McDougall's bird specimens still exist.

Since the death of Peter McDougall his name has been almost forgotten. This must be partly the fault of Montagu who, when naming the tern, left out the prefix 'Mc', so that the person honoured would appear to have been named 'Dougall'. Indeed, the Dutch and French names for the Roseate Tern are, respectively, 'Dougall's Stern' and 'Sterne de Dougall'. In 1842 William MacGillivray patriotically tried to correct the omission by naming the species 'MacDougall's Tern', but with no lasting effect.

William MacGillivray (1796-1852)

MacGILLIVRAY'S WARBLER *Oporornis tolmiei* (Townsend)

Sylvia Tolmiei J.K.Townsend, 1839. *The Narrative of a Journey Across the Rocky Mountains ... &c.*, p. 343: the Columbia [= Fort Vancouver, Washington]

In *The Birds of America* Audubon named two species after William MacGillivray but his desire to pay a lasting tribute to his good friend nearly foundered. The bird he called MacGillivray's Shore Finch had already been described and is now known as the Seaside Sparrow. His MacGillivray's Warbler had been described a few years earlier by John Kirk Townsend, who named it *Sylvia tolmiei* after Dr William Tolmie. Townsend's scientific name has precedence but Audubon's vernacular name has persisted despite some opposition from western birdwatchers who resented the eponym being applied to a man who had never even set foot in North America.

From Audubon's point of view there was good reason to honour MacGillivray as they had laboured together for nine years on the *Ornithological Biography* (1831–39), the text for Audubon's illustrations which were bound together separately as *The Birds of America* (1827–38). In October 1830, when he was about to begin work on the *Ornithological Biography*, Audubon wrote in his Edinburgh journal:

"I know I am a poor writer, that I scarcely can manage to scribble a tolerable English letter and not a much better one in French, though that is easier to me. I know I am not a scholar, but, meantime, I am aware that no man living knows better than I do the habits of our birds; no man living has studied them so much as I have done, and, with the assistance of my old journals and memorandum books, which were written on the spot, I can at least put down plain truths which may be useful and perhaps interesting, so I shall set to at once. I cannot, however, give scientific descriptions and here must have assistance."[1]

By the end of the year he had engaged MacGillivray to supply that assistance as scientific and general editor and he could hardly have found a better qualified or more conscientious helper.

MacGILLIVRAY. *SCOTTISH ORNITHOLOGIST, LATTERLY PROFESSOR OF NATURAL HISTORY AT ABERDEEN UNIVERSITY. HE WAS THE AUTHOR* OF RAPACIOUS BIRDS *AND* HISTORY OF BRITISH BIRDS *AND CO-WROTE AUDUBON'S* ORNITHOLOGICAL BIOGRAPHY.

Eleven years younger than Audubon, MacGillivray was born on 25 January 1796 in Aberdeen, on the north-east coast of Scotland. At the age of three he was taken to the Outer Hebrides to live with an uncle who was tacksman of the large farm of Northtown, near Obbe, in the south of Harris, since at that time MacGillivray's father was serving as a surgeon with the Cameron Highlanders. During those formative years in Harris young MacGillivray loved wandering by himself along the sandy beaches and in the windswept hills, but at the age of eleven he was sent off to Aberdeen to continue his education. After only a year at school there he entered the university—not then unusual at such a young age—and after following the arts curriculum and taking his M.A. he began to study medicine, working for a time as dissector to the lecturer on comparative anatomy at Marischal College.

His father was killed at the battle of Corunna in 1809 so MacGillivray continued to return to Harris each year for the long summer holidays, always walking from Aberdeen across the mainland to the west coast. In 1816 he made the return journey in September via Blair Atholl and upper Deeside, in order to investigate their plant communities, and the next year he took up zoology. By the age of twenty-one, MacGillivray had realized that he wanted to be a naturalist, not a doctor. At the beginning of August he left Aberdeen, planning to spend a year in the Hebrides.

When he arrived at Northtown he found that he was not the only visitor there. His uncle had recently married Mary MacAskill from North Harris and her younger sister, Marion, was staying for a time. Three years later she became William's wife. In the meantime there were other distractions, for his medical experience, limited though it was, was often sought by the locals and he also noted in his journal that he had been an enthusiastic participant at all the ceilidhs and other social events. Nevertheless he still spent a good deal of time studying the natural history of the islands. At the end of December he heard that a Walrus had been killed on the east coast of Harris and he arrived there in time to measure the corpse and tusks before the islanders cut it up for meat and blubber. For MacGillivray it provided the information for his first publication, which appeared in the *Edinburgh Philosophical Journal* of 1820. During this period he began to consider the possibility of some day writing a book about all of Scotland's birds.

In the autumn of 1819 MacGillivray walked from Aberdeen to London and arrived there footsore and tattered, having walked, by his own reckoning, over 830 miles. After collecting a parcel of clean clothes that he had sent ahead, he spent a week in the city, mostly at the British Museum where he studied the collection of British birds. He then took a steamboat back to Aberdeen.

In September 1820 he and Marion MacAskill (then aged seventeen) were married in Harris and their son John, the eldest of thirteen children, was born in Aberdeen in December 1822. The following year the family moved to Edinburgh, where MacGillivray became secretary and assistant to Professor Jameson and keeper of the Edinburgh University Museum. After a few years he resigned and supported himself by writing, translating and lecturing so that he could spend more time in the field—and there were many rewarding hot spots within easy reach of the city. In 1831 he returned to the indoor life when he was appointed Conservator of the Museum of the Edinburgh College of Surgeons and worked there for the next ten years.

When he contracted to work with Audubon at the end of 1830 the two naturalists had much in common. MacGillivray had explored his native countryside and its wildlife with as much enthusiasm as the American had his own backwoods, creeks and marshes. But the Scot also had a thorough education and a broad knowledge of science which Audubon lacked. In editing the *Ornithological Biography* he was able to appreciate Audubon's vivid accounts, correcting his still imperfect English only as much as was necessary; yet he was also able to add the essential detailed descriptions of each species as well as ninety-eight anatomical drawings. One wonders though if his careful drawings and descriptions of digestive tracts really interested any of the subscribers.

That winter Audubon stayed closeted in his rooms in George Street, scribbling furiously day after day, often from four in the morning until after ten or eleven at night; his wife, Lucy, made whatever corrections she could, then passed the manuscript on to MacGillivray to complete. The first volume went to press in February of the new year and in August the Audubons sailed to the United States because more specimens and more paintings were needed for *The Birds of America*. During the next three years the artist collected in Florida, South Carolina and the Gulf of St Lawrence, so delaying the second volume of the *Ornithological Biography* until he returned to Scotland.

During the next two years Audubon was able to write a little less frenetically and found time to accompany MacGillivray on some of his outings. In the middle of August 1835 they went to the Bass Rock in the outer reaches of the Firth of Forth, where the precipitous cliffs, crowded with Gannets, reminded Audubon of his day at Bird Rocks. After following the path to the summit they returned to the boat and rowed round the island, shooting at the Gannets and Kittiwakes, though Audubon was disappointed that the Cormorants had all been shot out in the spring.

In September 1836 Audubon returned to America, where he made the important acquisition of many west coast birds collected by Nuttall and Townsend. In the spring he sailed along the Gulf Coast with Edward Harris and in the summer crossed the Atlantic with Lucy, their two sons and his daughter-in-law Maria. For the next year he worked on *The Birds of America*

MacGillivray's Warbler

in London—where it was being engraved—and then went up to Edinburgh in the late summer. With MacGillivray as their guide the family had a week-long holiday touring the Trossachs and Loch Lomond, returning via Glasgow and the Falls of Clyde. 1838 saw the appearance of the final volume of *The Birds of America*. The *Ornithological Biography* was completed in 1839 and that same year *A Synopsis of The Birds of North America* was published as an index to both works, largely compiled by MacGillivray. In the autumn Audubon returned to the United States for the last time and MacGillivray was greatly saddened by his departure. Their friendship had been deep and lasting and MacGillivray, who made few close friends, valued those he had.

Now MacGillivray was able to devote more time to his own projects. While working at the Royal College of Surgeons, and on the *Ornithological Biography*, he had produced a number of publications of his own: in 1832 *The Travels of A. von Humboldt*; in 1834 *Lives of Eminent Zoologists, from Aristotle to Linnaeus* (both part of the Edinburgh Cabinet Library) and in 1836 *Descriptions of the Rapacious Birds of Great Britain*. The last-named was dedicated to John James Audubon "In admiration of his talents as an ornithologist, and in gratitude for many acts of friendship". It revealed MacGillivray's familiarity with the Scottish raptors, particularly the Golden Eagle, White-tailed Sea-Eagle, Peregrine Falcon, Merlin, Kestrel and Sparrowhawk—and also his blunt honesty. He readily admitted ignorance of many British species and quoted other naturalists, such as Montagu, Temminck and Audubon, giving them full credit when it was due.

The first part of MacGillivray's five-volume *A History of British Birds* (1837–52) had been published in 1837. This was his 'great work', as he called it and it took up much of his energy during the last fifteen years of his life. Like the *Rapacious Birds* it contained much in the way of interesting first-hand observations, but it was not a success and criticism of it deeply hurt him. The reasons for its failure were many: his nomenclature was unpopular; his unorthodox classification was based on the structure of the digestive organs; it was produced over fifteen years (so some of the original purchasers lost interest) and it had to compete with William Yarrell's work of the same name, which first appeared in the same year but was completed in less than half the time and contained better illustrations.

In 1841 MacGillivray returned to Aberdeen University and became Professor of Natural History at Marischal College. In the winter he taught zoology, in the summer botany, his lectures proving so popular that many students from other disciplines came to hear him. He took them out for long rambles and easily walked them all into limp helplessness. Many years later, one of his students remembered him well:

"Under medium height, spare in form, shy and reserved in manner, he walked swiftly along the street, generally alone, with his head inclined downwards and his eyes bent towards the ground, wrapt in his own thoughts. Celt of the Celts, he was singularly courteous and polite, with fine quiet dignity, but when offended he could use sharp words which left their sting".[2]

Now that he was back in the north-east, MacGillivray rekindled his love for the Grampian Mountains and the valleys of the Don and the Dee. In the late summer of 1850 he spent nearly two months walking in Deeside, following the river westwards through Aboyne, Ballater, Balmoral and Braemar, right up to its source between Cairn Toul and Ben MacDui. His eldest daughter, who shared his botanical interests, joined him for a time, as did his son Paul, but often he walked alone, jotting down botanical and geological notes, descriptions of the scenery and records of the birds of the high tops such as Dotterel, Ptarmigan and 'Brown Ptarmigan'—his name for the Red Grouse. These notes formed the basis of *The Natural History of Deeside*, which was published by the command of Queen Victoria three years after the naturalist's death.

Throughout his life MacGillivray frequently slept out in the hills at night with little protection apart from some extra clothing, but this last excursion seriously strained his health. During the following term he became so ill that he had to take a prolonged period of leave. To avoid the bitter winter in Aberdeen he went to Torquay in the south of England, but while he was there his wife died unexpectedly at home. In the spring he moved back to Aberdeen and completed the last volume of *A History of British Birds*, but he had not fully recovered and died at the beginning of September, at the age of fifty-six. He was buried in Edinburgh beside his wife and two of his children in the New Calton Cemetery, which lies between Calton Hill and Arthur's Seat.

MacGillivray's Petrel *Pterodroma macgillivrayi* (G.R. Gray) is not named after William MacGillivray, but for his eldest son John who served as naturalist on the voyage of H.M.S. *Herald* to South America and the South Pacific, between 1852 and 1855. The new petrel was discovered during the course of the expedition by Dr Rayner on Ngau Island, in Fiji, and this specimen is still the only skin in scientific collections. John MacGillivray left the *Herald* at Sydney and spent the rest of his life visiting various Australasian islands, studying the behaviour of the aboriginal peoples and the indigenous wildlife. He died at Sydney in 1867. His brother Paul also settled in Australia, at Bendigo in Victoria. At first he worked as a surgeon, but later devoted his time to gathering a large collection which was acquired by the Museum of Natural History in Melbourne. Paul inherited his father's journals, but all except two volumes (covering August 1817 to May 1818 and the journey to London in 1819) were destroyed by fire.

Charles Leslie McKay
(1855-1883)

McKAY'S BUNTING *Plectrophenax hyperboreus* Ridgway

Plectrophenax hyperboreus Ridgway, 1884. *Proceedings of the United States National Museum* 7, p. 68: St. Michael's, Alaska

When Robert Ridgway described McKay's Bunting in 1884 he named it "in memory of Mr. Charles L. McKay, who sacrificed his life in the prosecution of natural history investigations in Alaska, and in whose collections the new species was first noticed".

Charles Leslie McKay was raised on a Wisconsin farm near Lake Butte des Morts, to the west of Lake Winnebago. Born 21 April 1855, his horizons extended no further than the nearby city of Appleton until 1874, when David Starr Jordan became principal of the Appleton Collegiate Institute where McKay and his two brothers were students. Jordan immediately recognized McKay's remarkable aptitude for zoology and took him along on field excursions. By the end of the year McKay was familiar with the names and habits of all the birds and most of the fishes along the Fox River. Encouraged by Jordan, he decided to go to college in preparation for a career as a naturalist. After a year or so of

McKay's Bunting

McKAY. *SON OF A SCOTTISH FARMER IN WISCON-
SIN, HE BECAME DAVID STARR JORDAN'S BEST
ZOOLOGY STUDENT. HE DIED IN ALASKA TWO
DAYS BEFORE HIS TWENTY-EIGHTH BIRTHDAY.*

farm work he entered the Natural History Department of Cornell University at Ithaca and during the summer vacations he worked in the wheatfields of western New York.

From 1878 to 1879 he again studied under Jordan, at Butler University, Indianapolis, and when Jordan moved to Indiana University at Bloomington, McKay went with him, graduating with honours as a Bachelor of Science in the class of '81. During the early months of that year he spent some time in Washington D.C. as an assistant to the U.S. Fish Commission and prepared his only publication, a 'Review of the Centrarchidae' in which he outlined his conclusions about the freshwater sunfishes (*Proceedings of the U.S. National Museum*, 1881).

At that time Professor Baird was responsible for choosing Signal Officers for the remoter stations, and he reserved such positions for young men with some scientific training who were willing to study the plants and animals as well as record the weather. McKay registered with the Signal Corps in Washington on 28 March 1881 (for a term of five years) and the records give the following details: age 25 years 11 months; place of birth Appleton, Wisconsin; occupation student; height 5 feet 11½ inches; complexion fair; eyes grey; hair brown. In June he left for San Francisco, then immediately embarked on the last part of his journey to Fort Alexander, now Nushagak, on the north side of Bristol Bay, Alaska. For the next two years he collected locally as much as his duties allowed and among the places he reached were Lake Aleknagit, Ugashik, Iliamna Lake and Iliamna Village.

McKay had been instructed to observe the indigenous peoples and part of a long letter that he sent to a college friend, recording some of his impressions, was circulated in the *Indiana Student* in November 1882. To Baird he sent native artefacts, minerals, fishes, 123 species of plants, 23 species of mammals and about 340 specimens of birds. No analysis of the birds has been published but in the collection were two of the co-types of McKay's Bunting, a male and female in winter plumage, that he had shot at Nushagak in November and December 1882. Ridgway compared them with another two specimens, in summer plumage, obtained by Edward W. Nelson at St Michael, on Norton Sound, where he also was serving as a Signal Officer.

McKay was eager to search for new birds in a little known region west of the Chigmit Mountains and in his last letter to Baird, written 26 January 1883, he confided:

"No white man has ever been there. It is very probable that many species of birds that are not found here will be found there, the warblers especially. The mountain sheep and little 'chief hare' are very abundant there, but it is impossible to get the natives that live around here to kill the latter as they have some superstition about it. The natives also say that there is a goat that lives in these mountains ...

I have no doubt that if I could spend one summer in that region, I could do good work there."[1]

He proposed handing over his meteorological duties to a willing friend for the duration of the expedition, but it was not to be. During the summer his father, Hector McKay, received the following letter:

*ALASKA: WITH PLACES VISITED BY ADAMS, COOK, KENNICOTT, KITTLITZ, McKAY,
STELLER AND C.H. TOWNSEND*

"It is my sad duty to inform you that your son left this place on the 17th of April,
to make a short trip for the purpose of making collections, and that he never
returned. He left in company with a native, each of them in a single canoe and
passed the night in an Indian village, sixteen miles from the station. The following
day was very stormy and they lay over in the village. On the morning of the third
day (19th,) it being calm weather, they left the village to cross over the bay, a
distance of twelve miles. They were accompanied from the village by three other
native canoes. When about two-thirds of the way across a strong adverse wind
sprang up. In some manner, he was left behind and that was the last that was seen
of him. On the 22d, the report reached me and the same day we began to search
for him. We found broken pieces of his canoe, a gun, his rubber boots, hat and

various little articles on the beach about a mile on this side of the village they left that morning. We continued the search for over three weeks, but could not find the body …

I can readily understand with what feelings you will receive this letter, and believe me that if the sympathy of a stranger can serve to mitigate your grief in the slightest degree, you have mine. Being my sole companion for two years, I had learned to appreciate him and to esteem his manly, upright character."[2]

The letter was written by John W. Clark, the agent of the Alaska Commercial Company at Nushagak. He may have told McKay's father everything he knew, but he may not have done. In 1902 Wilfred H. Osgood, an assistant with the Biological Survey, made a reconnaissance of the base of the Alaska Peninsula, sometimes travelling in McKay's footsteps. Between Lake Clark and Swan Lake he was guided by a native who had led McKay through the same pass twenty years earlier and he heard a rumour that McKay's death had been due to foul play. At Nushagak, Osgood lodged in the old log cabin that had been McKay's home and found some of his meteorological records as well as several pounds of arsenic (used for preparing skins) still lying on the shelves.

MASSÉNA. *DUCHESS OF RIVOLI AND PRINCESS OF ESSLING. THE PICTURE SHOWS THE EMPRESS EUGÉNIE SURROUNDED BY HER LADIES-IN-WAITING; ANNA MASSÉNA IS AT THE FAR LEFT, LOOKING AT EUGÉNIE.*

Anna Masséna, Duchess of Rivoli
(1802-1887)

ANNA'S HUMMINGBIRD *Calypte anna* (Lesson)

Ornismya Anna Lesson, 1829. *Histoire Naturelle des Oiseaux-Mouches*, p. "xxxj" [= xxxi]; 1830, p. 205, Pl. 74: La Californie [= San Francisco, California]

Anna's Hummingbird was discovered by Dr Paolo Botta, a young Italian surgeon, during the three-year round-the-world voyage of the French trading vessel *Le Héros*. In 1827 and 1828 *Le Héros* sailed up and down the Californian coast for more than twelve months. Since Botta's responsibilities allowed him plenty of time ashore he stayed at all the larger Spanish missions between Sonoma and San Diego, visiting some of them several times. Although his natural history collection was small it included the type of the Greater Roadrunner as well as that of Anna's Hummingbird.

Immediately after Botta returned to France in the summer of 1829 he gave some of the birds to Victor Masséna, Duke of Rivoli and Prince of Essling (1799–1863), who in turn showed them to his friend René Primevère Lesson. Lesson described the roadrunner in volume six of his *Complément des oevres de Buffon* (1829) and the hummingbird in his *Histoire Naturelle des Oiseaux-Mouches* (1829), calling the latter Oiseau-Mouche Anna after Masséna's wife. He explained that she shared her husband's interest in the collection and that he was indebted to them both for placing the birds at the disposal of visiting naturalists. Since neither of the type specimens carried labels he knew only that they had been acquired in 'La Californie', but Anna's Hummingbird is assumed to have been taken around San Francisco, since Botta collected there for longer than at any other place and, in his narrative of the voyage, mentioned that he had shot hummingbirds at San Francisco in February 1827.

It is remarkable that Victor also had a hummingbird named in his honour. In 1829 Lesson wrote about both Anna's and Rivoli's Hummingbirds but an earlier description of the latter had been provided by William Swainson so Lesson's scientific name of *Ornismya Rivoli* became obsolete. Victor Masséna was not just a wealthy collector; he took a serious interest in natural history and with his nephew, Charles de Souancé, described a number of conures and other parrots from South America. His title 'Duc de Rivoli' was won by his father, André Masséna (1758–1817), who was Napoleon's greatest general and a Marshal of France. He won battle after battle leading up to his crowning victory at Rivoli in northern Italy.

Much less is known about Anna and her family. She was the daughter of General Jean François Joseph DeBelle, who died in June 1802, the year of her birth, while serving as Commander in Chief of the French Artillery on Saint Domingue (Dominican Republic). Anna married Victor Masséna on 23 April 1823, becoming Duchess of Rivoli and Princess of Essling. The following year she gave birth to the first of their four children, two daughters and two sons. Many years later she became Mistress of the Household of the Empress Eugénie.

Audubon recorded an interesting meeting with the Massénas on 10 September 1828 when he was in Paris seeking new subscribers to his *Birds of America*:

"... I called on Monsieur L.C. Kiener, bird stuffer too the Prince of Massena (or Essling), who wished me to call on the Prince with him at two, the Prince being too ill to leave the house. Mr. and Mrs. Swainson were to go with me to see the collection he had made, of many curious and beautiful things, and when we reached the house we were shown at once to the museum, which surpasses in magnificence and number of rare specimens of birds, shells, and books, all I have yet seen. This for a while, when I was told the Prince would receive me. I took my *pamphlet* in my arms and entered a fine room, where he was lying on a sofa; he rose at once, bowed, and presented his beautiful wife. As soon as I had untied my portfolio, and a print was seen, both exclaimed, 'Ah! c'est bien beau!' I was asked if I did not know Charles Bonaparte, and when I said yes, they again both exclaimed, 'Ah! c'est lui, the gentleman of whom we have heard so much, the man of the woods, who has made so many and such wonderful drawings.' The Prince regretted very much there were so few persons in France able to subscribe to such a work, and said I must not expect more than six or eight names in Paris. He named all whom he and his lady knew, and then said it would give him pleasure to add his name to my list; he wrote it himself, next under that of the Duke of Rutland. This prince, son of the famous marshal, is about thirty years of age, apparently delicate, pale, slender, and yet good-looking, entirely devoted to Natural History;

Anna's Hummingbird

his wife a beautiful young woman, not more that twenty [she was twenty-five or twenty-six], extremely graceful and polite. They both complimented me on the purity of my French, and wished me all success."[1]

Unfortunately, this account tells us more about Audubon than it does about Anna Masséna.

In 1846 Masséna sold his bird collection to Dr T.B. Wilson who presented it to the Academy of Natural Sciences of Philadelphia. Dr Wilson did not visit Paris himself but sent an agent, the zoologist John Edward Gray of the British Museum (Natural History) who later wrote an account of his triumph:

"On my arrival in Paris, I put up at Meurice's, and at once sent a messenger with a note to the Prince Massena, saying that I was willing to purchase the collection of birds at the rate of four francs per specimen, and that I was prepared to pay for it in ready money. While sitting at dinner at the table d'hote, an aide-de-camp came in, all green and gold, with a cocked hat and a large white feather, to enquire for me, with a message from the Prince to enquire what I intended by ready money ... [At seven o'clock the next morning] the prince met me, declared the collection agreed with the catalogue, on which I gave his highness a cheque ... and he gave me a receipt and handed me the keys of the cases, and I sealed them up, the affair being settled in a few minutes.

Having finished my work sooner than I expected, and it still being early, I went to call on my dear old friend Prof. DeBlainville and had breakfast with him. He asked me what brought me to Paris. I said, among other things, to purchase the Prince Massena's Collection of Birds, which I had done; on which he became much excited and said that the French Government had intended to purchase it and that he must take measures to prevent its leaving France. I said I was not aware that the Government wanted it for I knew it had been for several years in the market, and it was now too late, as I had paid for the collection, which was now in my possession ..."[2]

By buying this important collection of 12,500 birds (from all over the world) Dr Wilson made many type specimens available to American ornithologists. Among them were the two types taken by Botta in California: the Greater Roadrunner and Anna's Hummingbird.

Ernesto Mauri[†]
(1791-1836)

WESTERN SANDPIPER *Calidris mauri* (Cabanis)

Ereunetes Mauri Cabanis, 1857. *Journal für Ornithologie* 4, No. 24 (1856), p. 419: South Carolina

Ernesto Mauri was born in Rome on 12 January 1791 and died there on 13 April 1836. He was the director of the botanical gardens in Rome and the author of several papers on the flora of the region. He gave considerable assistance to Charles Bonaparte when he was working on his *Iconografia della Fauna Italica* (1832–41) and after Mauri's death Bonaparte described a new species of fish *Smaris maurii* in the third volume, dedicating it to the memory of "my dear professor of practical botany". In 1838, while compiling his comparative list of the birds of Rome and Philadelphia Bonaparte also named the sandpiper after his friend. However he did not publish a description and several synonyms were coined during the next two decades.

In May 1840 the German ornithologist Jean Cabanis shot several Western Sandpipers in South Carolina where he was a preparator for two unhappy years. Afterwards he worked at the Berlin Zoological Museum under Hinrich Lichtenstein who labelled the specimens *cabanisi*, but like Bonaparte, he failed to publish a description. Bonaparte visited the Berlin collections shortly before his death and persuaded Dr Cabanis to use the name *mauri* in his original description, which appeared in 1857.

Western Sandpiper

†No portrait traced

George Campbell Munro (1866-1963)

AKIAPOLAAU *Hemignathus munroi* Pratt

Heterorhynchus wilsoni (not *Himatione wilsoni* Rothschild, April 1893) Rothschild, November 1893. *The Avifauna of Laysan and the Neighbouring Islands*, p. 75: Hawaii
Hemignathus munroi Pratt, 1979. *Dissertation Abstracts* 40, p. 1581. New name for *Heterorhynchus wilsoni*, preoccupied

The extirpation of many of the forest birds of the Hawaiian Islands has been caused by the destruction of habitat, the introduction of ground predators and by mosquito-borne diseases, three factors that continue to be a problem and are likely to lead to further extinctions. The collecting activities of the early naturalists had little effect on the bird populations and but for their efforts some species would not be known to us at all.

The first Hawaiian birds to arrive in Europe were collected in 1778 during Captain Cook's last voyage. Following his discovery of the islands a succession of naturalists arrived but little progress was made in working out the avifauna.[1] Andrew Bloxam collected twenty-five specimens of nine species in 1825 when H.M.S. *Blonde* brought back the embalmed bodies of King Liholiho and Queen Kamamalu who had both died from measles on a visit to England; but the natural history appendix to the voyage was so bad that it caused much subsequent confusion and was dubbed 'a disgrace' by Alfred Newton. Ferdinand Deppe, John Kirk Townsend and Thomas Nuttall were there in the 1830s but no full accounts of their work were published, only scattered bird descriptions by other authors. In 1840 Titian Peale and Charles Pickering collected at several locations during Wilkes's U.S. Exploring Expedition but the specimens were lost when one of the ships was wrecked. Around 1860, J.D. Mills, a storekeeper at Hilo, made an important collection of birds on Hawaii which went to the Bishop Museum, and some early ranchers such as Francis Gay and Aubrey Robinson developed small private collections. Sanford B. Dole, in his early years before becoming the first Governor of the Territory of Hawaii, published the first useful lists of Hawaiian birds in 1869 and 1879 (see Dole). The *Challenger* expedition called in 1875 but made scarcely any inroads on the land birds.

MUNRO. *NEW ZEALANDER WHO SPENT MOST OF HIS LIFE RANCHING ON THE HAWAIIAN ISLANDS. THIS SNAPSHOT WAS TAKEN FROM DEAN AMADON'S PRES-ENTATION COPY OF MUNRO'S* BIRDS OF HAWAII *(1944). MUNRO SAID IT WAS TAKEN IN HIS EIGHTY-SIXTH YEAR.*

It was not until 1887 that organized and systematic collecting began in earnest, reaching a peak in the 1890s and continuing until just after the turn of the century. Those most active during the collectors' heyday were Scott Barchard Wilson, Henry C. Palmer and Robert C.L. Perkins who operated at much the same period as Brother Newell, Henry Henshaw, Francis Gay, Theo Ballieu and Valdemar Knudsen.

Amongst the closet naturalists especially keen to get information on the birds of the Hawaiian archipelago was Alfred Newton, Professor of Zoology at Cambridge University. The professor convinced his former student Scott Wilson that the islands would be an exciting prospect for new discoveries and the young man subsequently became one of the most successful bird collectors to work there.

The eldest of three children, Scott Wilson was born on 29 May 1864 at Wandsworth, near London, but grew up at Weybridge a few miles to the south-west. His father George Wilson was a wealthy retail manufacturer who founded Wisley Gardens and was Treasurer of the Royal Horticultural Society. Scott went to Magdalene College, Cambridge, and after graduating continued his early interest in birds by travelling to Portugal with Professor Hans Gadow, a distinguished Cambridge-based ornithologist. After making a trip to Switzerland Wilson completed twenty pages of 'Notes on Swiss Birds' for *The Ibis* which were published in 1887. On 24 February of that year he took up Newton's suggestion and left Liverpool bound for the Hawaiian Islands. On the way he called at the Smithsonian Institution to confer with Leonhard Stejneger who had experience of the Pacific and was working through Knudsen's collections from Kauai. Arriving at Honolulu on 8 April, Wilson stayed until the end of the following year collecting throughout the main island group.

With the assistance of Arthur H. Evans, who appears to have acted as general editor, Wilson produced the classic monograph *Aves Hawaiienses: The Birds of the Sandwich Islands*, which came out in eight parts between 1890 and 1899. Dedicated to Newton, it contained sixty-four meticulous and pleasing colour illustrations by F.W. Frohawk as well as the written results of Wilson's strenuous and difficult collecting. At the time, fourteen of the birds were said to be new but modifications in classification have eroded this figure to only the following species: Laysan Finch, Kona Grosbeak, Akiapolaau, Hawaii Creeper, Maui Creeper, Molokoi Creeper, Kauai Akepo and the Crested Honeycreeper. His Laysan Finch was one of about fifty collected by a passing schooner and purchased by Wilson. Concerning the Akiapolaau Wilson observed:

"In the island of Hawaii, to which, as far as we know at present, it is peculiar, this bird is decidedly rare, and I obtained only three specimens during a stay of some five weeks in June in Kona, where it frequents the koa trees alone, running up their great smooth trunks and along their limbs in search of insects. In the mamane woods near Mana, I subsequently found it in considerable numbers in the month of January, when these trees are in full flower, resembling laburnums with their golden clusters. Its movements are very rapid, and the quickness with which it slips from one side of a limb to the other is surprising . . .

Its vertical range seems to be from 3500 to 5000 feet, as I never met with it in the lower forest-zone."[2]

He did not see the Akiapolaau feeding on flowers and correctly surmised that it used its curious bill (a short straight lower mandible and longer curved upper one, unique among birds) for extracting grubs and insects from dead wood. Wilson gave the bird the scientific name *H. olivaceus* believing that it was the Nukupuu described by Baron Lafresnaye. Lord Rothschild looked at Lafresnaye's type, which Wilson had omitted to do, concluded that it was new and generously named it *H. wilsoni* in 1893. This name remained in use until 1979 when H. Douglas Pratt reclassified the Hawaiian honeycreepers and for complicated reasons of nomenclature had to find a new name for it.[3] He called it *H. munroi* after George C. Munro—of whom more later.

Wilson's *Aves Hawaiienses* was well received, though Newton had some reservations about its success: "He brought back such a collection as had never before been made there; but, rich as it was in some respects, defects became apparent as it was gradually worked out, and some of these defects were so grave that, until they were remedied, no complete list of the avifauna could be formed. However, he had done a great deal more than anybody before him."[4] Newton tried to persuade Wilson to return to the Pacific but he was unable to go again so soon. Newton then urged the British Association for the Advancement of Science to fund someone but though this was agreed there were some tedious delays and it was over a year before Robert Perkins could be sent out. He arrived in March 1892 and remained until autumn 1894, returning in 1895 for two more years at the expense of the Bernice P. Bishop Museum and followed this with many more years there, chiefly studying the entomology of the islands. In December 1894 the veteran American bird collector Henry Henshaw arrived at Hawaii for a ten-year spell to recover his health. In 1902 Henshaw brought out a 146 page octavo volume on the *Birds of the Hawaiian Islands*—the first work on the subject that was moderately priced. Perkins's ornithological results were published a year later in part four of the first volume of his *Fauna Hawaiiensis*.

Scott Wilson eventually got back to Hawaii in 1896 but by then little remained to be discovered because he, Henshaw and Perkins had been preceded by Henry Palmer, a very thorough collector indeed. Perkins's only novelty was the Black

Akiapolaau

Mamo. Henshaw was given the type of Newell's Shearwater. Wilson found no more new species.

Palmer had skimmed off the cream for Lord Rothschild, another former student of Newton's who was now an arch-rival. Rothschild had begun by hero-worshipping his professor but Newton's jealousy of his wealth and ability seems to have caused a permanent rift between them. Even so, Rothschild did not send out Palmer to deliberately spite Newton, obtaining birds from the Hawaiian Islands was just part of his great plan to acquire representative collections from all parts of the world—an ambition in which he was very largely successful.

Palmer arrived at Honolulu in December 1890, fresh from a collecting trip to New Zealand's Chatham Islands. He was a determined and skilled collector who proved his worth to Rothschild by being the first ornithologist to set foot on Laysan, discovering several new birds there and elsewhere in the archipelago until he left in August 1893 (see Palmer).

Rothschild subsequently wrote *The Avifauna of Laysan and the Neighbouring Islands* (1893–1900), an even more sumptuous book than Wilson's. Only 250 copies were issued, with eighty-three colour illustrations of superb quality by J.G. Keulemans and F.W. Frohawk. Rothschild now became the expert on Hawaiian birds, clearing up some of the deficiencies of Wilson, as for example when he changed *H. olivaceus* to *H. wilsoni* and added more information on females, juveniles and distribution. He included extracts from Palmer's diaries, which though of great historical interest make little or no mention of George C. Munro, his young and able assistant who worked for him for fifteen months.

Munro was born at Clevedon, near Auckland, New Zealand, on 10 May 1866 and was twenty-four years old when he started work with Palmer. Sometimes Munro was left alone in suitable locations, at other times they travelled together, such as when they took a forty-five ton schooner out to Midway and Laysan. In June 1892 Munro married a New Zealand girl and took a job as overseer on Gay and Robinson's Makaweli Ranch on Kauai, remaining there for seven years. He then transferred to the Molokai Ranch for the American Sugar Company for another seven years. In 1906 he went farming in New Zealand but he came back in 1911 to be ranch manager for the whole island of Lanai, which had been turned into a 141 square mile cattle spread. After six years he went to work for H.A and F.F. Baldwin. In 1923 he began work with the Hawaiian Pineapple Company and in 1934 became their advisor on forestry and land reclamation. During all these years he continued birdwatching and in 1920 was made an Honorary Associate in Ornithology with the Bernice P. Bishop Museum, to whom he gave a bird report each year. Between 1935 and 1937 he conducted bird surveys of all the forested islands of the main group, spending at least a month on each. Starting in 1937 he pioneered the banding of seabirds on various local islands later inaugurating a programme extending to eight of the outlying islands. When the opportunity arose he continued his collecting of birds as well as plants, insects and land shells. Latterly he became very interested in the conservation of endemic plants and in 1958 the Governor of Hawaii set up a special arboretum to support his efforts.

To ornithologists, Munro is best known for his *Birds of Hawaii* (1944), the first such book to appear for over forty years. He was able to look back on his early experiences with Palmer and tie these in with observations from his recent surveys, bird banding expeditions and the other work of modern ornithologists. He knew the Akiapolaau from 1891 and was able to comment on its peculiar mode of feeding. Though he missed them on his later survey of Hawaii they do still exist there in much reduced numbers, perhaps totalling only 1500 individuals. Especially interesting are his accounts of species that are now extinct, because, as we have seen, he was part of the privileged band of naturalists fortunate enough to have seen some of them in the wild. One of these birds was the Lanai Hookbill *Dysmorodrepanis munroi* which is known only from a single specimen collected by Munro:

> "I discovered this bird on February 22, 1913, in the Kaiholena Valley [on Lanai]. It flew from a tree with a little chirp and alighted a short distance away, and thinking it was an ou I shot it ... It was molting and hard shot, and its sex could not be ascertained on dissection ... Its striking peculiarity was its beak. Both mandibles were curved towards each other so that the tip of the lower is the only part that touched the upper leaving an opening between them like the Hawaii nukupuu, only more exaggerated.
>
> I saw a bird of this kind on two occasions afterwards, once on March 16, 1916, farther up the Kaiholena Valley, and again on August 12, 1918, at Waiakeakua at the southwest end of the forest. On both occasions I was sure that it was not one of the known birds of the Lanai forest ...
>
> I tried for several years to obtain another specimen so as to allay any doubts that might arise that this specimen is a sport but I failed to do so."[5]

Munro's book was revised and reissued in 1960 but despite some improvements it was rather let down by the substandard illustrations, most plates being conglomerate copies of Frohawk's birds. It is now largely superseded by Pratt, Bruner and Berrett's *Field Guide to the Birds of Hawaii and the Tropical Pacific* (1987) but much of Munro's text still remains informative and useful.

Dean Amadon visited Munro at his home in Honolulu in June 1962 and "found the old gentleman, if not as hale as when we first became friends during the war, at least with the same hearty laugh. When I left he walked me to the corner, chugging along with a cane in each hand."[6] Munro died the following year on 4 December, at the age of ninety-seven, leaving behind some twenty grandchildren.

Not all the other collectors fared so well. Palmer was murdered after leaving Hawaii and Wilson committed suicide at his home in the south of England. Wilson had collected in the Pacific on the Society Islands in 1904 but later dropped out of the ornithological scene and had been living with his wife at Everton, Hampshire. At the subsequent inquest, his sister said that "He was very subject to extremes, was easily depressed over quite trivial matters, and was always abnormally sensitive."[7] He had become over-worried about financial matters and on 20 January 1923 he had shot himself in the head with his revolver.

Adolphe Simon Néboux†
(born 1806)

BLUE-FOOTED BOOBY *Sula nebouxii* Milne-Edwards

Sula Nebouxii A. Milne-Edwards, 1882. *Annales des Sciences Naturelles. Zoologie*, Sér. 6, 13, p. 37, Pl. 14: la côte Pacifique de l'Amérique [= Pacific coast of America, perhaps Chile]

On 22 February 1806 Dr Jean Néboux rejoiced in the birth of the son he had helped to deliver at his home in the small French town of La Châtre, about 160 miles south of Paris. As he grew up the boy was encouraged to follow his father's profession and as soon as he could, at the age of sixteen, he left home to study for a year at the naval hospital at Cherbourg. Just over a year later the young surgeon was posted to his first ship, the corvette *L'Isis*, which immediately set sail for the Spanish campaign. After six months, young Néboux returned for a short spell ashore at Brest naval hospital. For the rest of his career this pattern was repeated again and again: each time he was posted to a different ship, each time he returned to Brest. Service in Algeria and Spain was followed by voyages to St Dominique, Martinique, Brazil, the Antilles, Lisbon, Martinique and the Antilles. In 1834 he married for the first time. He won promotion in 1836 becoming First Class Surgeon—a man with considerable experience in dealing with the infinite number of tropical maladies and the inevitable ship-board accidents.

In November 1836 he was appointed to the frigate *La Vénus*, and thus became connected with a little of the early ornithological history of the Pacific. *La Vénus* was to make a thirty-one-month tour of the world to investigate especially the whale fisheries in the North Pacific in the hope of extending French interests in the region. Apart from his obvious medical duties thirty year old Dr Néboux was directed to make natural history collections at sea and on the various stopovers.

La Vénus left Brest on 29 December 1836 under Captain Abel du Petit-Thouars who took the usual westerly course to the Pacific via the Canary Islands

†No portrait traced

and Rio de Janeiro. From Valparaiso a course was steered up the Pacific coast to Calao and across to the Galapagos and Hawaiian Islands before making for Kamchatka and the Bering Sea, part of their main objective. Returning southward in the autumn of 1837 *La Vénus* called at Monterey on the California coast. Scurvy had long since become evident amongst the three hundred men on board and Surgeon-Major Néboux insisted that they remain in port for some time to rest and recuperate and gather fresh supplies to curtail the outbreak.

The sea off Monterey was of course a wonderful area for scientific investigation: rich in seabirds as well as the sea mammals they sought. During the twenty-seven days spent there between 18 October and 14 November, when time away from his medical duties was permitted, he added new material to the natural history collections, sometimes helped by his aid or the chief of marines. When Néboux later wrote the first description of the Swallow-tailed Gull he said that the bird had come from Monterey. Perhaps he did shoot it there but the species has never been recorded in the area since, nor anywhere else in North American waters. It seems more likely that some specimens and labels became confused and that the bird was from more southerly waters, perhaps the Galapagos Islands.

Another specimen of another species is also of doubtful origin. Néboux was the first to collect the tiny Costa's Hummingbird that was described by Jules Bourcier in the *Revue Zoologique* for 1839 soon after the end of the voyage. Bourcier failed to give a precise location for its discovery but from its present distribution in late autumn, the most likely place at which Néboux collected it was in Baja California at Magdalena Bay,[1] the next port of call for *La Vénus* after Monterey.

From Magdalena Bay they proceeded to Mazatlán, San Blas, Acapulco and other points to the south, arriving back at Brest on 24 June 1839. Néboux wrote up his usual medical report, supplementing it this time with an additional one on the outbreak of scurvy. After three months leave to his home town to deal with family matters, Néboux was sent to Paris for three and a half months to work on the natural history collections he had brought back and to generally assist in the compilation of the reports ensuing from the voyage. Even after he returned to Brest it was stipulated that he was not to be posted far away since his presence might be required at Paris at any time; indeed in July of the following year he had to go back to clear up a few matters.

The captain's final report appeared as the *Voyage autour du monde sur la frégate La Vénus pendant 1836–39* in eleven volumes of text and four volumes of plates published between 1840 and 1864. The *Atlas de Zoologie* was issued in 1846 containing thirteen hand-coloured plates of mammals, ten of birds, three of reptiles, ten of fish and thirty-eight of molluscs and other small marine creatures. One of the most pleasing and colourful plates is of the *Kurukuru Nebouxii*, a pigeon from Tahiti, but this scientific name is not now recognized. Other birds illustrated include Prévost's Ground-Sparrow, Swallow-tailed Gull and one of the noddies. The text for the zoology was not published until 1855. The ornithological section was by Florent Prévost and Oeillet Des Murs who quote some of Néboux's notes and praise him for the dozen or so new species found during the expedition.[2] In retrospect this seems an abysmally poor record considering the date of the expedition and the number of places visited.

Under his own name, Néboux published a three-page paper describing eight supposed new species in the *Revue Zoologique* for October 1840.[3] Only his scientific name for the White-capped Fruit Dove of the Marquesas is still valid: he named it *[Ptilinopus] dupetithouarsii* after his captain. Though he reported on the Swallow-tailed Gull he did not give it a proper scientific name until the 1846 *Atlas* appeared.

In the meantime much had happened to Néboux. In March 1841 he was sent to Brazil on the frigate *La Gloire* on his last and longest assignment, of almost three year's duration. It ruined his health and he was never fully fit again. Returning to France on 4 July 1844 Néboux celebrated with a three-month period of leave which was extended for a further three months, partly because he was still not well but also because of his second marriage (to Antoinette-Louise Noël) just before the end of the year. Néboux remained on the inactive list for two and a half years and was pensioned off towards the end of 1847, after twenty-five years service in the French navy.

He went to live in Paris and three years later published a book in which he outlined his recommendations for providing medical care for the poor in the city. After this nothing more is known about him other than the fact that a letter of his sent from Paris to Brest, dated 10 January 1854, indicates that he at least reached the age of forty-eight.

In 1882, forty-three years after the return of *La Vénus*, Alphonse Milne-Edwards wrote a short review of the gannets and boobies in the *Annales des Sciences Naturelles*. During the preparation of this he came across a new species from the *La Vénus* expedition, which he named *Sula Nebouxii*. It had probably

Blue-footed Booby

been overlooked because the specimen was not quite in adult plumage. The label attached to it vaguely indicated that it came from the Pacific coast of America. The Blue-footed Booby ranges along the Pacific coast from central Peru northwards to Baja California and occasionally into southern California. It also frequents the Galapagos Islands. Edwards suggested that the specimen came from the Chilean coast since *La Vénus* spent an extended period there but he was only guessing. If Néboux had still been alive in Paris, Edwards could have asked him whether he remembered where it came from. But by then Néboux had probably been dead for many years.

Thomas Nuttall (1786-1859)

COMMON POORWILL *Phalaenoptilus nuttallii* (Audubon)

Caprimulgus Nuttallii Audubon, 1844. *The Birds of America* (octavo edn), Vol. 7, p. 350, Pl. 495: upper Missouri [= between Fort Pierre and mouth of the Cheyenne River, South Dakota]

NUTTALL'S WOODPECKER *Picoides nuttallii* (Gambel)

Picus Nuttalii [sic] Gambel, 1843. *Proceedings of the Academy of Natural Sciences of Philadelphia* 1, p. 259: near the Pueblo de los Angelos [sic], Upper California [= Los Angeles, California]

YELLOW-BILLED MAGPIE *Pica nuttalli* (Audubon)

Corvus Nutalli [sic] Audubon, 1837. *The Birds of America* (folio), Vol. 4, Pl. 362, Fig. 1: Upper California, around the village of Sta Barbara

The first American organization to be devoted to studying and watching birds was established in 1873 by William Brewster, Henry Henshaw, Ruthven Deane and other like-minded friends. They held their meetings at Brewster's home in Cambridge, Massachusetts, and called their new society the Nuttall Ornithological Club. None of these young men had ever met Thomas Nuttall, for they were still only schoolboys when the elderly English naturalist died at his home near Liverpool, England, in 1859, but his hard-earned reputation as one of America's great ornithologists, and his local connections, made their choice particularly apt.

For ten years Nuttall had taught natural history at Harvard and during that time produced the first compact guide to North American birds. Then in 1834 he had resigned from the university and travelled across the Rocky Mountains with John Kirk Townsend—a journey on which many new birds were discovered. Yet this outstanding ornithologist who wrote so knowledgeably about the

NUTTALL. *ENGLISH BOTANIST, ORNITHOLOGIST AND EARLY EXPLORER OF THE NORTH AMERICAN WILDERNESS. HE WROTE A MANUAL OF THE ORNITHOLOGY OF THE UNITED STATES AND OF CANADA (1832–34) AND CROSSED THE ROCKIES WITH J.K. TOWNSEND IN 1834.*

American birds was primarily a botanist, indeed he was one of the greatest who ever worked in the United States.

Nuttall's dedication to botany in particular and natural history in general has often led to him being portrayed as the archetypal eccentric naturalist. In *Astoria*, Washington Irving described him as a kind of bumbling idiot:

"Whenever the boats landed at mealtimes, or for any temporary purpose, he would spring on shore, and set out on a hunt for new specimens. Every plant or flower of a rare or unknown species was eagerly seized as a prize. Delighted with the pleasures spreading themselves out before him, he went groping and stumbling along among the wilderness of sweets, forgetful of everything but his immediate pursuit, and had often to be sought after when the boats were about to resume their course. At such times he would be found far off in the prairies, or up the course of some petty stream, laden with plants of all kinds.

The Canadian voyageurs ... used to make merry among themselves at his expense, regarding him as some whimsical kind of madman."[1]

Unlike his friend Audubon, Nuttall never portrayed himself as a macho frontiersman. In his journals he admitted occasions when he got lost and had to be rescued by Indians or fellow travellers. He never learnt to swim. He was a poor marksman and therefore unable to live off the country. On one occasion when his party was threatened by Indians the others were shocked to discover that Nuttall's gun-barrel was tightly packed with dirt; he had either decided that it was the safest place to store his seeds or else used it as a probe to dig up his plants! It was hardly surprising that ignorant boatmen considered the man to be a fool but for biographers to accept this image of him is unpardonable. Nuttall travelled further through the wilds of America than any other naturalist of his generation—down the Ohio; through Kentucky, Tennessee, North and South Carolina, Georgia, northern Florida and Alabama; across the Great Lakes and down the Mississippi to New Orleans; up the Missouri from St Louis to the Yellowstone; along the Red River and the Arkansas River; across the prairies by way of the Kansas and the North Platte Rivers, over the Rocky Mountains and down the Columbia River to the Pacific and even to Hawaii and along the Californian coast. Often he travelled alone. No matter how terrible the weather, how short the food supplies or how heavy his pack, Nuttall made his collections. Even when fainting with ague he still searched for the plants that were always his first priority.

The values that made Nuttall something of a misfit amongst the rough trappers and voyageurs of the West also led to him sometimes being considered rather odd. in the civilized East. The prim Miss Lawson who coloured some of his botanical plates said that: "Nuttall himself was the least attractive of the Genus Homo I ever met. I could never imagine any Englishman so dirty and disorderly in his dress and appearance ..."[2] Other acquaintances were more appreciative. Elias Durand, who met Nuttall in Philadelphia, recorded that:

"His appearance and manner made a lasting impression upon those who approached him. He was a remarkable-looking man: his head was very large, bald, and bore the marks of a vigorous intellect, his forehead expansive, but his features diminutive, with a small nose, thin lips, and round chin, and with grey eyes under fleshy eyebrows. His complexion was fair, and sometimes very pale from hard labor and want of exercise. His height was above the middle; his person stout with a slight stoop; and his walk peculiar and mincing, resembling that of an Indian ...

Nuttall was naturally shy and reserved in his manners in general society, but not so with those who knew him well. If silent or perhaps morose in the presence of those for whom he felt a sort of antipathy, yet, when with congenial companions, he was affable and courteous, communicative and agreeable.

From long solitary study the cast of his mind was contemplative and abstracted; but when doubts and difficulties were solved, he was apparently light and buoyant— 'at the conclusion of a piece of work,' says one who has been most intimate with him, 'I have seen him rise from his chair, approach the stove, and, in his own peculiar way, put his hands behind his back, and, for an hour or two, pour forth a torrent of narrative and scientific facts on which was the cast of his own philosophical thoughts and conclusions. I have frequently seen him in social circles, when he was the delight of the company, from his cheerful and natural replies to all interrogatories and his voluntary details upon the subject of his travels and adventures.'"[3]

Nuttall first came to the United States at the age of twenty-two. Born at Long Preston in Yorkshire on 5 January 1786 he left the village to serve a seven-year apprenticeship with his uncle, Jonas Nuttall, who was a printer in Liverpool. This rather unpromising start to his future career was ameliorated by a period of convalescence at home when he was in his late teens. During this short but important time in his life he met a zealous young plant collector of his own age who took him rambling in Ribblesdale and over the surrounding moors. When the apprenticeship ended, Thomas's childless uncle was desperately disappointed to learn that the boy had no interest in entering the family business. After long and bitter arguments Nuttall eventually sailed from Liverpool in March 1808 and landed at Philadelphia where he decided to stay for a while. He wasted no time in meeting the right people. After trying unsuccessfully to buy a copy of Benjamin Smith Barton's *Elements of Botany* (1803) he called on the author and was enthusiastically received. Barton was looking for a capable young plant hunter to replace Frederick Pursh who had recently left him; and through Barton, Nuttall met another mentor, William Bartram, whose home and gardens became an ever-welcoming haven where he could always find friendship and wise advice. For the next two years Nuttall worked six days a week as a printer but he gave all his spare hours to studying the American flora and on excursions to the Great Cypress Swamp of Delaware and the Niagara Falls he proved his worth to Barton, who then sponsored a much longer journey.

Nuttall set off alone for the Mid West in the spring of 1810. He travelled overland to Lake Erie, then crossed the Great Lakes with French-Canadian voyageurs and boated down the Mississippi to St Louis. After passing the winter there he joined a party of Astorian trappers and ascended the Missouri as far as the Yellowstone River. By December he was in New Orleans with an impressive hoard of dried plants and seeds, many new beetles that he donated to his friend Thomas Say and some fossils, minerals, snakes and lizards. His second apprenticeship, as a field collector, had now been served. Since there was a ship at New Orleans about to depart for Liverpool and since trade with Great Britain was already prohibited because war was brewing up between the nations, Nuttall

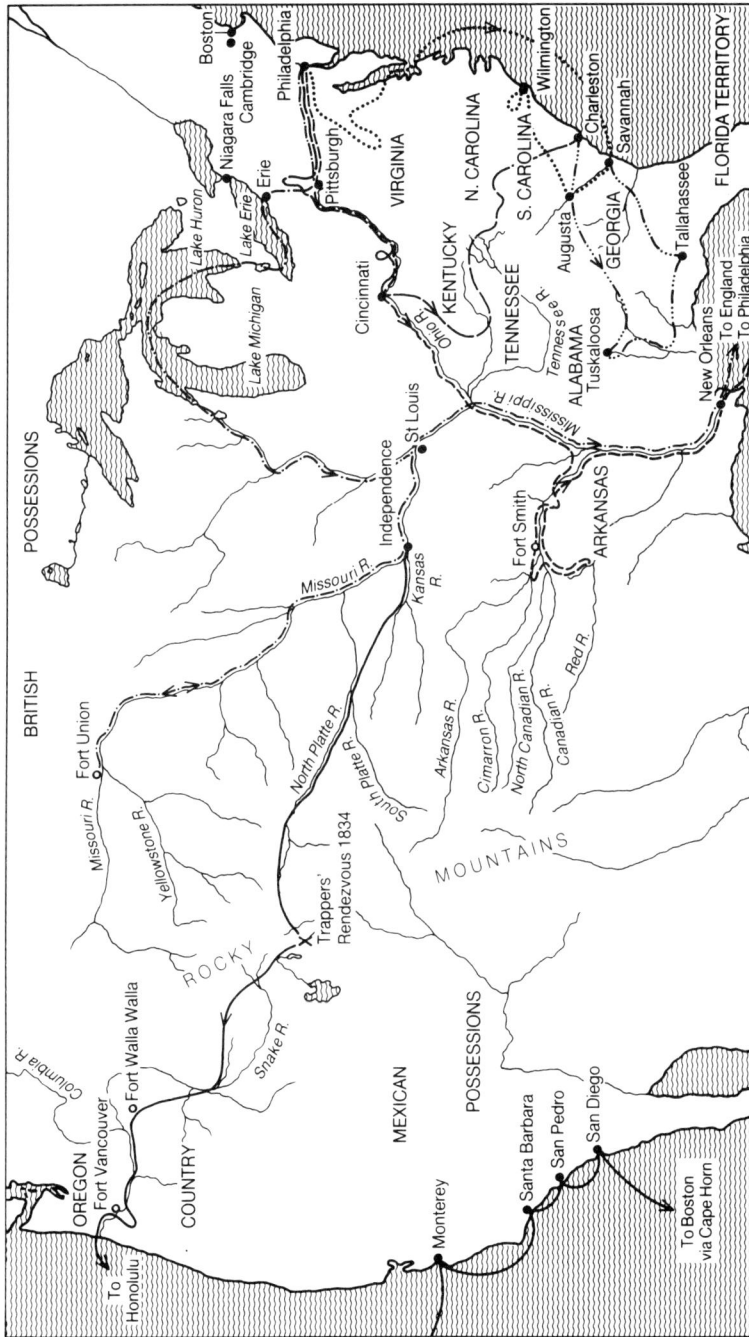

NUTTALL'S EXPEDITIONS

1810–11
1815–16
1816–17
1818–20
1830
1834–36

sent the major part of his collection and his diaries to Barton, then went aboard. By the beginning of 1812 he was back in England where he stayed until May 1815.

When he returned to the States he was twenty-nine years old and he made it his home for the next twenty-seven years. He was now a full-time naturalist earning sufficient income for his thrifty lifestyle by selling plants and seeds, often to gardeners in England. Using Philadelphia as his base he made two southern trips, the first to the Carolinas and the next to South Carolina by way of Kentucky and Tennessee. During this time he worked on his *Genera of North American Plants*, which was published in two volumes in 1818, earning him international acclaim. By then his talents had been rewarded locally by election to the American Philosophical Society and the Academy of Natural Sciences of Philadelphia. The latter group quickly appointed him to the committee responsible for the production of the first issue of their *Journal*. Funds were limited so Nuttall and Say set most of the type and printed it on a second-hand press. Since Nuttall's plant collection was housed at the Academy he spent most of his time there and often worked throughout the night or slept on the premises.

With the completion of his *Genera* Nuttall longed to travel west again but although Say was appointed to Major Long's Yellowstone expedition Nuttall was unable to acquire a place on any government survey. Instead he found sponsorship from four fellow members of the American Philosophical Society for a botanizing journey in Arkansas. In late April 1819 he reached Fort Smith where he found the surgeon to be a congenial companion and together they explored the valley meadows, full of colourful spring flowers. During these few weeks at the fort Nuttall discovered about a hundred new species of plants. In mid-May he joined a small group of soldiers who were travelling to the Red River to evict white settlers from Osage land. After crossing the Ouachita Mountains they came down into the bottomlands where Nuttall was in raptures over the beauty of the flowers and the novelty of the birdlife. Reluctant to leave such a choice site he lingered too long on the last day, missed his rendezvous and then followed the wrong trail. The platoon returned without him but he found shelter with some kind settlers and two weeks later made the return trip to Fort Smith with other travellers.

For a long time he had cherished hopes of reaching the Rockies and at the beginning of August he took the opportunity of travelling west with a trapper. After just a few days Nuttall began to suffer from one of his recurrent attacks of fever and was soon desperately ill. At the end of a nightmarish journey of five weeks they arrived back at the fort with Nuttall still in a feverish and debilitated state.

Once back in Philadelphia he began work on *A Journal of Travels into the Arkansas Territory*, which is full of the difficulties and dangers of wilderness travelling, detailed descriptions of the scenery and interesting accounts of some of the human inhabitants, but there is a disappointing lack of references to the bird life. Published in 1821 it was well received but like his *Genera* it was not a lucrative venture. During the next two years Nuttall was paid to give a series of lectures, which were extremely popular, but finances remained as tight as ever. A fellow academician, curious to know where he lodged, once followed him back to "the poor oyster-cellar of a colored man".

Financial security arrived in 1823 when Nuttall was offered the job of lecturer in natural history at Harvard University and curator of the Botanical Garden. He accepted and lived at Cambridge for the next ten years, but later, he summed up this academic phase of his career as a time of "vegetation amongst vegetables" without any important contribution to science. However this was hardly a fair assessment of his achievements and the ornithologist especially has to disagree.

It was during his Harvard period that Nuttall found time to concentrate for a while on the American birds that he had so often observed on his travels and which he studied in and around Cambridge. The idea that he should produce a guide to the birds was made by his close friend James Brown of the University Bookstore. Brown was a keen bird hunter and he pointed out that there was no modestly priced book available. (Wilson's *American Ornithology* and Audubon's *The Birds of America* were for most people prohibitively expensive.)

Nuttall's *A Manual of the Ornithology of the United States and of Canada* was published in two parts, the first covering the land birds and the next the water birds. In his research for volume one he used his own travel diaries and consulted the works of Wilson, Bonaparte and various European observers. He also visited Peale's Museum to study the skins, some of which had belonged to Alexander Wilson. Early in 1830 he made a 1200 mile birding trip on foot through South Carolina, Georgia, Alabama and northern Florida. The first part of Audubon's *Ornithological Biography* was published early in 1831, enabling Nuttall to refer to it before 'The Land Birds' was finished at the end of that year (dated 1832 it was on sale by December 1831).

In the summer of 1832 Audubon visited Boston and went to Harvard in search of Nuttall. They went for a walk together and Audubon shot a specimen of the Olive-sided Flycatcher that he was needing for his illustrations. For a long time it was known as *Nuttallornis borealis* but the generic name has now reverted to *Contopus*. On 14 August the artist wrote to his friend Edward Harris mentioning that he had made a drawing of the 'rare' Sedge Wren (which he had vainly searched for near Salem) from a specimen that Nuttall had shown him. To Dr Richard Harlan he wrote the same day saying: "Nuttall is a Gem—a most worthy, agreeable man—quite after our heart, and I am very happy to know him *as such*."[4]

During the winter of 1832–33 Nuttall took more than eight months leave to go to England, where he visited relatives and spent some fruitful days examining the collection of American birds in the British Museum.

In the summer of 1833 Audubon sailed to the coast of Labrador and on his return sent Nuttall many notes on the seabirds, which were duly used and acknowledged in 'The Water Birds', published the following year. Nuttall was greatly relieved at the completion of his *Manual of Ornithology* for he had become eager to leave the confines of Harvard and the routine of teaching. Nathaniel Wyeth had just returned from the West with some interesting plants for him and a proposition that he could not refuse. In 1832 Wyeth had travelled overland to the Columbia River on a commercial reconnaissance and on his way

back east had attended the Green River rendezvous and made a contract with
the Rocky Mountain Fur Company to haul in their supplies the next year. When
he returned to Boston he asked Nuttall to come along. It was just the sort of
opportunity Nuttall had desired for so long and despite a recent rise in salary
he handed in his resignation at the university and asked his young friend
Townsend to come with him as zoologist.

Nuttall and Townsend left Philadelphia together in March 1834 and soon
Nuttall was making his third descent of the Ohio. They joined Wyeth's brigade
at Independence and at the end of April the great cavalcade of men and horses
marched out of the town. Everyone was in high spirits, especially Nuttall for
he discovered a new sparrow on the prairies on the very first day! He christened
his new find the Mourning Sparrow because of its plaintive call but Audubon
later renamed it Harris's Sparrow and this name has stuck. In mid-June they
reached the trapper rendezvous and from there headed north-west to the
Columbia River where they arrived in September. During the fall Nuttall and
Townsend collected around Fort Vancouver then sailed to Hawaii, returning to
the fort in mid-April. In September Townsend decided to spend another year
on the Columbia but Nuttall began the return journey to the east coast. For
both naturalists it had been a demanding but supremely satisfying adventure
(details of which can be found in J.K. Townsend). Nuttall took his colleague's
collections with him as well as his own, which by now included the striking
Pacific Dogwood *Cornus nuttallii* and some new seashells, among them Nuttall's
Cockle *Clinocardium nuttallii*. Nuttall had become the first full-time naturalist
to cross North America and no other naturalist knew the country better.

On leaving the Columbia he sailed to Honolulu to spend the winter of 1835–36
in the Hawaiian Islands. In March he went to Monterey and from there took a
boat farther down the California coast to Santa Barbara. He found accommodation
in a house with a tree-bordered garden and happily observed the local birds.
Anna's Hummingbirds were beginning to lay and he managed to catch a female

Yellow-billed Magpie

with his hat as she sat tightly on her two little eggs, but as he did not have a gun with him her mate escaped. Later he must have acquired a weapon as he obtained two new species here: the Tricolored Blackbird and the Yellow-billed Magpie. Of the magpie he wrote: "In the month of April they were everywhere mated, and had nearly completed their nests in the evergreen oaks of the vicinity (*Quercus agrifolia*). The only one I saw was situated on a rather high tree, towards the summit, and much concealed among the thick and dark branches. Their call was *pait, pait*; and on approaching each other, a low congratulatory chatter was heard. After being fired at once, it seemed nearly impossible again to approach them within gun-shot."[5]

At San Diego Nuttall spent some time wandering along the beach and was spotted by a sailor who had formerly been a student at Harvard. This was Richard Henry Dana whose *Two Years before the Mast* became a popular classic. His reaction has appeared in virtually every biography, short or long, that has ever been written about Nuttall and since it seems to be compulsory, we shall repeat the story here:

> "I had left him quietly seated in the chair of botany and ornithology in Harvard University; and the next I saw of him was strolling about San Diego beach, in a sailor's pea-jacket, with a wide straw hat, and bare-footed, picking up stones and shells. He had travelled overland to the North-west Coast, and come down in a small vessel to Monterey. There he learned that there was a ship at the leeward, about to sail for Boston; and, taking passage in the *Pilgrim*, which was then at Monterey, he came slowly down, visiting the immediate ports, and examining the trees, plants, earths, birds, etc., and joined us at San Diego shortly before we sailed. I was often amused to hear the sailors puzzled to know what to make of him, and to hear their conjectures about him and his business ... The *Pilgrim's* crew christened Mr N. 'Old Curious', from his zeal for curiosities; and some of them said that he was crazy, and that his friends let him go about and amuse himself in this way. Why else a rich man should leave a Christian country, and come to such a place as California, to pick up shells and stones, they could not understand."[6]

Nuttall and Dana rounded Cape Horn on the *Alert* and arrived safely in Boston harbour on the evening of 20 September 1836. After his absence of more than two and a half years Nuttall was warmly welcomed by old friends in the city but no-one was more pleased to see him than Audubon. This was because some of the Rocky Mountain bird skins that Nuttall had shipped on ahead of him had already arrived in Philadelphia. Richard Harlan and Edward Harris had immediately contacted Audubon with the news so he had rushed to the Academy and been permitted to handle the specimens but he had not been allowed to draw them. Having now finished about 350 plates for *The Birds of America* he needed these new western species to complete his magnum opus. Since he could not get hold of them then and there he went on to Boston to seek more subscribers and by happy chance the *Alert* anchored in the harbour just a day or two later. Audubon wrote in his journal:

> "Called on Dr. Storer & heard that our learned friend Thomas Nuttall had just arrived from California. I sent Mr. Brewer after him, and waited with impatience for a sight of the great traveller, whom we admired so much when we were in

this fine city. In he came, Lucy, the same Thomas Nuttall, and in a few minutes we discussed a considerable portion of his travels, adventures, and happy return to this land of happiness. He promised to obtain me duplicates of all the species he had brought for the Academy at Philadelphia ... and we parted as we have before, friends, bent on the promotion of the science we study."[7]

Next morning, after the two friends met for breakfast, Nuttall presented Audubon with the skins of five new birds. Meanwhile, Harlan and Harris were applying pressure on Audubon's behalf and in October Harris bought ninety-three of Townsend's duplicate skins. After Nuttall returned to Philadelphia he also handed over some birds' nests, plants and moths which Audubon incorporated into his pictures and Nuttall promised to give him notes on many of the birds. Thus it was through the generosity of Nuttall and Harris that Audubon was able to figure and give original descriptions for such birds as the Black-footed Albatross, Western Gull, Black Oystercatcher, Yellow-billed Magpie, Tricolored Blackbird, Townsend's Solitaire and Green-tailed Towhee.

Another previously unknown species that Nuttall reported to Audubon was a small nightjar, now known as the Common Poorwill *Phalaenoptilus nuttallii*. Audubon mentioned it briefly in volume five of the *Ornithological Biography* (1839), called it Nuttall's Whip-poor-will *Caprimulgus Nuttallii* and explained: "it was frequently seen by him [in the Rocky Mountains], often within a few feet, but was not procured, probably because he is not in the habit of carrying a gun on his rambles." Later, in volume seven of the octavo edition of *The Birds of America* (1844), Audubon was able to give a proper physical description of the Common Poorwill because John G. Bell had shot one on 8 September 1843 during the latter part of the Missouri River expedition.

Nuttall now faced the all-important task of working through his botanical collections and distributing the rest to other specialists. Since he needed a lot of space he was given a room at the Academy where he could lay out his dried plants into related groups for study and comparison. Throughout the winter he worked steadily but when summer came he visited New York and Boston to see John Torrey and Asa Gray who were planning their *Flora of North America*

Common Poorwill

(1838–43). Though Nuttall considered such an ambitious undertaking to be premature he agreed to collaborate and contributed a great mass of information on western genera and species for the first volume, but the high-handed editing of the two desk-bound botanists infuriated Nuttall so much that he later withdrew his help.

In October Nuttall's Uncle Jonas died at *Nutgrove*, near Liverpool. His nephew was to be left a life interest in the estate on the death of his Aunt Frances, but there was a proviso—he could not inherit if he was absent from England for more than three months in any calendar year. Fortunately Nuttall's aunt lived on until 1841, allowing him a few more precious years to enjoy his American projects. In the meantime, he continued to live on a very limited income from his savings and sales of his specimens.

Nuttall's love of travelling was still as strong as ever and late in 1838 he left Philadelphia for the winter, accompanied by the young William Gambel who became one of his most promising disciples. Where they went is unrecorded but some of their specimens from this period originated from North Carolina.

Early in 1840 they went to Cambridge together where Nuttall began working on a new edition of his *Manual of Ornithology*, most of his extra material being observations made on the trail to Oregon and on the Pacific coast. He added two new birds: Harris's Sparrow and Gambel's Sparrow, the latter now demoted to the northern race of the White-crowned Sparrow *Zonotrichia leucophrys gambeli* (the coastal form is *Z.l. nuttalli*). He also incorporated information from Richardson and Swainson's *Fauna Boreali-Americana* (1832) and notes that Audubon supplied from his trip with Harris along the coast of Texas.

Towards the end of the year Nuttall began a new project, an appendix to F.A. Michaux's *The North American Sylva* written about thirty years before. When news came in the summer of 1841 that Nuttall's aunt had died he was forced to labour round the clock in order to complete this work on the trees and put all his other affairs in order so that he could go to England to comply with the terms of his uncle's will. At the very end of the year Nuttall reluctantly left the New World for the Old, realizing that the exciting and scientifically productive part of his life had passed.

Nuttall's new home at *Nutgrove* was run efficiently by his sister Elizabeth but though he owned considerable farmland and property, taxes on the estate were so high that he had only a small cash income. For a man who still thrived on exploration and discovery the prospects were depressing, but he found solace in his greenhouses and orchards and in correspondence with other botanists in Britain and America.

Early in 1842 Gambel wrote from California enclosing a manuscript describing a number of new birds, the first of which he had taken pleasure in naming *Picus Nuttalii*. He had found Nuttall's Woodpecker around the Pueblo de los Angeles and later at Santa Barbara he had discovered a nest containing young in the dead stump of an oak. A few years later he sent his collection of western plants to Nuttall and the task of examining them provided his friend with purposeful employment for some time.

In October 1847 Nuttall suddenly decided to visit his old haunts and sailed on a Cunarder steamship to Boston. After seeing his friends there and travelling in North Carolina he returned to Philadelphia and read his paper on Gambel's plants at the Academy of Natural Sciences. When he left the United States for the last time, in March 1848, he took with him many living plants for cultivation in his own greenhouse and garden. From this time onwards his travelling was vicarious. One of his great talents had been the ability to inspire and train young naturalists and his nephew, Thomas J. Booth, willingly became the next candidate. After learning the basics of the collector's craft he went out to the mountains of Assam in search of rhododendrons, orchids and other exotica that might be grown successfully in Britain. Among his most exciting discoveries was *Rhododendron nuttallii*, one of the most beautiful of all the rhododendrons with clusters of huge, fragrant snow-white flowers tinged with pink and gold. Nuttall found great satisfaction during his last years in propagating and tending the seeds and plants that Booth had brought him. *R. nuttallii* was unfortunately a difficult species to cultivate in the English climate but in May 1859 a specimen the

Nuttall's Woodpecker

old man had gifted to Kew Gardens came into flower for the first time.

By then Thomas Nuttall was fading away. He complained of rheumatic pains and felt feeble and slow. On 10 September he succumbed to bronchitis and an illness of six months' duration that the doctor termed 'acute gastritis'. He was seventy-three.

Harry Church Oberholser
(1870-1963)

DUSKY FLYCATCHER *Empidonax oberholseri* Phillips

Empidonax oberholseri Phillips, 1939. *Auk* 56, p. 311: Hart Prairie, San Francisco Mountain, Arizona

Harry Church Oberholser was born in Brooklyn, New York, on 25 June 1870 of Swiss and French-German-English extraction, his surname having originated in the small hamlet of Oberholz in the northern part of Switzerland. When he was seven years old the family moved to a farm at Newman Springs on the Navesink River, near Red Bank, New Jersey, and there he began watching birds. Since an ornithological career was not considered feasible by his parents he worked for a New York grocery firm until January 1890 when he moved to Wooster, Ohio, to look after his father's interest in a dry goods store. Though the hours were long he made frequent field excursions and became determined to make ornithology his life's work.

In the spring of 1894 he returned to New York and in the autumn applied for a post at the Division of Economic Ornithology and Mammalogy in the U.S. Department of Agriculture in Washington D.C. He was taken on as an ornithological clerk—retiring, forty-six years later, as Senior Biologist. During that time the organization changed to the Division and then the Bureau of Biological Survey, and after transfer to the Department of the Interior, became the Fish and Wildlife Service.

The departmental bird collection was housed in the U.S. National Museum, although at that time it was kept separate from the main collection of which Robert Ridgway was curator. When Oberholser arrived, at the age of twenty-four, Ridgway had been working there for twenty-five years and he willingly guided the newcomer through the maze of technical ornithology.

Oberholser's first assignment was the collation of data on the geographical distribution and migration of birds, which meant gathering records from published and unpublished sources, preparing maps and publishing his studies. Early in 1900 he was told to provide a comprehensive report on the birds of Texas and began the fieldwork in March of that year, making his first base on

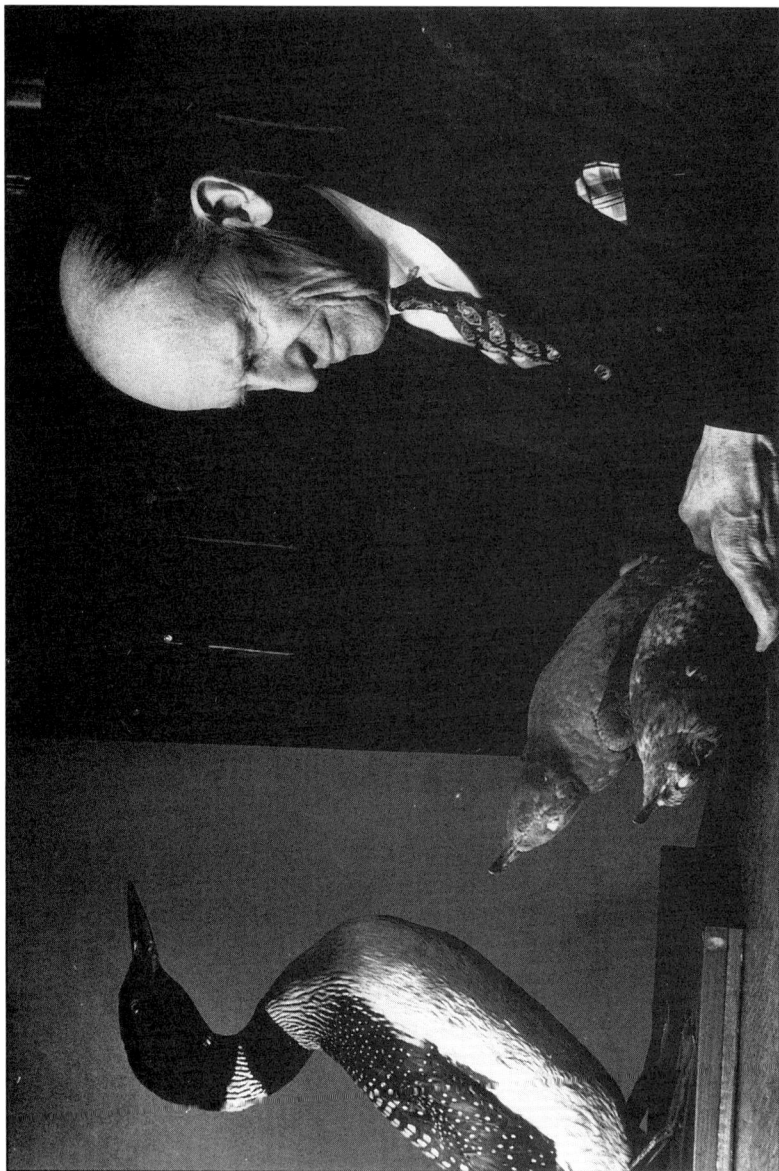

OBERHOLSER. PROFESSIONAL ORNITHOLOGIST WITH THE FISH AND WILDLIFE SERVICE. PIONEERING FIELDWORK WITH LOUIS AGASSIZ FUERTES AND VERNON BAILEY LED TO THE BIRD LIFE OF TEXAS (1974). LATTERLY HE MOVED TO THE CLEVELAND MUSEUM OF NATURAL HISTORY AND IS SHOWN HERE EXAMINING SPECIMENS OF POMARINE AND PARASITIC JAEGERS COLLECTED AT LAKE ERIE.

the coast at Port Lavaca, then travelling north to Henrietta where he worked until mid-August. (For a map see p. 498.) The next year he concentrated on the birds of the Rio Grande and the Chisos Mountains, accompanied by his senior colleague, Vernon Bailey, and the artist Louis Agassiz Fuertes. Initially Fuertes was irritated by Oberholser's "knowall calmness" but the relationship improved—especially after Oberholser rescued Fuertes from a narrow ledge on a cliff near Tornillo Creek, where he had got stuck while trying to retrieve a Zone-tailed Hawk. Oberholser later loved to tell the story of his dash back to the camp for a rope and his return with Bailey in time to pull the impetuous collector, and his hawk, to the top in safety. Fuertes had shot the bird after hunting it for three days and knowing that it would be a first record for Texas he had been unwilling to sacrifice his prize.

Oberholser returned to Texas for a third season in 1902 and the following year wrote the first draft of *The Bird Life of Texas*. He made many other trips to Texas and over the years rewrote the text three times. It grew to three and a quarter million words and he was still revising it when he retired. The two volumes eventually appeared in 1974, condensed and updated by Edgar B. Kincaid and illustrated with habitat photographs and distributional maps as well as Fuertes's paintings, commissioned by the Biological Survey at the beginning of the century. After more than seventy years in the making, *The Bird Life of Texas* came on the market eleven years after the author's death. *The Bird Life of Louisiana*, which Oberholser wrote after making several visits to the state, was published in 1938.

During his long career Oberholser visited every Canadian province and all but six states of the Union. He reckoned that the greatest sight he ever saw was the spectacle of over a million waterfowl on Utah's Great Salt Lake in 1927. At this time there was a growing awareness of a general reduction in duck numbers and in 1928 he organized a national waterfowl census, on which the Biological Survey based its waterfowl management policies. In recent years it has been continued as the Winter Waterfowl Survey. Often when Oberholser travelled it was to survey potential wildlife refuges as part of the Biological Survey's conservation programme. Other duties involved the development of bird banding, lecturing on wildlife protection and the identification of bird remains in court cases enforcing federal law. During his vacations he enjoyed teaching at summer schools in Pennsylvania and North Carolina.

Though he made frequent field excursions most of Harry Oberholser's work was done in Washington D.C. In his office he set high standards of hard work and attention to detail that were not easily met by his subordinates. He forbade smoking in his office and he was such an ardent prohibitionist that he inspired the nickname of 'H$_2$O'. A thrifty man, he carried his folding money in a used envelope and his right trouser pocket was usually full of pennies that he had checked for special issues. His other pocket was always bulging with overloaded keyrings, rubber bands, paper clips and coloured pencils. His leisure activities included stamp and coin collecting, choral singing and baseball; he was an enthusiastic follower of the Washington 'Senators'. He also took a keen interest in politics, closely followed the stock exchange and from 1899 to 1913 led the Adult Bible Class at the Metropolitan Memorial Methodist Episcopal Church.

On 30 June 1914 he married Mary Forrest Smith in Washington. Later that year he graduated from George Washington University with the degrees of B.A. and M.S. and two years later he obtained his doctorate.

Part of his ongoing research was taxonomic and resulted in the naming of eleven new families and subfamilies, ninety-nine genera and sub-genera and 560 species and subspecies of birds from all over the world. His propensity for describing new subspecies on the basis of very slight differences was frequently derided by his colleagues, but they had to admire his phenomenal memory for the minute morphological differences between geographical populations.

In November 1937 he gave advice to Allan R. Phillips of Flagstaff, Arizona, when he visited the U.S. National Museum to study the skins of the *Empidonax* flycatchers. In 1939 Phillips's paper, clarifying the confusion between the Gray and Dusky Flycatchers was published in *The Auk*. He named the latter *Empidonax oberholseri*, remarking that: "It gives me great pleasure to dedicate this species to Dr. Oberholser, whose unfailing assistance has made possible the preparation of this note." Oberholser's own publications eventually numbered approximately nine hundred.

On 30 June 1941, just after his seventy-first birthday, Oberholser retired from the Fish and Wildlife Service and moved to Ohio, where he served as Curator of Ornithology in the Cleveland Museum of Natural History until August 1947. His association with the museum was a long-standing one, since he had been advisor to the Department of Ornithology since 1928. In 1945, still an able field worker, he took part in an expedition to Arizona from April to July as the museum's representative.

After his second retiral he reluctantly parted with his very fine ornithological library by selling it to the University of Illinois. He continued to live in a suburb of Cleveland, working on his publications and assisting with those of others, one of which was *South Carolina Bird Life* (1949) by A. Sprunt and E.B. Chamberlain. To the end of his life Oberholser remained an active officer of the Cleveland Audubon Society, and shortly before his death was presented with the society's Award of Merit for the second time. He died at the age of ninety-three on Christmas Day 1963 and was survived by his wife.

Dusky Flycatcher

Henry C. Palmer
(fl. 1889-1894)

PUAIOHI *Myadestes palmeri* (Rothschild)

Phaeornis palmeri Rothschild, 1893. *The Avifauna of Laysan and the Neighbouring Islands*, p. 67: Halemanu, Kauai

CRESTED HONEYCREEPER *Palmeria dolei* (Wilson)

Palmeria Rothschild, 1893. *Ibis*, p. 113. Type, by monotypy, *Palmeria mirabilis* Rothschild = *Himatione dolei* S.B.Wilson
Himatione dolei S.B.Wilson, 1891. *Proceedings of the Zoological Society of London*, p. 166: Maui

Lord Walter Rothschild was a naturalist of great genius, vision and eccentricity. At his Tring Museum in Hertford, England, he assembled the biggest natural history collection ever made by one man, dispatched to him by a world-wide network of well over four hundred collectors. Rothschild was a twenty-two year old student at Cambridge when he sent Henry Palmer to the Chatham Islands (four hundred miles east of New Zealand) and he was so pleased with the consignments that he instructed him to proceed to Hawaii.

Between December 1890 and August 1893 Palmer visited the eight main Hawaiian Islands and devoted nearly three months to sailing among the islets and atolls of the north-western chain. His thorough coverage resulted in a remarkably complete series of 1832 skins, though this was not a large collection, given that he always worked with an assistant and also employed native birdcatchers. It seems that Palmer collected selectively and intelligently in accordance with Rothschild's desire to save for science species that were already on the road to extinction, without hastening their fate. The collection included ten new species: Laysan Albatross, Laysan Duck, Laysan Rail, Millerbird, Puaiohi, Greater and Lesser Koa-Finch, Greater Amakihi, Maui Parrotbill and Bishop's Oo and several first records for the islands. It led to Rothschild's first major work *The Avifauna of Laysan and the Neighbouring Islands: with a complete history to date of the birds of the Hawaiian possessions*, published in

PALMER. *LORD ROTHSCHILD'S COLLECTOR IN THE HAWAIIAN
ISLANDS FROM 1890 TO 1893. NO PORTRAIT OF PALMER HAS
BEEN TRACED, ONLY THIS SKETCH BY F.W. FROHAWK (FROM
A BROKEN GLASS NEGATIVE) WHICH WAS INCLUDED IN ROTHS-
CHILD'S* AVIFAUNA OF LAYSAN *(1893–1900).*

three parts (1893–1900) and superbly illustrated by J.G. Keulemans and F.W. Frohawk. Since Palmer's own diaries were wantonly destroyed by the British Museum (Natural History), the following account has been pieced together from the two summaries in *The Avifauna of Laysan* and from *Birds of Hawaii* (1944) by George C. Munro, who was Palmer's assistant for fifteen months.

Palmer and Munro reached Honolulu in December 1890 and on Christmas Eve began their exploration of the islands by embarking for Kauai where they stayed four months. At the Makaweli ranch on the south-west coast they were given hospitality and advice by Francis Gay and Aubrey Robinson who were already in possession of sizeable bird collections. They owned large tracts of virgin forest where the Apapane, Omao, Iiwi, Kauai Oo and Kauai Amakihi were common, but Palmer soon realized that many other forest birds, such as the Nukupuu, would be hard to come by. It was a delight to collect along the open fringes where the undergrowth had been grazed by cattle and pigs, and brilliantly coloured birds flashed across the glades in the sunshine; but hunting at higher elevations on steep terrain was another matter. Few tracks existed and fallen, rotting tree-trunks, creepers, tangled vines, mud and broken ground made progress desperately slow, while mist and rain obscured birds and landmarks.

On 21 March the men were working in the Alakai Swamp country, forced onto the low ground by a heavy fog, when they killed a new thrush. A few days later Palmer discovered that the precious corpse had disappeared: "You may imagine my feelings!" he wrote in his diary, "I at once suspected rats, so I crawled all under the cottage on my stomach, and finally succeeded in finding it in a rat's hole, although much damaged, yet partly preserved. They had indeed not touched one of the common birds! I am glad nobody heard the prayers uttered for the benefit of the rats."[1] Rothschild endowed this rat-bitten specimen with the name *Phaeornis palmeri*. Endemic to Kauai, it was formerly known as

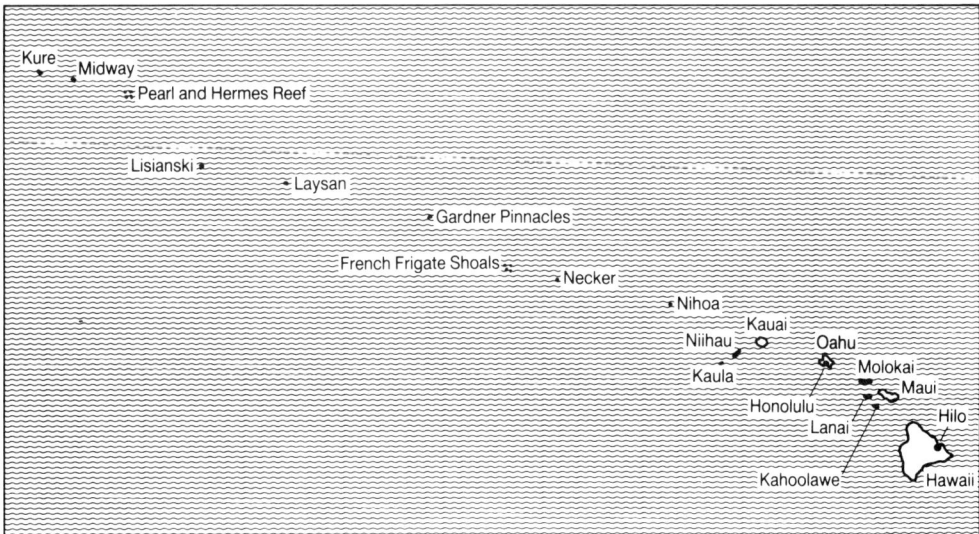

THE HAWAIIAN ISLANDS

the Small Kauai Thrush until the native name of Puaiohi was adopted. It is believed that there are now only a few dozen living individuals.

The lagoons at Mana were then a paradise for shorebirds and in March the collectors enjoyed stalking flocks of Bristle-thighed Curlew, Black-necked Stilt and Sanderling. On the 14th Palmer shot a Bonaparte's Gull there, the first record for the archipelago. At the end of April they left Kauai and returned to Honolulu in order to begin their cruise to the north-western islands. While waiting to embark they visited the new Bernice P. Bishop Museum to examine J.D. Mills's bird collection, made around Hilo, Hawaii, and dating back to the 1850s.

The voyage began on 23 May (1891), on a schooner commanded by a Captain Walker who frequently tried Palmer's patience beyond its limits. They had many arguments, usually because of Walker's refusal to allow his passengers ashore when conditions, to them, seemed ideal. The collectors were unable to land on Necker Island but at French Frigate Shoals they camped out on the low sandbanks for two nights, surrounded by breeding Sooty Terns. On turning over turtle shells discarded by ship-wrecked sailors, Palmer discovered that each one sheltered a Bulwer's Petrel incubating its single, pure white egg. He took both birds and eggs, but the extreme heat made preservation of the skins very difficult; and Munro went down with 'La Grippe'.

For nearly a week they beat about off the Gardner Pinnacles, defeated by the rough seas, but on 16 June they reached Laysan, where Palmer and Munro became the first ornithologists to set foot on the island. They stayed for twelve days by permission of the North Pacific Phosphate and Fertilizer Company who were lease-holders of the island from 1890 to 1904. Captain George Freeth, the governor of the island and manager of the works, afforded special protection to all of the breeding birds. His domain was a low, sandy island, luxuriantly covered with coarse beach grass and scrub, two miles long and about one mile broad with a large, saltwater lagoon in the centre. Palmer was staggered by the almost unbelievable abundance of bird life and exclaimed: "to my great delight, the island appears quite alive with birds, especially the Rail, which, although I had been told was tolerably plentiful, to my intense astonishment literally covers the island. Everywhere one walks this little creature hurries out of your way".[2] Almost all of the birds, including this tiny, flightless Laysan Rail were so tame that Palmer could catch them with his handnet. He sent a cage of the rails to Tring and, amazingly, they arrived alive. Rothschild invited Frohawk to observe

Puaiohi

the birds running about in his house and the artist produced an attractive painting of four adults and three black, downy chicks. He also wrote up the original description and named the rail *Porzanula Palmeri.*

Another new species the collectors found on Laysan was the Millerbird, a thin-billed brown and white warbler which frequented all the vegetated areas. While Palmer and Munro prepared their specimens in the laboratory that Freeth had put at their disposal, the Millerbird would hunt among the rafters. At night, when they sat talking by lamplight in Freeth's room, one would come in under the unsealed roof to pick off the large white miller moths which formed an important part of their diet.

During June Laysan was one vast, sprawling seabird rookery, a place of din and stench and frenzied activity. Wherever they went, the collectors had to pick their way carefully amongst the nests of Brown Noddies, Sooty Terns and Common Fairy-Terns or circle around the dense throngs of young Laysan Albatrosses. Christmas Shearwaters incubated their clutches on the sand, both adults sometimes sitting side by side, while Red-tailed Tropicbirds refused to leave the nest when approached and kept up a hideous harsh screaching until lifted up or left alone. Overhead, Great Frigatebirds chased Red-footed Boobies and terns, forcing them to disgorge their fish, and Palmer was shocked to see the frigatebirds killing and eating the nestlings of their own kind.

Palmer and Munro little realized that they were experiencing the unique avifauna of Laysan on the very verge of its destruction; only a very few other ornithologists ever enjoyed their extraordinary privilege. In 1909 Japanese feather hunters landed and killed more than 200,000 birds, mostly albatrosses. Ironically, their deliberate destruction caused far less damage than the well-intentioned, but blinkered act of Captain Max Schlemmer, manager of the guano works at the beginning of the twentieth century. He liberated rabbits and guinea pigs, which destroyed all the vegetation, as a consequence of which thousands upon thousands of petrels, shearwaters and other seabirds were buried alive in the drifting sands, the Laysan Rail became extinct and the Millerbird vanished from the island (though a different race, sometimes classified as a distinct species, inhabits Nihoa).

From Laysan the collectors sailed further west to Lisianski, where they found another huge colony of Laysan Albatrosses, then carried on to Pearl and Hermes Reef where Palmer fumed because Walker refused to allow him ashore even though the sea was calm. However, when they reached the outermost limit of their cruise, at Midway, they were able to explore again and found nesting Red-footed, Brown and Masked Boobies.

During the long four-week voyage back to Honolulu, Palmer and Munro waged war against small dermestid beetles which had got into many of the bird skins. Soon after their arrival on Oahu on 18 August Palmer went down with 'flu and two weeks passed before he was strong enough to set off on his next expedition, to Hawaii, where he stayed for nearly ten months from early September 1891 to the end of June 1892.

On Hawaii, as on Kauai, the collectors found that their strength was taxed to the uttermost. In addition to the usual problems caused by the rugged terrain, the rarified mountain air and the almost incessant rain, they had to cope with hostility from some of the local people who suspected them of being government spies, sent to investigate the illegal production of alcohol. In 1900 Alvin Seale testified to the appalling conditions to be surmounted when he wrote:

"The difficulty of collecting in these islands, with their dense tropical jungles and knife-like mountain ridges, has been mentioned by all former collectors, and I can only add, that while I have collected in difficult places before, including the boggy tundra of Siberia, the high mountains of Alaska, the Tamarack swamps of Michigan, and the Everglades of Florida, I have found nothing that could discourage an Ornithologist so much as one of these islands."[3]

Though Palmer had been preceded by many other naturalists on Hawaii he took three new species, as well as many rarities. With Munro he began in the Kona forests on the western side of the island, searching for the Kona Grosbeak which Scott B. Wilson had discovered four years earlier. After a week of roaming over the jagged lava flows they learnt to locate the grosbeaks by listening for the cracking noise made when the birds were breaking open the hard, dry seeds of the Naio tree with their huge bills. Later in September they discovered the Greater and Lesser Koa-Finches which, like the Kona Grosbeak, have now vanished forever. The koa-finches were so similar in their habits and appearance that the hunters, watching them feed in family parties on the green beans of the Koa tree, asssumed that they were all of the same species. When Rothschild delineated them separately in 1892 he named the larger kind *Rhodacanthis Palmeri*.

By December they were on the north-western slope of Hualalai stalking Nene across the lava fields. Since the geese were nesting they took only nine specimens, even though it was the open season and other hunters were killing large numbers. They then headed north into the Kohala Mountains to search for the Ula-ai-hawane *Ciridops anna* (see Dole) which had first been collected by Mills around 1859. They only secured one specimen, brought to them by a native, but it was the last ever to be taken. Forty-five years later, Munro thought that he glimpsed an Ula-ai-hawane in the same area, but none has been seen since.

Crested Honeycreeper

By March the collectors were at Hilo where Munro departed and was replaced by Wolstenholme. Soon afterwards, Palmer was kicked by his horse and had to rest for several weeks while collecting was continued by his new assistant who worked with a local birdcatcher. On 23 April, Palmer was hunting through the dense rain forests on the coastal slopes of Mauna Kea when he heard a call which was new to him; the bird which fell to his gun became his third undescribed species from Hawaii, the Greater Amakihi, now in all probability extinct. During May, Palmer's primary quarry was the Hawaiian Rail. Rothschild had sent a specially trained dog out from England to help him in the search, which was centred at Olaa, in the Puna district. Mills had collected specimens there in 1864 but Palmer failed to find any—probably they had already died out, predated by the rats, cats, dogs and mongooses against which they were defenceless.

At the end of June Palmer and Wolstenholme took a steamer to Maui where they spent the next three and a half months, interrupted by a six-day trip to the nearby island of Kahoolawe which Gay ran as a sheep ranch and was, consequently, denuded of most native birds. On Maui they obtained a series of skins of the Crested Honeycreeper, until then only known from a single juvenile specimen shot by Wilson in 1888 and named by him *Himatione dolei*. Rothschild changed the generic name to *Palmeria*. He also erected the genus *Pseudonestor* for one of Palmer's discoveries, the Maui Parrotbill.

After returning to Hawaii, Palmer was injured again, this time when his horse reared over backwards on top of him. Within a week he was trekking up the slopes of Mauna Kea, but on the back of a mule. In November he took a boat to Lanai for the first time, then visited Molokai where the first week of January 1893 was devoted to searching for Bishop's Oos in heavy rain and gales. On 25 February the collectors sailed to Oahu and stayed until 13 June. Here they met up with Robert C.L. Perkins who had come out from England the previous year to collect for the British Association for the Advancement of Science at the suggestion of Professor Alfred Newton. Though Newton and Rothschild were rivals with regard to the Hawaiian birds, Palmer struck a deal with Perkins and on the understanding that any new species would go to Rothschild, the three collectors shared a house, and later a tent, in the mountains of Waialua for several weeks. But there were no new discoveries and at the end of April Palmer wrote gloomily to the curator at Tring that his dog had broken a leg and would be useless for ever. He complained that he would rather have lost ten horses, for in the tangled jungles where he spent so much of his time a good retriever was indispensable.

However, Palmer's sojourn in Hawaii was drawing to an end. In mid-June he returned to Kauai to search for more Puaiohi to add to the single specimen that had been chewed by rats. Since lepers had in the meantime moved into the hills he was unable to search the same region as before, but took two young Puaiohis and many commoner species. In mid-July he made his last excursion, to Niihau, where he hunted shorebirds and waterbirds along the lagoons.

Palmer's exertions and injuries had taken their toll on his health, and in August he was warned by his doctor not to return to a wet district. He decided to leave and later that month sailed for England.

With his departure from Hawaii Palmer all but vanished from the pages of ornithological literature. We have been able to find only one tantalizing reference to his fate, in *The Auk* (1964).

By searching through the remnants of the Rothschild correspondence in the British Museum (Natural History) we have discovered a little about Palmer's activities during the year after he left Hawaii. On 17 October 1893 he wrote to Tring, from Cambridge, reporting that his health had improved, although his doctors wanted him to continue their treatment for another two months. On 5 April 1894 he was still at Cambridge but informed Rothschild that he had booked a passage to Sydney on the P&O liner *Rome*, due to sail in a week's time, and that the fare would be £40. Since Palmer expected Rothschild to send a cheque for that amount it is highly probable that Palmer was an Australian (or possibly a New Zealander) and that payment of his return fare was part of their contract.

Palmer's last extant letter to Rothschild is undated but it arrived at Tring on 3 September 1894 and was probably posted about two months earlier. Writing from 296 Parramatta Road, Petersham, on the outskirts of Sydney, Palmer explained that since his arrival he had been investigating the possibility of collecting on some islands, but had found the expense prohibitive. He was still hoping to represent Rothschild in some way but at present was settled at a nearby news agency.

Some information about subsequent events got back to George Munro, who told his friend Dean Amadon that Palmer had been obscurely murdered in the gold fields of Australia. Dr Amadon mentioned this fact in his obituary of Munro in *The Auk*, but he has been unable to provide us with any further details. 'Obscurely murdered' seems a good choice of phrase, since our efforts to track down any record of Palmer's death at the Registry Offices of each Australian State have proved totally fruitless.[4] The State Libraries have likewise been unable to supply us with any useful information. Somewhere—in a Registry Office or a newspaper—there must be more information about Palmer's demise; perhaps someone in Australia will take up the search.

'The brothers Paris'†
(fl.1837)

SCOTT'S ORIOLE *Icterus parisorum* Bonaparte

Icterus Parisorum Bonaparte, 1838. *Proceedings of the Zoological Society of London* (1837), p. 110: Mexico

In 1837 Charles Bonaparte prepared a paper for the London Zoological Society in which he described thirty-five birds from Mexico, thirty-nine birds from Guatemala, and a further twenty from Brazil. At the start of the Mexican section he said:

> "Messrs. Swainson and Wagler have, as far as their material would allow them, ably described the Birds of Mexico. Through the kindness of the Messrs. Paris I have been allowed to examine a small collection from that country, a list of which, with descriptions of new or interesting species, I shall subjoin; hoping thereby to add a little to our acquaintance with the ornithology of that interesting part of North America."

Amongst the descriptions were the original citations for the Pyrrhuloxia, Blue-hooded Euphonia and Scott's Oriole. The latter he described in Latin, commented in English that it was similar to the Black-cowled Oriole, and then added:

> "I have much pleasure in naming this bird after the brothers Paris, who, not withstanding the arduous nature of their professional engagements in Mexico, allowed no opportunity of furthering the interests of science to pass unimproved. I quite agree with the opinion, that in a country whose commercial transactions are so extensive as they are in this, the captain of a trading-vessel bringing home 'a "curious bird," which may prove to be new, has no claim to have his name immortalized;' but the same rule I would not apply to the Roman state, where a person crossing the sea is a rare occurrence."

Some writers have claimed that the Paris brothers were French but the above reference to the Papal States points to an Italian provenance. This seems all the more likely when Bonaparte's long association with Rome is considered. He lived near Rome on and off for much of his life, especially between 1828 and

†No portraits traced

1849; and when he visited London in 1837 he had just come from there. Recent enquiries in Italy have, however, failed to throw any light on the brothers.

From Bonaparte's paper it is impossible to assess their calibre as naturalists. He does not tell us whether they collected the birds themselves and we do not even know whether they supplied much in the way of accompanying notes. Bonaparte was never inclined to add extra or anecdotal material and all he has left us with is one abstruse statement, on the House Finch, for which he comments that "it is very common in the city of Mexico where according to Mr. Paris it takes the place of our common sparrow, provoking the science of the professors in the very yard of the university."

Scott's Oriole female

Robert Ridgway
(1850-1929)

RIDGWAY'S WHIP-POOR-WILL *Caprimulgus ridgwayi* (Nelson)

Antrostomus ridgwayi Nelson, 1897. *Auk* 14, p. 50: Tlalkisala, Guerrero, Mexico

AZTEC THRUSH *Ridgwayia pinicola* (Sclater)

Ridgwayia Stejneger, 1883. *Proceedings of the United States National Museum* 5 (1882), p. 460. Type, by original designation, *Turdus pinicola* P.L.Sclater. *Turdus pinicola* P.L.Sclater, 1859. *Proceedings of the Zoological Society of London*, p. 334: Pine-forests of the tableland above Jalapa, [Veracruz,] Southern Mexico

Robert Ridgway was one of the giants of American systematic ornithology. But he was far more than that: as a field worker *par excellence* he did sterling work in the western United States, Florida and Costa Rica; as a bird-lover and conservationist he preached and practised the protection of birds and their habitats; as a botanist he conducted important early surveys of the forests of the Wabash River valley and as an artist he not only drew and painted birds but standardized the nomenclature of colours.

Childhood experiences had an important influence on these various aspects of his subsequent career. Born in Illinois at Mount Carmel, Wabash County, on the Indiana border, he developed an intense love for the area, so much so that throughout his long career in Washington D.C. he never lost the desire to return to his native state. Born 2 July 1850 he was the eldest of ten children and both of his parents shared and encouraged his interest in nature. Every Sunday his father took him on long walks and taught him the local names for the birds, animals, trees and flowers they saw. They also enjoyed hunting trips together, as Robert later recounted:

"In those days Wild Turkeys, Passenger Pigeons, Ducks, and in fact game of all kinds were abundant (only deer were becoming scarce), and rarely indeed was our hunting trip unsuccessful. My father hunted with a long-barreled Kentucky

RIDGWAY. *CONSERVATIONIST, FIELD NATURALIST AND SYSTEMATIC ORNITHOLOGIST. THIS PHOTOGRAPH SHOWS HIM AT WORK IN HIS OFFICE IN THE SOUTH TOWER OF THE SMITHSONIAN INSTITUTION, AUGUST 1884.*

squirrel-rifle ... I with a shotgun. ... March and early April was 'gobbling time' during which we made several hunting trips together each year. One of these— the first, I believe—I have particular reason to remember. There had been rumors of a panther having been seen in the Hanging Rock Hills about three miles from the town, and we boys were of course afraid to visit that region by ourselves. Father and I started for the hills—a great place for Wild Turkeys—about three A.M. Part of our route led along a railway embankment that had been constructed a few years before, but the work being abandoned the fill was used only by pedestrians living in that direction. On each side of the embankment was a solid wall of virgin forest extending continuously for a mile or more. It was still dark as we were following the trail along the embankment. I was about ten feet behind my father when a blood-curdling scream from the woods on one side nearly paralyzed me with fear, though sufficient strength was left for me to reach my father's coat-tails in one jump. I had never before heard the wailing screech of the Barred Owl, and was sure the panther was coming to attack us."[1]

Robert was only ten years old when his father entrusted him with a shotgun and thereafter he was able to collect birds instead of just nests and eggs. Since he had then no knowledge of taxidermy, he made watercolour paintings of his quarry, grinding his own pigments at his father's drugstore.

Like many another young ornithologist of his generation, Ridgway came under the influence of Spencer F. Baird. Their relationship began in 1864 when Robert was unable to identify a colourful finch that had appeared locally in large flocks during the winter. A friend's mother suggested that he send a letter and a painting of the bird to the commissioner of patents in Washington and the package was eventually redirected to the Smithsonian Institution. Professor Baird immediately recognized the bird as a Purple Finch and impressed by the

Aztec Thrush

painting's realism he began to correspond with the young artist, sending advice and ideas, instructions on the preparation of skins and requests for the eggs of particular species.

Early in 1867 Baird offered his protégé the appointment of zoologist on Clarence King's survey of the Fortieth Parallel. Though Ridgway was then just sixteen, Baird assured him, "You need not have any apprehension of your fitness to discharge the duties; you know more than enough to begin, and you will soon grow up to all the possible requirements." He then thanked him for an American Robin and two hawk eggs, adding, "The albino robin was very neatly prepared; the only defect being in a want of finish about the head where the eyes and cheeks were not properly stuffed out. When you come on if you accept the place we will soon teach you all the nicities of the art."[2]

In his youth, Baird had been invited to accompany Audubon on his Missouri River trip, but had been prevented by his mother's anxiety. Perhaps afraid that Mr and Mrs Ridgway would be as protective, he wrote enticingly, "There are no dangers whatever from hostile Indians ... The expedition will not be an exhausting one in any respect ... The country is the finest imaginable for air and scenery ... you will never regret the acceptance of the offer."[3]

Ridgway agreed to go, and with the rest of the surveying team he sailed from New York on 10 May 1867 en route to San Francisco via the Isthmus. During the first summer they went over the Sierra Nevada to the Truckee River and Pyramid Lake in Nevada. Winter was spent at Carson City and during the next season they worked their way across Nevada to Utah, escorted by a company of U.S. regular cavalry. That fall they returned to Washington but reassembled at Salt Lake City in May and Ridgway passed the summer in the Wasatch Mountains, making a brief foray further east into the Uintas with the botanists.

The two and a half year trip was not quite the leisurely picnic that Baird had implied—there were several Indian scares and some of the desert regions were formidable. On one occasion Ridgway was found unconscious beside his mule, having succumbed to fever, but perhaps the greatest danger to life and limb was a deliberate risk he took when he climbed the Pyramid, an immense cone-shaped rock in Pyramid Lake.

At the conclusion of the expedition Ridgway deposited 769 bird skins, 753 nests and eggs, plus reptiles, fish and representatives of other fauna at the Smithsonian Institution. He produced a thoroughly satisfactory report (not published until 1877) that has proved to be of lasting interest and value to ornithologists working in California, Nevada and Utah.

Baird was eager to retain Ridgway's services but there was no permanent position available at the Smithsonian. However, he installed him in an office in one of the towers with a salary of $600 and the technical assignment of preparing drawings and descriptions of the land birds for *A History of North American Birds* (1874), which Baird was writing with Thomas Brewer. The two veteran ornithologists were so pleased with the quality of their assistant's contribution that they made him a co-author and later they all collaborated on *The Water Birds of North America* (1884).

During his first few years at the Smithsonian Ridgway was offered the position of ornithologist at the American Museum of Natural History at New York with a salary starting at $1500, but he turned down this lucrative proposal out of loyalty to Baird. Instead he suggested Dr J.A. Allen who was duly appointed.

At the age of twenty-five Ridgway married Julia Evelyn Perkins whom he had first met during the early 1870s in the main hall of the Smithsonian when she was just fourteen. It had been love at first sight—on Robert's part at least—though photographs of the suitor show him to have been a lean, handsome young man and something of a dandy with an impressive moustache and slicked back wavy hair. After the wedding on 12 October 1875, the couple lived with Evvie's grandparents in Washington. Their only child, Audubon Whelock

Ridgway, was born on 15 May 1877 and despite the burdensome name he imbibed his father's love of birds. He died of pneumonia in 1901 while working for the Chicago Field Museum.

In 1880 Ridgway became curator of birds at the Smithsonian, the appointment infuriating Elliott Coues who had long hoped for such a position. Coues could never understand that Baird valued tact and loyalty as much as ability and experience. In many ways, Baird and Ridgway were alike. Both had a kind and generous nature (indeed Ridgway has been described as generous to a fault), both were ever ready to help aspiring ornithologists and both hated controversy. A few years later Frank Chapman was moved to comment on Ridgway's gentle demeanour after meeting him for the first time at an A.O.U. Congress:

> "Ridgway, a fluent, graceful writer of undoubted ability, but so great is his diffidence that during the entire session he has not opened his mouth. Entirely unassuming; a stranger at the meeting would probably consider him merely an interested spectator. About thirty-five years old and five feet eight inches tall, dark reddish hair, red moustache and blue eyes. Agreeable in conversation, extremely polite. You might advance views contrary to his published statements and known ideas and he would simply smile and not contradict you."[4]

A year after his appointment as curator, Ridgway's *Nomenclature of North American Birds* (1881) appeared, revising nomenclature and to a lesser extent classification. 1886 saw the publication of *A Nomenclature of Colors for Naturalists*. This was a novel scheme to introduce a universal colour index for the exact description of plumage, fur and other animal parts. One hundred and eighty-six named colours were shown but the methods of reproduction resulted in variation between copies, so Ridgway later revised and enlarged the work, publishing *Color Standards and Color Nomenclature* in 1912. This time 1115 colours were named and absolute uniformity was achieved. The system was welcomed and widely adapted not only by artists and scientists but also by the commercial world.

Ridgway's next major publication was *A Manual of North American Birds* (1887). This hefty volume included the Mexican birds and gave a complete and concise description of each species with outline drawings of the generic characters, some by the author, but most by his brother John and another artist. The *Manual* was about to go to press when Baird died and so Ridgway hastily added a heartfelt tribute to the man who had taken him under his wing twenty-three years earlier.

Twenty-six years after the Fortieth Parallel Survey Ridgway again took part in a collecting expedition. During the early months of 1895, 1896 and 1898 he travelled through southern Florida first visiting the Lake Kissimmee region, then Lake Okeechobee, and finally the Big Cypress Swamp, mainly in search of the famous Ivory-billed Woodpecker.

In 1899 he joined the scientific cruise to Alaska organized by the railroad magnate E.H. Harriman. Since Ridgway was unattracted to northern regions he was a reluctant participant and did not enjoy himself despite having John Muir, William H. Dall, Daniel G. Elliot, George B. Grinnell and Louis A. Fuertes among his distinguished fellow passengers. Nevertheless, he collected at every opportunity and took the types from which he later described several new races of Song and Fox Sparrow.

Ridgway's next two expeditions, to Costa Rica, were much more to his taste. The first lasted from 8 December 1904 to 27 May 1905 and the second from 7 February to 8 May 1908. Accompanied by his wife, he enjoyed the hospitality of his old friend Don José C. Zeledon, an excellent ornithologist who guided him to some of the best sites. A great lover of forests, Ridgway was as excited by the huge trees with their brilliantly coloured burdens of orchids, bromeliads and mistletoes as by the tantalizing, tropical birds he collected for his department. Ridgway had already studied the skins of a huge number of Costa Rican species as part of the research for his last, monumental treatise *The Birds of North and Middle America* (1901–19), a work of sheer drudgery that Harry Oberholser described as one of the finest large contributions to technical ornithology ever made by one man. Ridgway began the work in 1880 when he took up his curatorship and in 1894 he started to assemble the information. By the time of his death, in 1929, eight volumes had been published but it was never completed, even though Herbert Friedmann produced three more parts.

In addition to the major works already mentioned, Ridgway wrote many papers. In 1928 Harry Harris concluded a biography of the author with a forty-seven page bibliography listing 540 publications. All but a very few dealt with the avifauna of North and Central America and amongst those based on personal fieldwork were 'The Prairie Birds of Southern Illinois' (1873); 'Notes on Birds Observed at Mount Carmel, Southern Illinois, in the Spring of 1878' (1878); 'Spring Notes on Migratory Birds' [Observations made at Laurel, Maryland] (1889); 'Nesting of the Duck Hawk in Trees' (1895) and 'The Home of the Ivory-bill' (1898).

After retiring from the Smithsonian at the age of sixty-five Ridgway spent his last fourteen years in Illinois. Throughout his working life he had longed to return to Mount Carmel but on his regular visits there he had seen the haunts of his youth obliterated by felling, drainage and cultivation. On his retiral he therefore chose to set up home at *Larchmound* in Olney, forty miles to the north-west of Mount Carmel, where the changes had been less drastic. Another compelling reason for choosing Olney was that no botanical work had yet been conducted in Richland County, whereas the vicinity of Mount Carmel had been thoroughly surveyed. Ridgway was concerned that the virgin forests were being destroyed even before their composition had been recorded and during the subsequent years he gathered much original data.

A few miles from *Larchmound* was *Bird Haven*, an eighteen-acre tract of abandoned farmland which the Ridgways had purchased in 1906. It had developed into a choice site of meadows and woodlands containing seventy-four tree species and varieties—for a long time a record for the temperate zone. The Ridgways delighted in tree planting and improving the conditions for many kinds of wild birds but the cost of maintaining and managing the sanctuary became so high that in 1925 a committee, led by Harry Oberholser, was appointed to raise sufficient funds for endowing *Bird Haven* as a permanent memorial. In 1932, when $50,000 had been raised and ninety-six adjoining acres of farmland had been accepted as a gift, the preservation of the area was entrusted to the Robert

Ridgway Memorial Association "to be perpetually maintained". Six years later an oil well was drilled and some land was sold. The sanctuary was neglected until 1970 when it was sold to the city of Olney and inundated by a lake, despite a ten year fight by many people to 'Save Bird Haven'.

Only a remnant remains. On the north side of East Fork Lake about fifteen to twenty acres of woodland, dominated by Black Locust, provide good cover for birds. Of the original eighteen acres of *Bird Haven*, two and a half acres survive which have since been expanded to over ten acres in a restoration project. Three nature trails—named for each member of the Ridgway family—have been blazed through the trees and the Robert Ridgway trail leads to the hill he chose for his final resting place. After his death, which occurred on 25 March 1929, the grave was marked by a simple granite boulder with a bronze plaque reading 'Robert Ridgway 1850–1929'.

Many ornithologists paid tribute to Ridgway by naming American races and species for him. The genus *Ridgwayia* was created for the Aztec Thrush by Leonhard Stejneger in 1883 when he was Ridgway's assistant at the Smithsonian, (some authors merge this thrush into *Zoothera*). Ridgway's Whip-poor-will, also known as the Buff-collared Nightjar, was collected by Edward W. Nelson and Edward A. Goldman in November 1894 and described in *The Auk* in Nelson's paper on 'New Birds from Mexico and Guatemala'. Robert and John Ridgway had previously painted the plates for Nelson's *Report on Natural History Collections made in Alaska between 1877 and 1881* (1887).

Ridgway himself named new birds after Spencer F. Baird, William Gambel, Charles K. Worthen, José Zeledon, Lyman Belding, Charles Nutting and many others, thus demonstrating his approval of eponyms when they honoured outstanding naturalists and collectors. It was a habit he had caught from Baird.

Ridgway's Whip-poor-will

ROSS, B.R. *HUDSON'S BAY COMPANY CHIEF TRADER WHO INVESTIGATED THE ZOOLOGY OF THE CANADIAN BOREAL FORESTS.*

Bernard Rogan Ross
(1827-1874)

ROSS'S GOOSE *Chen rossii* (Cassin)

Anser Rossii Cassin, 1861. *Proceedings of the Academy of Natural Sciences of Philadelphia* 13, p. 73: Great Slave Lake

For many people, thoughts of Ross's Geese bring to mind winter field trips in California, New Mexico, Texas or some other location in the southern States. But instead of being discovered on its wintering grounds the species first came to light in the summer months in remote and sparsely populated areas of northern Canada. The first inkling of the existence of Ross's Goose, the arrival at museums of the first specimens and the comparatively late, indeed recent discovery of the first down-lined nest are historical events all inextricably linked with the Hudson's Bay Company. In this short chapter on Chief Trader Bernard Rogan Ross we have chosen to place him within the wider context of the history of the Company and its dedicated officers with their long tradition of geographical exploration and scientific discovery.

Some of the earliest zoological specimens from North America were collected by Hudson's Bay Company employees. Birds from Alexander Light and James Isham were illustrated in George Edwards's *A Natural History of Birds* (1743–51). Andrew Graham, Humphrey Marten and Thomas Hutchins were among those who sent more specimens to Britain and these were described by J.R. Forster in 1772. Samuel Hearne, the most observant of the early contributors, was chosen by the Company to search for copper deposits and bring back information on the North-west Passage, both rumoured to exist further to the north. Hearne, a rugged individual of unusual determination, with no qualms about existing on deer blood and entrails, flesh and skin boiled in fat, or buffalo fetuses, made three overland attempts to reach the Arctic. In July 1771 he succeeded in reaching the mouth of the Coppermine River thus becoming the first white man to reach the Arctic in those regions. When Hearne's *Journey from Prince of Wales Fort in Hudson Bay to the Northern Ocean in the years 1769, 1770, 1771 and 1772* was published in 1795 it contained a reference to the "Horned Wavey". He was undoubtedly describing the bird we now know as Ross's Goose but with no specimens and no scientific name his account becomes only a historical curiosity:

"This delicate and diminutive species of the Goose is not much larger than the Mallard Duck. Its plumage is delicately white, except the quill feathers, which are black. The bill is not more than an inch long, *and at the base is studded around with little knobs about the size of peas,* but more remarkably so in the males. Both the bill and feet are of the same color with those of the Snow Goose.

This species is very scarce at Churchill river, and I believe are never found at any of the Southern settlements; but about two or three hundred miles to the north west of Churchill I have seen them in as large flocks as the common Wavey or Snow Goose. The flesh of this bird is exceedingly delicate; but they are so small, that when I was on my journey to the North, I eat two of them one night for supper. I do not find this bird described by my worthy friend, Mr. Pennant, in his 'Arctic Zoology.' Probably a specimen of it was not sent home ..."[1]

It was ninety years from the time of Hearne's sightings until the first skins of the goose were sent to a museum so that the species could be properly described and named. In 1861 in the *Proceedings of the Academy of Natural Sciences of Philadelphia* John Cassin published some notes on four species of geese, the third of which appeared as 'ANSER ROSSII Baird' to honour Bernard R. Ross, a regular correspondent of the Smithsonian Institution who knew the goose from the Great Slave Lake region and had constantly insisted in his letters that it was a valid new species. Although Cassin was the author of the paper he said that he was proposing the goose name at the suggestion of Professor Baird, Assistant Secretary at the Smithsonian.

Ross became one of the most outstanding contributors to a second resurgence of Hudson's Bay Company natural history investigation which began in the mid-nineteenth century and far exceeded the first period in the quantity and quality of material. Some indication of the unique biota of the Arctic regions had been given by the earlier Company officers and still more had been gathered by Dr John Richardson and Thomas Drummond, both of whom frequently relied upon the goodwill of the Company for the success of their expeditions. Their results were published in the *Fauna Boreali-Americana* (1829–37) and in the companion volume on the flora (1833–40); each gave a good baseline from which future research could proceed but much more remained to be discovered.

Chief Trader Ross found himself in Canada following contacts made by Sir George Simpson, Governor of the Hudson's Bay Company, with his maternal uncle in Ross's home town of Londonderry, Ireland. Born on 25 September 1827, educated at Foyle College, Londonderry, Ross arrived in Canada at the age of sixteen and remained there for the rest of his life, enjoying almost thirty years with the Company. Starting as an apprentice clerk at Norway House he went on to serve, in various capacities, at Lac La Pluie, York Factory, Frances Lake, Fort Simpson, Fort Norman, Fort Liard and Fort Resolution, a series of posts extending from the shores of Hudson Bay to the Canadian Rockies. He was appointed chief trader in 1856 and from 1858 to 1862 was at Fort Simpson in charge of the Mackenzie River District. (For a map see p. 190.)

The bulk of his natural history work was carried out while based at Fort Simpson. In June 1860 he wrote to Dr Richardson informing him that he lived at the fort in a fine house with a thousand-volume library and that he knew of thirty-one mammal and ninety-four bird species in the region. Ross had spent the previous winter examining an immense number of skins and had completed

'A popular treatise on the fur-bearing animals of the Mackenzie River' as well as 'An account of the animals useful in an economic point of view to the various Chipewyan Tribes,' each of which were published in the *Canadian Naturalist* for 1861. 'An account of the botanical and mineral products, useful to the Chipewyan Tribes of Indians' and his 'List of Mammals, Birds, and Eggs observed in the Mackenzie River District' appeared in the same journal in the following year. In the latter paper, Ross brought the number of known mammal species up to forty-eight and the birds up to 192. This rather high figure for the birds was due to the tendency then to separate out races into more species than necessary. In his short account of the Ross's Goose he pointed out that the Slave Lake Indians had special names for the different species of white geese, their various colour phases and their immature plumages. Ross himself believed that the colour phases of the Snow Goose were different species.

He was greatly encouraged at this time by his wife Christina Ross, daughter of Donald Ross who was in charge of the Norway House District. At the time of their marriage in 1860, Christina was already sending lepidoptera to W.H. Edwards, a wealthy American industrialist with an avid interest in butterflies.[2]

Bernard Ross did not specialize but devoted himself to securing practically anything collectable that would shed some light upon the wildlife and peoples of the north. From around Fort Rae on the Great Slave Lake he collected artefacts from the Dogrib Indians; from Fort Simpson he gathered geological specimens and Indian costumes of the Dogribs, Slaves and Nahannis; from Fort Churchill he sent items of Eskimo dress and everyday domestic utensils; and from all these places he collected mammals and birds. The British Museum (Natural History) lists 139 specimens of birds, 39 eggs and 16 nests from the Mackenzie River with many more from Fort Simpson itself. The Royal Scottish Museum still possesses Indian costumes and a small number of birds collected by him. To the Smithsonian, Ross sent mammals, birds, fish, insects and ethnological material as well as Sir John Franklin's gun from his first overland expedition and a sword from the second journey, retrieved from the Eskimos.

At Fort Simpson during the winter of 1859–60 Ross was host to Robert Kennicott, then at the start of his three year stint in the north. Ross was eight years older and less excitable but they had much in common. Writing to Baird in November 1859, Ross promised to continue to send material and meteorological records southwards and confirmed his support for the new arrival: "In conclusion I will merely say that all that lies in my power will be done to oblige you in any way. Every facility will be given to Mr. R. Kennicott to collect and forward specimens of natural history; free passage will be allowed him from post to post throughout the district, and to all his plans the various officers under my command will, I am sure, gladly render assistance."[3] Indeed it was Kennicott who sent the Smithsonian the first fragments of the new goose species: head, wings, tail and legs, as well as a nearly complete skin.

Ross had already devised his own programme of investigation and he encouraged his colleagues Roderick MacFarlane, Julian Onion, James Flett and Nicol Taylor to supply him with material and observations. Kennicott in turn encouraged MacFarlane, Laurence Clarke, John Reid, Alexander McKenzie, James Lockhart, William Hardisty and others. MacFarlane especially turned out to be an excellent bird man who gathered together more than 10,000 birds and eggs. He compiled 'Notes on and List of Birds Collected in Arctic America, 1861–66' published in the *Proceedings of the U.S. National Museum* for 1891 and a 'List of Birds and Eggs Observed and Collected in the Northwest Territories between 1880 and 1894' which appeared in Charles Mair's *Through the Mackenzie Basin* (1908). High on MacFarlane's list of priorities was the whereabouts of the Ross's Goose breeding grounds. He had great plans to explore the Mackenzie Delta and islands to the north, but he was transferred and saved from much fruitless searching in that area.

MacFarlane often speculated as to where the breeding grounds really were but though he lived until 1920 he never knew the answer. Ross retired in 1871 and died at Toronto on 21 June 1874—so he never knew either.

The puzzle was not solved until June 1938 when, rather appropriately, a Hudson's Bay Company officer made the discovery on the mainland, south of Victoria Island. Angus Gavin found the first nests of Ross's Goose "on an island in a small lake about 12 miles up the Perry River, and 8 miles southeast along a tributary". Two years later he set off with a colleague to search for more nests in order to gather material evidence. They took photographs, specimens and eggs, returned triumphantly to their post, then sent off their trophies to Ottawa, where P.A. Taverner confirmed that they had indeed cleared up the long-standing mystery.

Ross's Goose

James Clark Ross (1800-1862)

ROSS'S GULL *Rhodostethia rosea* (MacGillivray)

L[arus] roseus MacGillivray, 1824. *Memoirs of the Wernerian Natural History Society* 5, p. 249: Igloolik, Melville Peninsula

In 1836 an anonymous fellow officer on the *Cove* described James Clark Ross as "without exception, the finest officer I have met with, the most perservering indefatigable man you can imagine."[1] Roald Amundsen, the great polar explorer of a later generation, considered Ross "one of the most capable seamen the world has ever produced."[2] While few would quibble with these assessments of Ross's professional capabilities, his position as an amateur naturalist is much less clear. He certainly knew a great deal about the larger forms of polar wildlife. He enjoyed shooting and preparing bird and mammal skins but he never really studied natural history in depth and his degree of scientific competence in his favourite branch, marine zoology, is impossible to assess because he never wrote up the results of his major collection.

Born in London on 15 April 1800, James Ross first went to sea at the age of twelve with his uncle, John Ross, and made a number of subsequent voyages with him. In 1818 they went on the *Isabella*, a hired whaler from Hull, to search for the North-west Passage. James Ross, then only a midshipman, formed a lifelong friendship with Edward Sabine, a young army officer who had been invited along as astronomer. Sailing northwards, accompanied by the *Alexander* under William Parry, they reached as far as Lancaster Sound then turned back because John Ross mistakenly believed the way ahead to be blocked by mountains. William Parry returned the following season, with James Ross and Sabine, and spent the winter well to the west of Lancaster Sound but could find no through passage. In 1821 Ross went again with Parry and overwintered twice, at Winter Island and Igloolik. In 1824 he went on Parry's ill-fated third attempt at the North-west Passage and in 1827 Ross made a fourth expedition with Parry, this time in an unsuccessful attempt to reach the North Pole from Spitzbergen. Two years later Ross teamed up with his uncle for yet another attempt on the North-west Passage which turned into an epic struggle for survival when ice conditions forced them to spend four consecutive winters in

ROSS, J.C. *ROYAL NAVY OFFICER WHO MADE SEVERAL VOYAGES IN SEARCH OF THE NORTH-WEST PASSAGE AND LED A FOUR-YEAR EXPEDITION TO ANTARCTICA.*

the north. They eventually had to abandon their ship and by transferring to smaller boats made their escape by dragging them across the ice back to Baffin Bay. "Dressed in the rags of wild beasts ... and starved to the very bones", they were by chance spied by the whaler *Isabella* and taken back to Hull.

Ross's experience of eight winters and fourteen navigable seasons in the Arctic made him the ideal choice to lead a four and a half year expedition to the Antarctic. Three seasons were spent exploring the fringes of Antarctica from winter bases in New Zealand, Tasmania and the Falkland Islands. Ross was knighted for his leadership of this exacting voyage, from which he returned a very tired man. Now in his mid-forties he wanted to settle down to married life but instead went north for one last voyage to assist in the unsuccessful search for the missing Franklin expedition.

In his early years in the Arctic, Ross often took on the role of hunter–naturalist, his primary aim being to secure fresh meat for the crews. The collection and preparation of specimens helped him to while away some of the more tedious hours on board ship. As early as 1828, he claimed that so much work had been done on the natural history of the far north that little more remained to be discovered, from which we can infer that he was not a great scientist. He was content to record and describe the animals he saw for the pure and simple

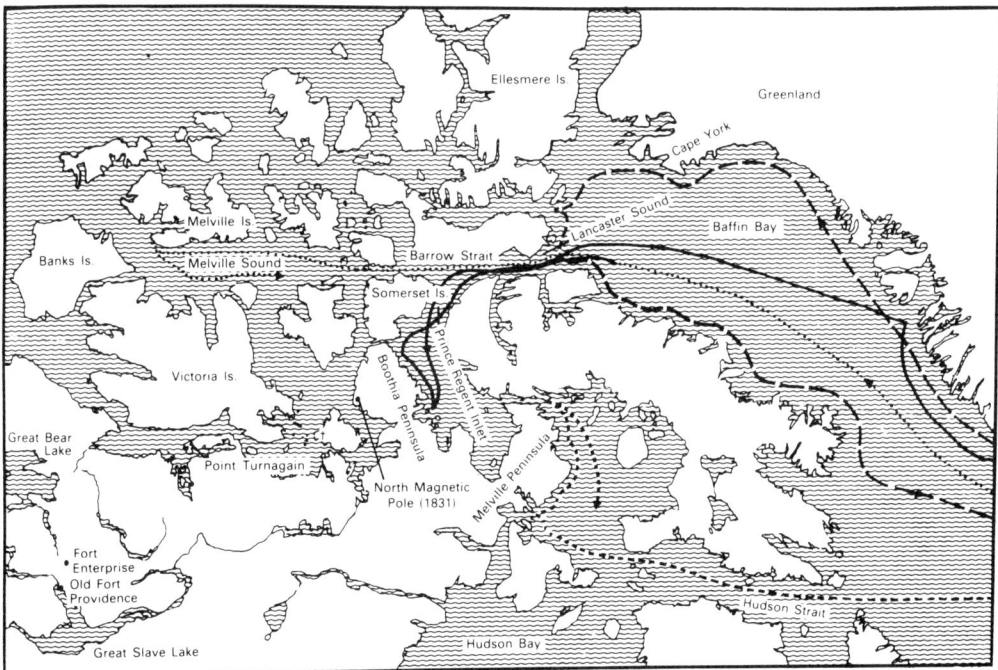

THE SEARCH FOR THE NORTH-WEST PASSAGE: ROUTES OF ROSS AND SABINE
‒ ‒ ‒ ‒ JOHN ROSS, 1818; WITH JAMES ROSS AND SABINE
············· PARRY, 1819–20; WITH JAMES ROSS AND SABINE
---------- PARRY, 1821–23; WITH JAMES ROSS
───── JOHN ROSS, 1829–33; WITH JAMES ROSS

enjoyment of it and even in his role as leader of the Antarctic voyage he continued his practice of occasionally making study skins.

Apart from some descriptive passages on birds in his *Voyage of Discovery and Research in the Southern and Antarctic Regions* (1847) Ross made only a small contribution to the zoological literature. He did not enjoy writing and confined himself to a few appendices for the official accounts of some of his Arctic expeditions. The first of these was for Parry's *Journal of a Third Voyage for the Discovery of a North-West Passage ... in the Years 1824–25* (1826) to which he added very brief, factual natural history notes. The second was for Parry's *Narrative of an Attempt to Reach the North Pole ... in the Year 1827* (1828) for which his notes were even more abrupt. For John Ross's *Narrative of a Second Voyage in Search of a North-West Passage During the Years 1829–1833* (1835) James added similarly short notes on the birds, mammals and fish. Other naturalists wrote better appendices for the remaining Arctic expeditions in which Ross took part. Edward Sabine and John Richardson, for instance, wrote up the zoology for Parry's 1819–20 and 1821–23 voyages.

It was during this last-mentioned attempt on the North-west Passage that Ross became associated with the bird now known as Ross's Gull. On 23 June 1823, while exploring the Melville Peninsula near Igloolik, Midshipmen Ross and Sherer each shot one of the gulls. Parry gives us the most interesting account of the event in the main text of his *Journal*:

> "Our shooting parties to the southward had of late been tolerably successful, not less than two hundred and thirty ducks having been sent in to the ships in the course of the last week. Mr. Ross had procured a specimen of gull having a black ring round its neck, and which in its present plumage, we could not find described. This bird was alone when killed but flying at no great distance from a flock of tern, which latter it somewhat resembles in size as well as in its red legs; but is on closer inspection easily distinguished by its beak and tail, as well as by a beautiful tint of most delicate rose-colour on its breast."[3]

The two gull specimens became part of the official collection; one went to Edward Sabine's brother Joseph and the other to Edinburgh University Museum. John Richardson was commissioned to work through the birds but, before he could finish, William MacGillivray saw the specimen at Edinburgh and mentioned it in his 'Descriptions, characters and synonyms of the different species of the genus *Larus*, etc.' He noted the unusual shape of the tail, called it *L. roseus*, and later used the name Ross's Rosy Gull. MacGillivray's paper was read before

Ross's Gull

the Wernerian Society in February 1824 and published that year. Richardson had already told the society about the gull in the preceding month, naming it the Cuneate-tailed Gull *Larus Rossii*, but his description had to be published in Parry's official account, which did not come out until later in 1824. Richardson was none too pleased when MacGillivray's scientific name won priority and he later complained that it was ridiculous that with only two known specimens there should be two different scientific names and two different common names.[4]

As the years went by, more Ross's Gulls were occasionally seen and collected. But the gradual increase in the number of records gave no clue as to the birds' breeding grounds and it was over eighty years before the main area was discovered, at an unexpectedly low latitude, on the Kolyma delta in eastern Siberia. Isolated breeding records have come from west and north Greenland, and from near Bathurst Island and Churchill in Canada.

Ross saw the gulls again on his trip north of Spitzbergen in 1827. He did not find any on his longest sojourn in the north between 1829 and 1833 but he did come back with three specimens of the Yellow-billed Loon. Ross thought that it was another new species but Edward Sabine convinced him that they were just very old individuals of the Common Loon. Ross gave one to Hull Museum and another to Audubon, who also failed to realize that it was a new species, so a published description did not appear until 1859 (see Adams).[5]

We do not know when or where Audubon met Ross but it was most likely in London, some time between May 1834 and September 1836 when the artist was in Britain to co-ordinate the production of *The Birds of America* and to work with MacGillivray on the *Ornithological Biography*. Audubon noted that his drawing of a male Lapland Longspur was "from a beautiful specimen in the collection of my esteemed friend Captain James Ross of the B.N." The Rock Ptarmigan, Red Phalarope and Hutchins's Goose were likewise drawn from Ross's skins.

In 1839, when Ross took command of the *Erebus* and the *Terror* for his Antarctic voyage, he had already developed an interest in marine zoology. In the months ahead he spent many hours searching the beaches of New Zealand and elsewhere for invertebrates and seems to have become slightly obsessive about dredging for marine organisms. He eventually amassed a substantial collection that he fully intended to work through when he returned home. Unfortunately, he had to give precedence to the official narrative (which took him four years to complete) and then he had to go north to search for Franklin and his two ships. By the time he got back the earliest marine specimens were already ten years old. The untimely death of his wife in 1857, leaving him responsible for four young children, put an end to any work on the samples and worse still, no one else examined them either. Shortly after Ross's death, on 3 April 1862, a huge pile of broken jars was found at the back of his house at Aylesbury, their contents all destroyed. It was a terrible waste of a deep-sea collection that could have stolen a little of the glory from the *Challenger* expedition of the 1870s.

During the southern voyage Ross had Robert McCormick and Joseph Hooker as his surgeon and assistant surgeon, both of them also acting as naturalists. McCormick was an old friend from the Spitzbergen voyage and though an ornithologist of sorts, in Hooker's opinion he was much less of a scientist than Ross (see McCormick). Hooker developed into one of the most prominent botanists of the late nineteenth century and was Director of Kew Gardens for twenty years, from 1865. Besides collecting plants, he helped with the marine work and in a letter to his father, he wrote: "No other vessel or collector can ever enjoy the opportunities of constant sounding and dredging and the use of the Towing-net that we do, nor is it probable that any future collector will have a Captain so devoted to the cause of Marine zoology, and so constantly on the alert to snatch the most trifling opportunities of adding to the collection."[6]

In the first volume of his *Flora Antarctica* (1847) Hooker thanked Ross for his encouragement, the use of his cabin and library, and for contributing some of the plants. After Ross's death he became much more critical, complaining that Ross "might have done a great deal to stimulate a love of scientific work in his officers, but they were treated as drudges and felt it keenly. I worked hard for him in his mania for collecting marine organisms but was never allowed an hour more of leave ashore than the idlest of my shipmates. Science was starved on board."[7] This seems most unfair, since Ross's primary objectives of geographical exploration and the accumulation of magnetic observations were exceeded to a remarkable degree, making it one of the most scientific of all nineteenth century voyages. Hooker's attitude seems to have been soured by the loss of the marine collection and the lack of time ashore—an age-old complaint from ship-board naturalists.

The Zoology of the Voyage of H.M.S. Erebus & Terror (1844–75) was edited by John Richardson and John Edward Gray. The best and most complete section, on the fishes, was by Richardson, based almost entirely on Ross's specimens. Gray's section on the seals of the southern hemisphere contains the first description of Ross's Seal, one of the rarest and least known of the Antarctic seals.

Gray handed over the ornithology to his younger brother, George Robert Gray, who concentrated on the birds from just one area. His *1. Birds of New Zealand* appeared in 1846 and dealt with about a hundred species. He listed six species of penguins, including the Emperor Penguin which Gray named *Aptenodytes forsteri* in the belief that J.R. Forster had seen this species when with Cook's second Pacific voyage. And just as Ross's name is associated with one of the most enigmatic birds of the north, so too was his name attached to a mysterious bird of the southern regions. The Auckland Island Shore Plover *Thinornis rossi* is known only from a single specimen, collected during the expedition in November 1840. It is perhaps an aberrant juvenile of the rare New Zealand Shore Plover but others have suggested that it is a similar, quite separate species that has now vanished.[8]

Lady Ross's Turaco *Musophaga rossae* is a central African bird named by John Gould in 1851 after Ross's wife, Anne.

Edward Sabine
(1788-1883)

SABINE'S GULL *Xema sabini* (Sabine)

Larus Sabini Sabine, 1819. *Transactions of the Linnean Society of London* 12, p. 522, Pl. 29: Sabine Islands, near Melville Bay, west coast of Greenland

Although he was a keen ornithologist Edward Sabine's main interests were terrestrial magnetism and astronomy. It was his capabilities in these latter fields that secured a place for him on John Ross's 1818 expedition to search for the North-west Passage (For a map of Sabine's voyages in 1818 and 1819–20, see J.C. Ross, p. 375.) The *Isabella*, under John Ross, and the *Alexander* under William Parry, set sail from England in the spring and headed northwards to the west coast of Greenland and Baffin Bay, in the hope of penetrating westwards between the islands of the Canadian Arctic to the Pacific Ocean. The Captain's nephew, James Clark Ross, was on board the *Isabella* with Sabine, and the two young men became lifelong friends.

By July the two ships were well beyond the Arctic Circle. On the 25th of the month, not far from Cape York, some low rocky islands, each about a mile across, were sighted about twenty miles offshore. Sabine, James Ross, several of the men and the assistant surgeon set out over the ice to investigate them. Amongst the hundreds of Arctic Terns breeding there, the explorers noticed a number of delicate tern-like gulls with forked tails, dark heads and yellow-tipped black bills. Their task of shooting some of them was made easier by the gulls' vigorous and again, tern-like defence of their newly hatched young. Several adult birds were quickly added to Sabine's collection but were later transferred to a southbound whaler and conveyed to Sabine's elder brother Joseph, in London. Joseph displayed the birds at a meeting of the Linnean Society in mid-December and the gulls, "hitherto unknown and undescribed ... in conformity with the custom of affixing the name of the original discoverer to a new species", he named *Larus Sabini*.

Sailing yet further northward, the explorers found the way between Greenland and Ellesmere Island blocked by ice, so they turned south and entered Lancaster Sound. John Ross thought he saw a range of mountains blocking their route westwards and decided to turn about and return home. It proved to be an

SABINE. *ENGLISH SCIENTIST WHO MADE TWO ARCTIC VOYAGES*
WITH HIS FRIEND JAMES CLARK ROSS.

unwise decision. Ross had concluded that no passage existed because there was no swell from the west and because there was no current with any driftwood. It is probable that all on board agreed with the decision but on the return journey some of the officers changed their minds. In the 'Narrative' of the expedition, published in the following year, John Ross tried to justify his action, noting that Sabine had agreed with him at the time. Ross also commented that he had been led to believe that Sabine would assist in compiling the zoological appendices. However, Sabine refused to cooperate and they were written by the surgeon and assistant surgeon, then edited and corrected by William Leach.

Instead, Sabine brought out his own account of the ornithology of the voyage. Entitled 'A Memoir of the Birds of Greenland', it was published in the *Transactions of the Linnean Society* (1819), in the same volume as his brother's description of the new gull. In Edward Sabine's thirty-two page account, he listed the twenty-four species of birds they had encountered during the course of the expedition, including Gyrfalcon, Peregrine Falcon, Rock Ptarmigan, Snow Bunting, Glaucous Gull, Northern Fulmar, King Eider and Ivory Gull. At 76° N, hundreds of Dovekies were killed daily for food. Of the shorebirds, they had seen Dunlin, Ringed Plover, both species of Arctic-breeding phalarope, Purple Sandpiper and Red Knot; failing only to observe the Ruddy Turnstone and Sanderling amongst the shorebirds known to breed in the areas visited.

Some more remarks on Ross's voyage were published in the same year by Sabine but they were of a discordant nature. His criticism of John Ross added still further to the ill feeling between them. But this animosity never extended towards James Clark Ross. When the Admiralty decided on a second expedition to the same area Sabine and the younger Ross gladly joined Parry, who had been selected as leader in preference to the elder Ross. In May 1819, the ships *Hecla* and *Griper* set out for Baffin Bay and sailed right through Lancaster Sound into open water beyond; the Croker Mountains which had been thought to bar the way westward did not exist.

Although winter was approaching they continued westward, extending their season a little longer than perhaps they might otherwise have chosen to do, because they all knew that by passing beyond 110° W, in that region, they would share a £5000 reward from the Admiralty. There was a great celebration

Sabine's Gull

on 4 September when they crossed the line, but winter closed in on them almost at once and they found it hard to locate a suitable anchorage. Eventually, with large saws, they managed to cut a 4000 yard channel through the ice and the ships were man-hauled into Winter Harbour, on Melville Island.

It was a long winter. The sun did not show itself from 4 November to 3 February, but they had plenty of provisions and settled in to pass the time as pleasantly as possible. Sabine began, and edited, a hand-written weekly journal entitled *The North Georgia Gazette and Winter Chronicle* which ran for twenty-one weeks. When daylight returned they made brief hunting trips and at the beginning of June Parry set out on a ten-day trek to the north shore of the island but he found it utterly desolate, almost devoid of life. In August they sailed fifty miles further west but the ice there was impenetrable. They were too far north to have any chance of finding a passage to the Pacific because of the permanent pack-ice to the west. They returned homewards after this disappointment and arrived in England in November 1820. Parry's *Journal* of the expedition had natural history appendices on the fish, mammals and birds, all written by Sabine. This time he was able to include the Sanderling and he also mentioned a single Sabine's Gull seen on the wing in Prince Regent Inlet.

The rest of Edward Sabine's life was dominated by his interest in the earth's magnetic field. In November 1821, in connection with this, he sailed on the *Pheasant* to the West Indies, Ascension Island and Sierra Leone, returning to England in January 1823, but in May of the same year he was back on the *Griper* continuing his magnetic observations in New York, Greenland, Trondheim and Spitzbergen. In July 1833 he was in south-west Ireland and found time to compile a *Notice on the Birds met with on the Coast of Ballybunian*, which appeared as an appendix to William Ainsworth's *Caves of Ballybunian* (1834).

In the following year, in conjunction with James Clark Ross and Professor Humphrey Lloyd, Sabine started the first systematic magnetic survey ever made in the British Isles. Starting in Ireland, it was extended to Scotland in 1836 and afterwards to England, Sabine being responsible for most of the written work. Not long after this, Humboldt urged the Royal Society to carry out a world-wide magnetic survey. The plan was quickly approved and a number of nations were invited to participate. Partly because of this, Ross was sent on a four-year expedition to Antarctica. When he first sighted land there he named the highest peak Mt Sabine in honour of "one of the best and earliest friends of my youth and to whom the compliment was more especially due, as having been the first proposer and one of the most active and zealous promoters of the expedition."[1]

To enable Sabine to cope with the mass of observations that were sent to him, he acquired a small clerical staff that assisted him at the War Office for the next twenty years. His wife, Elizabeth Leeves of Sussex, whom he had married in 1826, helped him in his scientific work for over fifty years. Eventually he was forced to discontinue his research which, perhaps inevitably, was never completed. He died without issue at Richmond, on 26 June 1883, at the age of ninety-five and was buried beside his wife's remains in the family vault in Hertfordshire.

Edward Sabine's cheerfulness and attractive personality had made him popular among most of his colleagues. His main scientific achievement was his

Contributions to Terrestrial Magnetism which he began in 1840. There were fifteen papers on the subject in the *Philosophical Transactions of the Royal Society*. His contribution to natural history was less significant but as President of the Royal Society (1861–71) he was in a position to advise and direct others. For instance, he was a friend and supporter of the botanist David Douglas and gave him instructions in surveying and in taking geographical observations which Douglas put to good use in his Canadian travels.

Sabine presented some birds he had obtained on his 1821–23 Atlantic voyages to the Natural History Department of the British Museum, but no record detailing the number of items or the identity of the species has survived. However, the collection almost certainly included Sabine's Puffback Shrike *Dryoscopus sabini* and Sabine's Swift *Chaetura sabini*; birds named by J.E. Gray and both native to the forests of Sierra Leone.[2] Shortly after his visit to West Africa, Sabine is reported to have shot a pair of Sabine's Gulls on Spitzbergen,[3] but there is no confirmation that the pair were breeding, which is unfortunate as proven instances from these islands are few in number.

After ending this brief outline of Sabine's scientific interests it is astonishing to realize that he also conducted an active military career. Edward Sabine was in the Royal Artillery for more than seventy years! When he retired from the army in 1877 he had risen to the rank of General, as his grandfather had done before him.

Edward was the fifth son and ninth child of Joseph Sabine of Tewin, Hertfordshire. Born at Dublin on 14 October 1788 and then educated at Marlow, Edward was no doubt encouraged to embark upon a military career. He entered the Royal Military Academy at Woolwich at fifteen and was made a first lieutenant in the following year. He was based at Gibraltar for a while before serving at various stations in Britain. In January 1813 he was sent to Canada, but the ship was only eight days out of Falmouth when it was attacked by an American privateer. There was a running battle for twenty-four hours during which Sabine and his soldier-servant helped to man the guns until forced to surrender. Not long after, the ship was recaptured by a British frigate and Sabine eventually landed safely in Canada. During that winter, as the American militia advanced on Quebec, Sabine was directed to garrison a small outpost. He later served on the border near Niagara and took part in an attack on Fort Erie when the British lost twenty-seven officers and over 300 men. He was again in action whilst on a sortie when the British suffered more heavy losses.

After his return to England in 1816 Captain Sabine devoted himself to his favourite scientific studies, for which he often received leave of absence. In 1830, after his aforementioned Arctic and Atlantic voyages, he was recalled for military duty in Ireland and for the next seven years managed to direct his magnetic surveys from there. In 1865 he was made Colonel Commandant of the Royal Artillery, and then, in the same year, was promoted to Lieutenant-General and later received a knighthood.

SAY. *FATHER OF AMERICAN ENTOMOLOGY. MEMBER OF MAJOR LONG'S ROCKY MOUNTAIN EXPEDITION (1819–20) FOR WHICH HE PREPARED THE REPORT ON THE BIRDS. THIS PORTRAIT WAS PAINTED BY CHARLES WILLSON PEALE IN 1819.*

Thomas Say
(1787-1834)

BLACK PHOEBE *Sayornis nigricans* (Swainson)

Sayornis Bonaparte, 1854. *Comptes Rendus de l'Académie des Sciences de Paris* 38, p. 657. Type, by monotypy, *Sayornis nigricans* Bonaparte = *Tyrannula nigricans* Swainson.
Tyrannula nigricans Swainson, 1827. *Philosophical Magazine*, new ser., 1, p. 367: Table land of Mexico [= Valley of Mexico]

EASTERN PHOEBE *Sayornis phoebe* (Latham)

Muscicapa Phoebe Latham, 1790. *Index Ornithologicus*, Vol. 2, p. 489. Based on the "Dusky Fly-catcher" Pennant, *Arctic Zoology*, Vol. 2, p. 389, and the "Phoebe Flycatcher" Latham, *A General Synopsis of Birds*, suppl., 1, p. 173: in America septentrionali, Noveboraco [= New York]

SAY'S PHOEBE *Sayornis saya* (Bonaparte)

Muscicapa saya Bonaparte, 1825. *American Ornithology*, Vol. 1, p. 20, Pl. 11, Fig. 3: Arkansaw River, about twenty miles from the Rocky Mountains [= near Pueblo, Colorado]

Thomas Say was a tall slim individual, fully six feet in height, with a dark complexion and a thick mop of black hair. His many friends respected him for his honesty, sincerity and kindness, and marvelled at his modesty and complete disinterest in money or possessions—with the exception of his natural history books which he cherished. There were two men who had a powerful influence on Say, the first of whom was his great-uncle, the botanist William Bartram. Bartram was forty-eight years old when Thomas was born on 27 June 1787 in Philadelphia, and as the boy grew up he frequently visited Bartram, taking with him boxes of butterflies and beetles that he wanted identified.

After leaving school Say worked in his father's apothecary shop and later, with fellow naturalist John Speakman, set up a similar business that went bankrupt because they were too willing to extend credit to their friends. In 1812 Say and Speakman became founder members of the Academy of Natural Sciences of Philadelphia and for some years Say lived in the rooms of the Academy, immersed in his entomological studies. In December 1817 the wealthy philanthropist William Maclure was elected president of the Academy and for years he furthered its objectives by generous gifts of money and books. He also took an interest in the young man who worked, ate and slept on the premises; Maclure approved of such dedication and became Say's other mentor and benefactor.

In the autumn of 1817 Maclure organized an expedition to Florida and took Say, George Ord and Titian Peale along with him. The trip was cut short by hostilities between the tribes of Spanish Florida and their American neighbours, so the naturalists spent most of the winter on the Georgian coast.

For the rest of his life Say continued his vocation as a naturalist, becoming

SAY'S TRAVELS
------- 1817–18
-·-·-·-·- 1819–20
----- 1823

North America's first great systematic entomologist. He also made his mark as a conchologist and took an interest in almost every other branch of zoology. Here we shall concentrate on the new bird discoveries with which he was associated, but it should be remembered that these were incidental to his main line of research. He contributed to the early history of American ornithology in two ways. Firstly, by describing nine new species discovered on an expedition he made to the eastern side of the Rocky Mountains in 1819–20; and secondly, by assisting Charles Bonaparte with his continuation of Wilson's *American Ornithology*.

The Rocky Mountain expedition, planned by the Secretary of War and underfunded by Congress, was part of a venture to establish military posts on the Upper Missouri. Say served as zoologist with the scientific section under the command of Major Stephen Harriman Long of the Topographical Engineers, and nineteen year old Titian Peale was appointed to help him collect and prepare specimens. They were the first members of a government expedition to be appointed specifically as naturalists.

On 5 May 1819 the group left Pittsburgh by steamboat and proceeded to St Louis for the beginning of the voyage up the Missouri. The first ornithological discovery was made four miles upriver at Bellefontaine, when many Lark Sparrows were shot. Say turned the problem of groundings and frequent mechanical breakdowns to his advantage by travelling overland for parts of the journey, at much the same speed as the boat. Early in August he was charged with leading a small exploring party on foot into the country between Fort Osage and the River Platte. He became sick, but the Kansas Indians welcomed them to their village and during four days of rest there he partially recovered and made notes about their lifestyle. The explorers were less fortunate in encountering a large band of Pawnees seven miles to the north, who stole their pack horses, plundered their supplies and took some of Say's personal belongings. Returning to the friendly Kansas they hired more pack animals and a mount for Say who was too weak to walk, then abandoning their original plan they headed back to the Missouri arriving at Cow Island, near Leavenworth, four days after the boat had left. Say remained on the island with members of the military expedition until he had recovered sufficient strength to follow on.

During late September and early October Long's soldiers built snug log cabins with huge fireplaces for their winter base, which they called Engineer Cantonment. It was located on the west bank of the Missouri five miles below the military camp at Council Bluffs. When all was completed Long returned to Washington for the winter.

Despite persistent poor health, Say made good use of his eight months at Engineer Cantonment and followed his instructions "to examine the country, visit the neighbouring Indians, procure animals &c ...". In the company of Dougherty, the translator, he went among the Pawnee, Oto, Iowa, Missouri and Omaha Indians who were camped in the vicinity and learned much about their traditions. Procuring animals was a responsibility that Peale enthusiastically shared and so it is not always possible to know which of them obtained which specimens. Peale spent a lot of time hunting, both to determine which species were wintering locally and to provide fresh meat for them all. In an effort to capture Coyotes he constructed various kinds of traps and soon found himself marvelling at their intelligence, for they proved to be superbly adept at seizing the bait and avoiding the trap, however carefully it was concealed. For a long time he only managed to catch shrews in the pitfalls—but two of these, the Least and the Short-tailed Shrew, were new to science. Eventually Peale succeeded in outwitting a Coyote, which Say also had the satisfaction of describing as new.

Say kept a bird list and noted that Snow Buntings, Black-capped Chickadees, American Goldfinches, Northern Flickers, White-breasted Nuthatches and a few Eastern Bluebirds were around the camp throughout the winter; Pine Grosbeaks and Carolina Parakeets were only seen occasionaly. By the end of March Ravens were down on eggs and Belted Kingfishers were back on the creeks. During April and May he enjoyed the almost daily arrival of summer visitors, and among those he mentioned for April were Turkey Vulture on the 2nd, American Robin on the 11th, Marsh Wren on the 15th, Common Nighthawk on the 19th and passage Blackpoll Warblers on the 26th. He also noted that mosquitos first appeared on April 12th!

May brought Summer Tanager, Passenger Pigeon, Ruby-throated Humming-bird, Willet, Osprey and an Orange-crowned Warbler "Shot at Engineer Cantonment early in May". Say was unable to find any description of this little bird and rightly assumed that it was a new species, passing through on its way north.

Another new species appeared among the many wader flocks that stopped off to feed on the marshy pools near the river, in spring and fall. This was the Long-billed Dowitcher and several in breeding plumage were shot at a pond near Bowyer's Creek and named by Say.

Major Long rejoined his command at the end of May, accompanied by Dr Edwin James in the capacity of botanist and geologist. Long's revised instructions were to ascend the Platte to its source in the Rockies, then head for the Mississippi by way of the Arkansas and Red Rivers, a journey for which they were extremely poorly equipped; there were no spare mounts and only eight mules to carry the inadequate provisions. Setting off on 6 June they visited some nearby Pawnee villages, but encountered no more natives on the westward journey.

For the naturalists, so long confined to the region around Council Bluffs, it was exhilarating to burst out onto the prairies that stretched ahead of them all the way to the mountains. They were awed by the huge herds of roaming Buffalo and fascinated by the prairie dog towns which sometimes extended for mile upon mile. Say spent hours watching these strange squirrels playing and grooming around the burrow entrances, ready to flee underground at the first sign of danger. He excavated some of the subterranean chambers and examined the fine, dry grasses that the animals used to make their nests. Wherever he saw prairie dogs he found Burrowing Owls as well, usually inhabiting the older, more ruinous abandoned passages.

Long followed the north side of the Platte to its fork where they crossed and continued along the South Platte. On 30 June they sighted the distant peaks of the Rockies and for the next week they followed the river up into the hills.

Here, where the cool, clear water gushed through narrow gorges and grassy valleys scattered with oak and hazel copses they found Rock Wrens; one of six new species that Say would describe from specimens taken during the strenuous and action-packed days between 6 and 20 July.

Now on their way southwards to the Arkansas River, the travellers spent two nights at a campsite by Plum Creek (near Castle Rock, Colorado). There were so many Beaver dams thrown up across the creek that it looked more like a succession of ponds than a continuous stream. Small willows and cottonwoods grew along the banks, providing building material for the Beavers, firewood for the campers and seeds for small flocks of House Finches that Say watched working their way through the treetops. Some of the more energetic members of the party set off to climb a steep hill to the east, and from the summit found that Pike's Peak was in sight to the south-west. Others followed the creek for as far as they were able and Peale made sketches of the more interesting geological features. He returned to camp carrying a large, plump pigeon he had shot, its plumage predominantly a mixture of soft

Eastern Phoebe

greys and purples, with a narrow white stripe on the nape and a dark band on the tail. Realizing that this was another new bird, Say named it *Columba fasciata*—the Band-tailed Pigeon. A Blue Grouse, taken on the same day, was also undescribed, although this species was one of those noticed by Lewis and Clark.

The Lesser Goldfinch was the next ornithological discovery, found at Boiling Spring Creek camp while Dr James and two men trudged up Pike's Peak. In the zoological notes that Say added to the expedition narrative he wrote of this "very pretty little bird" which "was frequently seen hopping about in the low trees or bushes, singing sweetly, somewhat in the manner of the American goldfinch, or hempbird, *Fringilla tristis*".

The explorers reached the Arkansas about twenty miles below present day Canon City. Here, they found two new mammals: the Golden-mantled Squirrel and the Colorado Chipmunk. Say and Peale examined the chipmunks' nests and watched them nimbly extracting pine seeds from fallen cones. They also took another two new birds: Say's Phoebe (of which more will be said later) and the Lazuli Bunting.

Cliff Swallows, which they had found to be common throughout all the rugged country near the mountains, were breeding here in great numbers. Say examined a colony that had built their nests under a large, overhanging rock and saw that they contained both eggs and young. He hoped that the swallows were yet another new species, but they had already been named by Louis Vieillot from Paraguayan specimens, in 1817. A rattlesnake and some other reptiles were captured at this spot and Say was of course continually adding to his collection of insects.

After spending only two weeks in the mountains, the explorers turned their backs on them and began the descent of the Arkansas River, viewing the thousand mile journey back to civilization with little enthusiasm. This retreat before they had properly mapped the sources of the Platte was forced upon them by the shortage of food. The following day, the 20th of July, was warm and fair and as they rode along the broad river valley between low, distant hills they shot a Western Kingbird, the last ornithological discovery of the expedition.

The next day they encountered native Americans for the first time since leaving the Pawnee villages and were informed that the greater part of six nations were fighting to the south-west, thus explaining why they had not met with any of the expected tribes. In the evening Long chose a campsite near the present location of La Junta and divided his men into two teams so that both the Red and Arkansas Rivers could be explored. He led the longer, more important investigation of the Red River, taking with him James, Peale and seven others while Captain J.R. Bell led Say, the artist, a cadet, three interpreters and five riflemen down the Arkansas, which had already been travelled by Zebulon Pike.

On 9 September Bell's contingent arrived at Fort Smith, the prearranged rendezvous, Say having suffered a sickening loss when three of the soldiers deserted. Each took with him one of the best horses and some of the packs, including the one that contained all of Say's natural history observations, his journal and three volumes of notes on Indian languages and ethnography. Fortunately the collections containing hundreds of precious type specimens of insects, birds, mammals and reptiles were safe. Long's party reached the fort only four days after Bell's, because the Major had mistaken the Canadian River for the Red River and realized his mistake too late to backtrack.

The reunited force left Fort Smith on 21 September and followed the Arkansas to the Cherokee villages on Illinois Creek, then went overland to Cape Girardeau where they were nearly all prostrated by fever. The expedition disbanded there, the members departing for different destinations as and when they recovered. Around the beginning of November, Say and Peale boarded a steamboat bound for New Orleans and returned by sea to Philadelphia.

During the winter Dr James wrote his *Account of an Expedition from Pittsburgh to the Rocky Mountains* (1823). Say supplied appendices listing the mammals, reptiles and birds observed at Engineer Cantonment, original descriptions of the new birds and mammals, meteorological records and native vocabularies as well as the account of Bell's descent of the Arkansas to Fort Smith.

In 1821 Say was made curator of the American Philosophical Society and as time allowed he wrote up the entomological results of the expedition, a lengthy and laborious undertaking as he had collected several thousand items. There were also fossils and molluscs to work through.

In 1823, at the age of thirty-six, Say travelled west again on another expedition led by Major Long, this time heading much further north into British territory around Lake Winnipeg. The journey of around four and a half thousand miles took only six months but as before Say collected insects, shells, birds and mammals at every opportunity. Throughout the summer he enjoyed much better health than on the previous expedition and he revelled in all the new experiences, which included frequent meetings with the natives and a return journey by canoe through Lake of the Woods to Lake Superior.

By the end of October he was back in Philadelphia, where he made the acquaintance of Charles Bonaparte who had arrived from Europe in the summer. Bonaparte had just begun to study the American avifauna and was very interested in the species Say had brought back from the West. Although Say had already described nine of them (Lark Sparrow, Long-billed Dowitcher, Orange-crowned Warbler, Rock Wren, Band-tailed Pigeon, Blue Grouse, Lesser Goldfinch, Lazuli Bunting and Western Kingbird) Bonaparte described two more birds in the collection that had been deposited at Peale's Museum. These were the Yellow-headed Blackbird, taken at Pawnee villages on the Platte (but which had long been confused with a similar inhabitant of South America) and Say's Phoebe, which Bonaparte named *Muscicapa saya*. He included both in his *American Ornithology; or, The Natural History of Birds Inhabiting the United States Not Given by Wilson*. Writing about Say's Phoebe, he said:

Say's Phoebe

"I dedicate it to my friend THOMAS SAY, a naturalist, of whom America may justly be proud, and whose talents and knowledge are only equalled by his modesty. The specimen now before us is a male, shot by Mr.T. Peale, on the seventeenth of July, near the Arkansaw river, about twenty miles from the Rocky Mountains . . .

We know nothing of the habits of this Flycatcher, except what has been communicated by Mr.T. Peale, from his manuscript notes. The bird had a nest in July, the time when it was obtained; its voice is somewhat different from that of the Pewee, and first called attention to its nest, which was built on a tree, and consisted chiefly of moss and clay, with a few blades of dried grass occasionally interwoven. The young birds were, at that season, just ready to fly."

Many of the species that Bonaparte added to Wilson's work were known to him only as skins, so Say's first-hand observations were of great value to him, especially for the sections on Cliff Swallow, House Finch, Burrowing Owl, Scissor-tailed Flycatcher and those species that Say had first described. The book was published in four volumes between 1825 and 1833 and many of the illustrations were by Titian Peale. Since Bonaparte had spent most of his life in

Italy and was unfamiliar with the English idiom, Say edited the first volume for him and also corrected his papers for the Academy's *Journal*.

The generic name *Sayornis* was not created by Bonaparte until 1854. Initially it was used for the Black Phoebe *Sayornis nigricans*, which he included in a list in his *Comptes Rendus de l'Académie des Sciences de Paris*. The genus now includes three species of phoebe: Black, Eastern and Say's. Only the last has any connection with Say as the Black and Eastern Phoebes were discovered by other naturalists.

Say's own great contribution to the natural history literature was his *American Entomology; or, Descriptions of the Insects of North America*. He had planned the work in 1816 and issued a prospectus the following year, but the first volume did not appear until 1824. After the publication of the second in 1825 he went on a collecting trip in Sussex County, New York, in the company of the French artist and zoologist Charles LeSueur and the Scottish-American geologist William Maclure—his old friend and advisor. Maclure was now planning to go into partnership with the social reformer Robert Owen, who had purchased an estate of 30,000 acres in Indiana, containing a village, factories, farms, orchards and vineyards. Here they intended to found a new Utopian society where all members would live in equality and work together for the good of the whole community.

Maclure persuaded Say and LeSueur to join his band of disciples and they travelled together to Indiana at the beginning of 1826. Among the group was Lucy Way Sistaire, whom Say married in January 1827. New Harmony, as the community was mis-called, was riven by dissension from the start and soon the two leaders were trying to sue each other. Mrs Say quickly became disillusioned with the communistic lifestyle and longed to return east. Her husband remained committed to the experiment but it is impossible now to judge how much Say's loyalty was due to his affection for Maclure, his financial dependence on him or a genuine attachment to his ideals.

During the winter of 1827–28 Say and Maclure travelled together in Mexico for the sake of the older man's health, visiting Vera Cruz and Mexico City. Say went home in the spring with good collections of insects and shells but Maclure remained behind after deciding that his enlightened views stood a better chance of acceptance in Mexico. Thereafter Say managed his friend's business affairs as well as continuing to carry responsibility for the promotion of education, literature and science at New Harmony.

Despite these priorities he found time to continue with his own studies. The third and last volume of the *American Entomology* was published in 1828. There were fifty-four high quality plates contributed by Titian Peale, Charles LeSueur and others, making it a very attractive book, but though widely praised it was too expensive to be popular. Through the scientific importance of this pioneering work, Say has become known as 'the Father of American Entomology'. According to one authority he described 1155 species of coleoptera, 225 species of diptera, 100 species of hemiptera and 100 species belonging to other groups. Through neglect, Say's entire collection was destroyed by dermestid beetles after his death.

Between 1830 and 1834 Say oversaw the printing and publishing of six parts of his *American Conchology* at New Harmony and the seventh was brought out posthumously. Like the *American Entomology*, it was dedicated to William Maclure. Of the sixty-eight plates, all but two were drawn by Lucy Say who shared her husband's interest in shells.

During his eight and three-quarter years at New Harmony Say suffered frequent bouts of sickness that usually began with bilious attacks and ended with fever and dysentery. According to some of his friends, his habit of fasting and eating frugally weakened his capacity to recover and his death on 10 October 1834, at the age of forty-seven, was not unpredicted. His widow moved to New York shortly afterwards, made her home there with her sister and continued her own work on molluscs. In 1841 Lucy Say became the first woman to be elected a member of the Philadelphia Academy.

Black Phoebe

SCLATER. *ENGLISH ORNITHOLOGIST WHO SPECIALIZED IN THE BIRDS OF CENTRAL AND SOUTH AMERICA. HE WAS A LONG-SERVING EDITOR OF* THE IBIS—*THE JOURNAL OF THE BRITISH ORNITHOLOGISTS' UNION.*

Philip Lutley Sclater
(1829-1913)

MEXICAN CHICKADEE *Parus sclateri* Kleinschmidt

Parus meridionalis (not Lilljeborg, 1852) P.L. Sclater, 1857. *Proceedings of the Zoological Society of London* (1856), p. 293: El Jacale in the State of Vera Cruz [or Puebla], Southern Mexico
Parus sclateri Kleinschmidt, 1897. *Journal für Ornithologie* 45, p. 133. New name for *Parus meridionalis* P.L. Sclater, preoccupied

Philip Sclater was one of the great nineteenth century ornithologists. He studied birds for over fifty years and was author or co-author of over 1300 titles, the majority concerning the avifauna of Central and South America.

Born at Tangier Park, Hampshire, on 4 November 1829, he remained based in the south of England throughout his life. As a student at Oxford he began preparing the skins of British birds for himself and on visits to the London dealers bought up foreign species. He was greatly stimulated and encouraged at this time by Hugh Strickland, a far-sighted ornithologist then living at Oxford who directed him towards the study of South American species.

Though Sclater practised as a lawyer in London for a few years, natural history was always uppermost in his thoughts. He was a frequent visitor to the European museums and zoos and at Paris became well acquainted with Charles Bonaparte. In July 1856 Sclater went to Saratoga, New York, to attend a meeting of the American Association for the Advancement of Science. Afterwards he made his way to Niagara, through the Great Lakes, and on foot to the St Croix River with two voyageurs, enjoying a canoe trip with them down to the Mississippi. He then spent a month at the Philadelphia Academy of Natural Sciences studying the bird collections.

In January 1859 Sclater made a three month collecting trip to Tunisia and eastern Algeria with E. Cavendish Taylor. In April, he became the new Secretary of the London Zoological Society and set about improving its finances, redeveloping the zoo and revitalizing the society's publications. Also in 1859 he became the editor of the first series of *The Ibis*, continuing as such until 1864; and from 1877 until 1912, he was either editor or co-editor with Osbert Salvin,

Howard Saunders or Arthur Evans. Among the more important books credited to Sclater are *Exotic Ornithology* (1866–69) and *Nomenclator Avium Neotropicalium* (1873) both with Salvin, and *Argentine Ornithology* (1888–89) with W.H. Hudson.

In 1862 Sclater published a *Catalogue of a Collection of American Birds belonging to P.L. Sclater* giving details on over 4000 specimens. But in the 1880s he began to dispose of his birds, through a series of instalments, to the British Museum (Natural History). By then there were nearly 9000 specimens representing over 3000 species even though he had restricted himself to those orders that were then least known, the passerines, woodpeckers and parrots.

Over the years, Sclater worked through the birds of many collectors describing their discoveries in a barrage of papers to the scientific journals. Sir Wyville Thomson asked him to review the birds collected by the *Challenger* expedition and, as two further examples, both of Mexican collections, Sclater wrote up the birds found by Matteo Botteri and Auguste Sallé. Amongst Sallé's 233 species from around Cordoba and Orizaba was the first known example of the Mexican Chickadee, now *Parus sclateri*. As one might expect for an ornithologist of Sclater's calibre this was by no means the only bird named for him, but it is the only one of them regularly breeding within the confines of the United States—where it is found only in the Animas Mountains of New Mexico and the Chiricahua Mountains of Arizona.

Mexican Chickadee

In 1884 Sclater went to North America for a second time, visiting Montreal and some of the eastern states. When he was sixty-nine years old the Smithsonian Institution honoured Sclater with a whole issue of the *Bulletin of the United States National Museum* (1896), giving a short biography of him together with a full list of his publications. Sclater's obituary notice and bibliography in *The Ibis* (following his death as a result of a carriage accident on 27 June 1913) remains the longest ever published by that journal.

Sclater had six children, the eldest of whom, William Lutley Sclater, adopted his father's interests. William married an American lady in 1896, and following a three year period at Colorado College Museum, from 1906 to 1909, he published *A History of the Birds of Colorado* (1912)—much of it written from the notes of Charles Aiken. Otherwise William tended to specialize in African birds. In London in 1944 he was fatally injured by one of Hitler's flying bombs.

SCOTT. *SOLDIER, PACIFICATOR AND PRESIDENTIAL NOMI-NEE. CALLED "OLD FUSS AND FEATHERS" BECAUSE OF HIS PUNCTILIOUSNESS IN DRESS AND DECORUM, HE IS BEST REMEMBERED FOR HIS SUCCESSFUL CAMPAIGN AGAINST MEXICO.*

Winfield Scott
(1786-1866)

SCOTT'S ORIOLE *Icterus parisorum* Bonaparte

Icterus Parisorum Bonaparte, 1838. *Proceedings of the Zoological Society of London* (1837), p. 110: Mexico

The striking black and yellow plumage of this smart oriole was first described by Charles Bonaparte. He examined skins collected in Mexico by the Paris brothers, whom he honoured in his scientific binomial. The name of Winfield Scott, the American hero of the Mexican War, was attached to the bird sixteen years later by Lieutenant Darius Nash Couch, who had served in the Mexican campaign. He found the oriole to be a common and melodious songster in Nuevo León and Coahuila when he collected in northern Mexico in April 1853. Thinking that he had discovered a new species he called it *Icterus Scottii* in the *Proceedings of the Academy of Natural Sciences of Philadelphia* (1854), with the explanation: "I have named this handsome bird as a slight token of my high regard for Major General Winfield Scott, Commander in Chief of the U.S. Army."

Scott was born on 13 June 1786 on the family estate near Petersburg, Virginia, the grandson of a Scottish Jacobite who had escaped to America after the battle of Culloden. By the age of twenty Scott was a big burly man weighing 230 pounds, six feet five inches in height and the strongest in the neighbourhood. He was also endowed with a large ego, abundant initiative, great self-reliance and an outspoken, generous nature.

His military career began in July 1807 when he joined the cavalry and a few weeks later helped to capture a small boat containing two British officers and six men. During the War of 1812 he fought on the Niagara frontier and rapidly rose to the rank of Major General but was seriously wounded at the battle of Lundy's Lane in July 1814. After recovering he studied military tactics in Europe then returned to become the most continually prominent public figure in the United States for the next forty-five years. Whenever there was imminent risk of war and a desire for peace he was sent to negotiate. He intervened successfully in the Canadian 'Patriot War' of 1837–38, the boundary dispute between Maine and New Brunswick in 1838 and the controversy with the British over San Juan Island in Puget Sound in 1859.

General in Chief of the U.S. Army for twenty years, from June 1841 until October 1861, his outstanding military achievement was the successful campaign against Mexico in 1847. By this time he was sixty years old and the veteran of many battles against the British and the native Americans. In March he besieged and captured Veracruz then marched his army inland for three hundred miles, up through the mountains to Mexico City. Despite a heroic defence he took control of the battered capital after the storming of Chapultepec. Mexico had now lost a third of its territory: California, New Mexico and the land east of the Rio Grande.

In 1852 the Whigs nominated Scott for the presidency but he was overwhelmingly defeated by Franklin Pierce. At the outbreak of the Civil War he was offered a commission by the Confederates but despite his Southern birth he retained command of the Union forces until 31 October 1861 when he resigned because of poor health. Clashes with Major General McClellan and the rout at Ball's Bluff probably contributed to his decision. Now aged seventy-five he began to write his *Memoirs of Lieut-General Scott* (1864). The reminiscences filled two volumes but there is no mention of his brief contacts with ornithologists. Couch goes unrecorded. So does Audubon, who visited the Department of State in July 1842 to collect letters of introduction signed by Scott and the President for his Missouri River expedition. In March 1853 Professor Baird met the general and was assured that army officers would be allowed to collect for the Smithsonian Institution, but we have only Baird's passing reference that Scott had "promised everything". There is nothing in the autobiography to suggest that Winfield Scott ever thought about natural history.

The old warrior lived to see the defeat of the South and died at West Point, New York, on 29 May 1866 shortly before his eightieth birthday.

Scott's Oriole male

Gideon B. Smith† (1793-1867)

SMITH'S LONGSPUR *Calcarius pictus* (Swainson)

Emberiza (Plectrophanes) picta Swainson, 1832. In W. Swainson and J. Richardson *Fauna Boreali-Americana*, Vol. 2 (1831), p. 250, Pl. 49: Carlton House, on the banks of the Saskatchewan [River]

In the 1830s and early 1840s a silk culture boom swept through some parts of the United States. Gideon B. Smith, a native of Maryland, was involved almost from the start: he invented a new silk reel in 1829, he produced and sold millions of silk worm eggs and he is credited with planting, at Baltimore, the first of a new strain of mulberry tree that grew rapidly with abundant foliage for feeding the larvae. From 1839 to 1840 he edited the earliest volumes of the *Journal of the American Silk Society*, which was produced at the society's headquarters in Baltimore, the focal point for the new industry. At the height of the silk craze he qualified as a doctor at the University of Maryland, at the age of forty-seven.

Smith was wealthy enough to be a subscriber to Audubon's expensive picture book *The Birds of America* and he acted as Audubon's agent in Baltimore, though neither of these points is proof that he was especially interested in birds; many subscribers were not. Their first known meeting was on 6 November 1840 when Smith received the artist at his office and put him up that night at his home. On 13 July 1842 Smith gave him hospitality again but had to minister to him in the night when he was ill. He also saw Audubon briefly in March 1843 when he passed through the city on his way to the Missouri River and afterwards received a series of letters from him outlining the expedition's progress. Smith was probably expected to relay the news to current and potential subscribers amongst Baltimore society. From these letters Smith learned how Audubon and his companions, Edward Harris, John G. Bell, Isaac Sprague and Lewis Squires, had travelled overland to St Louis to ascend the Missouri. One particular despatch was written in May when they were aground on a sand bar, a delay that gave Audubon the chance to write to his friend:

> "We have collected everything that was in bloom, and shall continue to do so, when in seed and ripe, for all our friends, far and near. In Zoology, we have done

pretty fair, in Ornithology better, as we have already four *new birds*, and will, no doubt, find more."[1]

They continued upriver and spent two months at Fort Union, then returned east in the fall. Smith was delighted when the octavo edition of *The Birds of America* was published in the following year. Amongst the new species were Harris's Sparrow, Bell's Vireo, Sprague's Pipit and Smith's Longspur—the latter species being dedicated by Audubon to his "good friend GIDEON B. SMITH, Esq., M.D., of Baltimore, Maryland, who has done much for science in several of its departments."[2]

'Smith's Lark Finches' were discovered by Harris and Bell and were best seen in the grasses from horseback, seldom taking flight except when directly disturbed. Audubon and the others "found these birds very abundant on the low prairie, near a lake in Illinois, about seven or eight miles distant from Edwardsville."[3] At the end of the expedition when the collection was divided up, Harris recorded in the back of his journal that his share was 159 specimens of about 40 species amongst which were seven 'Smith's Lark Finches', eight 'Sprague's Larks', two 'Harris's Finches' and two Bell's Vireos, indicating that the names had been thought up during the course of the expedition.[4]

While all this was going on the silk boom had begun to decline and had fizzled out almost entirely by the end of 1845. In 1848 Smith was "reported by the Board of Examiners [of the Baltimore Medical Society] ... as meriting expulsion for unprofessional conduct"[5] but no details are available. They may have been sufficiently embarrassing to cause him to move to Missouri where he evidently spent the last years of his life, still continuing his entomological studies. When he died in 1867, aged about seventy-four, he left an unpublished manuscript on the cyclical irruptions of cicadas in the state of Missouri.

Despite all the foregoing, the first specimen of Smith's Longspur was actually found in Canada. It was one of the birds collected by Dr John Richardson during John Franklin's second overland expedition, taken in April 1827 on the banks of the Saskatchewan River at Carlton House. It was named the Painted Bunting by Swainson and Richardson in the second volume of the *Fauna Boreali-Americana*, which was published in 1832. Their scientific name therefore has priority over the one proposed by Audubon twelve years later.

Smith's Longspur

Isaac Sprague
(1811-1895)

SPRAGUE'S PIPIT *Anthus spragueii* (Audubon)

Alauda Spragueii Audubon, 1844. *The Birds of America* (octavo edn), Vol. 7, p. 334, Pl. 486: near Fort Union [western North Dakota]

Isaac Sprague was born in Hingham, Massachusetts, on 5 September 1811, the eldest son of a box-cooper who died when Isaac was fourteen years old. At about that time the boy was apprenticed to his uncle, Blossom Sprague, a carriage painter who soon discovered that his nephew had a rare talent for artwork. He also had a great interest in nature and on becoming the owner of Nuttall's *A Manual of the Ornithology of the United States and of Canada* in 1834 he concentrated on studying and painting birds.

One day in August 1840 he went out shooting and on his return found that a visitor had called to see his work, leaving comments on some of the water-colours: below a Palm Warbler—"first rate, altogether, J.J.A."; and beside a Ruby-crowned Kinglet—"not quite finished and yet quite beautiful, J.J.Audubon." Audubon later recorded: "saw some very remarkable drawings of birds (far better than any ever made by the Immortal Alex[r] Wilson) by a young man named *Sprague*. Truly wonderful drawings, my dearest friends but this person was out shooting and I did not see him. I however wrote a few lines on several of them the purports of which, I trust, will not displease him."[1]

Sprague was not at all displeased, for two years later, when Audubon suggested that he work for him on his Missouri River expedition, the young man eagerly grasped this golden opportunity to visit the Mid West. In February of 1843 Sprague went to New York to stay with Audubon for a month and in March they left together, bound for the Missouri. During the long journey to St Louis Sprague became acquainted with the rest of the party; Edward Harris, a wealthy farmer from New Jersey who was one of Audubon's closest friends and co-sponsor of the expedition, John G. Bell, a New York taxidermist, and Lewis M. Squires who would be secretary, general assistant and hunter.

After some tedious delays at St Louis they began the 1300 mile journey upriver at the end of April, their steamboat halting several times on most days so that wood could be gathered or cut for the boilers. The motion of the boat as it

SPRAGUE. *BOTANICAL ILLUSTRATOR. IN 1843 HE SERVED AS AUDUBON'S ASSISTANT ARTIST ON THE MISSOURI RIVER EXPEDITION.*

fought its way against the current made sketching a frustrating occupation, so while the others went ashore to hunt, Sprague often stayed on board to concentrate on the more detailed parts of the plants, birds and mammals. He worked so quietly and industriously that Audubon was shocked when Sprague requested his permission to return to St Louis instead of staying with the party for the summer. Sprague had imagined that Audubon was dissatisfied with him, but the misunderstanding was soon cleared up and Audubon continued to be impressed by the output of his rather shy and reserved assistant.

After seven weeks on the river they reached Fort Union, which became their home for the next two months. Established fourteen years earlier by the American Fur Trading Company, the fort was situated close to the fork of the Yellowstone and Missouri Rivers, serving White and Indian hunters who worked the surrounding prairies, hills and creeks. Already Beavers and Muskrats had almost vanished from the region and other mammals were rapidly decreasing, but despite the depredations it was a safe and convenient base for the naturalists and gave Audubon the opportunity to hire experienced hunters.

The party was provided with just one room for their living quarters but it had the advantage of sufficient good light for the two artists to draw there undisturbed, an important consideration because the visiting Indians loved to watch the work in progress. While Bell, Harris and Squires hunted day after day, Audubon and Sprague usually remained behind, painting the animals while they were still reasonably fresh.

On 19 June Sprague drew all day as usual. The morning was so cold and wet that Harris and Bell did not go off hunting until it cleared up in the afternoon. On their way back to the fort they fired together at a wee brown bird that Audubon examined with great delight, realizing that it was an entirely new species. He named it Sprague's Missouri Lark—now Sprague's Pipit.[2] Three days later Audubon and Harris went out together for a morning's walk and heard a Sprague's Pipit singing high in the sky above them, its musical notes carrying a great distance across the prairie. On 24 June Sprague himself went out to search for the new pipits and found a nest, built of grass in a depression in the ground and containing five eggs of a greyish-purple hue. He also shot another specimen. Audubon was

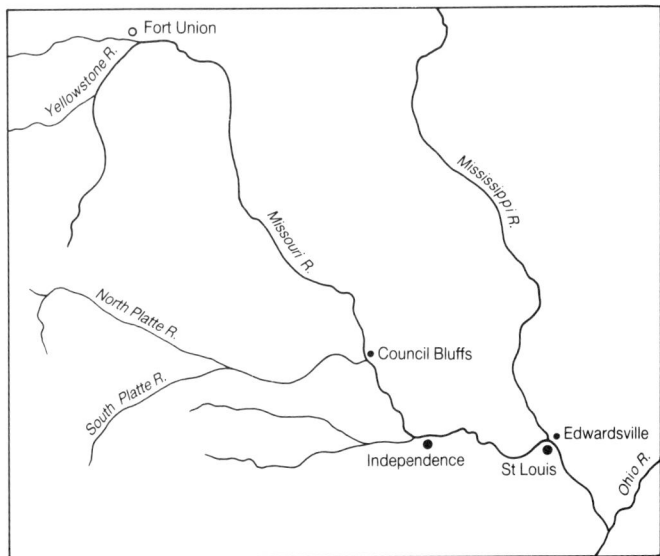

THE MISSOURI RIVER

elated, as he had rarely found a new species, its nest, its eggs and so much information about its habits in such a short space of time. He described the new bird in volume seven of the octavo edition of *The Birds of America*, which appeared the following year.

Throughout the summer Sprague worked on a variety of subjects. He had brought with him a camera lucida, an instrument that reflected an inverted image of the subject onto the paper. He often used it to draw outlines of living animals such as the Buffalo calves and young deer that were kept captive at the fort and he also used it to draw the landscape. Plant drawings took up much of his time and sometimes he made short trips to choose the specimens he wanted and to gather seeds. He produced a number of watercolours of Fort Union and the country nearby, helped Audubon to paint a portrait of the superintendant in charge of the trading post, and when not too busy went hunting with the others. However, he seems not to have got caught up in the lust for Buffalo hunting that channelled the energies of Harris and Bell. Sprague was so impressed by the wildlife of the area that for a while he toyed with the idea of overwintering at the fort, but he returned with his friends when they began the homeward journey in mid-August.

The experience of working under Audubon had been invaluable for Sprague, but Audubon incorporated his work into the illustrations for *The Viviparous Quadrupeds of North America* without giving any credit to his assistant so it brought him no artistic acclaim.

Sprague's reputation as North America's leading botanical artist of the period came later, through his association with Professor Asa Gray of Harvard

Sprague's Pipit

University—an association which began in 1844. Sprague was then employed as a summer clerk at a hotel in Nantasket and showed his artwork to one of the guests who knew that Gray needed a botanical draughtsman and recommended Sprague. Gray hired him to make large poster displays for a series of lectures and afterwards wrote to Dr John Torrey that "the pictures are worth something, [even] if the lecture was not"! Sprague's next assignments were the illustrations for several taxonomic papers and *The Botanical Text-Book* (1845). His knowledge of plants, his skill in making dissections and the beauty and accuracy of his work convinced Gray that together they could produce a guide to the American flora which would be every bit as good as the best in Europe. During the next few years Gray often complained about Sprague's slowness, but he appreciated the high quality of his work and his ability to make original observations: in October 1846 Gray informed Torrey that "Sprague has discovered some new quiddities about the position of the ovule in Ranunculaceae." The first volume of *Genera Florae Americae Boreali-Orientalis Illustrata. The Genera of the Plants of the United States* appeared in 1848, the second in 1849. It met with considerable acclaim and international recognition, as Gray had hoped it would.

There is no doubt that by providing Gray and Torrey with drawings that were both scientifically accurate and pleasing to the eye Sprague helped them to become accepted as world-class botanists. He provided the drawings for many of Torrey and Gray's papers on western plants, including 'Plantae Wrightianae' which dealt with Charles Wright's collections from Texas and New Mexico. He also illustrated botanical appendices to the Pacific Railroad Reports and made one hundred drawings of the flowering plants brought back by Wilkes's U.S. Exploring Expedition.[3] In 1853 Torrey paid tribute to Sprague's skills by naming the genus *Spraguea* in his 'Plantae Fremontianae; or, Descriptions of Plants Collected by Col. J.C. Fremont in California' which Sprague illustrated.

Sprague's most prolific period was between 1845 and 1865. It began when he moved to Cambridge with his bride, Hannah Colbath of Middleton, New Hampshire, whom he had married on 10 December 1844. Hannah died in January 1849 leaving him with a young daughter, so his mother came to care for them. He remarried in December 1854 to Sarah Eaton of Roxbury, Massachusetts. She bore two sons; Isaac in 1859 and Marvin in 1862. For the sake of his wife and sons, who were consumptive, Sprague left Gray's employ in 1865 and moved a little way out of Cambridge to Wellesley Hills, near Needham, but the country air had little effect on Sarah's health and she died five years later. Sprague then sent the boys to live with their maternal aunts; his brother, Lincoln, came to live with him and this arrangement lasted until Lincoln's death in 1891.

In the fall of 1870 Sprague was visited by the ornithologist Henry W. Henshaw whose father had recently moved to Wellesley Hills. Henshaw admired a pencil sketch of Audubon that was hanging in the parlour and learnt that it had been drawn just prior to the Missouri River trip. Audubon had declared it to be the best likeness to himself, and in a typical gesture, had taken the pencil from Sprague and added a few strokes to improve it further. Though Sprague no longer actively studied birds, he occasionally joined Henshaw for tramps in the nearby woods where they enjoyed birding together. He was greatly pleased when Henshaw took him to the nest of a Prairie Warbler, a species that had previously eluded him.

During the 1870s Sprague began to produce botanical illustrations of a less technical and more decorative nature for popular works. He painted for pleasure to the end of his life, producing pictures of plants and birds for himself, his family and friends. His last years were spent at the home of his son Marvin and he died there on 15 March 1895, at the age of eighty-three.

Georg Wilhelm Steller†
(1709-1746)

STELLER'S EIDER *Polysticta stelleri* (Pallas)

Anas Stelleri Pallas, 1769. *Spicilegia Zoologica*, Pt. 6, p. 35, Pl. 5: Kamchatka

STELLER'S SEA-EAGLE *Haliaeetus pelagicus* (Pallas)

Aquila pelagica Pallas, 1811. *Zoographia Rosso-Asiatica*, Vol. 1, p. 343 and plate: in Insulis inter Camtshatcam et Continentem Americes, praesertim in infami naufragio et monte Beringii insula [= Tauisk, on Sea of Okhotsk]

STELLER'S JAY *Cyanocitta stelleri* (Gmelin)

Corvus Stelleri J.F.Gmelin, 1788. *Systema Naturae*, Vol. 1, Pt. 1, p. 370. Based on "Steller's Crow" Latham, *A General Synopsis of Birds*, Vol. 1, Pt. 1, p. 387: in sinu Natka Americae borealis [= Nootka Sound, Vancouver Island, British Columbia]

On the morning of 31 July 1741[1] Georg Wilhelm Steller leapt onto the shore of Kayak Island and became the first European naturalist to stand on Alaskan soil. After a sea voyage of some two and a half thousand miles he was only allowed one frenetic, tantalizing day to explore the rocky shores and spruce forests. By the evening Salmonberry, Crowberry, Whortleberry and Upland Cranberry plants lay among his botanical samples and his Cossack hunter, Thoma Lepikhin, had brought him a dozen different kinds of birds. On examining the corpses Steller found only two that were familiar to him—the Black-billed Magpie and the Raven. He was struck by the brightly coloured plumage of the rest of the birds and as he turned over a blue-feathered jay with a blackish head and crest, he was reminded of a picture he had seen several years before in the library of the Academy of Sciences at St Petersburg: Mark Catesby's painting of a Blue Jay in *The Natural History of Carolina* (1731). The similarities between the two species were sufficient for him to comment, "This bird proved to me that we were really in America." The bird he handled was that bold and raucous inhabitant of western forests now known as Steller's Jay.

†No known portrait

Steller was a member of Vitus Bering's Second Kamchatka Expedition which had been sent by the Empress Anna to search for North America. After this trip ashore at Kayak Island the *St Peter* sailed south-westwards on the return voyage to Siberia and Steller was confined to the ship until 10 September when a shore party landed on Nagai Island (one of the Shumagin group) to refill the water casks. This time Steller had two days to explore. He noted that seabirds were abundant and listed gulls, swans, sandpipers, ptarmigan and Snow Buntings. However, he had to devote time to gathering the antiscorbutic plants which protected his health throughout the following weeks; had he been able to convince the naval

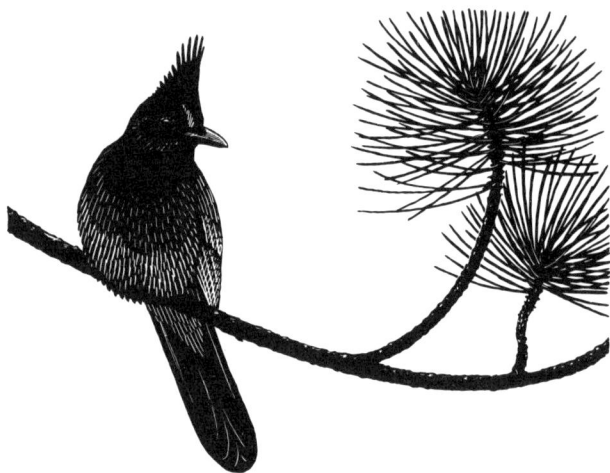

Steller's Jay

officers of their efficacy, subsequent events would have been very different. Those three Alaskan days on Kayak and Nagai Island were his only forays onto the unknown land.

During the next two months as the *St Peter* ploughed her way past the Aleutian Islands she was hit by westerly gales and several times was driven far off course. Many of the crew were rendered weak and helpless by scurvy and some died. When mountains were sighted to the west, Bering and his officers decided to steer towards them hoping that the land was Kamchatka. They were providentially swept into a calm channel beyond some rocky reefs and three weeks later, in a storm, after all the men had moved ashore, the ship was tossed high up onto the beach. The company was now stranded at the Russian end of the Aleutian Chain on a narrow, mountainous, uninhabited island, about fifty-five miles in length. For over nine months, through the winter, spring and summer of 1741–42, Steller was at last able to study some of the mammals, birds and plants of the archipelago.

The resultant pioneering fieldwork ensured Steller's fame, but we would have access to few details about his life history were it not for the obsession of one man. In his preface to *Georg Wilhelm Steller, the Pioneer of Alaskan Natural History*, Leonhard Stejneger began as follows:

"Authors who contemplate writing a biography are warned by a recent literary critic that the first condition for success is to select a proper subject. In the present instance, however, I had no choice as to subject, nor had I originally intended to write a biography. The subject itself took possession of me; it haunted me for more than fifty years, and before I became aware of it I had already the better part of a biography on my hands."

Stejneger's quest began in 1882 and like many another story in this volume it started when Professor Baird decided to send a promising young naturalist to a remote and potentially rewarding study area. On this occasion the naturalist was the thirty year old Norwegian who had recently arrived in the United States and the location was Bering Island, where Steller had been marooned. From the day of Stejneger's landing on 7 May 1882, until his departure eighteen months later, he regarded himself as being on classical ground.

Here Steller had observed the Northern Fur Seal, the Sea Otter, Steller's Sea Lion and Steller's Sea Cow, all previously unknown to scientists, and provided information about their behaviour as well as precise anatomical descriptions. When Stejneger arrived, 140 years later, the biology of the Northern Fur Seal was a matter of enormous interest, since the income from their pelts brought revenue to the U.S. Government. Their value was so great that following the purchase of Alaska from the Russians in 1867 (for $7,200,000) in just a few years the profits from the sale of Sea Otter and Northern Fur Seal pelts alone had recouped the cost.

One of Stejneger's primary objectives was to investigate the demise of Steller's Sea Cow. Endemic to Bering and Copper Island (the Commander Islands) it had been assumed extinct since 1768, slaughtered for food by the hunters who followed in the wake of Bering's expedition. Yet in 1879 the Arctic explorer N.A.E. Nordenskiöld had brought back a skeleton and reported rumours that living sea cows had been seen around Bering Island as late as 1854. Since Steller was the only naturalist to see them alive he was the only source of information about these huge manatees which grew up to thirty feet in length. From the door of his hut in the dunes he had watched them daily, browsing on seaweeds in the shallow water, sometimes with gulls perched on their backs.

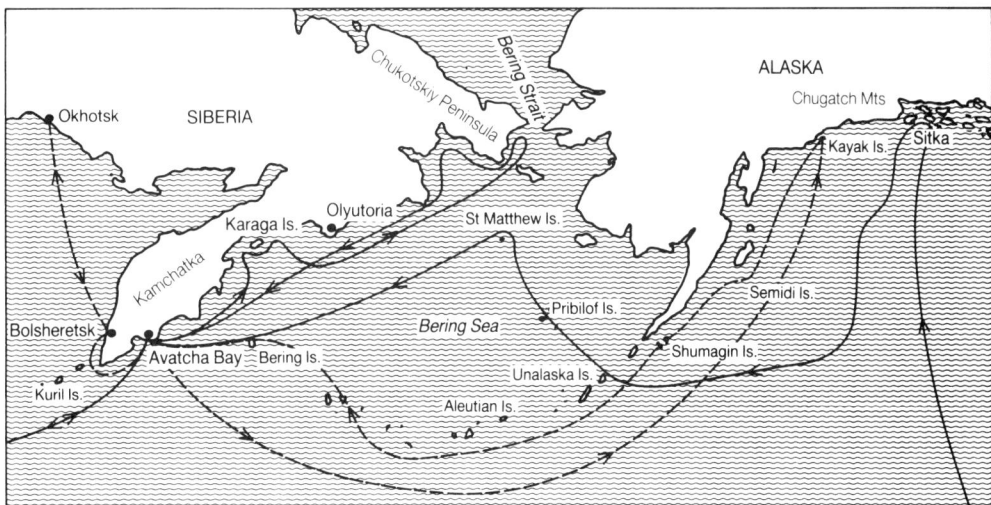

VOYAGES OF THE ST PETER *AND THE* SENJAWIN
----- ST PETER *(STELLER)* 1741–42
——— SENJAWIN *(KITTLITZ)* 1827–28

In July 1742 Steller measured and dissected a beached female, estimating that it weighed about 8000 pounds; the heart alone tipped the beam at over 36 pounds. He recorded: "The stomach of the sea-cow is of amazing size, 6 feet long, 5 feet wide and so stuffed with food and seaweed that four strong men with a rope attached could scarcely move it from its place and with great effort drag it out".[2] His examination was made on a cold and rainy day while packs of Arctic Foxes, which had tormented the men daily from the moment of their shipwreck, tore at the flesh and organs as soon as they were exposed. His human companions were of little help, their only interest in the docile beast being its usefulness as a source of tender, savoury cutlets. He had to pay for their grudging assistance with tobacco.

Another species that Steller regarded as common, but which had disappeared by Stejneger's day, was the Spectacled, or Pallas's Cormorant. This large, almost flightless bird was completely unafraid of man and was eaten with relish by the crew of the *St Peter*, and later by the Aleuts who were imported to Bering Island by the Russian–American Company in 1826. Stejneger was told that the birds had been seen until about 1850, their last refuge being a small islet off the coast.

With a few exceptions, such as the Spectacled Cormorant and the Bald Eagle, Steller judged the avifauna of the island to be very similar to that of the east Kamchatkan coast where he had passed the winter and early spring of 1740–41. Steller's Eider, which he had first seen and collected there, was abundant at sea off Bering Island and in April vast rafts amassed offshore prior to their northward migration. In 1769, soon after Peter Simon Pallas arrived at the Academy of Sciences at St Petersburg, he

Steller's Eider

named the duck *Anas Stelleri* after its discoverer. Pallas found that many of Steller's other bird skins had survived, and from these he added the original descriptions of Tufted Puffin, Parakeet Auklet, Crested Auklet and Steller's Albatross.

Steller's Albatross (more usually known as the Short-tailed Albatross) is the largest North Pacific albatross. Though once plentiful it was brought to the verge of extinction by Japanese plume hunters. Since the 1950s the population has slowly increased and in recent years wandering birds have been seen off the Hawaiian Islands, Alaska, California, British Columbia and Oregon.

In 1811 Pallas described Steller's Sea-Eagle. He called it *Aquila pelagica* and used Steller's notes:

"In the highest rocks overhanging the sea, it constructs a nest of two ells in diameter, composed of twigs of fruit and other trees, gathered from a great distance, and strewed with grass in the centre, in which are one or two eggs, in form, magnitude and whiteness, very like those of a Swan. The young is hatched in the beginning of June, and has an entirely white woolly covering. While Steller was cautiously viewing such a nest from a precipice, the parents darted with such unforeseen impetuosity as nearly to throw him headlong; the female having been wounded, both flew away, nor did they return to the nest which was watched for two days. But, as if lamenting, they often sat on an opposite rock. It is a kind of bird, bold, very cunning, circumspect, observant, and of savage disposition. Steller saw a Fox (*Vulpes lagopodus*) carried off by one and dashed upon the rocks, and afterwards torn in pieces. It lives also on dead substances cast up by the sea, and various offscourings of the ocean."[3]

Pallas assumed that Steller had found the eagles on Bering Island, but Stejneger, who knew that the species was only an occasional visitor there, considered that Steller must have become familiar with the birds on Kamchatka, or some other part of its stronghold around the Sea of Okhotsk.

Stejneger left Bering Island at the end of 1883 and in 1885 his 'Results of Ornithological Explorations in the Commander Islands and in Kamtschatka' was published as *Bulletin of the United States National Museum*, No. 29. From 1884 until his retiral he worked at the U.S. National Museum, first as Curator of Birds and later as the Head Curator for Biology. He made four more trips to Alaska and Kamchatka, visiting Kayak Island in 1922.

During those years he endeavoured to locate every possible scrap of information about Steller from the cradle to the grave, a task compli-

Steller's Sea-Eagle

cated by the fact that much of the material was written in Russian, German or Latin. From Steller's own journals and scientific papers (published posthumously), from documents hitherto buried in the archives of the St Petersburg Academy and from official expedition reports and miscellaneous sources he pieced together Steller's experiences on the voyage of the *St Peter*, his escape from Bering Island and the last four years of his life. In so doing, he revealed a tragic saga.

In the spring of 1742 Steller's companions broke up the wreck of the *St Peter* and began to build a smaller vessel. The sailors were determined to take great bundles of Sea Otter pelages with them and so Steller was forced to abandon his infinitely more precious collection of skins, skeletons and dried plants. When the forty-six survivors set sail in August, Steller only took his manuscripts, the palatal plates of one sea cow and some seeds, together with his share of the otter pelts. Thirteen days later they reached Avatcha Bay on the Kamchatkan coast.

Steller crossed the mountains and sent off his reports to St Petersburg from Bolsheretsk, then spent the next two years exploring the peninsula south to the Kuril Islands and north to at least as far as Olyturia, sending more specimens to the capital from time to time.

In the summer of 1744 he was recalled to St Petersburg, which he had left in January 1738. He sailed to Okhotsk, then began the long westward trek. After passing through the Urals he lingered in the mountains around Perm, botanizing with a friend, apparently reluctant to return to western civilization. In August he was arrested by a courier from St Petersburg and ordered to travel 2000 miles back to Irkutsk, to stand trial for an incident at Bolsheretsk where he had freed seventeen natives without proper authority. He had already been acquitted at Irkutsk, but this news had not yet reached the capital. Steller and his escort had gone about a third of the way when another courier caught up with them and ordered Steller's release. Travelling westwards again, in bitter weather, he took a fever and died at Tyumen on 23 November 1746,[4] at the age of thirty-seven, a victim of bungled Russian bureaucracy.

Since Steller died before Linnaeus introduced his binomial system of classification, all of his zoological and botanical discoveries have been renamed and redescribed by other naturalists—some by Pallas have already been mentioned. Another discovery was Steller's Jay and though the specimen taken on Kayak Island never reached St Petersburg, Steller's written description did get there and was known to John Latham in England. Working from a specimen taken on Vancouver Island during Cook's last voyage, Latham gave it the name Steller's Crow in his *General Synopsis of Birds* (1781–85). Since he only provided English names, the German zoologist J.F. Gmelin later copied Latham's description and attached Latin names of his own.

Forty-five years after Stejneger began his study of Steller he made a pilgrimage to Germany and visited Windsheim, near Nuremberg, where the naturalist had been born on 10 March 1709. He uncovered much new information about Steller's family, and schooling, and in 1930 he returned to make enquiries at Halle University where Steller had studied medicine and given botanical lectures for his fellow students. Steller was afterwards attracted to the newly-founded Academy at St Petersburg because he realized that there were few opportunities in his own country for physicians with a bent for exploration and the natural sciences. Three years later his hopes were realized when he was accepted as a naturalist on the Second Kamchatka Expedition.

Stejneger now possessed masses of biographical information from such diverse sources as log-books, journals, scientific reports, letters, official Russian documents, tombstones, church records and specimen labels. As well as uncovering his subject's activities and achievements he had found many pointers to his personality. Stejneger's respect for Steller's great courage, foresight and stamina did not blind him to his hero's shortcomings: the pride, obstinacy and explosive temper which were at times as obvious as the better qualities. *Georg Wilhelm Steller, the Pioneer of Alaskan Natural History* was published in 1936, when the author was in his mid-eighties; a rather rare kind of biography, written for naturalists, about a naturalist, by a naturalist.

Hugh Edwin Strickland (1811-1853)

STRICKLAND'S WOODPECKER *Picoides stricklandi* (Malherbe)

P.[icus] (Leuconotopicus) Stricklandi Malherbe, 1845. *Revue Zoologique* [Paris] 8, p. 373: du Mexique [= Mt.Orizaba massif, Veracruz]

On 14 September 1853 Hugh Strickland was carefully examining a railway cutting of the Manchester, Sheffield and Lincolnshire Railway near Clarborough in England. As he contemplated the newly exposed rock strata he saw a coal train approaching around the bend so he stepped onto the other line. The rumble of the passing train prevented him from hearing the approach of another and the driver reported that he seemed completely oblivious to the screeching brakes and frantic whistling as he was struck from behind and instantly killed.

So ended the life of one of Britain's most promising naturalists. He was born at Righton in Yorkshire on 2 March 1811. During school vacations in parts of England and northern France he developed a remarkable ability to quickly assess the main geological features of a district. In 1835, shortly after graduating from Oriel College, Oxford, he went to Asia Minor with William Hamilton, spending several months in Greece and western Turkey before coming back alone through Greece, Italy and Switzerland. He spent the next two years working through his geological results, afterwards concentrating more on the sandstone areas around his parents' house in Gloucestershire.

Ornithology was Strickland's second great interest and given a little more time would probably have become the dominant one. He had collected birds almost since childhood and his enthusiasm had increased under the influence of the renowned bird collector Sir William Jardine after Strickland married his daughter Catherine on 23 July 1845. The earliest specimen in Strickland's collection is dated August 1822; and by the time of his death he had amassed 6006 selected items. Only 301 of these birds had been collected in the field by his own hand but among them was the type specimen of the Olive-tree Warbler from his trip to Asia Minor. The largest contributions were bought from his cousins Nathaniel and Arthur, mostly derived from dealers and sea captains. His brother Algernon, a midshipman in the Royal Navy, brought back ninety bird skins for him from

STRICKLAND. *ENGLISH GEOLOGIST AND ORNITHOLOGIST WHO HELPED SET THE STANDARDS FOR STABILITY IN ZOOLOGICAL NOMENCLATURE.*

the Cape, Mauritius and southern India. The remainder were from dealers and a variety of ornithologists: British birds from Thomas Eyton, Indian birds from Thomas Jerdon and Brian Hodgson, African birds from J. Petherick and C.J. Andersson, Russian birds from J.F. Brandt and his brother, American birds from Baird and Audubon, and Mexican birds from T.G. Mann and a traveller by the name of Galeotti.

Strickland's Woodpecker was recognized as a new species by the French ornithologist Alfred Malherbe during a visit to London when Strickland showed him the skin of a female woodpecker with a completely brown back. In the *Revue Zoologique* for October 1845 Malherbe contributed a short paper entitled 'Description de trois espèces nouvelles du genre *Picus*'. One of the three was a detailed description of the Algerian race of the Great Spotted Woodpecker which Malherbe had published briefly elsewhere, the others concerned two of Strickland's specimens collected in Mexico by Mann. Malherbe called one woodpecker after Jardine (the name has since become obsolete), and the other he named after the "savant ornithologiste [Monsieur] H. E. Strickland". Strickland's Woodpecker is now split into two races, the nominate race *stricklandi* inhabiting open pine woods in the high mountains of lower eastern Mexico, the *arizonae* race preferring the oak or oak-pine woodlands of western Mexico northwards into south-east Arizona and the adjacent parts of New Mexico.

Strickland's widow presented the whole bird collection to Cambridge University in 1867 and his sister Frances endowed it with £150 per annum for its safe preservation. In 1882 Osbert Salvin completed *A Catalogue of the Collection of Birds formed by the late Hugh Edwin Strickland*—a publication of over 600 pages. The birds still remain at Cambridge alongside many of those formerly belonging to William Jardine, William Swainson and the Newton brothers.

By no means just a collector of birds, Strickland gained an international reputation for his serious ornithological work, particularly his attempt to devise universal guidelines for zoological nomenclature. In 1841 he outlined his proposals and circulated them to friends at home and abroad. In the following year the British Association for the Advancement of Science published a formal code drawn up by a committee that included Charles Darwin, Professor John Henslow, Dr John Richardson and Strickland. The Stricklandian Code, as it became known, was reprinted with some modifications by Jardine in 1863 and re-edited by Philip Sclater in 1878. By then it had gained widespread acceptance in Britain and was also used in North America by some zoologists, though others there preferred either the International, Dall or American Ornithologists'

Strickland's Woodpecker

Union codes. Eventually, all the various codes were replaced by the present International Code of Zoological Nomenclature.

Though sometimes involved in the minor details of ornithology such as describing new species, Strickland was more concerned about the general development of the subject. Two years after the work on nomenclature he wrote a 'Report on the recent progress and present state of Ornithology' for the British Association. In this masterly review he pointed out that more precision and uniformity were required in the designation of genera; that a mode of distinguishing real species from local varieties was urgently needed; and that information on behaviour, anatomy, oology and geographical distribution was still desperately wanting for the majority of exotic species. He also called for the centralization of the mass of scientific information that was thinly scattered between large numbers of journals—an idea he had suggested earlier when he unsuccessfully proposed the formation of a natural history publishing society to be called the [George] Montagu Society. A similar simultaneous idea by anothher naturalist eventually came to fruition as the [John] Ray Society, Strickland contributing to one of its first volumes a translation of Charles Bonaparte's 'Report on the State of Zoology in Europe'.

Following his marriage in 1845, Strickland and his wife visited many of the major cities of northern Europe so that he could examine museum collections. On their return they took a house at Oxford where he began a compendium of synonyms in an attempt to clear up the confusion caused by the profusion of bird names. In 1848, with Dr A.G. Melville, he produced a classic monograph entitled *The Dodo and its Kindred; or, The History, Affinities, and Osteology of the Dodo, Solitaire, and other Extinct Birds of the islands Mauritius, Rodriguez, and Bourbon.* It remains the authoritative work on the subject, though when first issued it brought Strickland some unjust criticism for reasserting an earlier suggestion (now widely accepted) that dodos were closely related to pigeons. The excellent illustrations were by Strickland and his wife, using a new lithographic technique.

His other published works include a variety of papers in the *Annals and Magazine of Natural History* and *The Proceedings of the Zoological Society*, one in the latter journal in 1836 first bringing his name to the attention of ornithologists at large: this was his 'List of the Birds Noticed or Obtained in Asia Minor in the Winter of 1835 and Spring of 1836.' His *Ornithological Synonyms* was never completed but one volume covering the birds of prey was issued in 1855 by his widow and his father-in-law.

It is perhaps worth mentioning that Strickland coined the generic name *Cyanocitta* for the Blue and Steller's Jay. Had he lived longer it is unlikely that he would have contributed specifically to North American ornithology but he was certainly the rising star of his generation, a world-class ornithologist capable of changing the international development of ornithology by a few select and influential publications.

William Swainson (1789-1855)

SWAINSON'S HAWK *Buteo swainsoni* Bonaparte

Buteo vulgaris (not Swainson, 1832) Audubon, 1837. *The Birds of America* (folio), Vol. 4, Pl. 372: near the Columbia River [= Fort Vancouver, Washington] *Buteo Swainsoni* Bonaparte, 1838. *A Geographical and Comparative List of the Birds of Europe and North America,* p. 3. New name for *Buteo vulgaris* Audubon, preoccupied

SWAINSON'S THRUSH *Catharus ustulatus* (Nuttall)

Turdus ustulatus Nuttall, 1840. *Manual of the Ornithology of the United States and of Canada,* 2nd edn, Vol. 1, pp. vi, 400, 830: forests of the Oregon [= Fort Vancouver, Washington]

SWAINSON'S WARBLER *Limnothlypis swainsonii* (Audubon)

Sylvia Swainsonii Audubon, 1834. *The Birds of America* (folio), Vol. 2, Pl. 198 (1834, *Ornithological Biography,* Vol. 2, p. 563): Edisto River, near Charleston in South Carolina

Writing in the *Ornithological Biography* (1834) under 'Swainson's Swamp Warbler' Audubon gave a lengthy preamble on the certainty of more and more birds remaining to be discovered in North America despite the doubts of other naturalists. He eventually got around to discussing the discovery of yet another new species, the bird now known more simply as Swainson's Warbler:

"The bird represented in the plate before you was discovered by my friend JOHN BACHMAN, near Charleston in South Carolina ... it was in the spring of 1832, when I was rambling over the rugged country of Labrador,[1] that my southern friend found the first specimen of this bird, near the banks of the Edisto river. I have been favoured by him with the following account of it.

'I was first attracted by the novelty of its notes, four or five in number, repeated at intervals of five or six minutes apart. These notes were loud, clear and more like a whistle than a song. They resembled the sounds of some extraordinary

SWAINSON. *ENGLISH ZOOLOGICAL ILLUSTRATOR AND AUTHOR. HE DESCRIBED BIRDS FROM MEXICO AND WAS CO-AUTHOR WITH DR JOHN RICHARDSON OF THE* FAUNA BOREALI-AMERICANA. *THIS LITTLE-KNOWN PORTRAIT IS IN THE POSSESSION OF THE SWAINSON FAMILY.*

ventriloquist in such a degree, that I supposed the bird much farther from me than it really was; for after some trouble caused by these fictitious notes, I perceived it near to me, and soon shot it.'"[2]

Bachman secured five skins but only four or five more were collected over the next forty years. Very gradually the haunts and habits of this uncommon and secretive bird were revealed, indicating that it was a summer visitor peculiar to the dense undergrowth of swamps and cane breaks in the low lying parts of the south-east United States. Then in 1935 it was also found to be a rare breeder at higher altitude in open woods in the Appalachians. Audubon would never have guessed at this as he knew only Bachman's assertion that these warblers are invariably found in "swampy muddy places".

Audubon bestowed Bachman's name on three species of birds but this new one he decided to name after William Swainson, with the following dedication:

"To none of my ornithological friends could I assuredly with more propriety have dedicated this species than to him, the excellent and learned author, whose name you have seen connected with it—to him, who has himself traversed large portions of America, who has added so considerably to the list of known species of birds, and who has enriched the science of ornithology by so many valuable works."[3]

Clearly Swainson had nothing whatsoever to do with the discovery of Swainson's Warbler. Nor had this Englishman "traversed large portions of America", though he had collected for about two years in eastern Brazil. But he had described new birds from many parts of the world and had written and illustrated a great many zoological books.

Other artist–naturalists such as Bewick, Wilson and Audubon have been written about over and over again, but Swainson, who was an innovative and skilful artist with as many natural history book titles to his name as the above-named trio put together, remains little known. While it is not unusual for early naturalists to lie forgotten as their texts become increasingly obsolete, it is rather rare for wildlife artists to be ignored. Swainson's subjects, usually without coloured backgrounds, lack the dramatic impact of Audubon's but they are pleasing, detailed and accurate. A full length biography with plenty of Swainson's illustrations is long overdue and would bring his works the attention they deserve. The following brief account concentrates on his work on North American birds and his association with Audubon.

William Swainson was born at Dover Place in the parish of St Mary Newington, London, on 8 October 1789. His mother died before he was three years old but he later gained a step-mother and half-brothers and sisters. At the age of fourteen he started work as a customs clerk in his father's office in Liverpool but longed to explore abroad. Intrigued by his father's collection of shells and insects he had become sufficiently knowledgeable by his late teens to be asked by the Liverpool Museum to compile some *Instructions for Collecting and Preserving Subjects of Natural History* (1808). Before it was published William was already in Sicily on the staff of the commissariat, an army post secured for him by his father. Based initially at Palermo he was at other times at Malta and Genoa. He was blessed with ample leisure to pursue his studies of shells, fish and plants, sometimes enjoying the company of Constantine Rafinesque-

Schmaltz, a like-minded naturalist who acted as interpreter when he wished to purchase specimens, for Swainson suffered from a speech impediment and never seems to have been proficient in any language other than English.

At the end of the Napoleonic Wars in 1815, Swainson retired from his position as Assistant Commissary General because of his poor health and returned to England. His half pay—the only steady source of income he received from the age of about twenty-six until his death—eased him through the difficult financial periods of his struggle as one of Britain's first professional zoologists and as an unsuccessful farmer in New Zealand.

In November 1816 Swainson judged himself fit enough to sail to Brazil to fulfil his early ambition to travel and collect in the tropics. His plans were a little curtailed by civil unrest but he journeyed inland from Recife and around Rio de Janeiro. At Rio he met up with several European naturalists including Prince Maximilian of Wied-Neuwied and Baron Langsdorff. In August 1818 Swainson returned home with 760 bird skins and large numbers of plants, fishes and insects. William Leach, a friend at the British Museum, now encouraged him to experiment with the new technique of lithography whereby subjects could be drawn directly on to stone, thus eliminating the need for engraving on to copper plates or wood. By this

Swainson's Warbler

method more lifelike representations of the subjects could be obtained, as it omitted the need for an intermediate professional engraver who inevitably affected the style and quality of the original drawing, adversely or otherwise. John Gould could be said to have perfected the process but as a pioneer Swainson had to overcome many technical problems to produce his *Zoological Illustrations* (1820–23), now considered his finest work. In three volumes, there were 182 plates, seventy of which were of birds. Speaking of these illustrations in a famous essay on ornithology Alfred Newton expressed the belief that Swainson was "an ornithological artist [who] had no rival in his time. Every plate is not beyond criticism but his worst drawings shew more knowledge of bird-life than do the best of his English or French contemporaries."[4] In 1822 Swainson completed his issue of *Exotic Conchology, or Coloured Lithographic Drawings of Shells*.

Some of his illustrations were taken from his Brazilian material and others may have been from items obtained at the sale of the Bullock Museum, which took place in London in 1819. There was so much to be disposed of that the sale lasted, with some interruptions, from 29 April to 11 June. It was so important that representatives from most of the major museums of Europe attended: Coenraad Temminck from Leyden, Baron Laugier from Paris, Hinrich Lichtenstein from Berlin, Captain Laskey from Glasgow, Dr Adams from Edinburgh and William Leach from the British Museum. Swainson was just one of many collectors quite unable to match the resources of the museum agents or the wealthy private buyers such as Lord Stanley. He was able to bid only small amounts for a hodge-podge of bird skins from the Far East, Australia, Central Africa and North America.

Not long after the sale William Leach was forced to resign his position as Assistant Keeper of Zoology because of a disturbing nervous complaint. Though lacking in scientific training Swainson believed himself well suited as a replacement. He was among the first to advocate the space saving practice of keeping birds as skins in drawers rather than as stuffed mounts in glass cases, and he knew a lot about the preservation of other types of specimens. Indeed he had only just published his *Naturalist's Guide for collecting and preserving subjects of Natural History* (1822). He applied for the vacancy with letters of support from several eminent naturalists, but the physicist J.G. Children was appointed. Swainson's friend Dr Thomas Traill immediately began publishing anonymous articles in the *Edinburgh Review* criticizing the trustees of the British Museum, in the hope that Swainson could still be appointed, but the articles only served to draw the public's attention to the fact that the national collections were suffering from neglect and decay. Swainson already knew this and later recalled that the museum basements resembled "the catacombs we have seen at Palermo, where one is opened every day of the year, merely to deposit fresh subjects for decay, and to ascertain how the progress has gone on."[5]

Unfortunately Swainson now became embroiled in the notorious Quinary System of classification first put forward by William MacLeay. Swainson once summarized the theory in five main principles of which it should be sufficient to quote only the first to show how bizarre the whole idea was: "1. That the series of natural animals is continuous, forming, as it were, a circle; so that, upon commencing at any one given point, and thence tracing all the modifications of the structure, we shall be imperceptibly led, after passing through numerous forms, again to the point from which we started."[6] Perhaps the series of diagrams of interconnecting circles would clarify the text but too much paper and print has already been wasted on this subject. Most scientists steered well clear of the theory; Swainson and Nicholas Vigors were the only notable British ornithologists who keenly supported it.

In 1823[7] Swainson married Mary Parkes of Warwick and they lived at *Elm Grove*, the Swainson family home in Liverpool. After a while they moved south to be closer to London, to Tittenhanger Green near St Albans, where they remained for the next ten years. Mary brought up children as William worked away on more and more projects, striving to make sufficient funds for the upkeep of himself and his family. In 1826 he sold off one of his shell collections and at some time invested heavily in a silver mine in Mexico, perhaps the one that belonged to his enterprising friend William Bullock who had gone out to Mexico after selling off his museum and had acquired an abandoned mine near Rincon de Temascáltepec, not far from Mexico City. Bullock came back to

England for a while in 1823 (with another collection of birds and curiosities) and may then have persuaded Swainson to invest in his venture.

Bullock's son was also a keen bird collector and was said to have been an excellent shot. He went out to Mexico with his father and remained behind to act as a mining agent. In April 1825 young Bullock met the German naturalist Ferdinand Deppe when he too arrived at Temascáltepec. In August they travelled together via Puebla and Tehuacán to Oaxaca. In October they continued south-eastwards to Tehuantepec and early in November went down to the Pacific coast at San Matteo and Santa Maria al Mar. By 22 November they were back in Oaxaca and Bullock went off to Mexico City to meet his father there. In July 1826 Deppe and Bullock met up again and collected around Temascáltepec until August when Bullock's father arrived again on business and took his son's bird collection with him when he returned to London in September. Deppe's 951 bird skins of over 300 species went to Lichtenstein, the Director of the Berlin Museum, who paid little attention to them. Many of the discoveries were not properly named until they were examined at Berlin by Johann Wagler in 1828, John Gould in 1833 and 1843, Charles Bonaparte in 1850, Christian Brehm in 1851, and Hermann Schlegel and Philip Sclater in 1868.

Bullock's birds were examined at once by Swainson and fourteen pages of descriptions were published in the *Philosophical Magazine* for May and June 1827 in a paper entitled 'A Synopsis of Birds discovered in Mexico'. While some of the species are endemic to Mexico and other parts of Central America, no fewer than nine are also common residents or summer visitors north of the border: Broad-tailed Hummingbird, Acorn Woodpecker, Black Phoebe, Violet-green Swallow, American Dipper, Curve-billed Thrasher, Black-headed Grosbeak, Canyon Towhee and Hooded Oriole. A further six species have a more southerly distribution and cross the border into only small portions of some of the states adjacent to Mexico: Groove-billed Ani, Green Violet-ear, Broad-billed Hummingbird, Magnificent Hummingbird, Lucifer Hummingbird and (once only) Blue Mockingbird. Another species, the Thick-billed Parrot, bred in Arizona until the 1930s, was wiped out by miners and settlers, and is currently being reintroduced to the area. Mention should also be made of Bullock's Oriole, which was named by Swainson in this paper but is now considered a race of the Northern Oriole.

In other journals and books which we will come to shortly, Swainson described a further six new birds from Mexico (from a variety of sources) whose range extends northwards: Cassin's Kingbird and Thick-billed Kingbird in the *Quarterly Journal of Science* (1826); Canyon Wren and Painted Redstart in his *Zoological Illustrations*, second series (1829); Western Bluebird in the *Fauna Boreali-Americana* (1832); and Phainopepla in *Animals in Menageries* (1837). Swainson was handling many western North American species before Townsend, Nuttall, Gambel and Heermann had set foot in the West. This explains why birds that are as widespread and familiar in the United States as the American Dipper, Western Bluebird and Canyon Wren have the scientific names *Cinclus mexicanus*, *Sialia mexicana* and *Catherpes mexicanus*.

While Swainson was working on Mexican birds Audubon was continuing *The*

Birds of America. In 1828 he played host to the American artist, who excitedly wrote back home on 2 June:

"Mrs. Swainson plays well on the piano, is amiable and kind; Mr. Swainson a superior man indeed; and their children blooming with health and full of spirit. Such talks on birds we have had together. Why, Lucy, thou wouldst think that birds were all we cared for in this world, but thou knowest this is not so. Whilst there I began a drawing for Mrs. Swainson, and showed Mr. Swainson how to put up birds in my style, which delighted him."[8]

In September Swainson and his wife accompanied Audubon to Paris; Swainson to work at the museums and visit French naturalists while Audubon hoped to win a few more subscribers. Calling at once on Baron Cuvier it was evident that the learned scientist knew of Swainson but had never heard of Audubon. Swainson was introduced to R.P. Lesson who had just returned from a voyage around the world aboard *La Coquille* and a few days later the Swainsons went with Audubon to admire the Rivoli bird collection (see Anna Masséna). When Swainson wanted to buy specimens from a dealer Audubon was easily able to act as interpreter, but the Swainsons' money was soon exhausted and after two months they returned home, doubtless loaded down with bird skins.

William Swainson was Audubon's first choice when he needed help to write the *Ornithological Biography*—the separately published text for *The Birds of America*—but negotiations broke down because Audubon would not pay the fee demanded and he refused to share the authorship with anybody. Swainson sent a testy letter to him in October 1830 regretting the impasse: "Your friends would tell you, if you enquired ... that even *my* name would *add* something to the value of 'The Birds of America' ... [I] must not suppose that you intended that I should give all the scientific information I have laboured to acquire during twenty years on ornithology—conceal my name,—and transfer my fame to your pages & to your reputation."[9] In the end it was William MacGillivray who had the honour of adding to Audubon's success with scant acknowledgement and for a much reduced fee.

Swainson also accused Audubon of withholding information from him: "Dr. Richardson's and my own volume on the Arctic Birds, is now in press. Not being able to refer to your plates, I have not had the power to quote your work, you know how repeatedly I have applied ... both to you and Mr. Havell [the engraver] in vain."[10] Audubon's reticence is understandable but contradicts his declaration that he cared only for the advancement of ornithology.

The volume on Arctic birds was Swainson's second major involvement in the North American avifauna and was a much more substantial and influential work than his earlier brief descriptions of Mexican birds. It was the first bird book funded by the British Government: *Fauna Boreali-Americana, or the Zoology of the Northern Parts of British America: part second, The Birds* (1832). Besides reviewing previous work it incorporated all the discoveries and observations made by Dr Richardson on his two overland journeys with Franklin to the Canadian Arctic. It seems that Richardson wrote the specific descriptions of the birds, but these were revised by Swainson who wrote the rest of the text and of course undertook the illustrations. Richardson generously allowed him to put his name to the descriptions of the new species despite the tedious delays his co-worker had inflicted upon him. The only new birds were the Black-backed Woodpecker, Smith's Longspur, Olive-sided Flycatcher, Western Bluebird (described in a four line footnote from a Mexican specimen) and Clay-colored Sparrow. There would have been more new birds but many of Richardson's specimens were left behind in the Arctic and others decayed en route to England or were otherwise lost.

The seventy-three colour plates of the birds were produced with little or no background. Those for the Evening Grosbeak and Steller's Jay are stunning and while most of the others are attractive it is worth singling out the plates of the Great Gray Owl, Eskimo Curlew and Smith's Longspur as being especially so. The text is also interspersed with about forty woodcuts of birds' heads, feet, bills and feathers. Swainson's Thrush is included in the work under the name of Little Tawny Thrush with Swainson expending nearly two full pages in an attempt to clear up the classification of a number of similar

Swainson's Thrush

thrushes. The earlier works of George Edwards, Mark Catesby and Alexander Wilson had not prevented over sixty years of confusion but it was still some time before the various species and subspecies were properly worked out. The name Swainson's Thrush now embraces the nominate race *Catharus ustulatus ustulatus* (Nuttall) and *Catharus ustulatus swainsoni* (Tschudi).

Some of the information from the *Fauna Boreali-Americana* was directly transcribed by Swainson into the fourth volume of Jameson's edition of Wilson and Bonaparte's *American Ornithology* (1831). Completely new books by Swainson followed on in rapid succession. A second series of *Zoological Illustrations* with 136 plates, including forty-eight of birds, appeared between 1829 and 1833. He then became involved in Dionysius Lardner's *Cabinet Cyclopaedia*, producing eleven octavo volumes between 1834 and 1840 which included his *Preliminary Discourse on the Study of Natural History* (1834), *A Treatise on the Geography and Classification of Animals* (1835), *On the Natural History and Classification of Birds* (1836–37), *Animals in Menageries* (1837) and *Natural History and Classification of Fishes, Amphibians and Reptiles* (1838–39). For Sir William Jardine's *Naturalist's Library* he wrote and illustrated two volumes on the *Birds of Western Africa* (1837) and another on *The Natural Arrangement and History of Flycatchers* (1838).

It is no surprise to learn that Swainson began a number of books that were never completed. Only five parts of *The Ornithological Drawings of William Swainson (Series 1): The Birds of Brazil* (1834) were issued, comprising sixty-two coloured plates without any text. He finished some page proofs for a *Mexican Zoology* and some sections evidently got as far as the printers; sixteen bird plates for it were eventually added to a reissue of the Brazilian project when it appeared as *The Birds of Brazil and Mexico* (1841). He also prepared a few drawings for *Tropical Ornithology* and twelve other colour plates were included in G.R. Gray's *Fasciculus of the Birds of China* (1871), a comprehensive work on the Chinese avifauna that never got past the early stages.

Another unfulfilled project, proposed by Charles Bonaparte, was a catalogue of all the known birds of the world. It is hard to imagine how Swainson would have found the time for this collaboration so it is perhaps fortunate that they failed to reach an initial agreement. Nevertheless, when Bonaparte had to think up a new name for Audubon's *Buteo vulgaris* (which was already in use for another species) he felt sufficiently gracious to call it *Buteo Swainsoni*. The new name appeared in Bonaparte's *A Geographical and Comparative List of the Birds of Europe and North America* (1838) but the author gives no special reason for his choice.

During this extremely busy period Swainson's wife died, leaving him with five children. As far back as 1830 he had confessed to being "sick of the world and of mankind, and but for my family would end my days in my beloved forests of Brazil."[11] He was obviously hankering after the carefree days of his youth, tired of all the petty squabbles he had become involved in. He had recently sided with Audubon against Charles Waterton, the English explorer–naturalist of South America, who attacked Swainson for his Quinary System and criticized Audubon for his distorted, unnatural bird poses as well as the supposed habits of some of the subjects he portrayed. Swainson's ideas

Swainson's Hawk

on classification exposed him to the ridicule of other naturalists besides Waterton and caused him to lose some of his scientific credibility. Despite the fact that he was fifty-one years old he decided that he would emigrate to New Zealand.

To raise money Swainson sold his bird skins and drawings to Cambridge University. He quickly finished off his final volume for the *Cabinet Cyclopaedia*, which was the *Taxidermy: with the Biographies of Zoologists* (1840), and knowing that he would soon be on the other side of the world he felt free to put down his true feelings about the naturalists he knew. He was kind to Waterton, but critical of Audubon of whom he wrote: "He can shoot a bird, preserve it, and make it live again ... but he cannot describe it in scientific ... perfectly intelligible terms ... A want of precision ... a general ignorance of modern ornithology sadly disappoint the scientific reader."[12] Swainson further slighted Audubon by only devoting one page to him, as an illustrator, and writing eight pages on Alexander Wilson, as an ornithologist. The longest of Swainson's biographical accounts was about himself, which may or may not be significant.

The youngest Swainson child was considered too young for the voyage and was left behind with friends who adopted him. Swainson then married his housekeeper, Anne Grasby, and sailed with the family on 27 November 1840 for his new start on a 300 acre land block already bought from the New Zealand Company. He took his main shell collection with him but most of his books were lost in a ship that foundered off the South African coast.

The Swainsons arrived in June 1841 and set up home at *Hawkshead* in the Hutt River valley near Wellington. Swainson had little experience of farming and soon found that he had scarcely any time for his former pursuits. The opposition of the local Maoris also worsened his position and the whole venture proved a costly failure for himself and other settlers. In May 1851 he sailed to Sydney to check on some land holdings and following a visit to his old friend MacLeay, he applied to the Victoria Government for a scientific position. Though hardly qualified for the post of Botanical Surveyor he spent more than a year studying the eucalyptus trees of Victoria, but his report was denigrated and ignored by subsequent botanists. From October 1853 to July 1854 he was in Tasmania continuing his work on the eucalypts but his field notes were never published. While in Tasmania he enjoyed collecting shells and wrote about some of them in the *Papers and Proceedings of the Royal Society of Van Dieman's Land*.

He returned to New Zealand some time in 1855 and died at his new home, *Fern Grove*, on 6 December. He was survived by his second wife, three of her children by him and four of the five children from his first marriage. His shell collection was split up among the family and though some items were deposited at the Wellington Museum they have been absorbed into the general collection without distinguishing marks. Another portion ended up on a rubbish tip. Many of Swainson's New Zealand and Australian drawings are preserved; most are of scenes, trees or shells with scarcely a single New Zealand bird amongst them.

Three years after Swainson's death Professor Baird named a North American vireo *Vireo swainsonii*. It was reclassified later as a race of the Warbling Vireo but full specific status and the name Western Warbling Vireo have been suggested for it.[13]

John Eliot Thayer
(1862-1933)

THAYER'S GULL *Larus thayeri* Brooks

Larus thayeri Brooks, 1915. *Bulletin of the Museum of Comparative Zoology, at Harvard College in Cambridge* 59, p. 373: Buchanan Bay, Ellesmere Land

John Thayer was born in Boston, Massachusetts, on 3 April 1862, the son of Nathaniel Thayer, a prosperous banker who built Harvard's Thayer Hall and sponsored Professor Agassiz on collecting expeditions for the Harvard Museums. John's childhood years were divided between the Thayer's town house and their farm at Lancaster, some thirty-five miles to the west. A year after he graduated from Harvard (in the class of '85) John married Evelyn Forbes, built a "large cheerful house" on George Hill in Lancaster and settled down there for the rest of his life, once writing in his diary "Had to go to Boston, how I hate it." Despite the wealth John Thayer inherited he worked hard throughout his life, seeking to put his money to practical use for the benefit of the community. Much of his time was spent on the farm where he grew apples, spearheaded horticultural improvements and bred dogs and horses.

His involvement with ornithology did not develop until the mid 1890s. By then his oldest son (one of five children) was old enough to join him in the field and together they hunted all the local birds, also taking nests and eggs. To house them, Thayer built a museum on the main street of Lancaster and in 1904 opened it to the public—during the first six years, more than 10,000 people signed the visitors' book. The lower floor contained his office, all his skin, nest and egg cabinets, mounted groups of local birds arranged according to habitat, and in the main hall several of Audubon's paintings. The upper floor housed his stuffed birds; some of these had been purchased at auctions, but more were brought back by collectors whom he sent out to the remoter and less explored parts of North and Central America. Thayer himself rarely travelled farther than Boston.

Early in the spring of 1906 Thayer despatched Wilmot W. Brown to Guadalupe, Mexico. Ornithologists had already worked the island but Brown's report was so disturbing that Thayer and Outram Bangs wrote 'The Present State of the Ornis of Guadaloupe Island' for *The Condor* (1908), drawing

THAYER. *WEALTHY AMATEUR ORNITHOLOGIST. HE ESTAB-
LISHED THE THAYER MUSEUM AT LANCASTER, MASSACHU-
SETTS, CONTAINING ONE OF THE LARGEST PRIVATE COLLEC-
TIONS OF BIRDS IN THE UNITED STATES. SEVERAL SPECIES
OF SHOREBIRDS WERE FIRST STUDIED ON THEIR NORTHERN
BREEDING GROUNDS BY HIS COLLECTORS.*

attention to the rapid destruction of the vegetation by thousands of goats and the threat to many endemic species from cats. Brown had found large colonies of Guadalupe Storm-Petrels in the montane pine forests but marauding cats were causing an appalling mortality—the last living individual was recorded only six years later. Thayer correctly predicted that the Guadalupe Flicker was also doomed to speedy extinction, for the birds were very unwary and easily caught by the cats—as well as by Brown, who shot twelve out of an estimated population of forty or fifty birds. The Guadalupe Caracara, last seen on the island in 1900, had already vanished.

A few years later Brown visited some of the islands off Cape San Lucas and sent Thayer the first account of the courtship behaviour of Heermann's Gulls, as well as new details about the breeding biology of Craveri's Murrelets.

Early in 1913 Thayer and some other Harvard graduates funded an expedition to Alaska and Siberia, with Joseph S. Dixon and W.S. 'Nick' Brooks serving as zoological collectors. It was intended that the voyage would last from April to September, but early in September the ship became trapped off the north Alaskan coast near the Canadian border and it was another ten months before the ice thawed. On 28 August 1913 Dixon shot a gull at Demarcation Point; when Brooks compared it with similar gulls in the Museum of Comparative Zoology at Harvard (labelled as Kumlien's Gulls and collected by J.S. Warmbath on Ellesmere Island in 1901) he decided that they were all of an undescribed species and named it *Larus thayeri*. Whether Thayer's Gull should be treated as a distinct species, or not, is still a matter of debate.

Thayer continued to send out collectors until the winter of 1927–28 when he became very ill and had to curtail his responsibilities. In 1931 he donated his 28,000 skins and 15,000 nests and eggs to Harvard, among them being the first clutches ever taken of Spoon-billed Sandpiper and Surfbird. The Surfbird eggs had been tracked down by Joseph S. Dixon and George M. Wright in May 1926, 1000 feet above the timberline near Mount McKinley, Alaska, on an expedition backed by Thayer. After Thayer's death, at his home in Lancaster on 29 July 1933, Harvard also received his magnificent collection of 3500 mounted birds.

Thayer's Gull

THOMSON. *SCOTTISH NATURALIST AND PIONEER MARINE BIOLOGIST, HE LED THE SCIENTIFIC STAFF ON THE CHALLENGER'S ROUND-THE-WORLD VOYAGE. THOMSON (AT CENTRE WITH GUN) IS SHOWN WITH THE SHIP'S CHIEF ENGINEER, PAYMASTER AND THREE CREWMEN ON KERGUELEN ISLAND IN JANUARY 1874.*

Charles Wyville Thomson
(1830-1882)

HAWAIIAN DUCK *Anas wyvilliana* Sclater

Anas wyvilliana P.L.Sclater, 1878. *Proceedings of the Zoological Society of London*, p. 350: Hawaiian Islands

Lithocoronis challengeri, Ipnops murrayi, Pentacrinus maclearanus, Calycopterix moseleyi, Lophius naresii, Umbellula thomsoni, Myzostoma wyville-thomsoni—and *Anas wyvilliana*—together with many other eponyms, serve as zoological reminders of the voyage of the *Challenger* and those who circumnavigated the world in her between the years 1872 and 1876.

This long voyage of scientific discovery was the brainchild of the eminent Scottish naturalist Charles Wyville Thomson[1] who had begun his career by studying medicine at Edinburgh University. His botanical genius was so marked that at the age of twenty he became a lecturer on the subject at Aberdeen University and three years later was Professor of Natural History at Queen's College, Cork. After a short time there he moved north to Queen's College, Belfast, where he taught for the next sixteen years, firstly on mineralogy and geology and later on all of the natural sciences. Within his broad and all-embracing knowledge of nature, Thomson was particularly fascinated by marine organisms and the possible limitations on life at great depths, about which there was then considerable speculation.

During the summer of 1868 Thomson and a colleague from the Royal Society of London borrowed the gunboat *Lightning* and dredged in the stormy seas between Shetland and The Faeroes. In the following years he continued his experiments at various sites off the British, Irish and French coasts, gathering new information about ocean currents, water temperatures and the distribution of animal life. In 1870, when Thomson became Professor of Natural History at Edinburgh University, he worked with the Royal Society to organize a prolonged voyage of scientific investigation, subsidized by the government. In December 1872 the *Challenger* put to sea. She was a three-masted, square-rigged wooden vessel of about two hundred feet in length, officially classed as a steam corvette but essentially a sailing ship. Commanded at first by Captain George Nares (and later by Captain Frank Turle Thomson) she carried twenty officers, over two hundred men and a scientific staff of six led by Wyville Thomson. Since the purpose of the expedition was to make a series of soundings and dredgings to assess the temperature and character of the water, to collect marine life from the surface and at all possible depths of the sea and also to study some of the remotest and rarely-visited islands, the vessel was specially adapted and equipped with workrooms, laboratories, storage space and trawling and dredging gear.

During the next three and a half years *Challenger*'s meandering itinerary took in Lisbon, Tenerife, Bermuda, Halifax, St Vincent, Fernando de Noronha, Cape Town, the Crozet Islands, Kerguelen Island, Melbourne, Wellington, Tonga, Cape York, the Molluccas, Manila, Yokohama, Hawaii, Tahiti, Juan Fernández Island, Valparaiso, the Falkland Islands, Ascension and numerous other ports and islands.

There were many memorable encounters with birds: Kerguelen Pintail satisfied the crew's cravings for fresh meat; Cape Pigeons and other petrels accompanied the ship while she dodged Antarctic icebergs; boobies and frigate birds perched on her yards off the New Hebrides; birds of paradise, exquisitely beautiful, lured the naturalists into the swamps and forests of the Aru Islands; Red-necked Phalaropes were found roosting on floating driftwood along the New Guinea coastline; steamer ducks paddling furiously away in clouds of spray amused the sailors as they navigated the tortuous Magellan Straits; and shoals of small cetaceans seen leaping through the sea off Tristan da Cunha proved on closer examination to be Rockhopper Penguins.

In this account it is the Hawaiian birds that deserve special mention, for it was in the Hawaiian Islands that *Anas wyvilliana* was added to the plethoric treasures of *Challenger*'s cargo. The ship reached Honolulu in August 1875 and the crew welcomed the King on board with his son and heir. King Kalahua delighted the naturalists by displaying a keen and intelligent interest in their activities and he peered through their microscope at a number of minute and curious organisms.

The next port of call was Hilo Bay, on Hawaii, where the *Challenger* anchored from 14 to 19 August so that the naturalists could ride inland to investigate the volcano of Kilauea. During this week they collected twenty-four bird skins of thirteen species, among them Wandering Tattler, Short-eared Owl and endemics such as the Hawaiian Hawk, Elepaio, Hawaiian Thrush, Ou, Iiwi and Apapane. A male specimen of the now extinct Hawaii Oo and two male Hawaiian Ducks (known locally as Koloas) were shot and examined. Three years later the ducks were described as a new species and named after Wyville Thomson by Philip Sclater in one of a series of papers dealing with the ornithological results of the voyage. Their status is rather uncertain and sometimes they are classified as a race of the Mallard. In his comments on the Hawaiian birds Sclater wrote:

> "The collection, although small, and containing nothing absolutely new except a single species of *Anas*, is of interest, as it enables us to record the actual island of the Sandwich group upon which the species contained in it were found, and as including an example of the little-known *Buteo solitarius* [Hawaiian Hawk] of Peale."[2]

In all, during the entire voyage, only 1172 specimens of birds' skins, skeletons and eggs were taken. The Admiralty later presented many of them to the British Museum (Natural History), thus adding thirty species that were not already held in their collections, including twelve type specimens. But the birds had only been an entertaining by-product of the main business of the voyage from which Thomson was rarely, if ever, diverted. After returning to Britain he was able to state:

"We always kept in view that to explore the conditions of the deep sea was the primary object of our mission, and throughout the voyage we took every possible opportunity of making a deep-sea observation. Between our departure from Sheerness on December 7th, 1872, and our arrival at Spithead on May 24th, 1876, we traversed a distance of 68,890 nautical miles, and at intervals as nearly uniform as possible we established 362 observing stations."[3]

Although the remarkable voyage was over, the scientific work had only just begun—now the cargo had to be disseminated among specialists for examination, comparison and description. Thomson, who was knighted on his return, was burdened with the responsibility of superintending the research and publishing the results, as well as recommencing his university duties. On top of all this, in 1877 he was invited to deliver the Rede lecture at Cambridge and the following year he presided over the geographical section at the meeting of the British Association in Dublin. Though he began a general account of the expedition, ill-health prevented him from completing the work and only two volumes of *Voyage of the Challenger, The Atlantic* were published (1877). The intimidating task was taken up by the Canadian naturalist Dr John Murray who had sailed with Thomson. He edited *The Report of the Scientific Results of the Exploring Voyage of HMS Challenger during the years 1873–1876* which appeared in fifty hefty volumes between 1885 and 1895.

In the summer of 1879 illness forced Thomson to retire from Edinburgh University and he spent his remaining few years at Bonsyde, his home near Linlithgow, Lothian, where he had been born on 5 March 1830. He died there, soon after his fifty-second birthday, on 10 March 1882, and was survived by his wife Jane (whom he had married in 1853) and their only son, Frank Wyville Thomson.

Hawaiian Duck

TOLMIE. SCOTTISH DOCTOR AND HUDSON'S BAY COMPANY OFFICIAL IN THE PACIFIC NORTH WEST. IN LATER LIFE HE SETTLED ON VANCOUVER ISLAND AND IS SHOWN HERE WITH HIS FAMILY AT CLOVERDALE IN ABOUT 1878. SIMON FRASER TOLMIE HAS A HAND ON HIS SHOULDER.

William Fraser Tolmie (1812-1886)

MacGILLIVRAY'S WARBLER *Oporornis tolmiei* (Townsend)

Sylvia Tolmiei J.K.Townsend, 1839. *The Narrative of a Journey Across the Rocky Mountains ... &c.,* p. 343: the Columbia [= Fort Vancouver, Washington]

MacGillivray's Warbler breeds mainly in the western parts of North America in Washington, Oregon, California, Idaho and Montana, extending northwards through British Columbia into south-eastern Alaska. It was first discovered near the Columbia River, so it is perhaps surprising that its common and scientific names should commemorate two men born thousands of miles away in the north of Scotland. One was born in Aberdeen, lived his entire life within the British Isles, and became a well-known ornithologist. The other was born sixteen years later, eighty-five miles away from Aberdeen at Inverness, and spent his adult life on the Pacific Slope of North America. The subject of this chapter, W.F. Tolmie, was the second of these men. Unlike William MacGillivray, he never wrote much about birds but he did at least have field experience of MacGillivray's Warbler and well deserves to be remembered in connection with the bird and its haunts in the Pacific north-west.

An early interest in plants had brought Tolmie into contact with William Hooker, then Professor of Botany at Glasgow University. He had also heard much about the Columbia River region from Dr John Scouler who had visited there in 1825 with a Hudson's Bay Company supply ship. In 1832 when the Company was looking for two medical officers to attend to their employees in the Columbia district, Hooker had no hesitation in putting forward the names of William Tolmie and another of his students, Meredith Gairdner.

Prior to this crucial event in his life, Tolmie had been educated in Inverness and Perth, then at the Faculty of Physicians and Surgeons at Glasgow. In early 1832 he helped out at an emergency cholera hospital in the city and having survived this experience he was ready to take up the position offered to him by the Hudson's Bay Company.

After travelling down to London he and Gairdner boarded the vessel *Ganymede* and were accompanied down the Thames to as far as Gravesend by the famous Arctic explorer and naturalist Dr John Richardson. Tolmie, who kept a journal for most of his early life, picked up a few tips from the great man (and guessed his age correctly): "He recommended me in packing birdskins to moisten the paper with Oil:Terebinth which preserves them from insects. The Doctor seems of a bold & energetic character, but does not possess much suavity of manner. He is of middle size, broad chested, strong limbed, well adapted in this respect for a traveller, appears to be about 45 years of age."[1]

Never one to be idle, Tolmie spent most of the outward voyage studying medicine, geography, literature, languages and natural history, but relations with Dr Gairdner became somewhat strained because Tolmie considered his colleague much too vain and proud. The *Ganymede* anchored at Fort George (Astoria) at the mouth of the Columbia on 1 May 1833. Three days later Tolmie arrived at Fort Vancouver and was set to work at once, though after a while he was allowed to explore and collect in the new country. (For a map see p. 74.) In characteristic style, he wrote in his diary: "In the grove bordering [the] river, the feathered songsters were warbling melodiously their note of praise to the Creator—saw one fine looking bird, the size of a thrush which I must endeavour to procure."[2] Over the years, as he made fresh and more detailed observations he was never afraid to record that he frequently fired and missed!

Tolmie was soon transferred to Fort Nisqually at the south end of Puget Sound. This proved to be an exceptional place for birds and also gave him the opportunity to ascend to the highest slopes of Mt Rainier, the first white man to do so. Setting off on 29 August with five Indians, the men struggled up the forested lower slopes, in pouring rain, and above the tree line reached a subsidiary peak (later called Tolmie's Peak) but the summit was inaccessible from there. During this trip Tolmie discovered several new species of plants and collected a few birds, amongst them the first MacGillivray's Warbler *Oporornis tolmiei*.[3]

After six months Tolmie was sent northwards up the Pacific coast for four hundred miles to Fort McLoughlin (Bella Bella, B.C.) where he arrived towards the end of December. To help pass away the winter Tolmie and a friend began to plan the first circulating library on the Pacific Slope and for the next decade it proved remarkably successful. Books were ordered from London and subscribing officers of the Company could then order them from Fort Vancouver.

Tolmie continued to note the birds he saw, recording hundreds on 7 April: "In the course of the forenoon about 5 or 600 surf ducks passed us in flocks of from 10 to 40—they feed on the herring spawn, which in some stations is abundantly deposited on the sea weed above low water mark. A few flocks of the Golden Eye Duck & of cormorants also passed—Mergansers, also hooded—Gulls very numerous. In the afternoon saw two cedar birds close to the Fort."[4] But his natural history observations unexpectedly declined during the summer because of a short but intense expedition. On 30 May Tolmie joined a party on the brig *Dryad* under Peter Skene Ogden heading further north beyond Fort Simpson. The intention was to establish a fort on the Stikine River to compete for furs that were being bought from the natives by the Russians. After a short time Ogden realized that he would be unable to make any headway against the Russian monopoly without starting a conflict, so he backed off. Later in the summer, Ogden moved Fort Simpson from its site on the Nass River southwards to McLoughlin's Harbour (Port Simpson, near Prince Rupert). Tolmie helped out during this great upheaval but was posted back to Fort McLoughlin at the beginning of November.

In the spring of 1836 Dr Tolmie moved south again to Fort Vancouver where he relieved John Kirk Townsend from his medical duties. Townsend had first arrived at the fort in September 1834 after crossing the Rockies with Thomas Nuttall. They had spent the fall on the Columbia, then wintered in Hawaii and returned to Fort Vancouver in the early spring. Nuttall had sailed back to the east coast in September 1835 but Townsend had stayed on for another winter, standing in as the fort doctor until Tolmie arrived, and remained in the area for a few more months. The two doctors became firm friends, eagerly discussing their travels and natural history discoveries. Tolmie usually sent his bird skins and plants back to Scotland but some of his specimens found their way into the collections of the Philadelphia Academy and the Smithsonian Institution via Townsend. One of these was the type of the Black Oystercatcher which Townsend intended to honour with Tolmie's name when he published the scientific description, but because of unforeseen circumstances this wish was denied. Townsend did manage to create the name Tolmie's Warbler *Sylvia Tolmiei* in the 1839 *Journal of the Academy of Sciences of Philadelphia*. In his 'Catalogue of the Birds of Oregon Territory' which appeared as an appendix to his *Narrative of a Journey Across the Rocky Mountains* (1839) he repeated his observations on the habits of the bird:

> "This pretty species, so much resembling the curious *S. philadelphia* [Mourning Warbler] of Wilson is common in spring on the Columbia. It is mostly solitary, and extremely wary, keeping chiefly in the densest and most impenetrable thickets, and gliding through them in a very cautious and suspicious manner. It may, however, sometimes be seen towards mid-day, perched upon a dead twig over its favorite place of concealment, and at such times it warbles a very sprightly and pleasant little song, raising its head until the bill is almost vertical, and swelling its throat in the manner of many of its relatives.
>
> I dedicate the species to my friend W.F. Tolmie, Esq., of Fort Vancouver."[5]

Townsend's name for the warbler was ignored by Audubon: a few months later, in his *Ornithological Biography*, he called it MacGillivray's Warbler *Sylvia Macgillivrayi* to honour his own faithful Scottish co-worker.[6] Audubon hoped

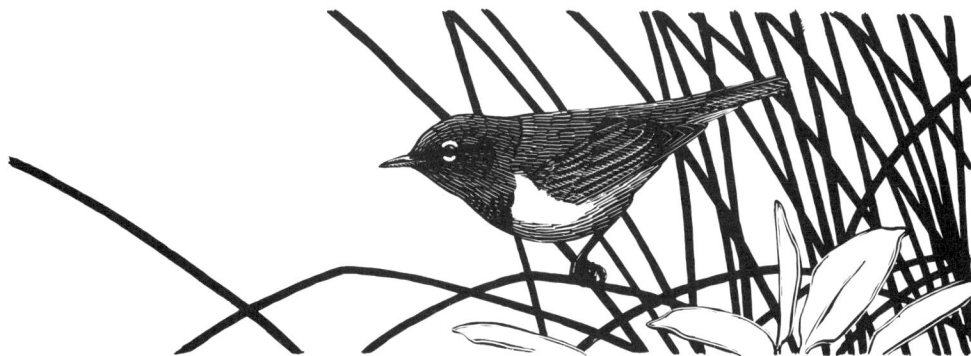

MacGillivray's Warbler

that his name would "endure as long as the species itself" but Townsend's scientific name has priority. The common name, Tolmie's Warbler, though surviving for several years (particularly in western America) has now been superseded by Audubon's proposal.

Tolmie was much more fortunate in his botanical work since several of his plant discoveries were named after him by Hooker. But Tolmie's interests were by no means confined to these two subjects; he studied Indian languages, sent Indian clothing, weapons and fish hooks back to Scottish museums and took a keen interest in livestock breeding. In 1841 he went home on leave and visited Paris to study at hospitals and other institutions. He also began to learn Spanish in anticipation of being posted to Yerba Buena in San Francisco Bay, but instead he was sent back to Fort Nisqually.

For the next twenty-five years Tolmie was involved in the Puget Sound area as medical officer, Indian trader and superintendent of agricultural operations. In 1846 Oregon and Washington were ceded to the United States so the Hudson's Bay Company moved its local headquarters to Victoria on Vancouver Island. Tolmie stayed on at Fort Nisqually and was appointed Chief Factor in 1855, then in 1859 he too moved up to Victoria where he spent the rest of his life on a large farm. Always a serious, hard-working and religious man, Tolmie supported a variety of causes and became increasingly involved in politics and educational reforms, so much so that he is one of the most esteemed pioneers of the north-west.

In February 1850 Tolmie had married Jane Work by whom he had five daughters and seven sons. The youngest son, Simon Fraser Tolmie, who was Premier of British Columbia from 1928 to 1933, wrote that his father "led a clean life and was a strict disciplinarian with his growing family. He was a profound student of religion, and though he did not attend church latterly on account of slight deafness, he insisted upon the family going and upon the younger members attending Sunday School. He obeyed the law and expected others to do likewise. He had a deep appreciation of the Pacific Northwest, and great faith in the future destinies of British Colombia, whose claims he never tired of backing."[7]

A description of Tolmie written in 1878 described him as "rather below medium height, broad-shouldered and stout … [with a] high forehead, coarse features, round deep-set eyes glittering from under shaggy brows, large round ruby nose."[8] Two years later when his wife died he became reclusive and insisted on taking all his meals alone in his library, except on Christmas Day. He died on 8 December 1886. Since he was born on 3 February 1812 he had lived to be nearly seventy-five.

Young Dr Gairdner who had set out with Tolmie on the *Ganymede* fared less well. In May 1835 he was allowed to make a trip to Fort Walla Walla and ventured into the Blue Mountains, but the hardships worsened his tuberculosis and only twenty-four months after his arrival on the Pacific coast he was forced to leave for Hawaii. Two years later he died in the islands aged twenty-nine. Gairdner's Woodpecker is now considered a race of the Downy Woodpecker *Picoides pubescens gairdneri*; named by Audubon, the subspecies was first collected by Townsend at Fort Vancouver.

Charles Haskins Townsend (1859-1944)

TOWNSEND'S SHEARWATER *Puffinus auricularis* Townsend

Puffinus auricularis C.H.Townsend, 1890. *Proceedings of the United States National Museum* 13, p. 133: Clarión Island, Revillagigedo Group

"My first book of birds, a happy discovery in our household library, was contained between the covers of a bulky report of the United States Department of Agriculture for the year 1856. It was a fifty-page chapter on *Birds Injurious to Agriculture* by Ezekiel Holmes, illustrated with thirty-two full-page woodcuts after Audubon. It supplied what I needed fairly well, the only other work in the house on natural history being J.G. Wood's *Bible Animals*, which unfortunately did not apply to my part of the world. When the old home was abandoned years later, I rescued this treasure of youth and had it bound, together with another chapter from the same ancient and battered report, almost as much prized, on the *Quadrupeds of Illinois* by Robert Kennicott, whose trail in northern Alaska I crossed years after."[1]

These reminiscences were written by Dr C.H. Townsend when he was in his late sixties and Director of the New York Aquarium. Most of the tomes in the old library were theological, his father being a minister in Parnassus, Pennsylvania. Townsend had been born there on 29 September 1859, into a family descended from an early immigrant who had arrived with William Penn on the *Welcome* in 1682. C.H. Townsend was once asked by Professor Baird if he was related to the ornithologist J.K. Townsend, but all he could suggest was that they probably shared the same Quaker ancestor.

The Reverend Townsend naturally hoped that his son would enter the ministry, but without any encouragement from either family or friends he determined to become a scientist. After persuading the local dentist to pass on his scanty knowledge of taxidermy Townsend practised on game birds that he shot locally. He must have shown some promise, for at the age of nineteen he was taken onto the staff of Professor Henry Ward's natural history establishment in Rochester, New York, where he was trained in museum methods. Three years later he moved to the Academy of Natural Sciences of Philadelphia where he might have remained as an all-round naturalist had he not, in 1883, met Professor

TOWNSEND, C.H. MARINE ZOOLOGIST, SHOWN HERE ON UNALASKA IN 1896 WITH THE JOINT BRITISH–AMERICAN COMMISSION FOR FUR-SEAL INVESTIGATION. FROM LEFT TO RIGHT: DAVID STARR JORDAN, GEORGE A. CLARK, JOSEPH MURRAY, CAPTAIN JEFFERSON MOSER, FREDERIC A. LUCAS, CHARLES H. TOWNSEND, D'ARCY WENTWORTH THOMPSON, JAMES MACOUN, LEONHARD STEJNEGER.

Baird, who counted amongst his responsibilities the Directorship of the Fish Commission. Baird offered Townsend a job and within a week sent him on his way to the McCloud River in northern California, where he worked at a salmon hatchery for two years. This was the beginning of Townsend's career with the Commission, specializing in fishery and oceanographic investigations, but the early influence of Holmes and Kennicott never faded—Baird made sure of that. As Secretary of the Smithsonian Institution Baird never wasted an opportunity to add to its collections and Townsend frequently received letters from him requesting particular species of birds. When Baird suggested that White-tailed Ptarmigan might occur in California, Townsend searched Mt Shasta and Lassen Peak without success, but sent 200 other species from the Golden State.

In 1885 Townsend was sent to Alaska on the Revenue Marine steamer *Corwin* with orders to explore the Kobuk River. (For a map of Alaska see p. 318.) The ship called at Golovnin Bay, Port Clarence and Cape Prince of Wales on the Seward Peninsula, then laboriously pushed its way through drift ice across the entrance to Kotzebue Sound. At the beginning of July the small party was put ashore near the forty-mile wide Kobuk delta where the river flows between thickets of dense willows. The group of five men was led by Lieutenant John Cassin Cantwell (nephew of John Cassin) who had explored part of the region during the previous summer. They ascended the broad river in a steam-powered launch and when the little boat got stuck on sand bars they often had to man-haul it through icy water that swirled around them up to their waists. In the upper reaches the Kobuk surges down a dangerous rock-strewn channel and the party had to make frequent portages, but eventually they reached the source at Walker Lake, the first white men to do so.

During the journey Townsend listed fifty-two species of birds and among the common breeders were Canada and White-fronted Goose, Sandhill Crane, Pintail, Bonaparte's Gull, Red-breasted Merganser, Lesser Yellowlegs, Least Sandpiper, Wilson's Warbler and Varied Thrush. He was surprised to see such familiar southern species as American Robin, Tree Swallow, Rusty Blackbird and Belted Kingfisher. As they were nearing the coast again at the end of August Townsend shot a Bristle-thighed Curlew and this individual constituted the third record for the United States. It was a young bird that must have recently left the nearby breeding grounds, but it was many more years before any nesting pairs were discovered.

As the *Corwin* sailed back southwards it called at the same places as before and also visited Hall Island on 8 September. As usual, Townsend took the opportunity to hunt ashore and thus became the first to discover the breeding grounds of McKay's Bunting. Robert Ridgway had published the original description during the previous year, speculating that "The summer home of this bird is probably the unknown region to the north of the Ar[c]tic mainland." It is now known to breed only on a few of the Bering Sea islands, primarily Hall and St Matthew. When the *Corwin* left St Paul Townsend stayed behind and spent a month collecting there before taking a homeward bound vessel, enjoying an unexpected two week delay at Unalaska that allowed him to further enrich his ornithological material.

His next appointment was to the Fish Commission's own steamer, the

Albatross, for a cruise among the Bahama Islands. Again he collected birds whenever duty allowed and Baird was delighted to receive several skins of the rare Kirtland's Warbler, which he had first described thirty-five years before.

In 1887 Townsend was sent to Honduras and since he was due to embark from New York Baird arranged for him to meet George Lawrence on his last evening. Lawrence was by then an octogenarian and had studied tropical birds for many years. He encouraged the young man to make the most of his opportunities—advice that was eagerly accepted. Sailing through the north-west Caribbean Townsend became the first naturalist to land on Swan Island and took thirty species of birds. On reaching Honduras he travelled by canoe, with native guides, up the Segovia River and was entranced by the birds of the tropical jungle. Their first camp was set up under a tree festooned with yard-long hanging nests that from the distance had looked to him like some strange, exotic tropical fruits, but they had been built by the remarkable Montezuma Oropendolas. Weeks later and much further upriver they entered an open pine forest where Scarlet Macaws gathered high in the trees.

Unfortunately Baird died in August that year and never had a chance to handle the colourful collection his young protégé brought back from Central America. Before his death Baird had made arrangements for the *Albatross* to visit the Pacific and so in the fall Townsend sailed with her round Cape Horn to the Galapagos Islands and wrote:

"The *Albatross* turned us loose on these classic islands for eleven arduous but wonderful days. There was help in the collecting, but I worked until midnight at the preparation of bird skins, leaving to others the preservation of reptiles and such miscellaneous zoological plunder as each returning boat dumped into our upper and 'tween-decks laboratories. Everybody was drafted including the surgeon, who helped to press plants, while boat crews swept the beaches with seines. There were eighteen tortoises crawling about the deck when we sailed, one of them big enough for the sailors to ride. The giant land tortoises for which the archipelago was named, had already disappeared from some of the islands and were scarce on the others. Ships had been carrying them away in large numbers for three centuries. I once made a study of whale-ship log-books to ascertain what the whalers of the nineteenth century did to the tortoises and found that seventy-nine ships carried away thirteen thousand tortoises."[2]

For several years the *Albatross* was stationed in the North Pacific and during the winter of 1888–89 cruised between San Francisco and the Gulf of California,

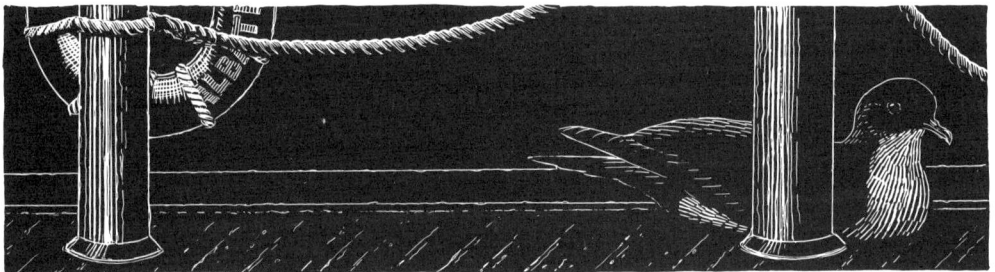

Townsend's Shearwater

anchoring at many remote islands. Townsend's Shearwater was discovered on 4 March 1889 when the *Albatross* was lying off Clarión Island, more than four hundred miles south-west of Baja California. In the darkness, attracted by the bright lights of the ship, more than twenty of these shearwaters came on board and Townsend took five specimens. In the morning he went ashore, the first naturalist known to have visited the rugged, volcanic island. Using a sharp machete he hacked his way through the cactus thickets and after a two hour struggle reached the central plateau about a thousand feet above the sea. His efforts were rewarded by another new species, the Clarion Wren. Burrowing Owls and Mourning Doves were also common on the rocky slopes and he watched Wandering Tattlers and a flock of Black-necked Stilts feeding along the shore.

As the *Albatross* steamed between the widely scattered islands of the Revillagigedo group Townsend frequently saw his new shearwater skimming low and fast over the waves but although he also went ashore on Socorro and San Benedicte, it was a little too early in the season for him to find occupied burrows. Ten years later the group was visited by the ornithologist Alfred W. Anthony, who found colonies of Townsend's Shearwaters on the higher parts of Clarión and San Benedicte Islands. The burrows were confined to the grassy slopes and at the end of May most pairs had newly hatched sooty-grey chicks. This shearwater is limited in its breeding range to the Revillagigedo and Hawaiian Islands and the subspecies that inhabits the latter group has been named *Puffinus auricularis newelli* (see Appendix).

Between 1890 and 1896 the *Albatross* remained in the North Pacific while the crew surveyed fishing grounds, charted part of the Aleutian chain, sounded the route for the Hawaiian cable, mapped the huge seal rookeries of the Pribilof and Commander Islands and made studies of the Northern Fur Seal. During this time Townsend enjoyed watching vast numbers of seabirds and two particular occasions forever remained vivid memories. One evening as the *Albatross* anchored at Yukon Harbor in the Shumagin Islands the boat was surrounded by Crested Auklets that covered the surface of the water and filled the air with their whirring wings. A long blast on the ship's whistle brought even more legions of auklets shooting out from the cliffs, startling and amazing everyone on board. The other experience concerned Sooty Shearwaters at sea, south-west of the Shumagin Islands:

"To discuss their numbers in commonplace terms of millions would not convey the impression made upon us at the time. The birds were no more to be numbered than 'the innumerable company of the stars'. The sea was smooth and the *Albatross* was making about eight knots through the bird-covered area, which extended as far on both sides as it was possible to see from the top-sail yard. It was 'Alaska weather'—inclined to be hazy and the horizon not very clear. Immediately ahead the birds kept rising and moving out of the ship's course. The rather narrow belts that were disturbed on either side rose and moved forward, joining the continuous flight of those ahead, which gradually tended to windward toward the left. The log of the *Albatross* might show how long we were crossing the avian galaxy. It would be worth looking up."[3]

Most of the birds Townsend shot on his wanderings were sent to Washington and described by Robert Ridgway (or some other ornithologist) long before the collector returned to headquarters. This led Townsend to comment: "It isn't safe to leave undescribed species where ornithological nomenclators are prowling about. Mr. Elliot, for instance, found some novelties in my collection of Aleutian ptarmigan; but I had at least the fun of shooting them on the mossy mountains of Attu, Aggatu, Kyska and Adak, to say nothing of Atka and Unalaska. And what joy for my setter, who was a sailor for years!"[4]

In 1897 Townsend was promoted Chief of the Division of Fisheries of the Fish Commission. After five years service in this capacity he was sent to Europe as fishery expert to the U.S. Government during its negotiations with the Russians at The Hague about conservation and fishing rights. On his return in 1902 Townsend took up the Directorship of the New York Aquarium and held the post for the next thirty-five years. While working there he wrote his previously mentioned autobiographical article for *The Condor*, which he finished with an explanation for his limited ornithological work: "It was not my destiny to go very deeply into ornithology. The long voyages with [Alexander] Agassiz, always illuminated with his enthusiastic talk in the ship's laboratory, amounted to a course in oceanography, which tended away from ornithology and eventually anchored me at the Aquarium. In this fishy atmosphere only the sea birds, pelicans, boobies, gulls, gannets and Galapagos penguins remain."[5]

Townsend may not have gone as deeply into ornithology as he would have wished, but his passion for birdwatching had been strong and lasting and he contributed in no small way towards a better understanding of the geographical distribution of many species, especially the pelagic birds. Inevitably, most of his publications were related to oceanography, whaling, fur seals and fishery matters, but he also wrote some notable papers on birds which included 'Field-notes on the Mammals, Birds and Reptiles of Northern California' in the *Proceedings of the U.S. National Museum*, (1887); 'Albatross Voyage. Tropical Pacific, 1899–1900. The Birds' which he wrote with Alexander Wetmore for the *Bulletin of the Museum of Comparative Zoology*, Harvard College (1919); and 'Albatross Voyage, Gulf of California. Birds collected in Lower California' in the *Bulletin of the American Museum of Natural History* (1923).

Townsend gave up his position at the Aquarium at the age of seventy-seven and spent the last seven years of his life in retirement. He died on 28 January, 1944 at Coral Gables, near Miami, Florida.

John Kirk Townsend (1809-1851)

TOWNSEND'S SOLITAIRE *Myadestes townsendi* (Audubon)

Ptilogony's [sic] *Townsendi* Audubon, 1838. *The Birds of America* (folio), Vol. 4, Pl. 419, Fig. 2 (1839, *Ornithological Biography*, Vol. 5, p. 206): Columbia River [= Astoria, Oregon]

TOWNSEND'S WARBLER *Dendroica townsendi* (Townsend)

Sylvia Townsendi (Nuttall MS) J.K. Townsend, 1837. *Journal of the Academy of Natural Sciences of Philadelphia* 7, p. 191: forests of the Columbia River [= Fort Vancouver, Washington]

In 1834 John Kirk Townsend was invited by Thomas Nuttall to join an expedition travelling to the Pacific coast. Crossing the Rocky Mountains in summer they spent the fall on the lower part of the Columbia River and when winter came they took ship to the Hawaiian Islands. The following spring they were back on the Columbia, Townsend collecting birds and mammals while Nuttall concentrated on his beloved plants. When Nuttall returned to the east coast, Townsend stayed on for another year, widening his field experiences and adding to his collections. When he too returned to the east he received a rapturous welcome from fellow naturalists and then settled down to write up his discoveries. He began to publish a comprehensive ornithology of North America that established him as one of the leading naturalists of his day. His book sold well and gave him the funds to study ornithology howsoever he wished.

Unfortunately, Townsend's return did not really work out like this. Though he had the privilege of exploring new parts of the American West he suffered several misfortunes. One of these was induced by another naturalist who made an impatient and determined bid to add to his own fame at Townsend's expense. This is partly why Townsend is less appreciated than he should be.

Starting again, from the beginning, John Kirk Townsend was born on 10 October 1809 in Philadelphia. His Quaker family encouraged him in his early

TOWNSEND, J.K. *PHILADELPHIAN NATURALIST WHO ACCOMPANIED NUTTALL ON HIS JOURNEY TO THE COLUMBIA RIVER AND HAWAII. HIS TALE OF THEIR ADVENTURES HAS BECOME A HISTORICAL CLASSIC.*

interests in natural history and he was further stimulated by his cousin, William P. Townsend of West Chester, Pennsylvania. Together they built up a fine collection of local birds, which were exhibited at the Westtown Boarding School in Chester County. After leaving the school John trained as a physician and pharmacist and continued with his passion for bird collecting. At the age of twenty-three he made his first permanent mark upon American ornithology by describing a strange bird. A bird so odd that no one has ever found another like it! On 11 May 1833 while shooting in a cedar grove at New Garden, Chester County, Townsend retrieved a grey and brown bunting, with white throat and white eyestripe, which he passed on to his friend Ezra Michener. The bird was subsequently described by Audubon as *Emberiza Townsendi* and is now known as Townsend's Bunting.[1] Although it could be a hybrid it may well be the very last remnant of a now extinct species. (It is placed in Appendix C of the 1983 A.O.U. *Check-list of North American Birds*.)

Later in 1833, Thomas Nuttall resigned from his position as teacher of natural history and curator of the botanic garden at Harvard. He wanted to join Nathaniel Wyeth's second geographical and commercial expedition across the Rockies via South Pass, Wyoming, to the shores of the Pacific. Nuttall was forty-seven years old and already had several thousand miles of wilderness travelling behind him. Knowing that botany would fully occupy his time and knowing too that there would be rich zoological discoveries to be made, and since he rarely carried a gun, Nuttall ensured that his young friend Townsend was invited along to specialize in the birds and mammals. Hearing of this the American Philosophical Society and the Academy of Natural Sciences of Philadelphia each gave Townsend $100 in advance for a share of the trophies.

Townsend could not have had a better companion with whom to discuss the birds for, though primarily a botanist, Nuttall had just published his classic *Manual of the Ornithology of the United States and of Canada* (1832–34). They made their way from Philadelphia to St Louis, then set out on foot to cross the next 300 miles, but at Booneville they met up with Wyeth and completed the rest of the journey to Independence by steamboat. By then they had already seen huge flocks of Passenger Pigeons, Sandhill Cranes, American Golden Plovers, Wild Turkeys and "vast numbers" of Carolina Parakeets. Soon the avifauna became much less familiar.

Besides Wyeth and the two naturalists, there were now assembled five missionaries, a fur trader and a further sixty men with some 250 horses. On 28 April 1834 the whole train set off across the prairies of north-east Kansas and eastern Nebraska. Nuttall very soon found Harris's Sparrow, a new bird. On the Platte River Townsend found another new species, adding a single specimen of the Chestnut-collared Longspur to the collection. But Townsend found it hard to come to terms with the hasty pace of the expedition. He often regretted that he did not have enough time in the best places, or the leisure to observe the habits of new species before collecting them:

> "On the afternoon of the 31st [May], we came to green trees and bushes again, and the sight of them was more cheering than can be conceived ... We encamped in the evening in a beautiful grove of cottonwood trees, along the edge of which ran the Platte, dotted as usual with numerous islands.

In the morning, Mr. N. and myself were up before the dawn, strolling through the umbrageous forest, inhaling the fresh, bracing air, and making the echoes ring with the report of our gun, as the lovely tenants of the grove flew by dozens before us. I think I never before saw so great a variety of birds within the same space. All were beautiful, and many of them quite new to me; and after we had spent an hour amongst them, and my game bag was teeming with its precious freight, I was still loathe to leave . . .

None but a naturalist can appreciate a naturalist's feeling–his delight amounting to ecstasy–when a specimen such as he has never before seen, meets his eye, and the sorrow and grief which he feels when he is compelled to tear himself from a spot abounding with all that he has anxiously and unremittingly sought for."[2]

Taking the north fork of the Platte into south-east Wyoming Townsend saw flocks of up to a hundred Lark Buntings, a distinctly new species, noting that "the piebald appearance of the males and females promiscuously intermingled, presented a curious, but by no means unpleasing effect." As they hurried across the tablelands near the Sweetwater Townsend saw a Mountain Plover and to get his only specimen felt "compelled to kill it without delay". While following this river up to South Pass, Common Poorwills *Phalaenoptilus nuttallii* were first seen on 10 June, and five days later at "a spot where we found a little poor pasture" beside the Sandy Creek, in Sweetwater County, Townsend discovered the Sage Thrasher—which he called the Mountain Mockingbird. To make room for the ever increasing number of specimens he had long since discarded his surplus clothing, soap and shaving gear.

At the beginning of the third week of June they reached the Green River rendezvous where Indians, trappers and hunters had gathered to sell their furs

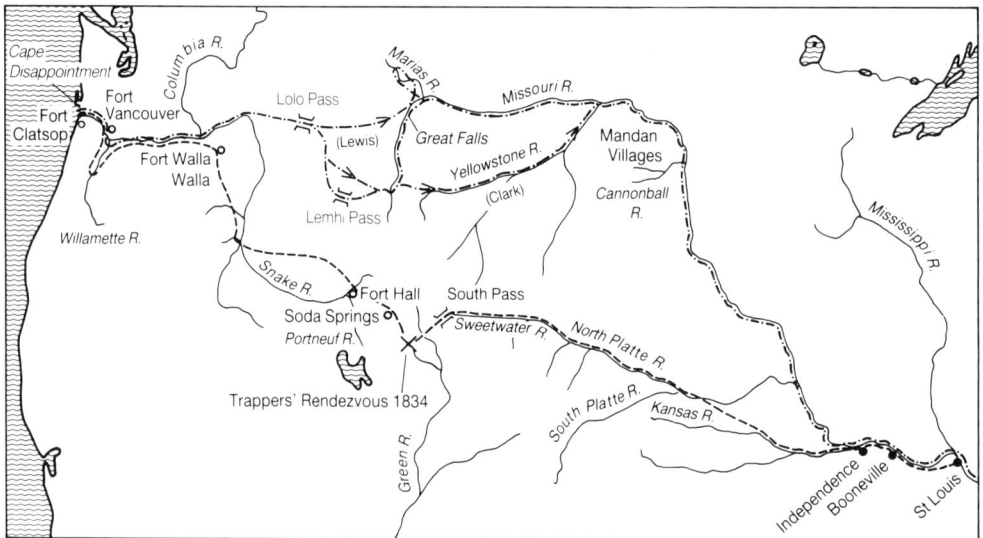

ROUTES OF TOWNSEND, NUTTALL, LEWIS AND CLARK
----- TOWNSEND AND NUTTALL 1834
·-··-··- LEWIS AND CLARK 1804–6

to traders and to stock up with provisions for another year. Wyeth had made a deal to sell some supplies but found to his dismay that another trader had already satisfied most of their needs. After ten days Wyeth left the wild gathering, now accompanied by the Scots adventurer Sir William Drummond Stewart, and continued westward up to Soda Springs in south-east Idaho. Here, amongst a thicket of Red Cedars, Townsend saw his first Clark's Nutcrackers and Lewis's Woodpeckers—a reminder that they had been preceded across the mountains by Lewis and Clark who had taken a more northerly route almost thirty years earlier.

On 9 July Townsend was able to report that White Pelicans were numerous and Canvasbacks and Northern Shovelers had broods of young. Three days later in Bannock County, Idaho, east of Pocatello, Townsend collected the type of the Green-tailed Towhee, another discovery that must have reinforced his belief that he had made a wise decision in accepting Nuttall's invitation.

On the south side of the Portneuf River Wyeth began to construct a trading post and the rough wooden building was considered to be complete on 5 August. On the next day Wyeth, Townsend, Nuttall and most of the men departed, leaving thirteen others in charge of the fort. The missionaries had already left with some Hudson's Bay Company hunters, and Stewart had likewise headed for the Columbia River. For the naturalists the next part of the journey proved to be the most demanding because of the high altitude, the rugged terrain and the lack of food, which combined to drain them of most of the surplus energy required for collecting. At one stage Townsend was so desperate that he was reduced to making a meal of rosebuds. He was more than a little put out when he returned to camp to find that Nuttall (who had become "exceedingly thin") had shared in a meal of an owl that Townsend had shot that morning and had wanted to preserve. He later complained that "the bird of wisdom lost the immortality which he might otherwise have acquired." Since this incident is mentioned in almost all the accounts ever written about the expedition, Townsend's statement would appear to be quite untrue!

Despite their hardships the party arrived safely at Fort Walla Walla on 3 September but there was so little food there that after a couple of days they continued on down the Columbia. Now travelling by canoe they suffered from gales, rough water and spray that soaked them thoroughly, and even penetrated Nuttall's precious plant collections. Townsend could only marvel at the older man's devotion: "... he has been constantly engaged since we landed yesterday, in opening and drying them. In this task he exhibits a degree of patience and perseverance which is truly astonishing; sitting on the ground, and steaming over the enormous fire, for hours together, drying the papers, and re-arranging the whole collection specimen by specimen, while the great drops of perspiration roll unheeded from his brow. Throughout the whole of our long journey, I have had constantly to admire the ardor and perfect indefatigability with which he has devoted himself to the grand object of his tour. No difficulty, no danger, no fatigue has ever daunted him, and he finds his rich reward in the addition of nearly *a thousand* new species of American plants ... My bale of birds, which was equally exposed to the action of the water, escaped without any material injury."[3]

At last, on 16 September, after six months travelling, Wyeth brought his contingent into Fort Vancouver where John McLoughlin, the Hudson's Bay Company's chief factor, welcomed them all and invited Townsend and Nuttall to stay at his home. Another Scot, Dr Meredith Gairdner, was also kind and friendly toward them. Gairdner knew the locality well and was able to give advice about where best to search for plants and birds since he had similar interests and had already sent a number of specimens back to Scottish museums.

Townsend and Nuttall now found themselves extremely well placed. They had a good base, they were in reasonable health and they were in excellent territory for making more new discoveries. Russian and British naturalists had searched the coastline but in the inland areas only David Douglas had travelled with the sole intention of collecting. Scouler, Menzies, Tolmie and Gairdner had investigated some small areas but much still remained unknown. Together the newcomers ransacked the foothill forests of cedar, hemlock and fir, the lowland stands of cottonwood, oak, ash, alder and maple, the thickets of hazel, elder and willow, and the open meadows, sloughs and lakes. Nuttall especially found many new species, since as usual he was casting a wide net for plants, birds, shells, crustaceans, butterflies and moths. Townsend continued to concentrate on the birds and mammals—especially after someone drank all the spirits from his two gallon jar of reptiles.

In October Wyeth began to establish another trading post on the lower Willamette River near where it joined the Columbia. The supply brig *May Dacre*, which he had sent round the Horn to the Columbia, now plied up and down the rivers and the two naturalists used it as a centre for their operations. In October Townsend was able to report:

> "5th.—Mr. N. and myself are now residing on board the brig, and pursuing with considerable success our scientific researches through the neighborhood. I have shot and prepared here several new species of birds, and two or three undescribed quadrupeds, besides procuring a considerable number, which, though known to naturalists, are rare, and therefore valuable. My companion is of course in his element; the forest, the plain, the rocky hill and the mossy bank yield him a rich and most abundant supply."[4]

Later they collected near Fort Vancouver but with the approach of winter their euphoria began to wane; botanizing was poor and after the fall migration bird life was much less evident. They therefore eagerly accepted a trip to Hawaii on the *May Dacre*, which was going there for three months to stock up with supplies for Wyeth. The brig started downstream in early December and after a few days gave Townsend a brief chance to collect seabirds on the beach at Cape Disappointment, a place that he determined to return to. On the 11th they crossed the treacherous sand bar at the mouth of the Columbia and delighted in the masses of cormorants, auks, gulls, petrels and marine mammals that surrounded them. Out at sea, flocks of shearwaters were intermittently abundant for the first few hundred miles and on Christmas Day, about half way to Hawaii, they added the Black-footed Albatross to the list of new species discovered.

They arrived at Oahu in early January 1835. On Kauai the following month Townsend was often helped by native children who brought him birds caught by hand. But despite their stay on the islands little has ever been made of their work there since neither they, nor anyone else, ever wrote a full account of it. The exotic plants were so far outside Nuttall's line of research that he seemed content to collect just a few ferns, though he greatly admired the sea shells and found seventeen unknown species. Most of the bird skins from this period (and Townsend's later visit to the islands) were passed on to the Academy of Natural Sciences of Philadelphia but others, as we shall see, went further afield in a rather haphazard manner.

Townsend and Nuttall left the Hawaiian Islands on 26 March and were back at the mouth of the lower Willamette on 17 April. Soon they were once again based in Fort Vancouver and were off into the woods after birds and plants, excited by the onset of spring in a territory that they knew only from the more dreary days of the preceding fall. Hordes of migrants poured in from the south. Black-throated Gray Warblers, 'Audubon's Warblers' and the resident 'Oregon Juncos' were abundant and easily gathered by Townsend. Vaux's Swifts were plentiful and he found them breeding in hollow trees. MacGillivray's Warblers, though common, kept to the thick undergrowth and were surprisingly elusive. The pair of Hermit Warblers he collected were the only ones he ever saw. He seems to have shot only one Townsend's Solitaire and he only managed to get two Townsend's Warblers, one fine specimen in the spring and another in duller plumage later in the year.

At the end of the summer season of 1835 Nuttall decided to return to the east coast after visiting California. Since there was no trade between Oregon and California it was easier for him to go there via Hawaii. Townsend elected to stay on for another year but took the opportunity to send 300 bird and 50 mammal skins back with Nuttall. The preservation of his skins was a constant problem for him and like all collectors he always feared for their safety, but he expected that they would be less vulnerable in the dry and solid buildings at Philadelphia than in the damp and comparatively primitive north-west.

Townsend's Warbler

In September Nuttall boarded the Hudson's Bay Company vessel *Ganymede* bound for Hawaii, together with the fatally ill Dr Gairdner. Townsend took over from Gairdner at the Fort Vancouver hospital until mid-March 1836 when Dr Tolmie came from the north to relieve him. Taking "the opportunity of peregrinating again in pursuit of specimens" Townsend collected locally and on 26 June headed upriver to Fort Walla Walla and the Blue Mountains, too late to do much useful work. In the fall he went back down to Cape Disappointment to concentrate on water birds. In early October, for example, he selected some Western Gulls for specimens and in November shot a single Surfbird, the only one he saw and an exceedingly desirable bird as far as eastern naturalists were concerned.

On 30 November Townsend left the Columbia bound for Honolulu, and on his arrival learned that the German collector Ferdinand Deppe was there. Writing up his journal on 15 January 1837 Townsend outlined his progress on Oahu: "Several days ago Mr. Deppe and myself visited Nuano valley, where we hired a native house, in which we are now living. Our object has been to procure birds, plants, &c., and we have so far been very successful. I have already prepared about eighty birds which I procured here."[5] Townsend stayed until March, spent about a month at Tahiti, then went on to South America, becoming so ill at Valparaiso that his absence from Philadelphia was extended for a further two months.

In the preceding year much had happened back home, little of it of benefit to the young naturalist. When all the boxes and barrels of specimens arrived at the Academy they aroused enormous interest. Dr Harlan wrote at once to his friend Audubon exaggeratingly exclaiming that there were a hundred new bird species! Audubon became frantic since he desperately wanted to include all the new birds in the last part of *The Birds of America*. He had no trouble persuading Nuttall to part with his duplicate skins because Nuttall was overloaded with plants, genuinely cared only for the advancement of science and worried scarcely at all about credit, honour or recognition. The bird skins collected by Townsend proved to be much more difficult to acquire since they were really the property of the Academy, many of whose members felt a certain loyalty to Townsend and for various reasons were disinclined to favour Audubon. But a man as determined as Audubon would hardly be put off by the Academy's initial unhelpfulness. He went to Philadelphia as soon as he could but after two days of campaigning was still not allowed to see them. His wealthy friend Edward Harris came to his aid by offering to pay for the bird skins but no decision could be made while Nuttall was in Boston. Audubon himself takes up the story:

"Dr TOWNSEND'S collection was at Philadelphia; my anxiety to examine his specimens was extreme; and I therefore, bidding farewell to my Boston friends, hurried off to New York, where, in a week, I added eighteen names to my list of subscribers ... Once more my son and I reached Philadelphia, where at once we placed ourselves under the roof of my ever staunch and true friend RICHARD HARLAN, Esq., M.D., with whom we remained several weeks ... Having obtained access to the collection sent by Dr TOWNSEND, I turned over and over the new and rare species; but he was absent at Fort Vancouver, on the shores of the Columbia River; THOMAS NUTTALL had not yet come from Boston, and loud murmurs were uttered by the *soi-disant* friends of science, who objected to my seeing, much less portraying and describing those valuable relics of birds, many of which had not yet been introduced into our Fauna. The traveller's appetite is much increased by the knowledge of the distance which he has to tramp before he can obtain a meal; and with me the desire of obtaining the specimens in question increased in proportion to the difficulties that presented themselves. Having ascertained the names of the persons best able or most willing to assist me on this occasion, and aided by THOMAS NUTTALL, who had now arrived, Drs PICKERING, HARLAN, S.G. MORTON, Secretary to the Academy of Natural Sciences, M'MURTRIE, TRUDEAU, and above all my friend EDWARD HARRIS, who offered to pay for them with a view of presenting them to me, I at length succeeded. It was agreed that I might *purchase duplicates, provided* the specific names agreed upon by Mr. NUTTALL and myself were published in Dr TOWNSEND'S name. This latter part of the affair was perfectly congenial to my feelings, as I have seldom cared much about priority in the naming of species. I therefore paid for the skins which I received, and have now published such as proved to be new, according to my promise. But, let me assure you, Reader, that seldom, if ever in my life, have I felt more disgusted with the conduct of any opponents of mine, than I was with the unfriendly boasters of their zeal for the advancement of ornithological science, who at that time existed in the fair city of Philadelphia."[6]

Audubon gloated unashamedly as he stole the glory from the absent Townsend, sending the Reverend John Bachman a very revealing letter:

"Now Good Friend, open your Eyes! aye, open them tight!! Nay place specks on your probosis if you chuse! Read aloud!! quite aloud!!! I have purchased *Ninety Three Bird Skins!* Yes 93 Bird Skins!—Well what are they? Why nought less than 93 Bird Skins sent from the Rocky Mountains and the Columbia River by Nuttall and Townsend!—Cheap as Dirt, too—only one hundred and Eighty-Four Dollars for the whole of these, and hang me if you do not echo my saying so when *you see them!!* Such beauties! Such rarities! Such Novelties! Ah my Worthy Friend, how we will laugh and talk over them!"[7]

Audubon took the skins to Charleston where he had arranged to stay with Bachman. Here, during the winter of 1836–37 he drew seventy-six more plates. Not all the species were entirely new but many were very rare in museums and difficult to track down. Townsend's birds make up very nearly a seventh of the species in Audubon's great work and helped make it truly *The Birds of America* instead of just an ornithology of the eastern United States.

The first text relating to Townsend's birds appeared in the *Journal of the Academy of Natural Sciences of Philadelphia* in 1837, written mainly by Nuttall under Townsend's name. Twelve birds were described briefly, without any extraneous detail on habitat or behaviour. Of these, the new species still recognized are the Black-throated Gray Warbler, Townsend's Warbler, Hermit Warbler, Chestnut-backed Chickadee, Bushtit, Chestnut-collared Longspur, Lark Bunting, Sage Thrasher and Mountain Plover. His Western Bluebird had already been described from a Mexican specimen by Swainson, the 'Oregon Junco' has recently been relegated to subspecific level, and 'Audubon's Warbler' is now lumped with the 'Myrtle Warbler' under the name Yellow-rumped Warbler.

In his folio plates Audubon claimed a further thirteen new species from the West of which only three, the Black Oystercatcher, Yellow-billed Magpie and Tricolored Blackbird are recognized, the latter two from Nuttall's spell in California. Later in his *Ornithological Biography* Audubon described nineteen more as new, of which only the Western Gull, Black-footed Albatross, Green-tailed Towhee and Townsend's Solitaire survive under his authorship. To all

Townsend's Solitaire

these new discoveries may be added Vaux's Swift and MacGillivray's Warbler (which Townsend dealt with in the appendix to his *Narrative*) and one more, Harris's Sparrow, that Nuttall kept back for the second edition of his *Manual*. These new birds and the nine described by Nuttall under Townsend's name add up to nineteen new species from their expedition. Besides the new species, they found many new races and extended the known range of a great many birds. Their observations, though of enormous value, are widely dispersed in the literature.

How different it might have been if Townsend had returned with his own material and worked through his discoveries, publishing all the scientific results together in a single volume, perhaps in conjunction with Nuttall. When Townsend did at last return in November 1837 it was too late for him to do much original work on his birds and he had so little money that he was forced to sell off his own portion of the collections to quieten his creditors. He was given a job as curator at the Academy, which voted him a second $100 for more duplicates but to further improve his finances he sold many of his Oregon birds to Audubon. He also arranged for the artist to sell the Hawaiian birds for him in Europe.

Townsend now concentrated on preparing *The Narrative of a Journey across the Rocky Mountains to the Columbia River and a Visit to the Sandwich Islands, Chili, &c.* (1839). It was lucidly and humorously written and remains an important historical document on Wyeth's route across the Rockies and pioneer days in the Pacific North West. The first edition of 300 copies sold out in three weeks but no reprint appeared during Townsend's lifetime. Birds are mentioned only in passing because he thought that they would be of little interest to his general readers. He put most of his observations and descriptions into his appendix: 'A Catalogue of the Birds found in the Territory of Oregon'—his best ornithological work. Much more detail is given here than in the 1837 paper by Nuttall and, as mentioned earlier, two new species, Vaux's Swift and MacGillivray's Warbler, are described. By this time Audubon had given names to many of Townsend's discoveries and they therefore had priority in scientific nomenclature. It must have been galling for Townsend to have to use them. For instance, under the Black Oystercatcher, named *Haematopus Bachmani* by Audubon, Townsend records that the bird was shot on the shores of Puget Sound by his friend Dr Tolmie who presented it to him. "I was anxious to give it the name of its discoverer, but I have been overruled by Mr. Audubon who has probably had good reasons for rejecting my proposed specific appellation, *Tolmiei*." The main reason was of course that Audubon had never heard of Tolmie, but Bachman was giving him hospitality while he painted the bird.

It is difficult to assess the relationship and the degree of bitterness that existed between Townsend and Audubon. Before the expedition they had been friendly but in 1839 when Audubon was working on *The Viviparous Quadrupeds of North America* a request for some mammals from Townsend received only "disappointing answers". In 1845, even as Audubon's powers were diminishing and his eyesight failing, he asked Bachman to try once more to get some mammals out of Townsend, who eventually parted with some new species including Townsend's Ground Squirrel, Townsend's Chipmunk, Townsend's Pocket Gopher, Townsend's Mole, Townsend's Vole and the Whitetail Jackrabbit *Lepus townsendi*. In the end so many birds and mammals were named after Townsend by Audubon and Bachman that one wonders if it was due to genuine gratitude, a real desire to honour the collector, or Audubon's guilty conscience. It cannot have been lack of imagination.

Audubon once called Townsend "Lazy and careless" but Townsend did not miss the opportunity to put pencil comments into the margin of the Academy's copy of the fifth volume of the *Ornithological Biography* to show up Audubon's own carelessness: Audubon twice mixed up descriptions given to him by Townsend so that they appear under the wrong birds, and one or two of Audubon's statements were pointed out as errors of fact. Townsend also complained that the name *tricolor* for the Tricolored Blackbird was Nuttall's invention for Nuttall's discovery and should have been credited as such. Audubon's absurd inclusion of a Chilean species in *The Birds of America* highlights the general lack of cooperation between them.

After completing his *Narrative* Townsend had ambitious plans to produce an *Ornithology of the United States of North America*. The title page was dated 1839 but only the first part was completed and was never distributed. Less than half a dozen copies now exist, making it one of the rarest items in American ornithological literature. In royal octavo size, it began with an illustrated account of the California Condor in order to attract the maximum amount of attention but Townsend had to abandon the whole idea because of lack of financial support and the sudden appearance of the small, octavo edition of Audubon's *The Birds of America* with which he had no chance of competing.

In 1839 Townsend published a 'List of Birds Inhabiting the Region of the Rocky Mountains, the Territory of the Oregon, and the North West Coast of America' in the *Journal of the Academy of Natural Sciences*. Unfortunately, it was only a list of names, of over 200 species, without any observations. At the end of the list, Townsend took the opportunity to publish a whole page note on the naming of MacGillivray's Warbler, the sentiments of which are open to interpretation and will be left to the reader to judge:

"When I first procured this bird on the Columbia River in the spring of 1835, I considered it an undescribed species, and have never since seen reason to renounce the opinion then formed. Specimens, in several varieties of plumage, were sent by me to the Academy in the following year, from among which Mr. Audubon selected several. A pair of these are figured in the 4th volume of his splendid work, under the title of SYLVIA *philadelphia* [Mourning Warbler], for which species Mr. A. always mistook it.

Being myself thoroughly convinced, that this usually accurate naturalist was for once mistaken, I ventured to insert a description of this bird in an appendix to my recently published work, 'Narrative of a Journey Across the Rocky Mountains, &c.' and honored it with the name of my friend Dr. W. Fraser Tolmie. At a subsequent period, Mr. Audubon, aware of his former error, relinquished the name of S. *philadelphia*, and in Vol. V. of his 'Ornithological Biography,' page 75, called the species S. *Macgilivrayi* [sic].

If I had been aware, before the publication of my appendix, of Mr. Audubon's wish to name this bird, I should have adopted his appellation with cheerfulness; but as his intention was never communicated to me, the name of S. *Tolmoei* [sic], which I have given it, having priority, must of course be retained."[8]

At about this time Townsend married Charlotte Holmes of Cape May Courthouse, New Jersey, and they had one son. Charlotte's sister married William Baird with whom Townsend spent many days birding around Cape May. By 1842 Townsend was in Washington working at the National Institute, a forerunner of the Smithsonian Institution which in turn became part of the U.S. National Museum. Here at last he began to carve out a worthwhile career for himself, collecting, purchasing and preparing birds and other zoological items. According to William Baird, who wrote to his better known brother Spencer, Townsend was buying up all sorts of birds in the market and becoming more and more expert in taxidermy:

"I have arrived at the same conclusion as yourself, that arsenical soap is not the best thing for preserving skins. Those put up with the powder look a great deal better and are much less trouble, even if the soap were as efficacious. At the end of this letter I will give you Townsend's receipt for preparing the powder, and which he always uses. It is safe, while there is a danger of pure arsenic injuring the lungs. I have noticed a great many little points in Townsend's manner of stuffing, which will be of great assistance to me when I prepare any skins myself. I looked at his tools and will try to get some like them. The value of proper instruments is very great in the saving of time as well as the appearance of the skins. Townsend can skin, stuff and sew up a bird, so as to make it look far superior to any I have ever seen, in five minutes ..."[9]

When the U.S. Exploring Expedition returned from its voyaging in the southern hemisphere, Captain Wilkes stored the zoological collections at the National Institute but a disagreement arose and for some reason poor Townsend was dismissed. He went back to Philadelphia in 1845 where he began to study dentistry but this too seems to have been a failure. By 1851 his health had seriously deteriorated, a fact that his brother-in-law, Dr Mahlon Kirk, attributed to the clouds of arsenic dust that he had inhaled over the years.

With hopes of recovery Townsend secured a place for himself on a government trip to the east coast of Africa but when the time came he was too ill to go. He died shortly after the ship left, outliving Audubon by only eleven days. Townsend's death in Washington on 6 February 1851, at the age of forty-one, robbed American ornithology of a man who was never able to demonstrate his full potential. Witmer Stone, writing in 1916, after speaking with people who had known Townsend, had this to say:

"John K. Townsend was evidently a genius whom force of circumstances prevented from reaching his proper place in ornithological annals. Had he had the financial backing that Edward Harris was ever ready to provide for Audubon, or had there been salaried scientific positions in those early days by which an ornithologist could make a living, the name of Townsend would have been among the leaders in American ornithology. I have talked with his cousin who remembers him dressed in the furs and skins that he brought from the far west, and with his brother-in-law who knew him in the intimacy of family relationship; and I have read the opinions of Cassin in his confidential letters to Baird and all testify to his high character and ability."[10]

Mahlon Kirk further praised him by saying that "His personality was most attractive. His courtesy, kindness of heart, and his brilliant conversational powers, fortified with a vivacious intellect and a fund of knowledge covering almost all subjects, made him a delightful companion and endeared him to every one who came within his influence."[11]

Lastly, the fate of Townsend's bird skins is worth relating in view of their convoluted history. Audubon was not an acquisitive person and once he had drawn the birds he gave them back to Harris or passed them on to Spencer Baird, with the result that most were eventually deposited at the Philadelphia Academy or the Smithsonian Institution. In 1937 some of the Harris collection was found mouldering away in an old barn at Moorestown, New Jersey (see Harris). Three of these birds had labels in Townsend's hand: two Snowy Egrets from Fort Walla Walla and a Gray Jay from Fort Vancouver. Eleven other unlabelled birds (including the possible co-type of the Hermit Warbler) were probably also derived from Townsend since they are birds characteristic of Oregon or Hawaii.[12]

As previously mentioned, some of Townsend's Hawaiian bird skins were sold by Audubon, to a dealer in Edinburgh by the name of Carfrae who sold them to Sir William Jardine. They remained in southern Scotland until a few years after Jardine's death. The British Museum (Natural History) tried to buy all the skins but their offer was refused. The collection was put up for auction in 1886 and the Hawaiian birds, still with Townsend's labels, were bought for Cambridge University Museum by Scott B. Wilson, an English ornithologist who was about to collect in the Hawaiian Islands and no doubt appreciated their true worth.[13]

TRAILL. *SCOTTISH MEDICAL DOCTOR WHO BECAME FRIENDLY WITH AUDUBON WHILE WORKING AT LIVERPOOL. IN THE PORTRAIT, PAINTED IN 1832, THE BIRD ON THE LEFT IS THE MAROON ORIOLE ORIOLUS TRAILLII (VIGORS) OF THE HIMALAYAS AND SOUTH-EAST ASIA.*

Thomas Stewart Traill (1781-1862)

WILLOW FLYCATCHER *Empidonax traillii* (Audubon)

Muscicapa Traillii Audubon, 1828. *The Birds of America* (folio), Vol. 1, Pl. 45 (1831, *Ornithological Biography*, Vol. 1, p. 236): woods along the prairie lands of the Arkansas River [= Fort of Arkansas [Arkansas Post], Arkansas

The Willow Flycatcher and the Alder Flycatcher were formerly recognized as a single species—Traill's Flycatcher. It was one of many birds that Audubon named after British friends who helped and encouraged him when he visited Europe in search of a publisher for *The Birds of America*.

Thomas Stewart Traill was born at Kirkwall, in Orkney, on 29 October 1781 and though he passed all his adult life on the British mainland he retained a strong affection for the northern isles. The son of the parish minister, he chose a medical career and studied at Edinburgh University, then settled in Liverpool for the next thirty years. By the time Audubon arrived at the port, in July 1826, Dr Traill had become a respected and influential member of the community; a founder of the Royal Institution, the Liverpool Mechanics' Institution and the Literary and Philosophical Society of Liverpool. Through his interest in the advancement of science and his lecture courses on a wide range of subjects related both to science and the arts he made many acquaintances and friends, among them William Scoresby, the Arctic explorer and surveyor. When Scoresby wrote his *Journal of a Voyage to the Northern Whale Fishery* (1823), Traill contributed a 'List of Animals met with on the Eastern Coast of West Greenland' and Scoresby named an island off the east coast of Greenland for Traill.

Traill was also a close friend of the Rathbones, a Quaker family interested in literature and the arts who imported baled cotton from the United States. Audubon visited them with a letter of introduction as soon as he arrived in Liverpool and they introduced him to Traill because of his zoological inclinations.

Traill and the Rathbones quickly decided that Audubon's venture deserved their support and they missed no opportunity to exert their influence for his benefit. Within two weeks they had arranged for a gallery at the Royal Institution to be made available for an exhibition of the paintings from his portfolio, and during the first two hours more than four hundred people rushed in. Traill then suggested that the next display should be at Manchester and he sent the curator from the Royal Institution with Audubon, to unpack and hang the paintings, but this second exhibition was not well received.

From Manchester Audubon travelled north to Edinburgh bearing letters of introduction from Traill to Professor Jameson and other possible benefactors. Here he met W.H. Lizars who agreed to engrave the book in the way Audubon had always envisaged it, as a double elephant folio. From the Scottish capital Audubon wrote to William Rathbone on 24 November: "I am sorry that some of my friends, particularly Dr. Traill, are against the pictures being the size of life, and I must acknowledge it renders the work rather bulky, but my heart was always bent on it, and I cannot refrain from attempting it. I shall publish the letterpress in a separate book, at the same time with the illustrations, and shall accompany the descriptions of the birds with many anecdotes and accounts of localities connected with the birds themselves, and with my travels in search of them."[1]

Though Audubon soon rejected Lizars in favour of Robert Havell of London, he made Edinburgh his base for long periods during the next thirteen years, as it was the home of William MacGillivray who wrote the scientific descriptions for the *Ornithological Biography*. In 1833 Dr Traill moved back to Edinburgh to take up the chair of Medical Jurisprudence at Edinburgh University and so his friendship with Audubon may have been resumed until the American finally returned home six years later. During his latter years Traill served as editor of the eighth edition of the *Encyclopaedia Brittanica* (1853–61) and wrote many of the articles but he is said to have trusted too much to his tenacious memory. Despite ill health he continued to teach at the university until a few days before his death, at the age of eighty, on 30 July 1862.

Willow Flycatcher

Henry Baker Tristram (1822-1906)

TRISTRAM'S STORM-PETREL *Oceanodroma tristrami* Salvin

Oceanodroma tristrami (Stejneger MS) Salvin, 1896. *Catalogue of the Birds in the British Museum*, Vol. 25, pp. xiv, 347, 354: Sendai Bay, [Honshu,] Japan

During the winter months Tristram's Storm-Petrel breeds on some of the western islands of the Hawaiian Chain and on the Volcano and Izu Islands south of Japan. The type specimen was obtained in Japan by Lieutenant G. Gunn of the Royal Navy, who sent it to Tristram. Tristram lent the skin to Leonhard Stejneger who referred to it as *Oceanodroma tristrami* in an unpublished manuscript and this name was repeated by Osbert Salvin in his volume on the petrels in *Catalogue of the Birds in the British Museum* (1874–98).

The Reverend Henry Baker Tristram was Salvin's cousin by marriage and a close friend. In 1858, together with the Newton and Godman brothers and Colonel Drummond-Hay they planned the formation of the British Ornithologists' Union, which was established at Cambridge that November. Their quarterly magazine, *The Ibis*, was the first exclusively ornithological journal published in Great Britain. It was another twenty-five years before a group of American ornithologists met together in New York with the purpose of founding an American Ornithologists' Union and named their journal *The Auk*.

During the intervening period the success of *The Ibis* was due, in no small measure, to the many papers contributed by Tristram. These were mainly of two kinds: short papers in which he described species new to science, often sent to him by his worldwide network of amateur collectors; and much longer, highly entertaining accounts of his expeditions and ornithological discoveries, sometimes published as a long series. The most important of the latter group were 'On the ornithology of Northern Africa' (1859–60) and 'On the ornithology of Palestine' (1862–68).

Throughout his adult life Tristram frequently travelled abroad and it could truly be said of him that he was blessed with poor health. While in his early twenties, tuberculosis led him to seek an appointment on Bermuda. The climate there having failed to effect a cure he went to Algeria where the hot, dry

TRISTRAM. *CANON OF DURHAM CATHEDRAL - HENCE HIS NICKNAME "THE GREAT GUN OF DURHAM." HE BUILT UP A HUGE COLLECTION OF BIRD SKINS AND EGGS FROM CORRESPONDENTS AND FROM HIS VISITS TO BERMUDA, ALGERIA, THE HOLY LAND AND CANARY ISLANDS.*

atmosphere dramatically improved his condition and he continued to avoid the English winters as frequently as possible. In the autumn of 1857 he joined yachting friends in the Mediterranean and in the spring toured Palestine for the first time. Thereafter the Holy Land held a special fascination for him and he returned in 1863–64, in 1872, in 1881, in 1894 and in 1897; in 1881 travelling also in Syria, Mesopotamia and Armenia. During the spring months of 1888 and 1889 he explored the seven major islands of the Canaries and in 1891 went out to Japan to visit one of his daughters, a missionary teacher.

Tristram authored a variety of books about the Middle East: *The Land of Israel* (1865), *The Natural History of the Bible* (1867), *Scenes in the East* (1870), *The Land of Moab* (1873) and *The Fauna and Flora of Palestine* (1884). He became the foremost authority on the natural history of Israel, writing extensively about every branch of the subject. His bird surveys are now much quoted by Israeli ornithologists as they provide the only reliable information on the former status of many species.

Another region of particular interest to Tristram was the South Pacific Ocean. From Fanning Island, Norfolk Island, New Caledonia, the Marquesas, the New Hebrides and the Solomon Islands missionaries and naval officers sent him many small collections which he discussed in *The Ibis*.

Tristram had no influence on the development of ornithology in North America but he became familiar with many Nearctic species while living on Bermuda, where he arrived on Christmas Day 1846, to serve as naval and military chaplain. He was fortunate that several other keen ornithologists—H.M. Drummond-Hay, J.W. Wedderburn and J.L. Hurdis—were then working on the islands and they all collaborated on a study of migration. When Tristram left Bermuda in May 1849 he took with him a small collection of bird skins which included a Red Phalarope found floating dead in Riddle's Bay; a Great Blue Heron once seen to seize a Common Ground-Dove and swallow it entire; a Northern Saw-whet Owl sponged out of the muzzle of a cannon on Ireland Island after a gale; and a far-flung Boreal Owl that was unfortunate enough to land on a schooner 150 miles north-west of Bermuda when Tristram was on board!

By the end of May Tristram was at Niagara where he hunted a little on each side of the border, adding Wood Duck, Summer Tanager, Cedar Waxwing, Solitary Vireo, Ruffed Grouse and a few other souvenirs to his collection. He

Tristram's Storm-Petrel

remained in America for six months but there is no record of his itinerary. It seems likely that he visited John G. Bell's shop in New York, since the catalogue of Tristram's collection lists 125 species attributed to the taxidermist, most of them dated 1846–49. Amongst other American species obtained by exchange or purchase were a few skins collected by Bendire, Coues, Dall, Heermann, Henshaw, Hepburn, Kennicott, Ridgway, Say and Xántus.

Between expeditions Tristram lived in north-east England where he had been born on 11 May 1822 at Eglingham Vicarage, near Alnwick, Northumberland. Educated at Durham and Oxford, his first clerical appointment was in Devonshire, but after returning from Bermuda he settled in the Durham area for the rest of his life; firstly as a rector at Castle Eden, then as a vicar at Greatham until 1873 when he was called to serve as Canon of Durham Cathedral.

At the age of twenty-seven Tristram married a childhood friend and in time they became the parents of seven daughters and a son. One of their many grandchildren wrote a lengthy tribute in which she recollected that "Grandfather was half Irish with an Irishman's quick sense of humour and irascibility, shown much more where principles were concerned than over people; and [had] a very great family affection and love of children. He married Eleanor Mary Bowlby ... [who] was as serene and equable as Grandfather was quick and irascible, and they were the most perfect example of married happiness with great complimentary and mutual gifts." Tristram hated the emptiness of a house without children so at least one young member of his family was almost always on a visit. During his last illness he sent for his grandson Archie, not then quite two years old, who came to stay for several weeks. On 8 March 1906, on the third anniversary of his wife's death, Tristram passed away at the age of eighty-three.

While in his early seventies Tristram had ensured that his beloved bird collection would not be carelessly disposed of when he was gone. He printed a catalogue listing his 17,000 skins of about 6000 species and in 1896 sold the collection to the Liverpool Museum where it remains to this day. Amongst his most prized specimens were some extinct species and 130 types. One of these types was a Nene, described by Nicholas Vigors while the bird was still alive and on show at the Zoological Society Gardens at Regent's Park.

During the last ten years of his life Tristram accumulated another 7000 skins which went to the Academy of Natural Sciences of Philadelphia, as did some duplicates from Liverpool. His huge collection of eggs was sold to Philip Crowley and was afterwards acquired by the British Museum (Natural History).

The informal, anecdotal style of writing that Tristram and his contemporaries produced for *The Ibis* was very different from the terse, statistical papers of modern journals. In 1959 Reg Moreau condemned Tristram for an article written in 1866 that showed "an ugly insensitiveness". After coming across a massacre by the shores of the Dead Sea Tristram had commented that "All the Vultures, Kites and Ravens in Arabia seemed to be rushing to the banquet" and regretted that it wasn't safe for him to remain and make use of the human bait to collect the birds. It is more than a little ironic that Moreau's criticism should have appeared in the centenary issue of *The Ibis*, since Alfred Newton believed that during a difficult period for the journal it had been Tristram's exciting papers on the Holy Land that had ensured its survival.[1]

William Sansom Vaux
(1811-1882)

VAUX'S SWIFT *Chaetura vauxi* (Townsend)

Cypcelus [sic] *Vauxi* J.K.Townsend, 1839. *The Narrative of a Journey across the Rocky Mountains...&c.,* p. 348: Columbia River [= Fort Vancouver, Washington]

> "This species, (which I dedicate to my friend, Wm. S. Vaux, Esq., of Philadelphia,) is common on the Columbia River; breeds in hollow trees, forming its nest in the same manner as the *pelasgius* [Chimney Swift], and lays four white eggs."

This dedication, written by John Kirk Townsend, followed his description of Vaux's Swift, which he called Vaux's Chimney Swallow. It was one of several species he discovered on the Pacific coast after crossing the Rocky Mountains with Thomas Nuttall in 1834.

Vaux's birth on 19 May 1811 made him two years younger than Townsend. Both were born and raised in Philadelphia and both became members of the Academy of Natural Sciences in their early twenties; Townsend in 1833 and Vaux in 1834. The Academy was socially and intellectually of enormous importance to Vaux for the rest of his life; for more than forty-eight years he served it in various capacities as auditor, curator and vice-president and in 1839, when new premises were being built at the corner of Broad and Sansom Streets, Vaux served on the building committee and contributed generously to the funds needed for its erection. He was also an active member of the Zoological Society of Philadelphia and one of the original members of the American Association for the Advancement of Science. He was a devoted mineralogist and after the end of the Civil War he made the first of many long journeys in Europe in search of specimens. By the time of his death, on 5 May 1882, his mineral collection was deemed to be the finest in the United States. He bequeathed it to the Academy, along with his library, much archaeological material and a large endowment for their maintenance.

Vaux's grandfather, James Vaux, had come to America from England in 1771 and bought a farm in Montgomery County, Pennsylvania, which he named *Vaux Hill*. In 1803 it became the home of Lucy Bakewell, whose father renamed it *Fatland Ford*, and here she met her neighbour and future husband, John James Audubon.

VAUX IN 1873. *EMINENT MINERALOGIST AND PATRON OF THE PHILADELPHIA ACADEMY OF NATURAL SCIENCES.*

Vaux's nephews and his niece, George, William and Mary Vaux, continued the family's scientific interests. The brothers made a special study of glaciers in British Columbia and the trio kept a thorough photographic record of their travels there. More than 2500 of their high-quality negatives, mostly processed by Mary, are now in the Archives of the Canadian Rockies in Banff. Mary also produced many beautiful botanical watercolours, later reproduced in five volumes by the Smithsonian Institution, entitled *North American Wildflowers* (1925).

During the 1920s George Vaux and the Academy of Natural Sciences co-sponsored Samuel G. Gordon to explore in South Africa, South America and Greenland; amongst the new minerals he discovered and named were vauxite, paravauxite and metavauxite. Eventually, George Vaux's collection rivalled that of his Uncle William who had first inspired his interest.

Vaux's Swift

JULES VERREAUX. *ELDEST OF THREE BROTHERS WHO COL-LECTED FOR THE FAMILY NATURAL HISTORY BUSINESS WHICH WAS BASED IN PARIS.*

Jules Pierre Verreaux
(1807–1873)

Jean Baptiste Edouard
Verreaux†
(1810-1868)

WHITE-TIPPED DOVE *Leptotila verreauxi* Bonaparte

Leptoptila [sic] *verreauxi* Bonaparte, 1855. *Comptes Rendus de l'Académie des Sciences de Paris* 40, p. 99: de la Nouvelle-Grenade [= Colombia]

Many of the exotic bird skins in the great museum collections of North America once passed through the hands of the Verreaux brothers. During the nineteenth century their *Maison Verreaux* at Paris was recognized as "one of the greatest, if not the greatest emporium of natural history that the world has ever seen."[1] Vast quantities of material were received from all over the world and redistributed to museums, private collectors, couturiers and milliners.

Trading in natural history specimens was a way of life to the Verreaux family. Jules Verreaux became enmeshed in his father's business when just a child and at the age of eleven he went with his explorer–naturalist uncle, Pierre Antoine Delalande, to southern Africa where they collected for two years for the Paris Natural History Museum. When Jules came home he studied at the museum under Baron Cuvier and Isidore Geoffroy St Hilaire for five years. He returned to the Cape and lived there from the end of 1826 until 1838, part of the time working at the Cape Town Museum. Following a four year spell back at Paris he sailed to Australia and Tasmania for five years of exploring and collecting. When he finally came back to France in 1848, he helped again at the *Maison Verreaux*, labelling and identifying skins. In 1864 he took over from Florent Prévost as assistant naturalist at the Paris Museum.

Edouard Verreaux went out to the Cape in 1830 to help his brother pack up and transport a large consignment to Paris. In 1832 Edouard returned to the Cape for a year before sailing to Sumatra, Java, the Philippines and Indo-China. After his father's death in 1834, he took control of the *Maison Verreaux*, continuing to receive shipments from Jules, as well as from many French naval and mercantile expeditions.

Unfortunately, when fulfilling orders it was not always possible for the brothers to be precise or accurate about the source of their stock. Often they dealt with unlabelled collections amalgamated from several ports of call and they were forced to make semi-informed guesses as to the likely place of origin. Naturalists were angered and frustrated by such skins, which were more trouble than they were worth. William Gambel, for instance, complained that two thrasher skins that he examined at Philadelphia were identical to ones he had collected in California, yet the labels erroneously stated "New Zealand, Verreaux."[2]

The Verreauxs were amongst the leading exponents of the art of taxidermy. One of their most ambitious and spectacular pieces was of an Arab courier mounted on a camel being attacked by two lions. It won a gold medal at the Paris Exposition of 1867 and was afterwards bought by the American Museum of Natural History at New York to help draw in the public.

When Dr Henry Bryant succeeded in buying Baron Lafresnaye's important bird collection for the Boston Society of Natural History in 1869, it was Jules Verreaux who prepared the catalogue. He spent three months at Lafresnaye's chateau making a 258 page list of the 8656 specimens, a work that is mainly of interest because it formed the basis of G.R. Gray's *Hand-list of the Genera and Species of Birds* (1869–71), which became invaluable for a whole generation of ornithologists.

The quantity of material published by the Verreauxs was comparatively modest in view of their extensive travels and immense general knowledge of the birds of the world. The groups that excited them most, partly for financial reasons, were the hummingbirds and sunbirds. In 1865 Jules and Edouard collaborated with Martial Etienne Mulsant on an *Essai d'une Classification méthodique des Trochilidés ou Oiseaux-mouches* and from 1874 to 1877 Edouard worked with Mulsant on their famous *Histoire Naturelle des Oiseaux-mouches*. Jules had intended to complete an account of the sunbirds but he died too soon; his valuable collection of sunbirds went to the Paris Museum.

The White-tipped Dove, a South and Central American species that is resident in small numbers on the lower Rio Grande, is just one of many species with the scientific name *verreauxi*. Charles Bonaparte in his succinct description of the dove gives no immediate clue as to whom the honour was intended for, except that elsewhere on the same page, and the one preceding, he refers to pigeons supplied to him by "MM. Verreaux"—the Messieurs Verreaux.[3] It seems likely therefore that the compliment was intended for Jules and Edouard, both of whom he knew well. He was probably not acquainted with the youngest brother, Joseph Alexis Verreaux, who went out to the Cape with Edouard in 1832 and died there in 1868. Alexis ran a gunpowder store in Cape Town and from time to time sent shipments of specimens back to the *Maison Verreaux*.

White-tipped Dove

WHITNEY. DIRECTED THE GEOLOGICAL SURVEY OF CALIFORNIA FROM 1860 TO 1874 AND HELPED FUND J.G. COOPER'S 'ORNITHOLOGY OF CALIFORNIA'. HE IS SHOWN HERE AT THE CENTRE OF A GROUP OF GEOLOGISTS. LEFT TO RIGHT, STANDING—WILLIAM M. GABB, JOSIAH DWIGHT WHITNEY, CLARENCE KING; SEATED—CHESTER AVERILL, WILLIAM ASHBURNER, C.F. HOFFMANN, WILLIAM H. BREWER.

Josiah Dwight Whitney
(1819-1896)

ELF OWL *Micrathene whitneyi* (Cooper)

Athene whitneyi J.G. Cooper, 1861. *Proceedings of the California Academy of Sciences*, Ser. 1, 2, p. 118: Fort Mojave, lat. 35° [N.], Colorado Valley [Arizona]

"For years our chief, Professor Whitney, has made brave campaigns into the unknown realms of Nature. Against low prejudice and dull indifference he has led the survey of California onward to success. There stand for him two monuments,— one a great report made by his own hand; another the loftiest peak in the Union begun for him in the planet's youth and sculptured of enduring granite by the slow hand of Time."[1]

When writing this flowery tribute, Clarence King forgot to mention the Elf Owl *Micrathene whitneyi*—a third memorial to Whitney and the one that naturalists most often associate him with. Whitney therefore has the distinction of having named after him the highest mountain in the contiguous United States and the smallest of all North American owls.

To discover how this came about it is necessary to recount some details of his life, making them from choice rather brief since his career details are of more interest to geologists than to ornithologists. Josiah Whitney was born on 23 November 1819, the eldest of thirteen children. The family home was at Northampton, Massachusetts, and Josiah attended schools at New Haven and Andover before entering Yale. At first he planned to study law but a summer spent as a geological assistant and a lecture he heard by Charles Lyell convinced him that he should devote his life to geology. To this end he studied in Europe at Paris, Rome and Berlin.

In May 1847 Josiah and his brother William took part in a survey of the northern part of Michigan, after which Josiah was allowed to co-author the main reports owing to the leader's sudden resignation. William went on to become a noted Sanskrit scholar and philologist, but Josiah also knew him as "a mighty hunter and collector of birds" who presented his trophies from New England to the Peabody Museum at Yale.

In June 1854 Josiah married Louisa Howe of Brookline and they had one child, a daughter. In the same year Josiah wrote *The Metallic Wealth of the United States*, which remained a standard reference for the next sixteen years.

From 1855 to 1858 he was actively involved in geological surveys in Iowa, Illinois and Wisconsin. From 1860 to 1874 he was State Geologist of California responsible for co-ordinating the state's first prolonged topographical and geological survey.

Clarence King was amongst the team, gaining his early field experience in the West before winning fame as the leader of his own survey along the 40th parallel. As an energetic twenty-two year old with the Whitney survey, King made his mark by naming the highest peak in the Sierra Nevada. He was near Bishop in July 1864 and in order to map the surrounding area and ascertain the heights of the various peaks he climbed Mount Tyndale with Richard Cotter. From the summit, far above the Owens Valley, there were stupendous views in all directions. "On setting the level it was seen at once that there were two peaks equally high in sight, and two still more elevated all within a distance of seven miles. Of the two highest, one rose close by ... The other, which we called Mount Whitney, appeared equally inaccessible from any point on the north or west side."[2]

Midway through the survey Whitney was appointed Professor of Geology and Director of the School of Mining and Practical Geology at Harvard, but he was given indefinite leave of absence to continue with his field work in California. When it ended he moved to New England, residing at Cambridge except when on active research. In 1882 he published his last great work, *Climatic Changes of Later Geological Times*, for which he drew heavily upon his western experiences. In the same year he suffered a personal tragedy, losing within four days of each other his daughter who had set up home in France, and his wife who had been an invalid for some years. Whitney's strong character was severely tested but he continued his geological work until his own death fourteen years

Elf Owl

later at Lake Sunapee, New Hampshire, on 18 August 1896. He lies buried beside his wife and daughter at Northampton, his grave being marked by a large glacial boulder of rose quartzite. His most important achievements were his reports on the natural resources of six States, his method of triangulation, which was used to map much of North America, and the two generations of geologists and topographers to whom he gave instruction and field experience.

A number of naturalists were also associated with Whitney including one of the most celebrated ornithologists of the western United States. In December 1860 Dr James G. Cooper joined the Geological Survey of California and was collecting all manner of natural history items through to April 1862 and also for much of 1863. The areas he explored were chiefly the Colorado River valley near Fort Mojave and along the route from the fort to San Diego, San Pedro and Santa Barbara but he also made trips out to some of the offshore islands. In 1864 he was in the Sierra Nevada as well as on the coast exploring the area from Bolinas southwards to Santa Cruz.

A few months after starting, Cooper sought Whitney's permission to describe at once two new birds so that priority would be won for the Geological Survey. He therefore read a short paper to the California Academy of Sciences in July 1861 in which he described Whitney's Owl *Athene whitneyi*—later renamed the Elf Owl *Micrathene whitneyi*. He gave only the basic details of the single specimen he had collected near Fort Mojave on 26 April but many years later when the survey report was completed he added some notes on the circumstances of the bird's discovery: "I found it in a dense thicket, on a very windy morning, where it may perhaps have taken a temporary refuge, after being blown down from some of the caverns in the barren mountains surrounding the valley. Its stomach contained insects and some small feathers."[3] This owl remained a unique specimen for eleven years until Major Bendire shot one in a willow thicket at Rillitto Creek near Tucson, Arizona, on 20 April 1872. Although the first specimens were both found in thickets, the Elf Owl, at least in the minds of many ornithologists, is more often associated with the giant Saguaro Cactus of the Sonoran Desert where it appropriates the nest holes excavated by Gila Woodpeckers.

The second new species described by Cooper was Lucy's Warbler. Again he found the bird near Fort Mojave, this time naming it for Professor Baird's daughter.

Some of Cooper's notes were incorporated into Baird, Brewer and Ridgway's *The Water Birds of North America* (1884). The bulk of his material appeared much earlier, in 1870, published by Whitney as part of the Geological Survey of California. It appeared under the title 'Ornithology. Volume 1. Land Birds', edited by Baird from the manuscript and notes of Cooper. Often known as Cooper's 'Ornithology of California' it included all the birds then known from west of the Rockies. Whitney had to finance much of the costs himself, hence his message to his more ornithologically minded brother William: "I take pleasure in sending you the first complete copy of the 'Birds'. *If you do not think it handsome please send it back!* Words cannot express how much labor (and money!) it has cost."[4]

WILLIAMSON. *TOPOGRAPHICAL ENGINEER WHO LED A NUMBER OF SURVEYING EXPEDITIONS IN THE WEST-COAST STATES. DURING THE CIVIL WAR HE FOUGHT FOR THE NORTH, LATER RETURNING TO CALIFORNIA.*

Robert Stockton Williamson (1824-1882)

WILLIAMSON'S SAPSUCKER *Sphyrapicus thyroideus* (Cassin)

Picus thyroideus Cassin, 1852. *Proceedings of the Academy of Natural Sciences of Philadelphia* 5 (1851), p. 349: California [= Georgetown, about twelve miles from Sutter's Mill, Eldorado County, California]

Robert Stockton Williamson graduated from West Point in 1848 at the same time as future generals Reno, Foster, Parks and Burnside. Assigned to the Topographical Engineers, Williamson is now best known for the part he played in some of the important early surveying expeditions that searched for a railroad route from the Mississippi to the Pacific Ocean. Instead of pointing to one particular route the surveyors identified a number of possible choices and often discovered valuable new agricultural or range lands far in excess of expectations. The situation was much more complicated than the Government had imagined and the hope of providing a single federally sponsored railroad to traverse the United States was still unfulfilled at the outbreak of the Civil War. Private enterprise on a grand scale eventually provided the railroads.

The twelve volumes of the Pacific Railroad Survey Reports give little indication of the competition and intrigue that developed as a result of the contrasting merits of each route but they have become valuable early records of natural history in the far West. With great foresight, Professor Baird succeeded in placing naturalists on almost all the surveys, each armed with a set of collecting instructions from the Smithsonian Institution giving directions for preparing their discoveries and advice on the necessary apparatus needed for preserving them. The results were written up by the field workers or, more often, by experts on the east coast, frequently by Baird himself. Though Williamson was not a zoologist he was associated with several well-known enthusiasts including one of the best ornithological field collectors of the day.

In February 1849, shortly after graduation, 2nd Lieutenant Williamson was ordered to California to join Captain William H. Warner who was an experienced frontier soldier. (For a map see p. 74.) On 27 June Warner received orders to lead an expedition from the upper Sacramento River across the Sierra Nevada to the Humboldt River to search for a railroad route to the east. In August Warner, Williamson and eleven civilians with an escort of eighty troopers and four officers headed into the notorious Klamath area. The terrain was so difficult that it was soon obvious that there was no chance of creating a suitable gradient for a railroad, so Warner left Williamson behind in charge of the sick and pressed ahead with only a few men. After travelling beyond Goose Lake (on the present border with Oregon) the party suffered an arrow attack by about twenty-five Pit River Indians that wounded several men and left Captain Warner and two others dead. Williamson heard the news, gathered up the survivors and headed back to Benicia. With the aid of notes recovered from Warner's body he was later able to write up his *Report of a Reconnaissance of a Route through the Sierra Nevadas by the Upper Sacramento* (1849–50).

It was Williamson's first experience of field surveying but the twenty-five year old[1] from New York had acquitted himself well. During 1851 and 1852 Williamson continued his military duties in northern California and Oregon, exploring eastwards from Yreka to the Klamath lakes and Fort Reading, and from Port Orford to the Rogue and Coquille Rivers. In 1853 he was involved in a second major expedition, this time to southern California where he was to investigate all the passes in the southern Sierras. He was also to extend his work as far south as San Diego and into the Mojave Desert to Fort Yuma, to survey again for an eastern route and to link up with work by other surveys conducted along the 35th and 32nd parallels by Lieutenants Whipple and Pope. Under his command Williamson had Lieutenant John G. Parke, two civil engineers, map maker Charles Preuss, mountain man Alexis Godey, and a young but experienced wilderness traveller acting as physician and naturalist. This last man was none other than Dr Adolphus Heermann who wrote up the zoology for this survey and went on to provide one of the best early bird lists for the state, chiefly drawn from his experiences with the Pacific Railroad Surveys under Williamson and Parke.

Travelling south from Benicia on 10 July 1853, Williamson and his men quickly ruled out the Walker Pass, which hitherto had been considered the best option, and instead recommended the passes of either Tehachapi or Cañada de las Uvas. Reconnoitring closer to Los Angeles and San Diego they discovered New Pass but were disinclined to favour Cajon Pass unless a long tunnel could be made. They reckoned the San Gorgonio Pass the best in the coast range as it took them easily across the mountains down into the San Bernardino Valley. Williamson and Heermann spent a little time at the site where Fort Tejon would soon be built and where, five years later, John Xántus would spend two important years ransacking the area for the Smithsonian.

Volume five of the Pacific Railroad Reports contained Williamson's *Report of Explorations in California for Railroad Routes to Connect with the Routes near the 35th and 32nd Parallels of North Latitudes* (1857). The zoological report appeared two years later in volume ten, with analyses of the reptiles by Edward Hallowell, fishes by Charles Girard, mammals by Baird and the birds by Heermann. Lieutenant Parke wrote up his own report for his separate travels in 1854 and 1855 between the Gila River and the Rio Grande and between Los

Angeles and San Francisco. Heermann, who left Williamson to be with Parke, again provided the ornithological reports.

In 1855 Williamson was assigned to yet another expedition, a reconnaissance for a line to connect Los Angeles and San Francisco with Oregon and Washington Territory. Assisted by Lieutenant Henry L. Abbot he headed north in July up the Sacramento Valley into the Pit River Canyon and Klamath lakes region where the team divided. Williamson explored the Cascades while Abbot made more directly for the Des Chutes River and down to Fort Dalles. Williamson went west to the Willamette River but was so ill that on reaching the Columbia he took a ship back to San Francisco. Abbot's party joined up with Williamson's men and together they explored a route west of the Cascades back up the Willamette to Sacramento via Yreka and Fort Reading. Dr E. Sterling was the chief zoologist but at Fort Dalles he was considered too sickly to carry on and much of the collecting work fell to the assistant surgeon and geologist John S. Newberry, an excellent all-round naturalist who had trained at the Smithsonian. He spent much of his time collecting plants, birds and mammals as well as carrying out his official tasks. On 23 August 1855, on the shores of Upper Klamath Lake, Oregon, he shot a black and white male woodpecker with a brilliant yellow belly and bright red chin, but no female:

> "The individual procured, when first seen, was creeping up the trunk of a large yellow pine, (*P. brachyptera*,) searching for insects in the bark. Its cry was very like that of *P. Harrisii* [Hairy Woodpecker]. When shot, though killed, he retained his hold of the bark of the branch on which he sat, as woodpeckers so often do, and I was compelled to dislodge him with the contents of my second barrel, by which he was somewhat mutilated."[2]

The expedition was concluded by the end of November, the surveyors having found two possible rail routes, one either side of the Cascades, which could link San Francisco with the settlements along the Columbia River. Williamson's poor health meant that the *Report upon Explorations for a Railroad Route from the Sacramento Valley to the Columbia River* (1857), which made up volume six of the Pacific Railroad Reports, had to be completed under Abbot's name. Newberry wrote the general notes on the geology and botany as well as the zoology, which included all but the fishes, land shells and reptiles, which were dealt with by other naturalists. Newberry's section on birds described the distinctive new woodpecker that he named *Picus williamsoni* after the expedition leader. Being badly shot up the woodpecker had been dumped in alcohol, which quickly extracted the yellow colour of the belly leaving it dull white—hence the inaccurately coloured plate that went with the text.

Williamson was recalled to the east at the onset of the Civil War. He served on General Burnside's staff with some distinction, being brevetted Major on 14 March 1862 for gallant and meritorious services at the Battle of Newbern, North Carolina, and was brevetted Lieutenant Colonel six weeks later following his actions during the siege of Fort Macon. He then joined the Army of the Potomac as Chief Topographical Engineer from 21 November to 21 December and then held the same post in the Department of the Pacific until March 1863 before transferring to the Corps of Engineers.

Williamson was not fully commissioned as Lieutenant Colonel until 1869, by which time the war was over and he was once more back on the west coast, working as an engineer with the lighthouse service. In 1870 he was involved in overseeing the contractor who blew up Blossom Rock in San Francisco Bay. Lying just below the water surface east of the Golden Gate the rock was a menace to shipping, but it required considerable engineering expertise to lay the charges inside the rock. Anticipation of the event provoked a spontaneous public holiday as hundreds of spectators lined the northern wharves or climbed Telegraph Hill to see the gigantic explosion.

Williamson's Sapsucker, found six years before the start of the Civil War, was by now well established in ornithological literature, appearing in all the major reviews of the period. Then, in 1873, Henry Henshaw made a startling discovery while based at Fort Garland, Colorado. One day in June, while collecting at the foot of Mount Baldy, he came across a pair of Williamson's Sapsuckers:

> "Chancing to shoot a female first, almost immediately I shot a male, and, laying them side by side, their relationship was at once apparent. Later I found mated pairs occupying the same cavity in live aspens, their favorite nesting tree, which of course was proof positive of their relationship. While it is true that, contrary to the rule that obtains among the woodpeckers, the male and female of the species are very differently colored, it is difficult to understand why the true facts of the case should so long have escaped the notice of closet ornithologists."[3]

A pair of Williamson's Sapsuckers, male at left

The most remarkable part of the discovery was that the female was a well-known bird that already had a specific name of its own! The Brown-headed Woodpecker had been collected by John G. Bell in Eldorado County, California, as long ago as 1849 and had been described by John Cassin in 1852 in the *Proceedings of the Academy of Natural Sciences of Philadelphia*. Although Bell's two specimens were females, when Cassin figured the birds under the name Black-breasted Woodpecker in his *Illustrations of the Birds of California* (1854) he called the females males and mistakenly described the females as "Similar to the male, but with the colors more obscure, and the black of the breast of less extent and not so deep in shade."[4] In 1860 Baird, Cassin and Lawrence in their *Birds of North America* gave a good description of a male Williamson's Sapsucker but said that the female was similar, with a white chin instead of red. So the adults of each sex were regarded as males; and the juveniles of each sex were thought to be females. No wonder Henshaw's discovery caused a stir. It had taken over twenty years for someone to observe the birds at their nests and correctly appreciate that the male and female Williamson's Sapsuckers were so remarkably different in plumage.

Cassin's scientific name of *Picus thyroideus* has priority over Newberry's *Picus williamsoni* but the common names of Brown-backed Woodpecker and Black-breasted Woodpecker were so obviously inappropriate for the magnificent male bird that Robert Williamson's name continues to be attached to the species.

Quite what the engineer made of all this is unrecorded. Little is known about his natural history interests or general character, and his name is often absent from the detailed biographical encyclopaedias of the Civil War. All his published works appear to be factual topographical or engineering reports or specialized accounts for improving surveying techniques. In June 1882 his persistent ill health seems to have forced him to retire from the lighthouse service since he died just a few months later in San Francisco, on 10 November, aged fifty-four. The funeral took place at the Masonic Temple, his pallbearers being selected two from the Oriental Lodge, two from the California Commandery Knights Templar, two from the Pioneers and two from the army.

WILSON. *AUTHOR AND ARTIST OF THE SEMINAL* AMERICAN ORNI-THOLOGY, *HE TRAVELLED THROUGH MUCH OF THE EASTERN UNITED STATES, OFTEN ALONE AND ON FOOT. THIS PORTRAIT IS BY REMBRANDT PEALE.*

Alexander Wilson
(1766-1813)

WILSON'S STORM-PETREL *Oceanites oceanicus* (Kuhl)

Pro.[cellaria] oceanica Kuhl, 1820. *Beiträge zur Zoologie*, Pt. 1, p. 136, Pl. 10, Fig. 1: No locality. South Georgia designated by Murphy

WILSON'S PLOVER *Charadrius wilsonia* Ord

Charadrius wilsonia Ord, 1814. In Wilson, *American Ornithology*, Vol. 9, p. 77, Pl. 73, Fig. 5: shore of Cape Island [= Cape May], New Jersey

WILSON'S PHALAROPE *Phalaropus tricolor* (Vieillot)

Steganopus tricolor Vieillot, 1819. *Nouveau Dictionaire d'Histoire Naturelle, Nouvelle édition*, Vol. 32, p. 136: Paraguay

HOODED WARBLER *Wilsonia citrina* (Boddaert)

Wilsonia Bonaparte, 1838. *A Geographical and Comparative List of the Birds of Europe and North America*, p. 23. Type, by subsequent designation, *Motacilla mitrata* Gmelin = *Muscicapa citrina* Boddaert (Ridgway, 1881)
Muscicapa Citrina, Boddaert, 1783. *Table des Planches Enluminées*, p. 41: Louisiana

WILSON'S WARBLER *Wilsonia pusilla* (Wilson)

Muscicapa pusilla Wilson, 1811. *American Ornithology*, Vol. 3, p. 103, Pl. 26, Fig. 4: southern States, ... lower parts ... of New Jersey and Delaware [= southern New Jersey]

CANADA WARBLER *Wilsonia canadensis* (Linnaeus)

Muscicapa canadensis Linnaeus, 1766. *Systema Naturae*, 12th edn, Vol. 1, p. 327: Canada

Writing to a friend, on 1 June 1803, Alexander Wilson declared "I am now about to make a collection of all our finest birds."[1] The letter was sent to Paisley, in Renfrewshire, where Wilson had been born nearly thirty-seven years earlier on 6 July 1766. In Scotland he had trained and worked as a weaver but since emigrating in 1794 he had only briefly followed his old trade; for seven years he had been a teacher and was now employed at Gray's Ferry School by the Schuylkill River, on the southern fringe of Philadelphia.

During the decade from 1803 until his death, Wilson dedicated himself to writing and illustrating the *American Ornithology* (1808–14). But firstly he had to learn to draw and here he was fortunate in having the venerable William Bartram as a neighbour and instructor. Bartram welcomed Wilson into his circle of eminent friends and put at his disposal all his resources—his gardens, his library, his influence and his own considerable ornithological knowledge. After he had made some progress with sketching, Wilson received further coaching from Alexander Lawson who later engraved all the plates for the *American Ornithology*.

The actual task of writing the manuscript was not a problem for Wilson. He was a communicator by profession and a poet by inclination. All through his life he had expressed everything that mattered most to him in the written word and the fluent prose of the *American Ornithology* bears witness to his literary vocation. It also reveals that he was as much a scientist as he was a poet, for he delighted in making ornithological discoveries, hated ignorance and superstition and carefully differentiated between hearsay and facts that he had observed himself.

By far the most formidable part of his undertaking was the gathering of the information. Many birds were completely undescribed and little had been written of a reliable nature about even the common species. He made full use of the available publications, but most of the early writers on American birds, such as George Edwards, John Latham and Thomas Pennant had not even set foot in America, for they had simply described specimens sent to them by collectors. Two notable exceptions were Mark Catesby, author of *The Natural History of Carolina, Florida and the Bahama Islands* (1771–43), who had spent over ten years in the New World, and Louis Vieillot, author of *Histoire Naturelle des Oiseaux de l'Amérique Septentrionale* (1807), who twice made brief visits to the United States. Frequently, the only way in which Wilson could establish the distribution, movements, diet or nesting habits of a species was through his own fieldwork. Here again, he was perfectly fitted for his mission by the dichotomous nature of his character, for his bookishness was more than equalled by his love for travel and exploration. Tall, strong and blessed with enormous energy, he believed that walking was by far the best mode of transport for the naturalist and from October to December 1804 he trekked 1300 miles to the Niagara Falls and back, accompanied by his nephew and a friend.

When the first volume of the *American Ornithology* came off the press in September 1808 he set out northwards in search of subscribers willing to commit $120 for his proposed ten-part work. Again he walked most of the way, travelling through New York, Connecticut, Massachusetts, Maine, New Hampshire and Vermont. He gained only forty-one subscriptions but met so many willing correspondents that he boasted "scarcely a *wren* or *tit* shall be able to pass along, from New York to Canada, but I shall get intelligence of it."[2]

Soon after returning to Philadelphia he was off again, this time on horseback, with the dual aim of studying the birds of the Carolinas and obtaining more subscribers. In March he wrote to Bartram from Savannah, Georgia:

"This has been the most arduous, expensive, and fatiguing expedition I ever undertook. I have, however, gained my point in procuring two hundred and fifty subscribers, in all, for my Ornithology; and a great mass of information respecting the birds that winter in the Southern States, and some that never visit the Middle States; and this information I have derived personally, and can therefore the more certainly depend upon it. I have, also, found several new birds, of which I can find no account . . .

On the commons, near Charleston, I presided at a singular feast. The company

WILSON'S TRAVELS
——— AUTUMN 1804
- - - - - WINTER 1808–9
················· JANUARY TO AUGUST 1810

consisted of two hundred and thirty-seven carrion crows *(vultur atratus)* [Black Vulture], five or six dogs, and myself, though I only kept order, and left the eating part entirely to the others. I sat so near to the dead horse, that my feet touched his, and yet at one time I counted thirty-eight vultures on and within him, so that hardly an inch of his flesh could be seen for them. Linnaeus and others have confounded this vulture with the turkey buzzard, but they are two very distinct species.

This journey will be of much use to me, as I have formed acquaintances in almost every place, who are able to transmit me information. Great numbers of our summer birds are already here; and many are usually here all winter."[3]

Having caught a fever and having no desire to retrace his steps, he returned home by sea.

By August he had completed the second volume, though delays in the colouring of the plates set back its publication until January 1810. It contained nine new species that Wilson had discovered in Pennsylvania: Marsh Wren, Solitary Vireo, Bay-breasted, Cerulean and Mourning Warbler, American Tree, Field and Song Sparrow and Pine Siskin. Though Wilson travelled for thousands of miles through the United States, it was eastern Pennsylvania and New Jersey that he knew best, especially around Philadelphia. He frequently patrolled along the Schuylkill and Delaware Rivers and crossed the wet meadows between, an area particularly good for raptors. Every winter he watched Red-tailed and Rough-legged Hawks hunting there and described how twenty or thirty of the latter regularly coursed over the fields below the city. Another favourite spot was a wooded swamp about seven miles south of Philadelphia where great numbers of Black-crowned Night Herons bred and where, one April, while wading out into the colony, he found a Long-eared Owl incubating four eggs in an old heron nest. Here too, he shot a Red-shouldered Hawk, the only female he ever obtained of that species.

Many of his accounts of the commoner birds were based on his local observations and one species whose habits he investigated with interest was the Brown-headed Cowbird:

"I had, in numerous instances, found in the nests of three or four particular species of birds, one egg, much larger, and differently marked from those beside it; I had remarked, that these odd-looking eggs were all of the same colour, and marked nearly in the same manner, in whatever nest they lay; though frequently the eggs beside them were of a quite different tint; and I had also been told, in a vague way, that the cow bird laid in other birds' nests. At length I detected the female of this very bird in the nest of the red-eyed flycatcher [Red-eyed Vireo], which nest is very small, and very singularly constructed; suspecting her purpose, I cautiously withdrew without disturbing her; and had the satisfaction to find, on my return, that the egg which she had just dropt corresponded as nearly as eggs of the same species usually do, in its size, tint, and markings, to those formerly taken notice of. Since that time, I have found the young cow bunting, in many instances, in the nests of one or other of these small birds ... I claim, however, no merit for a discovery not originally my own, these singular habits having long been known to people of observation resident in the country, whose information, in this case, has preceded that of all our school philosophers and closet naturalists, to whom the matter has till now been totally unknown."[4]

One problem of identification that Wilson had to address was the confusion regarding nightjars, as several authorities had considered the Common Nighthawk and the Whip-poor-will to be the same species. In order to eliminate all possible doubt he shot several of each after hearing them call, and took other individuals at the nest, then compared the skins.

The notion that swallows hibernate seemed ridiculous to Wilson, who had seen them on migration, but he nevertheless checked out the possibility one winter by digging into a Bank Swallow colony where he exposed hundreds of nest holes without finding a single swallow, "dead, living, or torpid".

In January 1810 Wilson began his longest "bird-catching expedition" after planning and looking forward to it for several years. Travelling alone, he reached Pittsburgh just as the ice was beginning to break up and purchased a small skiff which he launched into the Ohio on 24 February. After many short side trips up tributaries and excursions ashore he reached Louisville, 720 miles downriver, on 17 March. He then travelled overland until he came to Natchez, on the Mississippi, the most memorable event of the whole journey being occasioned by Passenger Pigeons that flew above him in their millions, hour after hour, darkening the sky for as far as he could see as he walked to Lexington. He bought a horse in the town and rode through Kentucky to Nashville, Tennessee, where he prepared for the risky journey ahead of him through Chickasaw territory. He told Lawson:

> "I was advised by many not to attempt it alone; that the Indians were dangerous, the swamps and rivers almost impassable without assistance, and a thousand other hobgoblins were conjured up to dissuade me from going alone. But I weighed all these matters in my own mind; and attributing a great deal of this to vulgar fears and exaggerated reports, I equipped myself for the attempt. I rode an excellent horse, on which I could depend; I had a loaded pistol in each pocket, a loaded fowling-piece belted across my shoulder, a pound of gunpowder in my flask, and five pounds of shot in my belt. I bought some biscuit and dried beef, and on Friday morning, May 4th, I left Nashville."[5]

The Indians he met posed no threat to him but he endured a violent storm and several soakings in deep creeks. The swamps with their lack of potable water proved to be the worst hazard and he arrived in Natchez considerably weakened by dysentery. After recovering at the home of a hospitable planter he travelled between plantations gaining many subscribers and two new birds, the Mississippi Kite and the Magnolia Warbler. He named the latter the Black and Yellow Warbler *Sylvia magnolia* after shooting a male feeding in the upper branches of a magnolia near Fort Adams, south of Natchez. It was a productive trip for warblers, since he had already discovered the Kentucky, Tennessee and Nashville Warblers as he travelled south.

From New Orleans Wilson sailed to New York, continuing his birdwatching at sea. Off the east coast of Florida and the Carolinas flocks of petrels followed the ship and he stayed on deck throughout one wet and windy night to watch them as they flew noisily about the rigging. He identified them as Storm-Petrels but in 1824 Charles Bonaparte pointed out that they were a different species, which he designated *Procellaria Wilsonii*. It turned out that they had already been described by Heinrich Kuhl so Bonaparte's specific name is lost but they are still known as Wilson's Storm-Petrel. In his lengthy chapter on this little bird, which he much admired, Wilson described a characteristic habit:

"... the most singular peculiarity of this bird is its faculty of standing, and even running, on the surface of the water, which it performs with apparent facility. When any greasy matter is thrown overboard, these birds instantly collect around it, and, facing to windward, with their long wings expanded, and their webbed feet patting the water, the lightness of their bodies, and the action of the wind on their wings, enable them to do this with ease. In calm weather they perform the same manoeuvre, by keeping their wings just so much in action as to prevent their

Wilson's Storm-Petrel

feet from sinking below the surface. According to Buffon, it is from this singular habit that the whole genus have obtained the name Petrel, from the apostle Peter, who, as Scripture informs us, also walked on the water."[6]

Wilson shot fourteen of the petrels in order to examine both male and female plumages and to learn about their diet by studying the stomach contents, but his writings indicate that he found little pleasure in killing birds. He regarded hunting as a means to an end, rather than a sport in itself. This antipathy towards the unnecessary destruction of life, though rare among naturalists of his generation, was much in keeping with his gentle and sensitive nature.

When volume three of the *American Ornithology* appeared early in 1811 it contained the new species from Wilson's southern travels and three more that Lewis and Clark had brought back from their epic journey: Lewis's Woodpecker, Clark's Nutcracker and Western Tanager. Also included was the bird now known as Wilson's Warbler, which Wilson called the Green Black-capt Flycatcher. The current eponym is derived from Bonaparte's name for it, *Sylvia Wilsonii*. Bonaparte was also responsible for creating the genus *Wilsonia*, which contains the Hooded Warbler, Wilson's Warbler and the Canada Warbler, all of which were known to Wilson.

The next three volumes, issued between September 1811 and August 1812, almost completed Wilson's studies of the landbirds, but his peace of mind was shattered when the United States declared war on Great Britain in June. His publisher insisted that he return to New England in the fall to ensure that subscribers in the maritime townships settled their accounts despite the financial

Hooded, Wilson's and Canada Warblers

uncertainties. During the journey Wilson experienced heart palpitations and in the winter his anxiety increased when the colourists deserted him. From the outset it had been difficult to find enough skilled artists to hand colour each plate and now he took it on himself to slave away at the task, becoming virtually housebound. An unexpected present cheered him up in December when he received a fine specimen of a Peregrine Falcon, since he had often seen them but had never managed to shoot one.

During the last few years of his life Wilson often went to the New Jersey coast, particularly to Cape May and Great Egg Harbor. One of the wonderful sights at Cape May, then as now, was the annual mating ritual of the Horseshoe Crab, which comes ashore in vast multitudes. Wilson watched huge flocks of gulls and shorebirds feeding on the crab eggs in amongst herds of hogs that were driven down each spring from the neighbouring farms. Along the salt marshes Clapper Rails bred so abundantly that Wilson knew of one man who took a hundred dozen eggs in a single day. Rail nests were also predated by Mink, foxes and crows and the weather likewise took its toll. During a storm, Wilson once found thousands of male Clappers in a single meadow, forced inland by the high seas which had drowned their incubating mates. During the summer he spent many hours watching Ospreys fishing along the creeks and described their behaviour at some length, concluding his account in verse, as he was occasionally wont to do when a species particularly impressed him.

In May 1813 Wilson spent four weeks at Cape May with George Ord, then returned to Philadelphia to push on with his work. On 6 July he wrote "Intense application to study has hurt me much. My 8th volume is now in the press and will be published in November. One volume more will complete the whole which I hope to be able to finish in April next."[7] His hopes were in vain. Worn out by the constant labour and stress of the preceding years he died of dysentery on 23 August and was buried next day in the graveyard of Old Swedes Church.

The ninth and last volume was compiled by Ord from Wilson's notes and drawings, supplemented by Ord's own records. Searching through Wilson's material he overlooked the drawing of an unknown phalarope that was subsequently described by Louis Vieillot in 1819. Joseph Sabine, ignorant of all the foregoing, described the bird as *Phalaropus Wilsoni* in 1823 simply because of his great admiration for Wilson. The name appeared in his zoological appendix to the narrative of Franklin's first overland Canadian expedition and

Wilson's Phalarope

though the Latin name is invalid, the English name, Wilson's Phalarope, has persisted. However, Ord did provide the original description of Wilson's Plover, along with the following notes:

> "Of this neat and prettily marked species I can find no account, and have concluded that it has hitherto escaped the eye of the naturalist. The bird from which this description was taken, was shot the 13th of May, 1813, on the shore of Cape Island, New Jersey, by my ever-regretted friend; and I have honoured it with his name. It was a male, and was accompanied by another of the same sex and a female, all of which were fortunately obtained. . . .
>
> That the species is rare we were well convinced, as we had diligently explored the shore of a considerable part of Cape May, in the vicinity of Great Egg Harbour, many times at different seasons, and had never seen them before."[8]

Wilson's Plovers, two males and a female

The *American Ornithology* was now complete and as surely as Lewis and Clark had opened up the West, Wilson had pioneered a scientific route for ornithologists to follow. His thorough and reliable fieldwork rendered all previous books obsolete and set a new standard for his successors; the illustrations, though quaint to modern eyes, were perceived as accurate and attractive renderings by his contemporaries. The high quality of Wilson's text can be judged not just from Bonaparte, Audubon and Nuttall's indebtedness to it during the next three decades but by its lasting value. Even during this century A.C. Bent quoted Wilson in nearly all of his volumes of *Life Histories of North American Birds* (1919–1968). It seems likely that as time passes and wild places become ever fewer, Wilson's lively descriptions of wetlands and woodlands teeming with birdlife will become increasingly precious glimpses into an America we can never know.

Wollweber† (fl. 1850)

BRIDLED TITMOUSE *Parus wollweberi* (Bonaparte)

Lophophanes wollweberi Bonaparte, 1850. *Comptes Rendus de l'Académie des Sciences de Paris* 31, p. 478: ex Mexico Zacatecas [= Zacatecas, Zacatecas]

Those wishing to see the Bridled Titmouse within the United States must travel to the mountains of southern Arizona and New Mexico to search among the oak, juniper and sycamore woodland. Once seen, the distinctive markings make the Bridled Titmouse hard to confuse with any other North American species.

The first person to describe their plumage was Charles Bonaparte. In the *Comptes Rendus* for 1850 he gave a two page article on two new species of *Paridae* from Mexico and discussed their relationship with other members of that family. The first species later turned out to be the Bushtit first noticed by John Kirk Townsend. The second species was the Bridled Titmouse, which

Bridled Titmouse

†No portrait traced

Bonaparte dedicated to the "intrépide et généreux voyageur" who had sent it to Dr Johann Jacob Kaup at the Darmstadt Museum. But no one has been able to discover anything about Wollweber other than the fact that he found the Bridled Titmouse in Zacatecas, Mexico, some time prior to 1850. Bonaparte was vague about the extent of Wollweber's collections, simply referring to the two species and "beaucoup d'autres" sent to Darmstadt.

In 1990 Dr Hanns Feustel of the Hessisches Landesmuseum of Darmstadt kindly searched the museum archives for us but could not find a single reference to Wollweber.

WRIGHT. *AMERICAN PLANT HUNTER. HE STARTED HIS CAREER COLLECTING IN TEXAS, NEW MEXICO AND ARIZONA, LATER CONCENTRATING ON CUBA.*

Charles Wright
(1811–1885)

GRAY FLYCATCHER *Empidonax wrightii* Baird

Empidonax wrightii Baird, 1858. In S.F.Baird, J.Cassin and G.N.Lawrence *Reports of Explorations and Surveys ... for a Railroad [Route] from the Mississippi River to the Pacific Ocean*, Vol. 9, p. 200 (in text): El Paso, Texas

Charles Wright was one of the best and most successful of the nineteenth century American plant hunters. Apart from a three year stint attached to the Rodgers and Ringgold Pacific Exploring Expedition most of his collecting was done in Texas, New Mexico, Arizona and Cuba.

Written accounts suggest that Wright was more than a little eccentric. To Mrs Asa Gray, wife of the famous Harvard Professor of Botany, he was "queer Charles Wright"[1] and Gray himself considered the slightly squint-eyed traveller rather strange: "... you do not know what a helpless odd fellow he is—good for nothing but to collect & dry specimens—but one of the most unselfish & good natured men I know ... [He] prefers exploring to anything else—will go to Santa Fe, or wherever we choose to send him. He is used to roughing it, and long been in the woods and prairies."[2]

Wright's association with Asa Gray was mutually beneficial. Gray enhanced his own career through Wright's discoveries; while Wright needed Gray to keep him supplied in the field with money, books and plant pressing equipment. They endured a long and mainly friendly relationship though the collector sometimes felt himself isolated and neglected as he struggled in difficult circumstances. For long periods Gray failed to write to him, yet Gray rated him extremely highly and regarded him with some affection:

> "Mr. Wright was a person of low stature and well-knit frame, hardy rather than strong, scrupulously temperate, a man of simple ways, always modest and unpretending, but direct and downright in expression, most amiable, trusty, and religious. He accomplished a great amount of useful and excellent work for botany in the pure and simple love of it."[3]

The son of a carpenter and joiner, from a long established Connecticut family, Wright was born at Wethersfield on 29 October 1811. He graduated from Yale in August 1835 after studying classics and mathematics and by October was on his way to Natchez, Mississippi, to become a tutor to a plantation owner's family. Two years later, when his employer's business collapsed, Wright drifted into Texas with a wave of other immigrants and spent most of the next fifteen years there. In his own estimation he "became a pretty fair deer-hunter" and "learned to dress deer-skins after the manner of the Indians, and to make moccasins and leggins."[4] He was also variously engaged as a teacher and land surveyor, at other times continuing his fascination for botany, which he had apparently first developed on solitary rambles at Yale. In 1844 he opened his long correspondence with Gray.

In 1845 he started a two year period as assistant principal at Rutersville College, afterwards teaching at Austin. During the Mexican War he took off for a few months to join (the botanist) Dr John Veatch whose company was defending the frontier at Eagle Pass. On Wright's return through San Antonio there was a letter from Gray offering him a curator's position at Harvard for the coming winter. Following this Gray won permission for him to accompany an army expedition through Texas.

Arriving by ship at Galveston on 24 April 1849 with more than a month to spare, Wright botanized slowly across the country to San Antonio before joining the army baggage train bound for El Paso. The unsympathetic officers would allow transport only for his gear and gave him no other assistance so he had to travel the 673 miles on foot for the next 104 days—often in terribly wet weather.

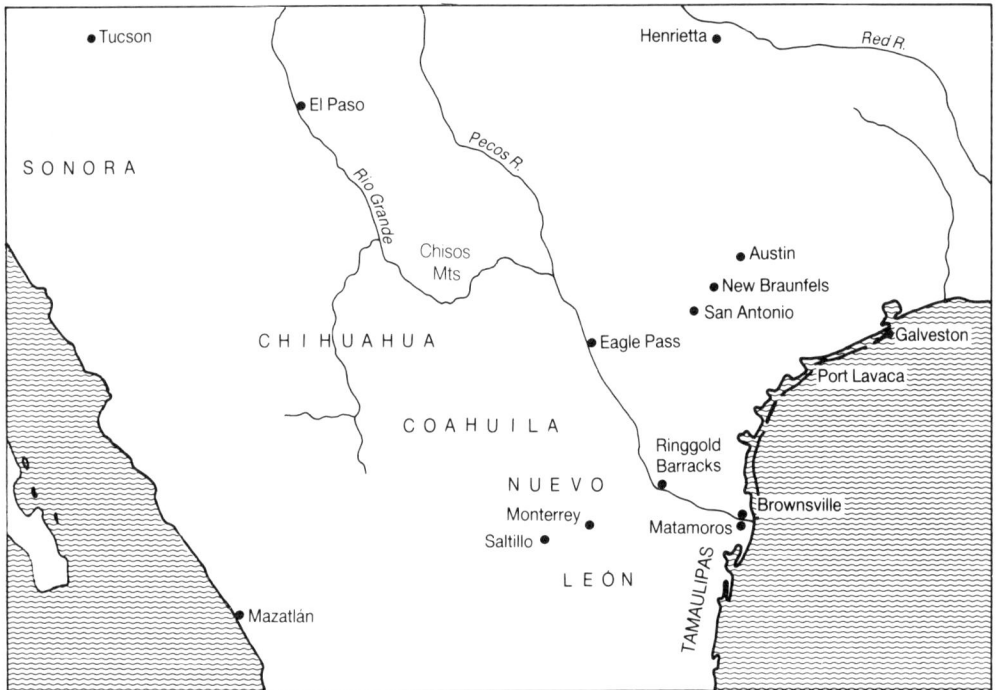

NORTHERN MEXICO AND TEXAS: WITH PLACES VISITED BY COUCH, McCOWN, OBERHOLSER AND WRIGHT

After exactly a month collecting around El Paso he made his way back with the train by another route in less than half the time, bringing with him hundreds of plants which he sent on to Gray.

For a while, Wright taught school at New Braunfels, near San Antonio, then rejoined the United States and Mexican Boundary Survey in the spring of 1851 under Colonel J.D. Graham. This time he worked as botanist and surveyor in circumstances much more conducive to collecting and pioneered botanical fieldwork across the southern parts of New Mexico and into mid Arizona, as well as through parts of Chihuahua and Sonora. At the end of the 1852 season he returned to San Antonio and thence to St Louis where he gave his cacti to George Engelmann before continuing to Cambridge with the remainder of his plants.

His collections from his two trips across Texas form the basis for Gray's classic two part 'Plantae Wrightianae' for the *Smithsonian Contributions to Knowledge* (1852–53). Some of Wright's results were also incorporated into the John Torrey and George Engelmann botanical sections of William Emory's *Report on the U.S. and Mexican Boundary Survey* (1859).

In May 1852 (in a letter to John H. Clark who was then at Frontera, Texas, with the Boundary Survey) Professor Baird commented that "Mr. Wright sends me some highly interesting notes on the birds observed by him last winter and it is a thousand pities he got no specimens, as several were no doubt new."[5] But Wright certainly collected some birds during his time in Texas because he added the Green Kingfisher from the Rio Grande to the United States list[6] and found at least two specimens of the Gray Flycatcher at El Paso, though Baird gave no date of collection for the type specimen when naming it *Empidonax wrightii*. The name Wright's Flycatcher, though derived from this scientific name, was used erroneously for many years for the bird now known as the Dusky Flycatcher.

Charles Wright spent the winter of 1852–53 back at Harvard. In the spring he joined the Rodgers and Ringgold Pacific Exploring Expedition as botanist. Sailing from Virginia in June, he had opportunities to collect at Madeira, Cape Verde, Cape Town, Sydney and Hong Kong, spending longer at the last location

Gray Flycatcher

than intended because of Ringgold's derangement. More chances to collect arose on the Bonin Islands, Japan and the western side of the Bering Strait where Wright, the zoologist William Stimpson and a few other men were set ashore for a month while the *Vincennes* went even further north. Wright left the expedition at San Francisco in February 1856 and went south to Nicaragua, returning via Greytown (San Juan del Norte) to New York. His collection of plants from Hong Kong was later used as the basis for George Bentham's *Flora Hongkongensis* (1861), and other plants, especially those from Japan, were of the utmost interest to botanists. Unfortunately the official expedition report was never completed because of Congressional antipathy and the Civil War. At least Wright's plants fared better than Stimpson's zoological specimens, which perished in the great Chicago fire of 1871.

Not one to be settled for long, Wright set off that same autumn of 1856 for Cuba where he collected for most seasons until July 1867, exploring all the island's habitats. From January to July 1859 he joined Johannes Gundlach in the region around Monteverde; and in the winter of 1861–62 they explored together around Cárdenas and penetrated the great marshy region to the south. In the summer of 1864 Wright came back to spend a year at home in Connecticut and at Cambridge, and then returned to Cuba for another two years.

A glance through Baird's *Revue of American Birds in the Museum of the Smithsonian Institution* (1872) reveals a surprisingly high tally of birds collected by Wright. They include such Cuban species as the Cuban Vireo and Cuban Solitaire as well as migrants from the north such as the Worm-eating Warbler, Black-and-white Warbler, American Redstart and Ovenbird. He wrote little about birds (and not much more about plants) but it was inevitable that he should learn a little about his avian field companions. It would be interesting to discover just how much time he devoted to birds and the extent of his knowledge. Perhaps Wright's manuscript autobiography, now lost, would have told us.

Wright made one more foreign trip after Cuba, a short government expedition to San Domingo in 1871 but it was a brief and hurried affair at the wrong season. Now aged sixty, his travelling days were over. He worked for a while at Harvard but the last ten years of his life were spent back at the house where he was born looking after his invalid brother and two ailing sisters, all of them unmarried. He died suddenly on 11 August 1885 while making his usual evening rounds to feed the livestock.

John Xántus
(1825-1894)

XANTUS'S MURRELET *Synthliboramphus hypoleucus*
(Xántus de Vesey)

Brachyramphus hypoleucus Xántus de Vesey, 1860. *Proceedings of the Academy of Natural Sciences of Philadelphia* 11 (1859), p. 299: Cape St. Lucas, Lower California [= 14 miles off the coast of Cape San Lucas, Baja California]

In late summer and fall, Xantus's Murrelets regularly wander up the west coast to as far north as Washington after breeding on rocky islands off the coasts of southern California and Baja California. The females lay their two large eggs in nooks and crannies among loose boulders and often leave the clutch unattended while away feeding. The young are never fed at the nest, but one or two nights after hatching they leave for the sea where they dive and swim with ease. The first known specimen of this seabird was shot fourteen miles off the coast of Cape San Lucas by John Xántus and described by him as being "considerably weatherbeaten". Taken on 14 July 1859 during his first few months in Baja California it was just one of many thousands of zoological specimens he collected there between April 1859 and August 1861.

During the previous two years Xántus had established himself as one of the great early western naturalists by sending the Smithsonian Institution a regular flow of barrels, boxes and bales from Fort Tejon in southern California. Through his zealous collecting Xántus earned lasting fame—through his publications he gained a unique infamy among the pioneer naturalists, for they revealed him as a liar and a plagiarist of the first order. Sometimes he merely exaggerated his accomplishments, sometimes he blended truth with fiction and sometimes he indulged in pure fantasy.

Because the autobiographical information about Xántus is so unreliable it is impossible to be certain about some aspects of his life but we are indebted to Henry Miller Madden who investigated all of the collector's writings and activities. Madden's painstaking biographical research in both Hungary and the United States was published in 1949 in *Xántus, Hungarian Naturalist in the Pioneer West*.

János Xántus, as this strange and fascinating individual was christened, was

XÁNTUS. *COLLECTED FOR THE SMITHSONION INSTITUTION AT FORT TEJON, CALIFORNIA, AND AT CAPE SAN LUCAS, BAJA CALIFORNIA. REPORTS OF HIS AMAZING EXPLOITS, PUBLISHED IN HIS NATIVE HUNGARY, WERE LATER EXPOSED AS PLAGIARISMS. HE IS SHOWN HERE IN THE UNIFORM OF THE U.S. NAVY—WHICH HE NEVER JOINED.*

born on 5 October 1825 at Csokonya, Somogy, in southern Hungary. His father, Ignác Xántus, was employed by a local family as solicitor, land agent and steward of their estates. His mother, Terézia, to whom he later wrote many boastful letters, gave birth to two other, younger children. Nothing is known about the boy's childhood but on leaving the Benedictine gymnasium at Győr he studied law in the same town, then obtained employment as a vice-notary in Somogy for the next three years. In 1847 he passed the bar examinations in Pest but on the outbreak of the Hungarian War of Independence in 1848 Xántus joined the artillery and fought as a sergeant at the battle of Pákozd. After transferring to the infantry he became a lieutenant in the 46th Battalion and was captured at Ersekújvár. The war ended in the autumn of 1849 with Hungary a conquered country, reduced to a province of the Austrian empire. Xántus was thereafter forced to serve as a private soldier until his mother bought his release in the summer of 1850. Later rearrested in Prague, Xántus managed to escape, fled through Germany to Hamburg and from there sailed to London and on to the United States, probably arriving some time in 1851, at the age of about twenty-five.

No-one knows how Xántus earned his living during his early years in America or where, exactly, he wandered in a futile and increasingly desperate search for the sort of prestigious employment he considered suitable for a man of his background, education and ability. In letters to his family written between December 1852 and June 1854 he portrayed himself successively as a heroic leader, a great traveller, a successful and scholarly writer and as a professor of languages at the University of Louisiana. His mother was so impressed by his exciting life that she showed the letters to István Prépost, a literary agent who considered them suitable for publication. Xántus gave his permission, made some revisions and supplied additional biographical details that tell a rather different story:

"[I] have been a jack of all trades—a newspaper boy, sailor, store clerk, bookseller, pharmacist, piano teacher, railroad cartographer, engineer, and teacher of German, Latin, and Spanish. If I had no money, I did not wait for a job as an artist ... but, according to the American custom, I went to work as a day-laborer so that I might eat. Up to the waist in water I dug for days in a canal, and this is what changed my luck; for people saw that I was willing and energetic, and helped me forward with my plans. Believe me, this part of my life was sometimes as dark as soot, but I shall always remember it with pride."[1]

Since the expeditions and literary efforts that Xántus referred to have been shown by Madden to be fictional and since there is no evidence that he ever taught at the university, it seems likely that the refugee tried at least some of the jobs that he mentioned to Prépost while living in or around New Orleans (where the letters were posted).

The first book by Xántus, entitled *Xántus János Levelei Éjszakamerikából* [John Xántus's Letters from North America] contained thirty-seven letters written between December 1852 and July 1857. Most of them were more or less products of his imagination but one, written from the fictitious address of Elk Creek, Kansas Territory, on 26 September 1856 was a plagiarism of Captain

Randolph Marcy's *Exploration of the Red River of Louisiana, in the Year 1852* (1853). The Hungarian version, supposedly written as a journal between 10 July and 9 September 1856 used Marcy's narrative with the names of places and personnel changed, with Xántus himself, of course, in command of the expedition. The *Levelei* was published in Pest by Lauffer and Stolp in 1858 and two years later they produced a second book by Xántus, *Utazás Kalifornia déli részeiben* [Travels in the Southern Parts of California].

Meantime, in August 1854 Xántus went to the Hungarian settlement at New Buda in southern Iowa where he stayed for about a year, filed a claim for land and made a collection of insects and reptiles for the National Museum of Hungary. Having no real interest in working as a farmer he departed after falling out with his compatriots.

Job prospects now looked so bleak that in October 1855 Xántus entered the U.S. Army as a private, under the assumed name of Louis de Vésey. Later, he explained this action to Professor Baird as follows:

"I came to this country amongst the first of my countrymen, in advance of Kossuth, and by order of President Fillmore I received a grant of land in Iowa, as the others of my fellow Refugees. But actually I never took possession of, but being a good piano player, and a tolerable draughtsman, I procured a honorable support by teaching for a short time; when I went successively with the Prince of Würtemberg, Dr Wagner & Scherzer, & Dr Kroyer as collector. At last I fitted out of my hard earnings an expedition into North Minnesota, which failed so entirely, that in a moment of utmost despair, & under circumstances completely beyond my control—I was forced to enter the American Army.

With a honest past, very respectable connexions, and a good many enemies at home, I think I was justified in changing my name, and not to carry it on the muster Rolls of the American Military Ranks. My name at home was John Xántus de Vésey; but as Véseys are good many, and John Xántus is only one, I took the former."[2]

In fact, de Vésey had never been a part of the family name and its adoption caused much confusion. His claim to have travelled with the above-named German, Austrian and Dutch naturalists was often repeated by him, but was complete fabrication. Perhaps the most truthful part of the letter was the admission that he joined the army in a moment of utmost despair, so humiliated by his situation that he never told his family anything about it. Yet it was in the army that he found his salvation—in the person of Dr William A. Hammond.

After enlisting at St Louis, Xántus was sent to Fort Riley in the Kansas Territory where he met Hammond, the assistant surgeon, one of many medical officers in the U.S. Army who enjoyed collecting for Professor Baird. Hammond urged Xántus to help him collect and prepare birds, mammals, reptiles and insects and used his influence to free the lowly private from the garrison chores that he hated. Xántus had first shown an embryonic interest in natural history while in Iowa. Now, with careful tuition, he gradually developed some essential skills in fieldcraft, identification and taxidermy. At last he had discovered a vocation that bolstered his self-esteem and brought him a degree of the recognition for which he craved. His first collections were sent to the Academy of Natural Sciences of Philadelphia, to which he was elected a life member on 30 December

1856 after being proposed by the distinguished entomologist Dr John L. LeConte and the herpetologist Dr Edward Hallowell.

Hammond suggested that Xántus should send some of his findings to Baird and these were gratefully accepted and acknowledged in the Pacific Railroad Reports. In February 1857, less than a year and a half after his enlistment, Xántus began to correspond with the professor and this mutually beneficial relationship continued over the next seven years. The letters from Xántus have survived and provide a fascinating insight into his character: his persuasiveness, his pride, his swings between optimism and pessimism, his hypersensitivity to slights either real or imagined and his resentment of those in authority over him. Though they highlight some of the flaws in Xántus's personality they are, on the whole, honest records of his activities, albeit a sometimes biased version of events. They also chronicle the frustrations and disasters common to all who collected on the frontier under conditions that were often less than ideal and demonstrate his transformation into a pioneer naturalist.

For some time Baird had been keen to exploit the opportunity afforded by a new military post in Tejon Pass. As the son-in-law of Brigadier-General Sylvester Churchill it was easy for him to arrange the transfer of Xántus from Kansas to California and on 18 May 1857 Xántus arrived at Fort Tejon.

The fort had been built in the most frequently used pass between the coast and the San Joaquin Valley in order to protect travellers from Indians and Mexican bandits and to suppress stock rustling. As the regimental headquarters of the 1st Dragoons it was an important military, social and political centre. The immediate surroundings were extremely rich in birdlife. White Oaks shaded the edges of the parade ground, a clear stream lined by maples and willows flowed past the barracks and all around were grassy, partly wooded hillsides. Had Baird himself been allowed to choose the perfect location for a collector's base in southern California he could hardly have found a more conveniently placed site where the juxtaposition of mountains, foothills and deserts provided a tremendous diversity of hunting grounds.

If Baird was hoping for ecstatic reports from Xántus then he was disappointed. The early letters were a litany of troubles and complaints; the breeding season was almost over when he arrived, the collecting box had gone missing en route to the fort, his gun had been left behind at Fort Riley so he had to use a rusty old carbine, and to cap it all, collecting was dangerous: "We have here grizlys [?Black Bears] in great abundance," wrote Xántus in his own peculiar style, "they are really a nuisance, you cannot walk out half a mile, without meeting some of them, and as they just now have their clubs, they are extremely ferocious to, I was already twice driven on a tree, and close by to the fort."[3]

A major problem for Xántus was his superiors' lack of interest. As well as his duties as hospital steward he had to serve in the bakery and the library so collecting had to be limited to his few off-duty hours. There was no friendly protector here, as there had been at Fort Riley. He complained bitterly, with an attitude that can hardly have endeared him to anyone:

"I had an immense trouble with the boxing of nests, eggs, labelling etc etc I had to make everything myself and lost much time with this manufacturing business.

Lost much time I say, because I am assisted by nobody here, I depend entirely and exclusively on myself. In F^t Riley not only Col. Cooke & Dr Hammond took interest in my pursuits, but I was assisted by *everybody*, citizens as well as soldiers. They brought me in a great number of specimens from all parts around. But here everybody is a gambler and drunkard, they sit day & night in whisky shops, or gambling holes; and instead of supporting me they ridicule my [sport?] and trow every obstacle in my way.

  I treat them of course with princely contempt, and go on with doubled step in my path."[4]

Nevertheless, Xántus did report enthusiastically about the wildlife. Apart from the bears there were Pronghorns, Mule Deer, Elk, Gray Foxes, Mountain Lions, Bobcats, Valley Pocket Gophers, California Moles and lots of California Ground Squirrels. But by 20 July Xántus's popularity had waned still further: "I have at present quite a menagerie; a fine grizzly cub, & 5 badgers (he, she, & 3 young ones) besides several ground Squirrels. I am sorry, that there is no opportunity whatever to send them to you; more yet because I am bound soon to kill the grizzly, he is getting from day to day more mischivious, he kills my hens as fast as possible, and tore to pieces already the Colonels dog."[5]

In the same letter Xántus mentioned that all his jars and bottles had been broken in an earthquake. Since reptiles and amphibians had to be preserved in alcohol this was causing him some anxiety. So did the high cost of freighting the collections to Washington. Still, he was dispatching huge numbers of specimens and in January Baird wrote encouragingly, and with a little flattery: "I have seen many collections of specimens but nothing within my experience comes up to the perfection of finish of your skins in every respect. The skins are all that could be asked. Just right, not too much or too little stuffing. May I say that they are much better than your Kansas collections. The packing of the [?] vials attracted especially my admiration, as well as of the nests. Everything was perfect."[6]

The end of January brought a new calamity. Lieutenant Beale arrived from Texas leading an expedition that used camels for transportation. He enlisted some of the local men to provide him with an escort and commandeered every weapon at the post for them. Poor Xántus lamented, "I am now entirely naked (scientifically speaking)!" Undeterred, he turned his attention to gathering nests and eggs, to the advantage of Thomas Brewer of Boston who had recently produced the *North American Oölogy* (1857) and still wanted more information. Baird had particularly requested specimens for him, but it would seem that Xántus was not an enthusiastic nester for in November he wrote:

"I hope you will find something interesting amongst my Reptiles of this year, although the collection is by far not such as I had devised. But I was very much engaged in the whole spring, with those confounded nests & eggs, and naturally enough I had to neglect somehow the more important branches, as birds reptiles etc . . .

  Dr Thos Brewer wrote me a letter last mail, asking the permission to describe & photograph the nests & eggs. In answer I told him, that I already informed you long ago, that the particular purpose of my collecting nests & eggs was to furnish some additional material for his work. Of course then he shall use my collection *for illustrations descriptions*. Dr Brewer seems to be a very polite & nice gentleman, and I am quite glad to have formed with him in this way an Aquaintance."[7]

One can only hope that Brewer appreciated the efforts made on his behalf. Xántus certainly made sure that Baird realized how much trouble he had gone to:

> "To illustrate the difficulty of taking some nests, I will tell you that the first *vireo* nest I discovered, took fully four days meditating & devising plans, as how to obtain it. The nest was situated on a limb of water oak, which was scarcely an inch tick, nearly twenty feet long, & projecting over a water, fully 4 feet deep, not including mud. The nest was at the furthest extremity of the limb, about twenty feet above water. Finally, after every device failed—I chopped logs, undressed, carried the logs neck deep in water under the nest, & built a regular lower work (in the log house architecture style), until I reached to about 12 feet of the nest; I mounted then, pulled down the limb with a hook, & got the nest. What you say to such feats?
>
> With the Muscicapa verticales [Western Kingbird] & several humming birds I had adventures of very similar character, the description of all together would make a very amusing volume, & might be entitled '*The Ramblings & climbings of a fool*'."[8]

During his twenty-month sojourn at Fort Tejon Xántus catalogued about 140 species of birds, several of which he described as new to science. In 1858 he named Hammond's Flycatcher for his friend and teacher and *Vireo cassinii* for John Cassin. Cassin's Vireo was long regarded as a race of Solitary Vireo, but it should perhaps be given full specific status. Another bird discovered and described by Xántus was the Spotted Owl, a western species almost entirely restricted to the old-growth forests and canyon lands, now becoming increasingly rare through the destruction of its habitat.

While stationed at the fort Xántus also wrote his second book, *Utazás* or 'Travels', which he dedicated to Agoston Kubinyi, Director of the Hungarian National Museum. It was in three sections, headed 'From Los Angeles to Fort Tejon', 'Tejon and the Tejon Indians' and 'The California Peninsula' which he had not yet visited! Madden concluded that the first section contained some plagiarisms but was largely imaginary. The second part began with Xántus's own account of the establishment of Fort Tejon, but the rest was a translation of Jonathon Letterman's *Sketch of the Navaho Tribe of Indians* (1855) with a few changes and additions. In the third chapter on Baja California Xántus plagiarised Major William Emory's *Notes of a Military Reconnoissance, from Fort Leavenworth ... to San Diego* (1848) and the 'Report of Lieutenant James W. Abert of His Examination of New Mexico in the Years 1846–1847' which was part of the same publication. Xántus even copied some of their illustrations!

Throughout his stay in California Xántus was irked by the restraints imposed on him. Most of his collecting had to be limited to the vicinity of the fort, with his longest trip being one of twenty miles to San Emigdio, where a soldier was in need of medical treatment. Army life proved to be so frustrating that he was soon agitating for pastures new. As always, Baird had the right connections.

He managed to wangle a position for his collector as a tidal observer with the Coast Survey at Cape San Lucas and so on 25 January 1859 Xántus rolled away down the pass in the mail coach, bound for San Francisco.

After three long years as an enlisted man he was exuberant at being set free. For the next six weeks he remained in the city, learning to manage the tidal gauge, laying in stores and enjoying the novelty of a social life. He made two useful contacts, one of them being Ferdinand Gruber, a taxidermist who offered to exchange birds and eggs from the Farallon Islands for ones from the Cape. The other was Andrew Jackson Grayson who had been studying and painting the western birds for about five years. Unable to give credit to any possible rival, Xántus wrote Baird that Grayson was unable to tell a flycatcher from a dove.

On 4 April 1859 Xántus was set ashore on the beach at San Lucas with all his gear, an assistant and a carpenter who was to construct a shed for the gauge—a large machine weighing about half a ton. Once the gauge had been installed, Xántus only had to visit it briefly each day to check the clock. Once a month he had to change the roll of recording paper and devote about two days to calculating his observations and writing up reports, but otherwise his time was his own. Since his assistant could make the daily visits for him he was free to ransack the broad, sandy, windswept shore and the cactus desert which lay inland for a distance of about six miles to the foot of the sierras. He was surprised to find that this harsh dry environment sustained a wealth of birds, mammals, snakes and lizards. Though his base was only a tent on the beach, unsheltered from the sun, and though all his drinking water had to come from a neighbour's brackish well at the other end of the bay, Xántus cared little about the inconveniences—he was in his element.

By the end of the first breeding season he had sent off forty-two different kinds of birds, among them such typical desert species as the Gila Woodpecker, Phainopepla, Cactus Wren, Verdin, Scott's Oriole and Greater Roadrunner. In later years Robert Ridgway made a careful examination of the Smithsonian records and concluded that Xántus had eventually collected 130 bird species at the Cape. Three of these were new: Xantus's Hummingbird, the Gray Thrasher and the aforementioned Xantus's Murrelet. The hummingbird, a resident species endemic to Baja California (with one 1980s record for California) was taken at San Nicolás about ten miles north-east of the collector's base. It was named by George Lawrence who doubtless expected Xántus to be pleased by the compliment. Instead he wrote indignantly to Baird: "I am much obliged to Mr Lawrence that he named the humming birds Xántusii, although I had preferred to describe myself. You know very well we agreed, that the birds shall be described *all* by me, or in my name, should you prefer not to await my arrival there."[9] Xántus succeeded in describing both his murrelet and the Gray Thrasher in the *Proceedings of the Academy of Natural Sciences of Philadelphia*. He gave Latin names to them both, but the English names were proposed later by other naturalists.

During his spell at the Cape Xántus made a number of excursions into the mountains and along the coast up to seventy-five miles from his base. One of his most productive sites was at San José del Cabo where the river flooded the bottomlands, creating a large lake, freshwater swamps and a brackish lagoon. He camped there in November 1859 and shot huge quantities of ducks, particularly Mallard, Pintail, Cinnamon Teal and Northern Shoveler as well as diverse shorebirds. When a flock of seven Wood Storks turned up he killed the lot but five Sandhill Cranes that hung around for a few days proved to be far too wary for him.

As at Fort Tejon, Xántus collected very thoroughly, though here he found a much greater diversity of animals and hunted birds, nests and eggs, reptiles, amphibians, fish, echinoderms, insects, crustaceans, spiders, molluscs, polyps, corals, plants and rocks. As the biota of Baja California was then almost entirely unknown to scientists Baird was more than pleased since every observation and discovery was of value and he passed most of the material on to the relevant specialists. Even the first instalments demonstrated some of the region's biological peculiarities and its zoological affinity with the Gila River and the Rio Grande, rather than the rest of Mexico. When John L. LeConte studied the beetles he was so struck by the close relationship of one species to another from the Colorado Desert that he wrote about "the propriety of including [Baja California] within the zoological, as it will eventually be within the political, boundaries of the United States."!

The crustaceans were worked through by William Stimpson, whose relationship with Xántus got off to a bad start. The collector wrote to Baird on 5 July complaining about Stimpson's insultingly simplistic advice, that "crustacea are found at low water mark under stone, or by digging in sand". Xántus then went on to describe his favourite method of searching the inter-tidal zone:

"I discovered among the other *tricks* quite a novel method of collecting crustacea, viz I trained my pointer dog for the purpose, and he at present digges out for me everything, I have only to watch close by with a pipe or cigar, & bottle the specimens. This method will prove an immense facility for collectors as a poor biped has to dig sometimes a dozen holes until he finds one crab; when a pointer *never* digges an empty hole, but smells always the mouth and assures himself of the presence of the resident before he commences digging. My *Jack* catches now even the largest size crabs and with such canine caution, that the crab can never hurt his nose, and still he never damages specimens, but brings to me, and drops in the tin can, which I open to him."[10]

Xantus's Murrelet

His unorthodox approach was clearly successful for Stimpson later grumbled that he was receiving so many new species of crabs from the Cape that he could hardly think up enough names for them all! Several times he solved the problem by naming the species after its collector.

By the end of his second year at the Cape Xántus was growing weary of the isolation and he poured out his unhappiness to Baird:

> "I am quite sick indeed of this place, every day seems a long year, and every one with the same monotonous desolation around me, not affording the least pleasure, variety, or enjoyment of any kind. I have no reading matter whatever, and the latest San francisco paper is that which announces by Pony [Express] Lincolns election. I wonder what all happened since in the world!!
>
> I am now of Gods grace nearly two years perched on this sandbeach, a laughing stock probably of the Pelicans & Turkey buzzards, the only signs of life around me. To the E & SE the eternally smoky Gulf, to all other points of the compass the sandy desert, covered with white salina and ornamented with cactuses in every forms, sticking out like candlesticks on a white cloth ... there is not a blade of grass in the country, & not a green leaf."[11]

In June he sailed east to Mazatlán and visited Grayson who was working on *The Birds of Western Mexico*, an excellent project that was never completed. Grayson's stimulating company, the change of scenery and the exotic birds soon restored his spirits. On returning to the Cape he found orders instructing that he immediately dismantle the gauge and return to San Francisco. In haste he packed up, said farewell to friends, arranged the shipment of his final collections and left on 7 August 1861.

Though Baird was delighted with his efforts, the officers of the Coast Survey were not so impressed. The observations for the first year had been far less accurate than they should have been and so Xántus was not offered another appointment. He therefore returned East and at the beginning of October visited Baird. After explaining that he had decided to return to Hungary but lacked the necessary funds he offered his duplicate collection of skins as security for a loan. Baird not only arranged various small loans from himself and other friends such as D.G. Elliot, J.L. LeConte and Baron Osten-Sacken but also organized a free cabin on the *Bremen* in exchange for some Californian birds for the Bremen Museum.

On returning to his homeland after an absence of ten years Xántus was acclaimed as a distinguished traveller and scientist. Posing in the uniform of a captain in the U.S. Navy he had his portrait painted by a well-known artist and a reproduction of this picture was offered by the newspaper *Gjöry Közlöny* as a gift to new subscribers. On reaching Pest, Xántus was met at the station by several professors from the university, serenaded by the academic choral union, and in January made a corresponding member of the Hungarian Academy of Sciences, an honour he found deeply gratifying. In typical Xántusian style he gave a lecture at the academy that was wholly imaginary, telling of how, as an American naval officer, he had collected thousands of marine protozoa at every latitude, longitude and depth, down to a mile and a half, between North America and Australia. Having sent regular fallacious reports to Hungary for many years, Xántus seems to have been caught up in such a web of deceit that it was impossible for him not to elaborate on earlier accounts, but it was unfortunate that having made genuinely important discoveries he preferred to expound his fantasies.

Despite the triumphant welcome, offers of work failed to materialize so Xántus returned to the United States with a commission from the editor of the *Gÿory Közlöny* for more letters from America. In Washington he visited Dr Hammond who had recently become Surgeon General of the U.S. Army and was immediately taken on as his secretary. While enjoying this sinecure in the war-time capital Xántus had his reputation attacked by some of the Hungarian settlers in New Buda. Having been irritated by their compatriot's self-glorifying tales in the past they suspected the falseness of the claims that were now being recounted in the Hungarian papers and wrote to the Secretaries of War, the Navy and the Interior to ask if Xántus had really led surveying parties in the West or conducted marine research in the Pacific. On discovering the mendacity of some of the claims they then wrote an exposé for a daily newspaper in Pest. It is amazing that despite this attack Xántus emerged with his reputation intact, supported as ever by no lesser persons than the Assistant Secretary of the Smithsonian Institution and the Surgeon General of the U.S. Army, who witnessed to their friend's great honesty and scientific accomplishments—though none of the specific allegations were refuted. One wonders whether Baird and Hammond acted out of loyalty or real conviction.

As a naturalized citizen of the United States of America Xántus had long aspired to become a consul in some suitably unexplored location and in November 1862 he was appointed to the post of Manzanillo, on the coast of Colima in western Mexico. Since a consul had to pay his own travel expenses and Xántus was in his usual impecunious state, George Lawrence lent him $300. Xántus retained the post for only six months as he made an error of judgment that led to his sudden dismissal, but he stayed on in Colima and began to rake in money by representing the underwriters of a treasure ship wrecked on the nearby coast. He also explored the forests around the towns of Colima and Manzanillo but concentrated on the more flamboyant and conspicuous specialities instead of making a thorough study. In poor health he left in March 1864 and went to New York to visit Lawrence, before departing for Washington to argue with Baird over the disposal of his duplicates. In Philadelphia he called on John Krider and John Cassin, then returned to New York and sailed for Europe once more.

Despite paving the way for himself with gifts to the Hungarian National Museum, numerous letters to the press, two volumes of travels and considerable publicity, Xántus had to wait two years for employment as a naturalist in his own country. When a zoological garden was opened at Pest in August 1866 he became its director but resigned after two years in order to join the Austro-Hungarian East Asiatic Expedition. It was organized to form commercial links with the nations of the Far East but scientists and collectors also took part. Xántus travelled independently to Ceylon, Penang, Malacca and Singapore where he joined the others in April and accompanied them to Japan, Siam, Cochin China and China, but his ardent patriotism was too fiercely and tactlessly expressed for happy relations to flourish with the Austrians. He particularly criticized Karl Ritter von Scherzer—with whom he had formerly claimed to have travelled in Louisiana. After six months Xántus went his own way, visited Formosa and Singapore and collected in Borneo, Java and Sumatra, not returning home until 1871.

The next year he began his final career as keeper, and later director, of the ethnographical department of the National Museum. In his late forties he married the actress Gabriella Doleschall and they had one son, Gábor, born in 1874, but they were later divorced.

In 1891 Xántus took part in organizing the Second International Ornithological Congress which was held at Budapest, but during the next few years he declined mentally and was admitted to a private asylum. After he suffered an attack of pneumonia his second wife, Ilona Steden, took him in April 1894 to Volosca on the Adriatic coast but on returning to Budapest in July he failed to recover his strength and died there on 13 December, aged sixty-nine.

It was many years before biohistorians in America questioned Xántus's integrity. While Xántus was at Colima, Professor Josiah D. Whitney, the state geologist in California, had called attention to what he regarded as "curious errors" in the German translation of the *Utazàs* but he assumed that Xántus had confused his notes and concluded, "It is evident that Mr Xantus' notice of rich gold, lead and copper mines on the peninsula must be taken with many grains of allowance."

When Harry Harris wrote 'Notes on the Xantus Tradition' for *The Condor* in 1934 he was obviously puzzled by some of the contradictions in Xántus's writings but he failed to reach the appalling conclusion that such an important naturalist had been a deliberate liar.

The first published exposé appeared in 1942 when E.E. Hume included a chapter on Xántus in his book on the *Ornithologists of the U.S. Army Medical Corps*. Hume got the truth from Madden who was then preparing his thesis and so he was able to contrast Xántus's general unreliability with his real scientific accomplishments. Madden's own definitive biography appeared in 1949.

A further slant on Xántus's skill as a collector was revealed by the Californian novelist John Steinbeck, who heard of his local reputation when he visited Cape San Lucas in 1940:

> "Speaking to the manager of the cannery at the Cape, we remarked on what a great man Xantus had been. Where another would have kept his tide charts and brooded and wished for the Willard Hotel, Xantus had collected animals widely and carefully. The manager said, 'Oh, he was even better than that.' Pointing to three little Indian children he said, 'Those are Xantus's great-grandchildren,' and he continued, 'In the town there is a large family of Xantuses, and a few miles back in the hills you'll find a whole tribe of them.'"[12]

Bibliography and Notes

General Bibliography

Original descriptions of bird species and associated texts were consulted and have often been quoted—references are at chapter headings. Various ornithological journals were examined, especially the early issues of *The Auk, The Condor* and *The Ibis*. The following works are not mentioned elsewhere in this bibliography but were invaluable for specific information or historical background:

Alden, R.H. and Ifft, J.D. 1943. Early Naturalists in the Far West. *Occasional Papers of the California Academy of Sciences, No. XX,* 60 pp. San Francisco.

American Ornithologists' Union. 1933. *Fifty Years' Progress of American Ornithology 1883–1933.* Revised edition.

American Ornithologists' Union. 1983. *Check-list of North American Birds*, 6th edition [and supplements Nos. 35–37].

Hall, E.R. 1981. *The Mammals of North America*, 2nd edn, 2 vols. John Wiley and Sons, New York.

Jobling, J.A. 1991. *A Dictionary of Scientific Bird Names.* Oxford University Press, Oxford.

Meissel, M. 1924–29. *A Bibliography of American Natural History; the Pioneer Century, 1769–1865*, 3 vols. Premier Publishing Company, Brooklyn.

National Geographic Society. 1987. *Field Guide to The Birds of North America.* 2nd edn.

Palmer, T.S. *et al.* 1954. *Biographies of Members of the American Ornithologists' Union.* [Reprinted from *The Auk*, 1884–1954]. Washington, D.C.

Pratt, H.D., Bruner, P.L. and Berrett, D.G. 1987. *A Field Guide to The Birds of Hawaii and the Tropical Pacific.* Princeton University Press, Princeton.

United States Pacific Railroad Survey. 1855–59. *Reports of explorations and surveys, to ascertain the most practicable ... route for a railroad from the Mississippi River to the Pacific Ocean*, 11 vols [in 12]. Washington, D.C.

Wood, C.A. 1931. *An Introduction to the Literature of Vertebrate Zoology.* Oxford University Press, London.

Wynne, O.E. 1969. *Biographical Key—Names of Birds of the World—to Authors and those Commemorated.* Wynne, Fordingbridge, Hants.

Introduction

See e.g. Thomson, A.L. 1985. Nomenclature, *in* Campbell, B. and Lack, E. (Eds) *A Dictionary of Birds*, pp. 397–399. Calton (Poyser).
1. Coues, E. 1874. *Birds of the Northwest*, pp. 367–368. Government Printing Office, Washington, D.C.
2. *Quoted in* Cutright, P.R. and Brodhead, M.J. 1981. *Elliott Coues. Naturalist and Frontier Historian,* p. 321. University of Illinois Press, Chicago.

Abert

Abert, J.W. 1882. Zoological Miscellany. *Journal of the Cincinnati Society of Natural History* **5**: 57–59.
Bailey, F.M. 1928. *Birds of New Mexico*, pp. 17–18 [Abert and Emory in New Mexico], 718 [Abert's Towhee]. New Mexico Department of Game and Fish, Washington, D.C.
Davis, J. 1951. Distribution and Variation of the Brown Towhees. *University of California Publications in Zoology* **52**: 110.
Emory, W.H. 1848. *Notes of a Military Reconnoissance, from Fort Leavenworth, in Missouri, to San Diego, in California, including parts of the Arkansas, Del Norte, and Gila Rivers.* [Includes: Appendix No. 6. Notes of Lieutenant J. W. Abert, pp. 386–405]. House Executive Document No. 41, 30th Congress, 1st Session, 1849–50. Washington, D.C.
Ewan, J. and Ewan, N.D. 1981. *Biographical Dictionary of Rocky Mountain Naturalists, 1682–1932,* p. 1. Frans A. Stafleu, Utrecht.
Galvin, J. (Ed.) 1966. *Western America in 1846–1847: The Original Travel Diary of Lieutenant J.W. Abert.* J. Howell-Books, San Francisco.
Geiser, S.W. 1958. Men of Science in Texas, 1820–1880. *Field and Laboratory* **26**: 89.
Goetzmann, W.H. 1959. *Army Exploration in the American West, 1803–1863,* pp. 144–149. Yale University Press, New Haven.
National Cyclopaedia of American Biography 1897, Vol. IV, pp. 395–396. J.T. White and Co., New York.
1,2,3. Galvin, J. (Ed.) 1966. *op. cit.,* entries for 29 June 1846, 1 December 1846, 7 March 1847.
4. *Proceedings of the Academy of Natural Sciences of Philadelphia* 1847, **3**: 221–222 [Note on Scaled Quail by J.W. Abert]; Deane, R. 1905. John James Abert, to John James Audubon. (Hitherto unpublished letters). *Auk* **22**: 172–175.

Adams

Adams, E. 1878. Notes on the Birds of Michalaski, Norton Sound. *Ibis*, pp. 420–442.
Bury and Norwich Post 17.12.1856. Obituary.
Collinson, R. 1889. *Journal of H.M.S. Enterprise on the expedition in search of Sir John Franklin's ships by Behring Strait 1850–55.* Samson Low & Co., London.
1. These drawings by Adams were said to have been deposited at the British Museum but their present whereabouts are unknown. They may be in private hands (*see*

Günther, A. (Ed.) 1904–12. *The History of the Collections Contained in the Natural History Departments of the British Museum*, Vol. II, p. 251. BM publication, London).
2. At least two publications erroneously connect the White-billed Diver with Royal Navy Surgeon Arthur Adams who also visited the Arctic and who also made a number of zoological sketches: Warren, R.L.M. 1966. *Type Specimens of Birds in the British Museum (Natural History)*, Vol. I; and Coomans de Ruiter, L., *et al.* 1947. *Beteekens en Etymologie van de Wetenschappelijke namen der Nederlandsche Vogels*. Club van Nederlandsche Vogelkundigen.
3. Seebohm, H. 1885. On the occurrence of the White-billed Diver, Colymbus adamsi, on the British coasts. *Zoologist*, pp. 144–145.
4. Captain Cook's expedition collected a Bluethroat "on the ice" north of Bering Strait between August–September 1778 and July 1779 (Pearse, T. 1968. *Birds of the Early Explorers in the Northern Pacific*, pp. 177, 202. T. Pearse, Comox, British Columbia).

Alexandre

No biographical information traced; sources checked include the Bibliothèque Interuniversitaire de Lyon, the Musée Guimet d'Histoire Naturelle at Lyon, and the Société de Sciences Naturelles Loire-Forez at St Etienne (which holds Mulsant's MS notes). Adolphe Boucard named a supposed new species of hummingbird *Aphantochroa alexandri* after "Mr. Alexander, of New York, who discovered it [in "Demarara, B. Guinea"] in 1878" (*The Humming Bird* 1891, **1**: 18). There is no reason to suppose that this is the same man who discovered the Black-chinned Hummingbird *Archilochus alexandri*.

Allen

Mailliard, J. 1931. Charles Andrew Allen. *Condor* **33**: 20–22.
1,2. *Quoted in* Bent, A.C. 1940. *Life Histories of North American Cuckoos, Goatsuckers, Hummingbirds and Their Allies*, Part II, pp. 416, 412. *Bulletin 176, United States National Museum*. Washington, D.C.
3. Mailliard, J. 1924. Autobiography of Joseph Mailliard. *Condor* **36**: 14.

Anderson

Anderson, Mary White. Unpublished notes on Mary Virginia Childs Anderson. [Includes correct date of birth of Ann Catherine Anderson—1864, not 1861 as in Hume]
Dictionary of American Biography 1928–36, Vol.IV, p. 71 [Thomas Childs]. New York.
Hume, E.E. 1942. Chapter I: William Wallace Anderson (1824–1911). *Ornithologists of the U.S. Army Medical Corps*, pp. 11–21. John Hopkins University Press, Baltimore.
Parler, J.P. 1938. Afterglow. A Record of Two Centuries of Living at Borough Plantation. *South Atlantic Quarterly* **36**: 73–85.
Additional information supplied by Mrs Richard K. Anderson and the South Caroliniana Library, University of South Carolina.
1. *Quoted in* Hume, E.E. 1942. *op. cit.*, p. 17.

Audubon

Audubon, J.J. 1831–39. *Ornithological Biography*, 5 vols. Edinburgh.
Audubon, M.R. (Ed.) 1960. *Audubon and His Journals*, 2 vols. Dover Publications Inc., New York.
Ford, A. 1965. *John James Audubon*. University of Oklahoma Press, Norman.
Herrick, F.H. 1917. *Audubon the Naturalist*, 2 vols. D. Appleton and Co., New York.
Stone, W. 1906. A Bibliography and Nomenclator of the Ornithological Works of John James Audubon. *Auk* **23**: 298–312.
1. Stone, W. 1906. *op. cit.*

Bachman

Bachman, C.L. (Ed.) 1888. *John Bachman D.D., LL.D., Ph.D.* Walker, Evans and Cogswell Co., Charleston, S.C.
Ford, A. 1965. *John James Audubon*. University of Oklahoma Press, Norman.
Hamel, P.B. 1986. *Bachman's Warbler. A Species in Peril*. Smithsonian Institution Press, Washington, D.C.
1. Ford, A. 1965. *op. cit.*, p. 285.
2,3. *Quoted in* Bachman, C.L. 1888. *op. cit.*, pp. 95–96, 102.
4. Audubon, J.J. 1967. *The Birds of America*, Vol 2, p. 93. Dover Publications Inc., New York.
5,6,7. *Quoted in* Bachman, C.L. 1888. *op. cit.*, pp. 112, 156, 250.

L.H. Baird

Dall, W.H. 1915. *Spencer Fullerton Baird. A Biography*. J.B. Lippincott Co., Philadelphia and London.
Palmer, T.S. 1943. Lucy Hunter Baird. *Auk* **60**: 483–484.

S.F. Baird

Dall, W.H. 1915. *Spencer Fullerton Baird. A Biography*. J.B. Lippincott Co., Philadelphia and London.
Deiss, W.A. 1980. Spencer F. Baird and his collectors. *Journal of the Society for the Bibliography of Natural History* **9**: 635–645.
Herber, E.C. 1963. *Correspondence between Spencer Fullerton Baird and Louis Agassiz— Two Pioneer American Naturalists*. Smithsonian Institution, Washington, D.C.
Ridgway, R. 1888. Spencer Fullerton Baird. *Auk* **5**: 1–14.
1. *Quoted in* Dall, W.H. 1915. *op. cit.*, pp. 441–442.
2. Elliot, D.G. 1896. In Memoriam. George Newbold Lawrence. *Auk* **13**: 6.
3. Henshaw, H.W. 1920. Autobiographical Notes. *Condor* **22**: 3.
4. *Quoted in* Deiss, W.A. 1980. *op. cit.*, p. 643.
5,6. *Quoted in* Dall, W.H. 1915. *op. cit.*, pp. 57–59, 166.

7. *Quoted in* Cutright, P.R. and Brodhead, M.J. 1981. *Elliott Coues. Naturalist and Frontier Historian*, p. 37. University of Illinois Press, Chicago.
8. [Baird, S.F.] 1859. *Directions for Collecting, Preserving, and Transporting Specimens of Natural History*, 3rd edn, p. 27. Smithsonian Institution, Washington, D.C.
9. *Quoted in* Hume, E.E. 1942. *Ornithologists of the U.S. Army Medical Corps*, p. 17. Johns Hopkins University Press, Baltimore.
10. *Quoted in* Dall, W.H. 1915. *op. cit.*, p. 330.
11. *Quoted in* Zwinger, A.H. 1986a. *Xántus. The Letters of John Xántus to Spencer Fullerton Baird from San Francisco and Cabo San Lucas 1859–1861*, p. 220, note 2. Dawson's Book Shop, Los Angeles.
12. *Quoted in* Harris, H. 1928. Robert Ridgway. *Condor* 30: 18.
13,14,15,16. *Quoted in* Dall, W.H. 1915. *op. cit.*, pp. 276, 291, 283, 342.
17. *Quoted in* Peck, R.M. 1990. *Land of the Eagle*, p. 262. Guild Publishing, London.
18. *Quoted in* Zwinger, A.H. 1986b. *John Xántus: The Fort Tejon Letters 1857–1859*, p. 81, note 5. University of Arizona Press.
19. *Quoted in* Zwinger, A.H. 1986a. *op. cit.*, p. 230, note 4.
20. *Quoted in* Deiss, W.A. 1980. *op. cit.*, p. 642.
21. *Quoted in* Cutright, P.R. and Brodhead, M.J. 1981. *op. cit.*, p. 107.
22. *Quoted in* Zwinger, A.H. 1986b. *op. cit.*, p. 30.
23. *Quoted in* Cutright, P.R. and Brodhead, M.J. 1981 *op. cit.*, p. 58.
24. *Quoted in* Dall, W.H. 1915. *op. cit.*, pp. 297–298.
25. *Quoted in* Ridgway, R. 1888. *op. cit.*, pp. 5–6.
26. *Quoted in* Cutright, P.R. and Brodhead, M.J. 1981. *op. cit.*, p. 285.

Ballieu

Biographical information from service records supplied by the Chef de la Division Historique, Ministère des Affaires Étrangères, Paris.
1. *Bulletin de la Muséum d'Histoire Naturelle, Paris* 1909, p. 562. Other catalogues and labels at the museum usually spell the name Bailleu and once Baillieu (C. Jouanin *in litt.*). The French Consular Service gives the spelling Ballieu.
2. Wilson, S.B. and Evans, A.H. 1890–99. *Aves Hawaiienses. The Birds of the Sandwich Islands*. Unpaged, see text for *Loxioides bailleui*. R.H. Porter, London.
3. C. Jouanin *in litt.*, 30 September 1991.

Barrow

Barrow, J. 1801. *An Account of Travels into the Interior of Southern Africa in the years 1797 and 1798*, Vol. I. T. Cadell & W. Davies, London.
Barrow, J. 1847. *An Autobiographical Memoir of Sir John Barrow, Bart., Late of the Admiralty*. J. Murray, London.
Dictionary of National Biography 1885–1900. Vol. III, pp. 305–307. London.
Lloyd, C. 1970. *Mr. Barrow of the Admiralty: A Life of Sir John Barrow 1764-1848*. Collins, London.
1. Ross, J.C. 1847. *A Voyage of Discovery and Research in the Southern and Antarctic Regions, during the years 1839–43*, Vol. I, p. 187. J. Murray, London.

2. Richardson, J. and Swainson, W. 1832. *Fauna Boreali-Americana*, Vol. II, pp. 456–457. J. Murray, London.

3. *Quoted in* Lloyd, C. 1970. *op. cit.*, p. 90.

Bartram

Bartram, W. 1988. *Travels* [Introduction by J. Dickey]. Penguin Books Ltd., New York.

Dictionary of American Biography 1928–36, Vol. II, pp. 26–28 [J. Bartram], 28-29 [W. Bartram].

Earnest, E. 1940. *John and William Bartram*. University of Pennsylvania Press, Philadelphia.

Ewan, J. 1968. *William Bartram. Botanical and Zoological Drawings, 1756–1788. Memoirs* of the American Philosophical Society, Vol. 74. Philadelphia.

Hunter, C. (Ed.) 1983. *The Life and Letters of Alexander Wilson. Memoirs* of the American Philosophical Society, Vol. 154. Philadelphia.

Mearns, B. and Mearns, R. 1988. *Biographies for Birdwatchers. The Lives of Those Commemorated in Western Palearctic Bird Names*, pp. 50–53, 435. Academic Press, London.

Stone, W. 1913. Bird Migration Records of William Bartram, 1802–1822. *Auk* **30**: 324–358.

Wennerstrom, J. 1990. A Founding Father. An appraisal of Alexander Wilson: early American ornithologist. *Birders World* **4**: No. 4—August, pp. 36–39.

1. *Quoted in* Earnest, E. 1940. *op. cit.*, p. 175 [Written by William Dunlop, the painter and playwright who visited Bartram in 1797].

2,3. *Quoted in* Hunter, C. 1983. *op. cit.*, pp. 261–262, 317.

Bell

Audubon, M.R. (Ed.) 1960. *Audubon and His Journals*, 2 vols. Dover Publications Inc., New York.

Chapman, F.M. 1890. John G. Bell. *Auk* **7**: 98–99.

Chapman, F.M. 1933. *Autobiography of a Bird-Lover*, pp. 29–31. D. Appleton-Century, New York.

Cutright, P.R. 1956. *Theodore Roosevelt the Naturalist*, p. 7. Harper and Brothers, New York.

McDermott, J.F. (Ed.) 1951. *Up the Missouri with Audubon. The Journal of Edward Harris*. University of Oklahoma Press, Norman.

1. Audubon, M.R. (Ed.) 1960. *op. cit.*, Vol. 2, p. 137.

2. *Proceedings of the Academy of Natural Sciences of Philadelphia* 1850, **5**: 105.

3. Chapman, F.M. 1933. *op. cit.*, pp. 30–31.

Bendire

Bendire, C.E. 1875. Notes on seventy-nine species of Birds observed in the neighborhood of Camp Harney, Oregon, compiled from the correspondence of Capt. Charles Bendire. *Proceedings of the Boston Society of Natural History* **18**: 153–168.

Bendire, C.E. 1877. Notes on some of the Birds found in Southeastern Oregon, particularly in the Vicinity of Camp Harney, from November, 1874, to January, 1877. *Proceedings of the Boston Society of Natural History* **19**: 109–149.

Hume, E.E. 1942. Chapter II: Charles Emil Bendire (1836–1897). *Ornithologists of the U.S. Army Medical Corps*, pp. 22–37. John Hopkins University Press, Baltimore.

Knowlton, F.H. 1897. Major Charles E. Bendire. *Osprey* **1**: 87–90.

Merrill, J.C. 1898. In Memoriam: Charles Emil Bendire. *Auk* **15**: 1–6.

1. Bendire, C.E. 1892. *Life Histories of North American Birds*, Vol. 1, p. 232. Washington, D.C.
2. Bendire, C.E. 1875. *op. cit.*, p. 159.
3. Bendire, C.E. 1891. Instructions for collecting, preparing and preserving birds' eggs and nests, p. 3. *Bulletin 29, United States National Museum*, Pt. D. Washington, D.C.
4. *Quoted in* Bent, A.C. 1942. *Life Histories of North American Flycatchers, Larks, Swallows, and Their Allies*, p. 270. *Bulletin 179, United States National Museum*. Washington, D.C.
5. *Quoted in* Hume, E.E. 1942. *op. cit.*, p. 26.

Bewick

Bain, I. 1979. *Thomas Bewick. An illustrated record of his life and work*. The Laing Gallery, Tyne and Wear County Council Museums, Newcastle upon Tyne.

Bain, I. (Ed.) 1981. *Thomas Bewick. My Life.* Folio Society, London.

Dictionary of National Biography 1885–1900. Vol. IV, pp. 452–460. London.

Goddard, T.R. [1929] Early History of the Museum Collections. *A History of the Natural History Society of Northumberland, Durham and Newcastle upon Tyne 1829–1929*, pp. 12–26. A. Reid & Co. Ltd, Newcastle upon Tyne.

1. Audubon, J.J. 1831. *Ornithological Biography*, Vol. 1, p. 96. Edinburgh.
2. *Quoted in* Bain, I. 1981. *op. cit.*, pp. 20–22.

Bishop

Kent, H.W. 1965. *Charles Reed Bishop. Man of Hawaii*. Pacific Books, Palo Alto.

Nellist, G.F. (Ed.) 1925. *The Story of Hawaii and its Builders*, pp. 63–65. Honolulu Star–Bulletin Ltd., Honolulu.

Peterson, B.P. (Ed.) 1984. *Notable Women of Hawaii*, pp. 43–46 [Bernice P. Bishop]. University of Hawaii Press, Honolulu.

1. *Quoted in* Rothschild, W. 1893–1900. *The Avifauna of Laysan and the Neighbouring Islands*, p. 225. R.H. Porter, London.

Blackburne

Gentleman's Magazine 1787, **57**: 204 [John Blackburne].
Gentleman's Magazine 1794, **64**: 180 [Anna Blackburne].

Wystrach, V.P. 1974. A note on the naming of the Blackburnian warbler. *Journal of the Society for the Bibliography of Natural History* 7: 89–91.

Wystrach, V.P. 1975. Ashton Blackburne's place in American ornithology. *Auk* **92**: 607–610.

Wystrach, V.P. 1977. Anna Blackburne (1726–1793)—a neglected patroness of natural history. *Journal of the Society for the Bibliography of Natural History* 8: 148–168.

Additional material supplied by the Central Library, Warrington.

1. Swainson, W. 1840. *Lardner's Cabinet Cyclopaedia of Natural History, Vol.XI; Taxidermy: with the Biography of Zoologists, and notices of their works*, p. 245. London.
2. Sources differ on the number and order of the children. Wystrach, V.P. 1977. *op. cit.*, adds to the muddle by calling Anna *and* Ashton the fifth child, p. 150 and p. 165, note 8.
3. Pennant, T. 1784. *Arctic Zoology*, Vol. 1, unpaged, see "Advertisement." London.
4. Pennant, T. 1785. *Arctic Zoology*, Vol. 2, pp. 324–326. London.
5. Pennant, T. 1785. *Arctic Zoology*, Vol. 2, p. 412. London.
6. Wystrach (1974) outlines the great confusion that exists in the literature concerning the naming of the Blackburnian Warbler. He points out that Pennant was the first to name it but does not explain that there was no accompanying dedication and that it is impossible to know exactly for whom the honour was intended.
7. Wystrach, V.P. 1975. *op. cit.*, p. 609.
8. *Quoted in* Wystrach, V.P. 1975. *op. cit.*, p. 608.
9. Rylands, J.P. 1881. *Hale Hall, with notes on the family of Ireland Blackburne*, facing p. 34. Printed for private circulation, Liverpool.
10. *Gentleman's Magazine* 1794, **64**: 180.
11. Fairley, R. 1989. *Jemima—The Paintings and Memoirs of a Victorian Lady*. Canongate Publishing Ltd., Edinburgh.

C.L. Bonaparte

Brodhead, M.J. 1978. The Work of Charles Lucien Bonaparte in America. *Proceedings of the American Philosophical Society* **122**: 198–203.

Emerson, W.O. 1905. A Manuscript of Charles Lucien Bonaparte. *Condor* 7: 46–47.

Ford, A. 1965. *John James Audubon*. University of Oklahoma Press, Norman.

Mearns, B. and Mearns, R. 1988. *Biographies for Birdwatchers. The Lives of Those Commemorated in Western Palearctic Bird Names*, pp. 75–82, 436–437. Academic Press, London.

Müller, J.W. 1852. Biographische Notiz über Carl Lucian Bonaparte. *Naumannia* **2**: 90–94.

Proceedings of the Linnean Society of London 1859, **3**: 41–44. Obituary.

Stresemann, E. 1975. Chapter 9: Charles Lucien Bonaparte. *Ornithology, from Aristotle to the Present*, pp. 153–169. Harvard University Press, Cambridge.

1. *Quoted in* Burns, F.L. 1917. Miss Lawson's Recollections of Ornithologists. *Auk* **34**: 281.
2. *Quoted in* Brodhead, M.J. 1978. *op. cit.*, p. 202.
3,4. *Quoted in* Ford, A. 1965. *op. cit.*, pp. 351, 355.
5. *Quoted in* Brodhead, M.J. 1978. *op. cit.*, p. 202.
6. It is stated in the *New Catholic Encyclopedia* (1967) and elsewhere that Charles Bonaparte became director of the Jardin des Plantes in 1854. In an unpublished

letter to Sir Hugh Gladstone of Capenoch, in 1946, Le Directeur des Archives de France denied that Bonaparte had ever held the post, though he did apply for it.

7,8. *Quoted in* Stresemann, E. 1975. *op. cit.*, pp. 167, 168.

9. Stresemann, E. 1975. *op. cit.*, p. 169.

Z. Bonaparte

Aronson, T. 1964. *The Golden Bees.* Oldbourne, Book Co. Ltd., London.

Charles-Roux, F. 1952. *Rome, asile des Bonaparte*, pp. 198–201. Hachette, Paris.

Mailhol, D. de. 1896. *Dictionaire Historique et Héraldique de la Noblesse Française*, pp. 215–218. Paris.

1. Mailhol and some other sources give her names as Zénaïde Charlotte Julie but according to the Musée National des Châteaux de Malmaison et de Bois-Préau Zénaïde Laetitia Julie is correct.

Botteri

Anon. 1883. *Apuntes Biográficos del Sr. D. Mateo Botteri.* 21 pp. México.

Bonplandia 1853, **1**: 143.

Crosse, H. and Fischer, P. 1878. M. Mateo Botteri. *Journal of Conchyliology* **18**: 112.

Diccionario Porrúa. Historio, Biografía y Geografía de México 1986, 5th edn., Vol. 1, p. 388. México D.F.

Sclater, P.L. 1857. On a Collection of Birds made by Signor Matteo Botteri in the vicinity of Orizaba in Southern Mexico. *Proceedings of the Zoological Society of London*, pp. 210–215.

Brandt

Gebhardt, L. 1964. *Die Ornithologen Mitteleuropas*, p.48. Giessen.

Løppenthin, B. 1984. *Johann Friedrich von Brandt. Icones Avium Rossico—Americanarum Tabulae VII, Ineditae. With Comments on Birds, Expeditions and People Involved.* Scandinavian Fine Editions, Copenhagen.

Pearse, T. 1968. Chapters I–V: Russian Explorers. *Birds of the Early Explorers in the Northern Pacific*, pp. 28–126. T. Pearse, Comox, British Columbia.

Quäbicker, G. 1942. Joh. Friedrich von Brandt (1802–1879). *Journal für Ornithologie* **90**: 324–325.

1. Løppenthin, B. 1984. *op. cit.*, gives 1838 as the year that Brandt's Cormorant was described. Cf. 1837 in 1983 A.O.U. *Check-list.*

2. Since 1972 Brandt's Cormorant has bred in small numbers in Alaska's Prince William Sound, well north of its usual breeding and wintering areas. If it occurred there two hundred years ago the original specimen could have been collected by a wider range of Russian explorers, seamen, traders and merchants.

3. Günther, A. (Ed.) 1904–12. *The History of the Collections Contained in the Natural History Departments of the British Museum*, Vol. II, p. 318. BM publication, London.

Brelay

La Fresnaye, F. de. 1839. Quelques Oiseaux Nouveaux de la collection de M. Charles Brelay, à Bordeaux. *Revue Zoologique* **2**: 97–100.
Penard, T.E. 1945. Lafresnaye. *Auk* **62**: 227–233.
1. *Translated from* La Fresnaye, F. de. 1839. *op. cit.*, p. 98, note 1.
[Our recent enquiries at Bordeaux have not produced any biographical information: there is no reference to Charles or Aglaé Brelay in the Archives of the Bibliothèque Municipale; there is no mention of them in local biographical lists nor in the *Actes de la Société Linnéenne de Bordeaux* 1830–1842; and the present curator of the Bordeaux natural history museum has never come across any of their specimens.]

Brewer

Bulletin of the Nuttall Ornithological Club 1880, pp. 102–104.
Dictionary of American Biography 1928–36. Vol. III, pp. 24–25. New York.
Proceedings of the Boston Society of Natural History 1881, pp. 411–415.
Wood, C.A. 1923. A Letter from T.M. Brewer to Osbert Salvin. *Condor* **25**: 100–101.
1,2. *Quoted in* Audubon, M.R. (Ed.) 1960. *Audubon and His Journals*, Vol. 2, pp. 48–49; Vol. 1, p. 73. Dover Publications Inc., New York.
3. *Quoted in* Dall, W.H. 1915. *Spencer Fullerton Baird. A Biography*, p. 111. J.B. Lippincott Co., Philadelphia and London.
4. *Quoted in* Wollaston, A.F.R. 1921. *A Life of Alfred Newton*, p. 25. J. Murray, London.
5. Brewer, T.M. 1852. A few Ornithological Facts, gathered in a hasty Trip through portions of New Brunswick and Nova Scotia, in June 1850. *Boston Journal of Natural History* **6**: 297–308.
6. *Proceedings of the Academy of Natural Sciences of Philadelphia* 1867, **19**: 233.
7. Allen, J.A. 1880. Report on the Mammals and Birds, p. 19. *Annual Report of the Curator of the Museum of Comparative Zoölogy at Harvard College ... for 1879–80.* University Press, Cambridge.
8. Henshaw, H.W. 1919. Autobiographical Notes. *Condor* **21**: 171.
9. *Dictionary of American Biography, op. cit.*, p. 25.
10. For Coues's quotes and summary of the 'Sparrow War' *see* Cutright, P.R. and Brodhead, M.J. 1981. *Elliott Coues. Naturalist and Frontier Historian*, pp. 201–205. University of Illinois Press, Chicago.

Buller

Buller, W.L. 1887–88. *A History of the Birds of New Zealand*, 2nd edn, 2 vols. Published by the author, London.
Buller, W.L. 1905–06. *Supplement to the Birds of New Zealand*, 2 vols. Published by the author, London.
Dictionary of New Zealand Biography 1940. Vol. I, p. 119. Wellington.
Oliver, W.R.B. 1955. History of Ornithological Discovery in New Zealand. *New Zealand Birds*, pp. 18–25. Wellington, New Zealand.

Turbott, E.G. (Ed.) 1967. Editor's Introduction. *Buller's Birds of New Zealand*, pp. xi–xviii. Macdonald, London.
1. Bent, A.C. 1922. *Life Histories of North American Petrels and Pelicans and Their Allies*, p. 101. *Bulletin 121, United States National Museum*. Washington, D.C.
2. Buller, W.L. 1887–88. *op. cit.*, Vol. 2, p. 240.

Bulwer

Gladstone, H.S. 1946. British Birds named after Persons: Bulwer's Petrel. *Transactions and Journal of Proceedings of the Dumfriesshire and Galloway Natural History and Antiquarian Society 1940–44*. Ser. 3, **23**: 178–184, 188–189.
[Lyall, A.] 1827. *Rambles in Madeira and Portugal*. Dublin.
1. *Ibis* 1886, pp. 531–532. [Sale of the Jardine Ornithological Collection.]

Cassin

Brewer, T.M. 1869. [Obituary. J.Cassin] *Proceedings of the Boston Society of Natural History*, pp. 244–248.
Dictionary of American Biography 1928–36. Vol.III, pp. 568–569. New York.
Shelton, F.H. 1918. The Birthplace of John Cassin. *Cassinia* **22**: 3–5.
Stone, W. 1901. John Cassin. *Cassinia* **5**: 1–7.
Stone, W. 1918. Cassinia. *Cassinia* **22**: 1–2.
1. Sibley, C.G. and Munroe, B.L. 1990. *Distribution and Taxonomy of Birds of the World*, p. 454. [Described by Xántus 1858 not 1866]. Yale University Press, New Haven and London.
2. *Quoted in* Stone, W. 1901. *op. cit.*, p. 4.
3. *Proceedings of the Academy of Natural Sciences of Philadelphia* 1881, **5**: 153.

J.H. Clark

Dall, W.H. 1915. *Spencer Fullerton Baird. A Biography*, pp. 170–171, 260, 266–268, 276–277, 306–307, 332–333. J.B. Lippincott Co., Philadelphia and London.
Ewan, J. and Ewan, N.D. 1981. *Biographical Dictionary of Rocky Mountain Naturalists, 1682–1932*, p. 42. Frans A. Stafleu, Utrecht.
Geiser, S.W. 1948. *Naturalists of the Frontier*, 2nd revised edn, p. 272. Southern Methodist University Press, Dallas.
Geiser, S.W. 1958. Men of Science in Texas, 1820–1880. *Field and Laboratory* **26**: 113.
1. *Quoted in* Dall, W.H. 1915. *op. cit.*, p. 266.
2. Henshaw, H.W. 1919. Autobiographical Notes. *Condor* **21**: 217–218.
3,4. *Quoted in* Dall, W.H. 1915. *op. cit.*, 276–277, 307.
5. Herber, E.C. 1963. *Correspondence between Spencer Fullerton Baird and Louis Agassiz*, pp. 64–65. Smithsonian Institution, Washington, D.C.
6. *Quoted in* Dall, W.H. 1915. *op. cit.*, p. 333.
7. Bendire, C.E. 1892–95. *Life Histories of North American Birds*, Vol. 1, p. 174. Washington, D.C.

8. Smithsonian Institution Archives, Record Unit 7002. Spencer F. Baird Collection, 1793–1923.
9. Dall, W.H. 1915. *op. cit.*, p. 171 states that Clark was a native of Anne Arundel, Maryland, but there is no J.H. Clark listed in the Church Records Index to Births and Baptisms 1663–1967, the Church Records Index to Deaths and Burials 1663–1967, the Birth Records Index 1804–1877 or the Anne Arundel County Index to Inventories 1891–1900. Herber, E.C. 1963. *op. cit.*, p. 64 claims that Clark was a native of Virginia but the state library there has no record of him.

    In the 1890s there was a J.H. Clark living in Paterson, New Jersey, who was an ornithologist and had travelled in Mexico but this was Josiah Huntoon Clark (1873–1928) known particularly as an oologist.
10. *Quoted in* Dall, W.H. 1915. *op. cit.*, p. 276.

W. Clark

Cutright, P.R. 1969. *Lewis and Clark: Pioneering Naturalists.* University of Nebraska Press, Lincoln and London.
Dictionary of American Biography 1928–36. Vol. IV, pp. 141–144. New York.
Encyclopedia Americana 1978, Vol. 17, pp. 275–277 [Lewis and Clark Expedition]. Americana Corporation, Danbury.
Thwaites, R.G. 1959. *Original Journals of the Lewis and Clark Expedition*, Vols. 1–8. Antiquarian Press, New York.
1. Thwaites, R.G. 1959. *op. cit.*, Vol. 3, p. 17.

Cook

Pearse, T. 1968. Chapter VII: English Explorers [Cook, Ellis, Webber, Anderson, Clerke]. *Birds of the Early Explorers in the Northern Pacific*, pp. 151–212. Theed Pearse, Comox, British Columbia.
Stresemann, E. 1949. Birds collected in the North Pacific area during Capt. James Cook's last voyage (1778 and 1779). *Ibis* 91: 244–255.
Stresemann, E. 1950. Birds collected during Capt. James Cook's last expedition (1776–1780). *Auk* 67: 66–88.
1. The original description does not identify Cook. We have assumed, as others have done, that G.R. Gray intended to honour Captain James Cook.

Cooper

Baatz, S. 1990. Knowledge, Culture, and Science in the Metropolis: The New York Academy of Sciences 1817–1970. 1. Science in the Early Republic, 1817–1844. *Annals of the New York Academy of Sciences* 584: 9–56.
Emerson, W.O. 1899. Dr. James G. Cooper. A Sketch. *Bulletin of the Cooper Ornithological Club* 1: 1–5.
Fairchild, H.L. 1887. William Cooper. *A History of the New York Academy of Sciences*, pp. 70–73. New York.

Ford, A. 1965. *John James Audubon.* University of Oklahoma Press, Norman.
1. Cox, J.B. 1990. The Enigmatic Cooper's and Cox's Sandpipers. *Dutch Birding* **12**: 53–64.

Cory

Davis, W.E. 1987. Early Years, 1873–1919. *History of the Nuttall Ornithological Club 1873–1986*, pp. 9–30. *Memoirs of the Nuttall Ornithological Club No. 11.* Cambridge, Massachusetts.
Osgood, W.H. 1922. Charles Barney Cory. *Auk* **39**: 151–166.
1. *Quoted in* Cutright, P.R. and Brodhead, M.J. 1981. *Elliott Coues. Naturalist and Frontier Historian*, p. 267. University of Illinois Press, Chicago.
2. Chapman, F.M. 1933. *Autobiography of a Bird-Lover*, p. 53. D. Appleton-Century, New York.
3. Osgood, W.H. 1922. *op. cit.*, p. 159.
4. *Quoted in* Hanley, W. 1977. *Natural History in America*, p. 274. Quadrangle/The New York Times Book Co., New York.

Costa

Dictionaire de Biographie Française 1933– . Vol. IX. Costa de Beauregard No. 13, column 788. Paris.
Fleming, J.H. 1919. The Costa Collection of Birds. *Condor* **21**: 39.
Kofoid, C.A. 1923. A Little Known Ornithological Journal and its Editor, Adolphe Boucard, 1839–1904. *Condor* **25**: 85–89.
Palmer, T.S. 1918. Costa's Hummingbird—Its Type Locality, Early History and Name. *Condor* **20**: 114–116.
1. When Bourcier first named the hummingbird he did not identify Costa but he did so a year later (*Annales des Sciences Physiques et Naturelles, d'Agriculture etc. de Lyon* 1840, **3**: 225–226). F. Prévost and O. Des Murs erroneously claimed that the bird was named for "M. Costa, directeur du Musée de Naples." (*Oiseaux*, p. 194. *in* du Petit-Thouars, A.A. 1855. *Voyage autour du monde sur la frégate La Vénus.* Paris.) This Costa was Oronzio Gabrielle Costa (1787-1867), author of the *Fauna di Regno di Napoli* (1857).
2. Kofoid, C.A. 1923. *op. cit.*, p. 88.

Couch

Brown, J.H. (Ed.) 1900. *Lamb's Biographical Dictionary of the United States*, Vol. II, pp. 207–208. Boston.
Dall, W.H. 1915. *Spencer Fullerton Baird. A Biography*, pp. 294–297, 299, 321–322. J.B. Lippincott Co., Philadelphia and London.
Dictionary of American Biography 1928–36. Vol. IV, pp. 463–464. New York.
Faust, P.L. (Ed.) 1986. *Encyclopedia of the Civil War*, p. 187. Harper & Row, New York.

Geiser, S.W. 1948. *Naturalists of the Frontier*, 2nd revised edn, pp. 53–54, 272. Southern Methodist University Press, Dallas.
Geiser, S.W. 1958. Men of Science in Texas, 1820–1880. *Field and Laboratory* **26**: 117.
National Cyclopaedia of American Biography 1909, Vol. IV, p. 207. New York.
1. *Quoted in* Cassin, J. 1853–56. *Illustrations of the Birds of California, Texas, Oregon, British and Russian America*, p. 138. J.B. Lippincott, Philadelphia.
2,3,4. *Quoted in* Dall, W.H. 1915. *op. cit.*, 294–295, 322, 295.
5. Couch, D.N. 1854. Descriptions of new Birds of Northern Mexico. *Proceedings of the Academy of Natural Sciences of Philadelphia* **7**: 66–67.
6. *Quoted in* Baird, S.F. 1859. Birds of the Boundary, p. 8. *in* Emory, W.H. *Report on the U.S. and Mexican Boundary Survey*. Washington, D.C.

Coues

Boston Transcript 16 June 1909. [Dana Estes Dead].
Boston Evening Transcript 28 December 1925. [Grace D. Coues Estes Dead]
Coues, E. 1865. Ornithology of a Prairie-Journey, and Notes on the Birds of Arizona. *Ibis*, pp. 157–165.
Cutright, P.R. and Brodhead, M.J. 1981. *Elliott Coues. Naturalist and Frontier Historian*. University of Illinois Press, Chicago.
Dictionary of American Biography 1928–36. Vol. VI, pp. 188–189 [D.Estes]. New York.
Kilgour, R.L. 1957. *Estes and Lauriat. A History 1872–1898*, pp. 203–211. University of Michigan Press, Ann Arbor.
Additional information supplied by Professor M.J. Brodhead, University of Nevada-Reno, who possesses a transcript of Elliott Coues's 'Book of Dates'.
1,2. *Quoted in* Cutright, P.R. and Brodhead, M.J. 1981. *op. cit.*, pp. 59, 65.
3. *Quoted in* Bent, A.C. 1953. *Life Histories of North American Wood Warblers*, p. 363. *Bulletin 203, United States National Museum*, Washington, D.C.

Craveri

Denza, P.F. 1890. Federico Craveri. *Società Meteorologica Italiana, Bollettino Mensuale, Osservatorio Cent., Turino* (2nd ser.) **10**: 81–82.
Molinaro, E. 1980. Il Museo Civico Craveri di Bra di storia naturale. *Cassa di Risparmio di Bra*, pp. 5–45.
Molinaro, E. 1990. Celebrazioni centanarie di Federico Craveri. *Cassa di Risparmio di Bra* **6**: 1 Giugno pp. 6–9, 2 Dicembre pp. 27–30.
Palmer, T.S. 1928. Notes on Persons Whose Names Appear in the Nomenclature of California Birds. *Condor* **30**: 275.

Dole

Amadon, D. 1944. Sanford Ballard Dole: Early Hawaiian Ornithologist. *The Elepaio* **5**: 12–13.
Dictionary of American Biography 1928–1936, Vol. V, pp. 358–359. New York.

Nellist, G.F. (Ed.) 1925. Sanford Ballard Dole. *The Story of Hawaii and its Builders*, pp. 405–407. Honolulu Star-Bulletin Ltd, Honolulu.
Nellist, G.F. (Ed.) 1929. Anna Prentice Cate Dole. *Women of Hawaii*, pp. 94–95. E.A. Langton-Boyle, Honolulu.
Palmer, T.S. 1927. Sanford Ballard Dole. *Auk* **44**: 160–161.

Fischer

Gebhardt, L. 1964. *Die Ornithologen Mitteleuropas*, p. 95. Giessen.
Neue Deutsche Biographie 1952– . Vol. V, pp. 212–213. Berlin.

Forster

Dictionary of National Biography 1885–1900. Vol. XX, pp. 15–16. London.
Hoare, M.E. 1976. *The Tactless Philosopher. Johann Reinhold Forster (1729–98)*. Hawthorn Press, Melbourne.
Lysaght, A.M. 1971. *Joseph Banks in Newfoundland and Labrador, 1766. His Diary, Manuscripts and Collections*, pp. 93–97, 251, 397–400, 407–20. Faber and Faber, London.
Mearns, B. and Mearns, R. 1988. *Biographies for Birdwatchers. The Lives of Those Commemorated in Western Palearctic Bird Names*, pp. 154–159, 441. Academic Press, London.
1. Houston, C.S. 1983. Birds First Described from Hudson Bay. *Canadian Field-Naturalist* **97**: 95–98.
2. *Philosophical Transactions of the Royal Society of London* 1772, Vol. LXII, Article XXIX, pp. 421–422.

Franklin

Dictionary of National Biography 1885–1900. Vol. XX, pp. 191–196. London.
Houston, C.S. (Ed.) 1984. *Arctic Ordeal. The Journal of John Richardson, surgeon-naturalist with Franklin, 1820–1822.* McGill-Queen's University Press, Kingston and Montreal.
Houston, C.S. 1988. John Richardson: Deserving of Greater Recognition. *Canadian Field-Naturalist* **102**: 558–563.
Mearns, B. and Mearns, R. 1988. *Biographies for Birdwatchers. The Lives of Those Commemorated in Western Palearctic Bird Names*, pp. 160–167, 441. Academic Press, London.
1. Stresemann, E. 1954. Ferdinand Deppe's travels in Mexico, 1824–1829. *Condor* **56**: 86–92 (see pp. 91–92).
2. For more detailed account of the birds see Houston, C.S. 1984. *op. cit.*, Appendix A, pp. 223–244 [Bird observations by John Richardson during the first Franklin expedition].
3. *Quoted in* Hall, F.S. 1934. Studies in the History of Ornithology in the State of Washington. *Murrelet* **15**: 12.

4. *Transactions of the Linnean Society of London* 1829, **16**: 139.
5. Hodson, V.C.P. (Ed.) 1947. *List of the Officers of the Bengal Army, 1758–1834*, Part II, p. 214. London; Kinnear, N.B. 1952. The History of Indian Mammalogy and Ornithology, Part II: Birds. *Journal of the Bombay Natural History Society* **51**: 106.

Gambel

Beidleman, R.G. 1958. William Gambel. Frontier Naturalist. *Pacific Discovery* **11**: 10–14.
Campbell, H. 1978. William Gambel, Naturalist. *New Mexico Wildlife* Nov–Dec 1978, pp. 3–9.
Ewan, J. and Ewan, N.D. 1981. *Biographical Dictionary of Rocky Mountain Naturalists, 1682–1932*, pp. 81–82. Frans A. Stafleu, Utrecht.
Graustein, J.E. 1967. *Thomas Nuttall, Naturalist*, pp. 342–351, 354, 358–359, 366–367 *et passim*. Harvard University Press, Cambridge.
[Grinnell, J.] 1939. Notes and News. [Gambel-Wistar] *Condor* **41**: 129.
Harshberger, J.W. 1899. *The Botanists of Philadelphia, and their Work*, pp. 231–233. Philadelphia.
Pacific Coast Avifauna 1909, **5**: 10–11. [Bibliography of California Ornithology]
Palmer, T.S. 1928. Notes on Persons Whose Names Appear in the Nomenclature of California Birds. *Condor* **30**: 278.
Stone, W. 1910. William Gambel, M.D. *Cassinia* **14**: 1–8.
Stone, W. 1916. Philadelphia to the coast in early days, and the development of western ornithology prior to 1850. *Condor* **18**: 11–12.
Woods, D.B. 1851. Obituary. Dr. William Gambel. *American Journal of Science* (Ser. 2.) **11**: 143–144.
1. *Quoted in* Graustein, J.E. 1967. *op. cit.*, p. 349.
2. *Quoted in* Beidleman, R.G. 1958. *op. cit.*, p. 11.
3. *Proceedings of the Academy of Natural Sciences of Philadelphia* 1843, **1**: 260–261.
4. Additional details on the Mountain Chickadee are from later Gambel papers.
5,6. *Quoted in* Stone, W. 1910. *op. cit.*, pp. 2, 3.
7,8. *Proceedings of the Academy of Natural Sciences of Philadelphia* April 1846, pp. 47–48; April 1847, pp. 202–203.
9. *Quoted in* [Grinnell, J.] 1939. *op. cit.*, p. 129.
10. Woods, D.B. 1851. *op. cit.*, p. 144.
11. *Quoted in* Harshberger, J.W. 1899. *op. cit.*, p. 233.

Gray

Dictionary of National Biography 1885–1900. Vol. XXIII, pp. 7–11. London.
Günther, A. (Ed.) 1904–12. *The History of the Collections Contained in the Natural History Departments of the British Museum*, Vol. II, pp. 79–88 [General sketch of the history of the bird collections by R. Bowdler Sharpe]. BM publication, London.
Ibis 1872, pp. 340–342. Obituary.
Mearns, B. and Mearns, R. 1988. *Biographies for Birdwatchers. The Lives of Those Commemorated in Western Palearctic Bird Names*, pp. 173–178, 442 [J.E. and G.R. Gray]. Academic Press, London.
1. *Quoted in* Günther, A. (Ed.) 1904–12. *op. cit.*, p. 83.

Gundlach

Barbour, T. 1923. [Historical. Ornithological Collections in Cuba.] *The Birds of Cuba*, pp. 6–9. *Memoirs of the Nuttall Ornithological Club No. 6.* Cambridge, Massachusetts.
Cory, C.B. 1892. In Cuba with Dr. Gundlach. *Auk* **9**: 271–273.
Gebhardt, L. 1964. *Die Ornithologen Mitteleuropas*, pp. 128–129. Giessen.
Ramsden, C.T. 1915. Juan Gundlach. *Entomological News* **26**: 241–260.

Hammond

Dictionary of American Biography 1928–1936. Vol. VIII, pp. 210–211. New York.
Hume, E.E. 1942. Chapter XII. William Alexander Hammond (1828–1900). *Ornithologists of the U.S. Army Medical Corps*, pp. 176–189. John Hopkins University Press, Baltimore.

Harris

Audubon, M.R. (Ed.) 1960. *Audubon and His Journals*, 2 vols. Dover Publications Inc., New York.
Brannon, P.A. 1947. *Edward Harris, Friend of Audubon*. Birmingham Publishing Company for The Newcomen Society of England—American Branch. 36 pp.
Ford, A. 1965. *John James Audubon*. University of Oklahoma Press, Norman.
McDermott, J.F. (Ed.) 1951. *Up the Missouri with Audubon. The Journal of Edward Harris*. University of Oklahoma Press, Norman.
Morris, G.S. 1902. Edward Harris. *Cassinia* **6**: 1–5.
Street, P.B. 1948. The Edward Harris Collection of Birds. *Wilson Bulletin* **60**: 167–184.
1. *Quoted in* Brannon, P.A. 1947. *op. cit.*, p. 12.
2. *Quoted in* McDermott, J.F. (Ed.) 1951. *op. cit.*, p. 54.
3. *Quoted in* Ford, A. 1965. *op. cit.*, p. 411.
4. *Quoted in* Bachman, C.L. (Ed.) 1888. *John Bachman D.D., L.L.D., Ph.D.*, pp. 276–278. Walker, Evans and Cogswell Co., Charleston, S.C.

Heermann

Ewan, J. and Ewan, N.D. 1982. *Biographical Dictionary of Rocky Mountain Naturalists, 1682–1932*, p. 100. Frans A. Stafleu, Utrecht.
Field, M.C. 1957. *Prairie and Mountain Sketches*. [Edited by Gregg, K.L. and McDermott, J.F.]. University of Oklahoma Press, Norman.
Gieser, S.W. 1959. Men of Science in Texas, 1820–1880. *Field and Laboratory* **27**: 37–38.
Heermann, A.L. 1853. Notes on the Birds of California, observed during a residence of three years in that country. *Journal of the Academy of Natural Sciences of Philadelphia*, 2nd series, **2**: 259–272.
Hume, E.E. 1942. Chapter XIII. Adolphus Lewis Heermann (1827–1865). *Ornithologists*

of the U.S. Army Medical Corps, pp. 190–205. John Hopkins University Press, Baltimore.

Porter, M.R. and Davenport, O. 1963. *Scotsman in Buckskin. Sir William Drummond Stewart and The Rocky Mountain Fur Trade*. Hastings House, Publishers, New York.

Stone, W. 1907. Adolphus L. Heermann, M.D. *Cassinia* **11**: 1–6.

1. *Quoted in* Stone, W. 1907. *op. cit.*, pp. 4–5.
2. Stone was puzzled by Dresser's belief that Heermann travelled to the Rocky Mountains with Frémont. There was much coming and going between the two camps and though Heermann surely met Frémont he did not travel with him. When speaking of his journey to the Rockies Heermann no doubt mentioned Frémont and Dresser must have misunderstood the connection.
3. *Quoted in* Field, M.C. 1957. *op. cit.*, p. 53.
4. The report was not published until 1859, but it was written by 14 October 1854, when Heermann sent it with a letter of that date to Lt. J.G. Parke. Heermann must therefore have visited Switzerland some time prior to 1849, since his documented travels between 1849 and 1854 preclude such a trip.
5. Heermann, A.L. 1853. *op. cit.*, p. 268.
6. *Quoted in* Dall, W.H. 1915. *Spencer Fullerton Baird. A Biography*, p. 280. J.B. Lippincott Co., Philadelphia and London.
7. *Quoted in* Harris, H. 1941. The Annals of *Gymnogyps* to 1900. *Condor* **43**: 36.
8,9. Dresser, H.E. 1865. Notes on the Birds of Southern Texas. *Ibis*, pp. 316, 484.

Henslow

Brent, P. 1981. *Charles Darwin. A Man of Enlarged Curiosity*. Heinemann, London.
Dictionary of National Biography 1885–1900. Vol. XXVI, pp. 135–136. London.
Jenyns, L. 1862. *Memoir of the Reverend John Stevens Henslow*. London.

1. Audubon, M.R. (Ed.) 1960. *Audubon and His Journals*, Vol. 1, p. 290. Dover Publications Inc., New York.
2. Brent, P. 1981. *op. cit.*, pp. 89–90.

Hornemann

Gladstone, H.S. Unpublished MS, from: Christensen, C. 1936. Jens Wilken Hornemann. *Dansk biografisk Leksikon*, Vol. X, pp. 599–603. Copenhagen; Ostermann, H. 1936. Carl Peter Holböll. *Dansk biografisk Leksikon*, Vol. X, pp. 368–369. Copenhagen; and other sources.

Hutton

Palmer, T.S. 1928. Notes on Persons Whose Names Appear in the Nomenclature of California Birds. *Condor* **30**: 283.

1,2. *Quoted in* Stone, W. 1916. Philadelphia to the coast in early days, and the development of western ornithology prior to 1850. *Condor* **18**: 12–13, 13.
3. *Proceedings of the Academy of Natural Sciences of Philadelphia* 1851, **5**: 151.

4. *University of California Publication in Zoology* 1932, **38**: 298.
5. *Ibis* 1906, p. 361. [Obituary. F.W. Hutton].

Kennicott

Dall, W.H. 1915. *Spencer Fullerton Baird. A Biography*, pp. 231–232, 307–308, 333–335, 342–352, 357–363, 367–369, 371–378. J.B. Lippincott Co., Philadelphia and London.
Dictionary of American Biography 1928–36. Vol. X, pp. 338–339. New York.
[Stimpson, W.]. 1869. Robert Kennicott. *Transactions of the Chicago Academy of Sciences* 1: 133–226.
Zochert, D. 1980. Notes on a Young Naturalist [Kennicott]. *Audubon Magazine* **82**: 34–47.
1. *Proceedings of the Academy of Natural Sciences of Philadelphia* 1867, 19: 99–100.
2,3,4. *Quoted in* [Stimpson, W.]. 1869. *op. cit.*, pp. 146–150, 170–172, 175.
5,6. Dall, W.H. 1915. *op. cit.*, pp. 345, 359.
7. *Quoted in* [Stimpson, W.]. 1869. *op. cit.*, p. 218.
8. Dall, W.H. 1870. *Alaska and Its Resources*, p. 70. Sampson, Low, Son and Marston, London.
9. Preble, E.A. 1922. Roderick Ross Macfarlane, 1833–1920. *Auk* **39**: 203–210.

Kirtland

Dictionary of American Biography 1928–36. Vol. X, pp. 438–439. New York.
Gehr, A.R. 1952. Jared Potter Kirtland. *The Explorer* (The Cleveland Museum of Natural History) 2: No. 7, pp. 1–33.
Gifford, G.E. 1972. Doctors Afield. Dr. Jared Kirtland and His Warbler. *New England Journal of Medicine* **287**: 909–911.
Mackenzie, J.P.S. 1977. Kirtland's Warbler. *Birds in Peril*, pp. 159–168. Houghton Mifflin Co., Boston.
Mayfield, H.F. 1960. Chapter 2. History. *The Kirtland's Warbler*, pp. 5–8. Cranbrook Institute of Science, Bloomfield Hills, Michigan.
Mayfield, H.F. 1965. *Jared Potter Kirtland—Pioneer Ornithologist of Ohio*, pp. 3–8. Department of Natural History of the Ohio Historical Society, Columbus, Ohio.
National Cyclopaedia of American Biography 1909. Vol. XI, pp. 347–348. New York.
Newberry, J.S. 1886. Biographical Sketch of J.P. Kirtland. *Biographical Memoirs of the National Academy of Sciences*, Vol. 2, pp. 129–138.
1. *Quoted in* Mackenzie, J.P.S. 1977. *op. cit.*, p. 162.
2. *Quoted in* Mayfield, H.F. 1965. *op. cit.*, pp. 7–8.

Kittlitz

Gebhardt, L. 1964. *Die Ornithologen Mitteleuropas*, pp. 182–183. Giessen.
Greenway, J.C. 1967. *Extinct and Vanishing Birds of the World*, 2nd edn, pp. 74–77. Dover Publications Inc., New York.

Kittlitz, F.H. von 1861. *Twenty-four views of the vegetation of the coasts and islands of the Pacific.* Longman, Green, Longman and Roberts, London.

Litke, F. 1987. *A Voyage Around the World 1826–1829.* Volume 1—To Russian America and Siberia. [Supplemented with a parallel account by F.H. von Kittlitz]. Edited by R.A. Pierce. Alaska History No. 29. The Limestone Press, Kingston, Ontario.

Løppenthin, B. 1984. F.H. v. Kittlitz's Voyage on Board the Senyavin. *Johann Friedrich von Brandt. Icones Avium Rossico—Americanarum Tabulae VII, Ineditae. With Comments on Birds, Expeditions and People Involved*, pp. 29–32. Scandinavian Fine Editions, Copenhagen.

Neue Deutsche Biographie 1952– . Vol. XI, pp. 694–695. Berlin.

Pearse, T. 1968. Frederic Lutke. *Birds of the Early Explorers in the Northern Pacific*, pp. 116–123. Theed Pearse, Comox, British Columbia.

1. *Quoted in* Thayer, J.E. 1914. Nesting of the Kittlitz Murrelet. *Condor* **16**: 117–118.
2. Litke, F. 1987. *op. cit.*

Lawrence

Dictionary of American Biography 1928–36. Vol. XI, p. 49. New York.

Elliot, D.G. 1896. In Memoriam: George Newbold Lawrence. *Auk* **13**: 1–10.

Foster, L.S. 1892. The Published Writings of George Newbold Lawrence. [Biographical Sketch, pp. vii–viii]. *Bulletin 40, United States National Museum.* Washington, D.C.

1. Dutcher, W. 1891. The Labrador Duck. *Auk* **8**: 201–216, *see* p. 215.
2,3. *Proceedings of the Academy of Natural Sciences of Philadelphia* 1852 (1850) **5**: 105, 105.
4. Elliot, D.G. 1896. *op. cit.*, p. 7.
5. *Quoted in* Cutright, P.R. and Brodhead, M.J. 1981. *Elliott Coues. Naturalist and Frontier Historian*, p. 33. University of Illinois Press, Chicago.
6. Chapman, F.M. 1933. *Autobiography of a Bird-Lover*, p. 49. D. Appleton-Century, New York.
7. Elliot, D.G. 1896. *op. cit.*, pp. 9–10.

Leach

Dictionary of National Biography 1885–1900. Vol. XXXII, pp. 311–312. London.

Gladstone, H.S. Unpublished MS, from: Mullens, W.H. 1917–18. *Museums Journal* **17**: 51–56, 132–137, 180–187; Mullens, W.H. and Swann, H.K. 1917. *Bibliography of British Ornithology*, pp. 343–344. MacMillan & Co., London; and other sources.

Smith, E. 1980. Some early Nineteenth Century Devonshire Naturalists. *Transactions of the Devonshire Association* **112**: 8–13.

Swainson, W. 1840. *Lardner's Cabinet Cyclopaedia of Natural History, Vol. XI; Taxidermy: with the Biography of Zoologists, and notices of their works*, pp. 237–240. London.

1. Günther, A. (Ed.) 1904–12. *The History of the Collections Contained in the Natural History Departments of the British Museum*, Vol. II, pp. 208–245. BM publication, London.
2. Temminck, C.J. 1820. *Manuel d'Ornithologie*, 2nd edn, Vol. II, pp. 812–813. Paris.

3. *Quoted in* MacGillivray, W. 1837–52. *A History of British Birds*, Vol. V. p. 454. W.S. Orr & Co., London.
4. Swainson, W. 1840. *op. cit.*, p. 239.

LeConte

Dictionary of American Biography 1928–36. Vol. XI, pp. 89–90. New York.
Essig, E.O. 1931. *A History of Entomology*, pp. 680–685. MacMillan, New York.
Ewan, J. and Ewan, N.D. 1981. *Biographical Dictionary of Rocky Mountain Naturalists, 1682–1932*, pp. 130–131. Frans A. Stafleu, Utrecht.
Horn, G.H. 1883. John Lawrence LeConte. *Science* 2: 783–786.
Lesley, J.P. and Horn, G.H. 1884. [Memoirs of John L. LeConte]. *Proceedings of the American Philosophical Society* 21: 290–299.
1. Audubon, M.R. (Ed.) 1960. *Audubon and His Journals*, Vol. 1, p. 510. Dover Publications Inc., New York.
2. Dall, W.H. 1915. *Spencer Fullerton Baird. A Biography*, pp. 95–96. J.B. Lippincott Co., Philadelphia and London.

Lesson

Biographie Universelle, nouvelle édition [1843–65], Vol. XXIV, pp. 330–331 [R.P. Lesson]. Paris.
Bonnemain, H. 1989. [Revue of: Laguerenne, Claude Le Clerc, Grozieux de. 1989. *Étude de la matière médicale de 1833 à travers les écrits du pharmacien de la Marine René-Primevère Lesson*, 2 vols. Univ. de Nantes]. *Revue d'Histoire de la Pharmacie* No. 283, pp. 421–424.
Lesson, R.P. 1846. *Musée Anaïs ou Choix de Vues*, pp. vi–ix. Imprimerie de Henry Loustau et Cie, Rochefort.
Nouvelle Biographie Générale 1856–66. Vol. XXX, columns 972–974 [R.P. Lesson]. Paris.
Ronsil, R. 1948. [Publications by R.P. Lesson]. *Bibliographie Ornithologique Française*, Vol. 1, pp. 289–294. Paris.
1. Her year of birth is based on R.P. Lesson's statement (*Musée Anaïs*, p. ix) that Clémence was thirty-seven years old when she died. The Musée d'Art et d'Histoire at Rochefort gives 1800 as her year of birth.
2. *Revue Zoologique* 1839, p. 44.

Lewis

Cutright, P.R. 1969. *Lewis and Clark: Pioneering Naturalists*. University of Nebraska Press, Lincoln and London.
Dictionary of American Biography 1928–36. Vol. XI, pp. 219–222. New York.
Encyclopedia Americana 1978, Vol. 17, pp. 275–277 [Lewis and Clark Expedition]. Americana Corporation, Danbury.

Thwaites, R.G. 1959. *Original Journals of the Lewis and Clark Expedition*, Vols 1–8. Antiquarian Press, New York.
1. Thwaites, R.G. 1959. *op. cit.*, Vol. 5, pp. 70–71.
2. Wilson, A. and Bonaparte, C.L. 1831. *American Ornithology* (Edited by R. Jameson), Vol. 1, pp. 168–169. Constable and Co., Edinburgh.
3. Cutright, P.R. 1969. *op. cit.*, p. 100.
4. Jaeger, E.C. 1948. Does the Poor-will "Hibernate?" *Condor* **50**: 45–46.

F.L and F.J. L'Herminier

Charleston Museum Contributions 1923, No. 4. pp. 13–15. [F.L. L'Herminier].
Guibourt, G. 1834. Notice sur Félix-Louis L'Herminier. *Journal de Chemie Médicale* **10**: 221–224.
Gazzette Officielle de la Guadeloupe 13 Dec 1866 and 18 Dec 1866. [Obituary. F.J. L'Herminier].
Nouvelle Biographie Générale 1855–66. Vol. XXXI, column 74. Paris. [F.L. L'Herminier].
Urban, I. 1898–1913. *Symbolae Antillanae*. Vol. 3, pp. 74–75 [F.L. and F.J. L'Herminier]. Berolini.
1. *Charleston Museum Contributions* 1923, No. 4. p. 13.
2. *Quoted in* Weiss, H.B. and Ziegler, G.M. 1931. *Thomas Say. Early American Naturalist*, p. 57. Charles C. Thomas, Springfield and Baltimore.
3. *Encyclopaedia Britannica* 1885. 9th edn, Vol. XVIII, pp. 21–24. Edinburgh.
4. Lafresnaye, F. de 1844. Description de quelques Oiseaux de la Guadeloupe. *Revue Zoologique* **7**: 167–169.

Lincoln

Audubon, M.R. (Ed.) 1960. *Audubon and His Journals*. Vol. 1. Dover Publications Inc., New York.
Hobart, R.W. 1986. Chapter XIV. Homes and Families. *Dennysville 1786–1986 ... and Edmunds, Too!* pp. 83–86. The Dennysville Bicentennial Committee.
1. Audubon, J.J. 1834. *Ornithological Biography*, Vol. 2, pp. 437–438. Edinburgh.
2. *Quoted in Maine Sunday Telegram* 4 May 1969. "Maine's Lincoln" by La Rue Spiker.
3. *Quoted in* Audubon, M.R. (Ed.) 1960. *op. cit.*, Vol. 1, p. 365.
4. Audubon, J.J. 1967. *The Birds of America*, Vol. 3, p. 116. Dover Publications Inc., New York.
5. *Quoted in* Audubon, M.R. 1960. *op. cit.*, Vol. 1, p. 411.
6. *Quoted in* Durant, M. and Harwood, M. 1980. *On the Road with John James Audubon*, p. 484. Dodd, Mead and Co., New York.
7. *Quoted in* Deane, R. 1910. Dr. William Ingall's Recollections of the Labrador Trip. *Auk* **27**: 47.

McCormick

Dictionary of National Biography 1885–1900. Vol. XXXV, p. 11. London.

Günther, A. (Ed.) 1904–12. *The History of the Collections Contained in the Natural History Departments of the British Museum*, Vol. II, pp. 421–422. BM publication, London.

McCormick, R. 1884. *Voyages of Discovery in the Arctic and Antarctic Seas and Round the World*, 2 vols. Samson, Low, Marston, Searle and Rivington, London.

Ross, M.J. 1982. *Ross in the Antarctic*. Caedman of Whitby, Yorkshire.

1. Günther, A. (Ed.) 1904–12. *op. cit.*, p. 421.
2. McCormick, R. 1884. *op. cit.*, Vol. 2, p. 218.
3,4. Burkhardt, F.H. and Smith, S. (Eds.) 1984. *The Correspondence of Charles Darwin*, Vol. 1, pp. 176, 238. Cambridge University Press, Cambridge.
5. McCormick, R. 1884. *op. cit.*, Vol. 1, p. 145.
6,7. *Quoted in* Ross, M.J. 1982. *op. cit.*, pp. 44, 32.

McCown

Arkansas Democrat 22 Jan 1879, p. 4, column 5. [Death of Gen. McCown]

Brown, J.H. (Ed.) 1903. *Lamb's Biographical Dictionary of the United States*, Vol. V, p. 218. Boston.

Cullum, G.W. 1891. *Biographical Register of the Officers and Graduates of the U.S. Military Academy*, 3rd edn, Vol. 2, pp. 31–32. Houghton, Mifflin and Co., Boston and New York

Faust, P.L. (Ed.) 1986. *Encyclopedia of the Civil War*, pp. 457–458. Harper & Row, New York.

Geiser, S.W. 1959. Men of Science in Texas, 1820–1880. *Field and Laboratory* 27: 135.

Killgore, N.H. 1947. *History of Columbia County*, pp. 146–147. Magnolia Printing Company, Magnolia, Arkansas.

Lawrence, G.N. 1851a. Descriptions of new species of Birds ... together with a list of other species not heretofore noticed as being found within the limits of the United States. *Annals of the Lyceum of Natural History of New York* 5: 112–119.

Lawrence, G.N. 1851b. Descriptions of New Species of Birds. *Annals of the Lyceum of Natural History of New York* 5: 121–123.

Wakelyn, J.L. 1977. *Biographical Dictionary of the Confederacy*, pp. 295–296. Greenwood Press, Westport.

Warner, E.J. 1959. *Generals in Gray: Lives of the Confederate Commanders*, pp. 199–200, 383. L.S.U. Press, Baton Rouge.

1. Lawrence, G.N. 1851b. *op. cit.*, p. 123.
2,3. McCown, J.P. 1853. Facts and Observations from Notes taken when in Texas. *Annals of the Lyceum of Natural History of New York* 6: 9–14, see p. 11, 11.
4. *Quoted in* Cassin, J. 1865. *Illustrations of the Birds of California, Texas, Oregon, British and Russian America*, pp. 42–43. Philadelphia.
5. *Quoted in* Warner, E.J. 1959. *op. cit.*, p. 383.
6. *Quoted in* Faust, P.L. 1986. *op. cit.*, p. 458.

McDougall

Anon. 1814. *A Catalogue of a Choice Collection of British and Foreign Birds, comprehending many curious and rare specimens in high preservation. The entire collection of the late Dr. Peter M'Dougall.* J. Smith, Glasgow. [In the collection of Glasgow University Library].

Duncan, A. 1896. *Memorials of the Faculty of Physicians and Surgeons of Glasgow*, p. 144. Glasgow.

Glasgow Directory 1804–13. W. M'Feat & Co., Glasgow.

Glasgow Chronicle 5.5.1814.

Glasgow Courier 30.4.1814–11.6.1814.

Laskey, J. 1813. *A General Account of the Hunterian Museum Glasgow*, p. vi. Glasgow.

Montagu, G. 1813. *Supplement to the Ornithological Dictionary*. S. Woolmer, Exeter.

Parish Records: Kilsyth; Ramshorn and Blackfriars (Glasgow).

1. In the eighteenth and nineteenth centuries *Mc* and *Mac* were often interchangeable. There was no significance between the difference and both were frequently abbreviated to *M'*. The baptismal record refers to Peter's father as Alexander McDougall and this version has been adopted throughout.

    Two Glasgow University publications refer to the younger McDougall as Patrick or Patricius. All other documents, including his baptismal entry, death notices and burial record give his name as Peter.

2,3. Montagu, G. 1813. *op. cit.*, Letterpress for Roseate Tern.

4. Graves, G. 1821. *British Ornithology*, Vol. III, letterpress to pl. 39. London.

MacGillivray

Dictionary of National Biography 1885–1900. Vol. XXXV, pp. 90–92. London.

Ford, A. 1965. *John James Audubon*. University of Oklahoma Press, Norman.

MacGillivray, W. 1836. *Descriptions of the Rapacious Birds of Great Britain*. Maclachlan and Stewart, Edinburgh.

MacGillivray, W. 1901. *A Memorial Tribute to William MacGillivray M.A., LL.D., Ornithologist*. Printed for Private Circulation, Edinburgh.

MacGillivray, W. 1910. *Life of William MacGillivray*. J. Murray, London.

Mullens, W.H. 1909. Some Early British Ornithologists and their Works. IX.—William MacGillivray (1796–1852) and William Yarrell (1784–1853). *British Birds* 12: 389–399.

1,2. *Quoted in* MacGillivray, W. 1901. *op. cit.*, pp. 35, 60.

McKay

Jordan, D.S. 1883. Charles Leslie McKay. *Indiana Student* 10: 2–5.

Jordan, D.S. 1922. *The Days of a Man*, Vol. 1, pp. 120–122, 239. World Book Co., Yonkers-on-Hudson, New York.

McKay, C.L. 1882. [Letter from Alaska, dated 14 April 1882] *Indiana Student* 9: 18–19.

Osgood, W.H. 1904. *North American Fauna No. 24. A Biological Reconnaisance of the Base of the Alaska Peninsula*, pp. 25–26. Government Printing Office, Washington, D.C.

Service record from: Descriptive Book of the Signal Corps, 1860–1889. Record Group 111, Records of the Office of the Chief Signal Officer. National Archives and Records Administration, Washington, D.C.

1,2. *Quoted in* Jordan, D.S. 1883. *op. cit.*, pp. 4, 4–5.

Masséna

Encyclopaedia Britannica 1883, 9th edn Vol. XV, pp. 616–617 [André Masséna]. Edinburgh.

Mailhol, D. de 1896. *Dictionaire Historique et Héraldique de la Noblesse Française*, p. 1802. Paris.

Palmer, T.S. 1917. Botta's Visit to California. *Condor* **19**: 159–161.

Six, G. 1934. *Dictionaire Biographique des Généraux*, Vol. I, p. 299 [J.F.J. Debelle]. Paris.

Stone, W. 1899. Some Philadelphia Ornithological Collections and Collectors, 1784–1850. *Auk* **16**: 166–177.

1. Audubon, M.R. (Ed.) 1960. *Audubon and His Journals*, Vol. 1, pp. 313–314. Dover Publications Inc., New York.

2. Stone, W. 1899. *op. cit.*, pp. 174–175.

Mauri

Journal für Ornithologie 1856, pp. 419–421.

Palmer, T.S. 1931. The scientific name of the western sandpiper—who was Mauri? *Condor* **33**: 243–244.

Munro

Amadon, D. 1964. Obituary. George Campbell Munro. *Auk* **81**: 256.

Munro, G. C. 1960. *Birds of Hawaii*, revised edn. Charles E. Tuttle Co., Rutland, Vermont & Tokyo, Japan. [First published Honolulu, 1944]

Pan-Pacific Who's Who 1940–41, p. 509 [G.C. Munro]. Honolulu Star-Bulletin Ltd., Honolulu.

Venn, J.A. 1954. *Alumni Cantabrigienses*, pt. II, vol. VI, p. 528 [S.B. Wilson]. Cambridge University Press, Cambridge.

Weybridge census returns for George Wilson household: 1871, 1881.

Wilson, S.B. and Evans, A.H. 1890–99. *Aves Hawaiienses. The Birds of the Sandwich Islands*. R.H. Porter, London.

1. *See e.g.* Rothschild, W. 1893–1900. The Principal Literature. *The Avifauna of Laysan and the Neighbouring Islands*, pp. v–xii. R.H. Porter, London.

2. Wilson, S.B. and Evans, A.H. 1890–99. *op. cit.*, unpaged, see text for *Hemignathus olivaceus*.

3. Douglas Pratt's *Hemignathus munroi* applies only if the Akiapolaau and amakihis are placed in the same genus. Sibley and Munroe (1990) use a classification different from Pratt's in which the Akiapolaau and amakihis are not congeners so they use

the older name *Hemignathus wilsoni*. The 1983 A.O.U. *Check-list of North American Birds* favoured Pratt's classification.
4. *Quoted in* Wilson, S.B. and Evans, A.H. 1890–99. *op. cit.*, p. xvii.
5. Munro, G.C. 1960. *op. cit.*, pp. 122–123.
6. Amadon, D. 1964. *op. cit.*, p. 256.
7. *Lymington & South Hants Chronicle*, Thursday, January 25, 1923, p. 1, column 7. "Found Shot. Shocking Discovery at Everton."

Néboux

Biographical details from service record and archive material supplied by the Chef du Service Historique de la Marine and the Chef du Service des Archives et des Bibliothèque de la Marine, Vincennes, France.
1. Palmer, T.S. 1918. Costa's Hummingbird—Its type locality, early history and name. *Condor* **20**: 114–116.
2. Prévost, F. and Des Murs, O. 1855. Oiseaux, pp. 177–284. *in* du Petit-Thouars, A.A. 1840–60. *Voyage autour du monde sur la frégate La Vénus. Zoologie, mammifères, oiseaux, reptiles et poissons*. Gide et J. Baudry, Éditeurs. Paris.
3. Néboux, A.S. 1840. Descriptions d'oiseaux nouveaux recueillis pendant l'expédition de *La Vénus*. *Revue Zoologique* **3**: 289–291.

Nuttall

Ewan, J. and Ewan, N.D. 1981. *Biographical Dictionary of Rocky Mountain Naturalists, 1682–1932*, p. 163. Frans A. Stafleu, Utrecht.
Graustein, J.E. 1967. *Thomas Nuttall, Naturalist. Explorations in America 1808–1841*. Harvard University Press, Cambridge.
Kastner, J. 1977. Chapter 12. The Innocent Traveler [Nuttall]. *A Species of Eternity*, pp. 254–283 *et passim*. Alfred A. Knopf, New York.
Nuttall, T. 1821. *A Journal of Travels into the Arkansa Territory*. T.H. Palmer, Philadelphia. Reprinted in Thwaites, R.G. 1904–1907. *Early Western Travels*, Vol. XIII. Arthur H. Clark Co., Cleveland, Ohio.
1. *Quoted in* Graustein, J.E. 1967. *op. cit.*, p. 326.
2. Burns, F.L. 1917. Miss Lawson's Recollections of Ornithologists. *Auk* **34**: 282.
3,4. *Quoted in* Graustein, J.E. 1967. *op. cit.*, pp. 286, 242.
5. Audubon, J.J. 1831–39. *Ornithological Biography*, Vol. IV, p. 450. Edinburgh.
6. Dana, R.H. 1912. *Two Years Before the Mast*, pp. 223–224. T. Nelson and Sons Ltd., London. [Numerous other editions]
7. *Quoted in* Graustein, J.E. 1967. *op. cit.*, p. 319.

Oberholser

Aldrich, J.W. 1968. In Memoriam: Harry Church Oberholser. *Auk* **85**: 25–29.
Cleveland Audubon Society. 1964. In Memory of Harry C. Oberholser. 8 pp. Cleveland, Ohio.

Gilfillan, M.C. 1963. Ohio's Top Bird Man. *Ohio Conservation Bulletin*, January, pp. 3–5.

Oberholser, H.C. 1957. Autobiography. *The Biltmore Immortals*, Vol. II, December 5, pp. 152–163.

[Portrait and note] *Condor* 1904, **6**: 98.

Additional information supplied by the Cleveland Museum of Natural History, Ohio.

Palmer

Bryan, E.H. 1956. History of Laysan and Midway Islands, *in* Bailey, A.M. *Birds of Midway and Laysan Islands*, pp. 8–17. Museum Pictorial No. 12. Denver Museum of Natural History, Denver, Colorado.

Munro, G.C. 1960. *Birds of Hawaii*, revised edn. Charles E. Tuttle Co., Rutland, Vermont & Tokyo, Japan. [First published Honolulu, 1944]

Rothschild, M. 1983. *Dear Lord Rothschild. Birds, Butterflies and History*, pp. 73, 76–78, 139, 155–156, 185. Hutchinson & Co., (Publishers) Ltd., London.

Rothschild, W. 1893–1900. *The Avifauna of Laysan and the Neighbouring Islands*. [Includes: Résumé of Palmer's Diary (Jan. 1, 1891, to August 1893), pp. (Di.)1–(Di.)21; and Diary of Henry Palmer from May 5th to August 18th 1891, pp. vii–xiv]. R.H. Porter, London.

Additional information from the Rothschild correspondence—British Museum (Natural History), London.

1,2. Rothschild, W. 1893–1900. *op. cit.*, p. (Di.)2, pp. ix–x.

3. Berger, A.J. 1972. *Hawaiian Birdlife*, p. 185. University Press of Hawaii, Honolulu.

4. There was no record of the violent death of Henry C. Palmer in any of the Australian Registry Offices for the period 1893–1903. The State Libraries were likewise unable to trace any record of his death. Palmer may have died after 1903 or his death may be recorded under the name of Henry Palmer, without the middle name or initial.

Paris

No biographical information traced. Quotations and other details from: Bonaparte, C.L. 1838. [Notices and descriptions of new or interesting birds from Mexico and South America]. *Proceedings of the Zoological Society* 1838 (1837), pp. 108–114.

Ridgway

Anon. [no date] *Bird Haven. Robert Ridgway Memorial Arboretum and Bird Sanctuary*. Leaflet: available [in 1991] from Olney City Hall, 300 Whittle Avenue, Olney, Illinois 62450.

Behle, W.H. 1990. *Utah Birds: Historical Perspectives and Bibliography*, pp. 15–17. Utah Museum of Natural History.

Bullard, A.K. 1979. Robert Ridgway. Bird Haven. Larchmound. *Footprints, Past and Present* [Quarterly publication of the Richland County Genealogical and Historical

Society] **2**: 4–9. [And additional information on Larchmound supplied by Anne Katherine Bullard, Olney, Illinois—February 1991]

Harris, H. 1928. Robert Ridgway. With a bibliography of his published writings and fifty illustrations. *Condor* **30**: 4–118.

Oberholser, H.C. 1933. Robert Ridgway: A Memorial Appreciation. *Auk* **50**: 158–169.

Ridgway, R. 1905. A Winter with the Birds in Costa Rica. *Condor* **7**: 150–160.

1,2,3. *Quoted in* Harris, H. 1928. *op cit.*, pp. 9, 18–19, 18–19.

4. Chapman, F.M. 1933. *Autobiography of a Bird-Lover*, p. 44. D. Appleton-Century, New York.

B.R. Ross

Cartright, B.W. 1940. Where the Ross' Geese nest. I—The Story of the Search. *Beaver*, outfit 271 (Dec. 1940), pp. 6–8.

Dall, W.H. 1915. *Spencer Fullerton Baird. A Biography*, pp. 335, 342–350, 358, 362, 378. J.B. Lippincott Co., Philadelphia and London.

Deignan, H.G. 1947. HBC and the Smithsonian. *Beaver*, outfit 278 (June 1947), pp. 3–7.

Dictionary of Canadian Biography 1966–91. Vol. X, p. 629. Toronto.

Gavin, A. 1940. Where the Ross' Geese nest. II—The Discovery. *Beaver*, outfit 271 (Dec. 1940), pp. 8–9.

Gavin, A. 1947. Birds of the Perry River District, Northwest Territories. *Wilson Bulletin* **59**: 195–203.

Houston, C.S. 1983. Birds first described from Hudson Bay. *Canadian Field-Naturalist* **97**: 95–98.

Houston, C.S. 1989. Samuel Hearne. *Picoides* **3**: 6–9.

Kerr, R. 1953. For the Royal Scottish Museum. *Beaver*, outfit 284 (June 1953), pp. 32–35.

Preble, E.A. 1922. Roderick Ross MacFarlane, 1883–1920. *Auk* **39**: 203–210.

Ross, B.R. 1861. [Letter to John Richardson, dated 20 June 1860: Mammalia and Birds of Arctic Regions]. *Edinburgh New Philosophical Journal*, new series, **13**: 161–163.

Sherk, J.M. 1926. H.B.C. Pioneers. Bernard Rogan Ross (1827–1874). *Beaver*, outfit 257 (Dec. 1926), p. 25.

1. *Quoted in* Cassin, J. 1861. Communication in reference to a new species of Goose from Arctic America. *Proceedings of the Academy of Natural Sciences of Philadelphia* **13**: 72; and [with minor differences] in Cartright, B.W. 1940. *op. cit.*, p. 6.

2. Ewan, J. and Ewan, N.D. 1981. *Biographical Dictionary of Rocky Mountain Naturalists, 1682–1932*, p. 187 [Christina Ross]. Frans A. Stafleu, Utrecht.

3. *Quoted in* Deignan, H.G. 1947. *op. cit.*, p. 5.

J.C. Ross

Densley, M. 1988. James Clark Ross and Ross's Gull—A Review. *Naturalist* **113**: 85–102.

Dictionary of National Biography 1885–1900. Vol. XLIX, pp. 265–269. London.

Dodge, E.S. 1973. *The Polar Rosses. John and James Clark Ross and their Explorations.* Faber & Faber, London.

Mearns, B. and Mearns, R. 1988. *Biographies for Birdwatchers. The Lives of Those Commemorated in Western Palearctic Bird Names*, pp. 307–315, 449–450. Academic Press, London.

Ross, M.J. 1982. *Ross in the Antarctic*. Caedmon of Whitby, Yorkshire.

1. *Quoted in* Dodge, E.S. 1973. *op. cit.*, p. 182.

2. *Quoted in* Ross, M.J. 1982. *op. cit.*, no page number: between title page and contents list.

3. Parry, W.E. 1824–25. *Journal of a Second Voyage for the Discovery of a North-West Passage ... in the years 1821–3*, Vol. 2, p. 449. J. Murray, London.

4. For earlier undescribed specimens of Ross's Gull *see* Densley, M. 1988. *op. cit.*, pp. 89–90.

5. For an earlier specimen of the Yellow-billed Loon, and unpublished description by John Richardson dated 1 November 1820, *see* Houston, C.S. (Ed.) 1984. *Arctic Ordeal. The Journal of John Richardson, surgeon-naturalist with Franklin 1820–1822*, pp. 225–226. McGill-Queen's University Press, Kingston and Montreal. Richardson told Audubon about a "very large and handsomely crested Diver" which Audubon provisionally named *Colymbus Richardsoni* in 1838 in his *Ornithological Biography*, Vol. IV, p. 53.

6,7. *Quoted in* Ross, M.J. 1982. *op. cit.*, pp. 232; 215.

8. Fuller, E. 1987. *Extinct Birds*, pp. 92–93. Viking Rainbird, London.

Sabine

Dictionary of National Biography 1885–1900. Vol L, pp. 74–79. London.

Parry, W.E. 1821–24. *Journal of a Voyage for the Discovery of a North-West Passage from the Atlantic to the Pacific; performed in the years 1819–20* [with appendices on mammals, birds, fish by E. Sabine]. J. Murray, London.

Ross, J. 1819. *A Voyage of Discovery ... for the purpose of exploring Baffin's Bay*. J. Murray, London.

Sabine, E. 1819. A Memoir of the Birds of Greenland. *Transactions of the Linnean Society of London* **12**: 527–559.

1. Ross. J.C. 1847. *A Voyage of Discovery and Research in the southern and Antarctic Regions during the years 1839–43*, Vol. I, p. 183. J. Murray, London.

2. Günther, A. (Ed.) 1904–12. *The History of the Collections Contained in the Natural History Departments of the British Museum*, Vol. II, p. 460. BM publication, London.

3. Richardson, J. and Swainson, W. 1832. *Fauna Boreali-Americana*, Vol. II, pp. 428–429. J. Murray, London.

Say

Dictionary of American Biography 1928–36. Vol. XVI, pp. 401–402. New York.

Essig, E.O. 1931. *A History of Entomology*, pp. 750–756. MacMillan, New York.

James, E. 1823. *Account of an Expedition from Pittsburgh to the Rocky Mountains*. Philadelphia. Reprinted in Thwaites, R.G. 1904–07. *Early Western Travels*, Vols. XIV–XVII. Arthur H. Clark Co., Cleveland, Ohio.

Weiss, H.B. and Ziegler, G.M. 1931. *Thomas Say, Early American Naturalist*. Charles C. Thomas, Springfield, Illinois and Baltimore, Maryland.

Sclater

Goode, G.B. 1896. Biographical Sketch of Philip Lutley Sclater, pp. ix–xix *in* Bibliography of the Published Writings of Philip Lutley Sclater. *Bulletin 49, United States National Museum*. Washington, D.C.

Günther, A. (Ed.). 1904–12. *The History of the Collections Contained in the Natural History Departments of the British Museum*. Vol. II, pp. 464–469. BM publication, London.

Ibis 1913, pp. 642–654 [Obituary. P.L.Sclater].

Ibis 1945, pp. 115–121 [Obituary. W.L.Sclater].

Scott

Dictionary of American Biography, 1928–36. Vol. XVI, pp. 505–511. New York.

Encyclopaedia Brittanica 1886, 9th edn, Vol. XXI, p. 551. Edinburgh.

Faust, P.L. (Ed.) 1986. *Encyclopedia of the Civil War*, pp. 662–663. Harper & Row, New York.

Scott, W. 1864. *Memoirs of Lieut-General Scott, LL.D*, 2 vols. Sheldon, New York.

Smith

Baltimore Sun 19 September 1954. "Maryland's Big Silk Bubble."

Cordell, E.F. 1903. *The Medical Annals of Maryland 1799–1899*, p. 571. Williams and Wilkins, Baltimore.

Gifford, G.E. and Gifford, L.T. 1961. John James Audubon and Gideon B. Smith, M.D. *Bulletin of the History of Medicine* **35**: 475–477.

McDermott, J.F. (Ed.) 1951. *Up the Missouri with Audubon. The Journal of Edward Harris*. University of Oklahoma Press, Norman.

Watters, R.C. 1939. Audubon and his Baltimore patrons. *Maryland Historical Magazine* **34**: 138–143.

1. *Quoted in* McDermott, J.F. (Ed.) 1951. *op. cit.*, pp. 21–23.

2,3. Audubon, J.J. 1844. *The Birds of America*, 8vo edn, Vol. 7, pp. 337, 336. New York and Philadelphia.

4. McDermott, J.F. (Ed.) 1951. *op. cit.*, pp. 211–212.

5. Cordell, E.F. 1903. *op. cit.*, p. 571.

Sprague

Audubon, M.R. (Ed.) 1960. *Audubon and His Journals*, 2 vols. Dover Publications Inc., New York.

Garden and Forest 1895, **8**: 130. Obituary.

Gifford, G.E. 1975. Isaac Sprague. Audubon's Massachusetts Artist. *Massachusetts Audubon Newsletter* **14**: 7–10. [With self portrait drawn when Sprague was a young man].

Henshaw, H.W. 1919. Autobiographical Notes. *Condor* **21**: 167–168.
Rudolph, E.D. 1990. Isaac Sprague, "Delineator and Naturalist." *Journal of the History of Biology* **23**: 91–126.
1. *Quoted in* Rudolph, E.D. 1990. *op cit.*, p. 94.
2. Our account of the discovery of Sprague's Pipit is based on the journals of Audubon and Harris, written when the events were fresh in their memory. The contradictory statement made by Audubon in the original description—that the new pipit was first procured by Sprague on 19 June 1843—is assumed to be in error.
3. Rudolph, E.D. 1990. *op cit.*, pp. 116–126 gives the full references for 'Plantae Wrightianae' and 'Plantae Fremontianae' in a list of the books and papers which contain illustrations by Isaac Sprague.

Steller

Ibis 1943, pp. 354–355. [Obituary. L. Stejneger]
Stejneger, L. 1936. *Georg Wilhelm Steller, the Pioneer of Alaskan Natural History.* Harvard University Press, Cambridge.
1. The old Julian calendar was still in use by the Russians and Stejneger adhered to it throughout his narrative of the expedition. In the present work we have converted his dates (by the addition of eleven days) to make them correspond to our Gregorian calendar (but not in the chapter on Steller in Mearns, B. and Mearns, R. 1988. *Biographies for Birdwatchers. The Lives of Those Commemorated in Western Palearctic Bird Names*, pp. 347–355. Academic Press, London.)
2. *Quoted in* Stejneger, L. 1936. *op. cit.*, p. 365.
3. *Quoted in* Bent, A.C. 1937. *Life Histories of North American Birds of Prey* Part I, p. 350. *Bulletin 167, United States National Museum.* Washington, D.C.
4. Not 14 November, as in *Biographies for Birdwatchers* (1988), p. 355.

Strickland

Dictionary of National Biography 1885–1900. Vol. LV, pp. 50–52. London.
Jardine, W. 1858. *Memoirs of Hugh Edwin Strickland.* London.
Nature 19 February 1874, p. 312. [Endowment of Strickland collection]
Salvin, O. 1882. *A Catalogue of the Collection of Birds formed by the late Hugh Edwin Strickland, M.A.* Cambridge University Press, Cambridge.

Swainson

Dictionary of National Biography 1885–1900. Vol. LV, pp. 192–193. London.
Dictionary of New Zealand Biography 1940. Vol. II. pp. 350–352. Wellington.
Galloway, D.J. 1978. The botanical researches of William Swainson F.R.S., in Australasia, 1841–1855. *Journal of the Society for the Bibliography of Natural History* **8**: 369–379.
Günther, A. (Ed.) 1904–12. *The History of the Collections Contained in the Natural History Departments of the British Museum*, Vol. II, pp. 208–245 [The Bullock Collection]. BM publication, London.

McMillan, N.F. 1968. Biography of William Swainson. *in* Abbot, R.T. (Ed.) *Swainson's Exotic Conchology*, pp. viii–x. D. Van Nostrand Company, Inc., Princeton, New Jersey, for the Delaware Museum of Natural History.

McMillan, N.F. 1970. William Swainson's *Birds of Brazil, Mexican Zoology*, and *Tropical Ornithology*. *Journal of the Society for the Bibliography of Natural History* **5**: 366–368.

McMillan, N.F. 1980. William Swainson (1789–1855) and his shell collections. *Journal of the Society for the Bibliography of Natural History* **9**: 427–434.

McMillan, N.F. and Cernohorsky, W.O. 1979. William Swainson, F.R.S., in New Zealand with notes on his drawings held in New Zealand. *Journal of the Society for the Bibliography of Natural History* **9**: 161–169.

Stresemann, E. 1954. Ferdinand Deppe's travels in Mexico, 1824–1829. *Condor* **56**: 86–92.

Additional information from the Norah F. McMillan collection of Swainson material at the British Museum (Natural History), London.

1. Audubon was in Labrador in 1833.
2,3. Audubon, J.J. 1967. *The Birds of America*, Vol. 2, pp. 84, 85. Dover Publications Inc., New York.
4. *Encyclopaedia Britannica* 1885, 9th edn, Vol. XVIII, p. 13. Edinburgh.
5. *Quoted in* Allen, D.E. 1978. *The Naturalist in Britain*, p. 85. Penguin Books Ltd., London.
6. *Quoted in Encyclopaedia Britannica* 1885, 9th edn, Vol. XVIII, p. 15. Edinburgh.
7. The McMillan collection gives 1823. Many other sources give 1825.
8. *Quoted in* Herrick, F.H. 1917. *Audubon the Naturalist*, Vol. 1, p. 299. D. Appleton and Co., New York.
9,10,11,12. *Quoted in* Ford, A. 1965. *John James Audubon*. pp. 271, 266, 266, 374–375. University of Oklahoma Press, Norman.
13. Sibley, G.C. and Munroe, B.L. 1990. *Distribution and Taxonomy of Birds of the World*, p. 454. Yale University Press, New Haven and London.

Thayer

Anon. 1935. John Eliot Thayer. *The Fiftieth Anniversary Report of the Harvard Class of '85*, pp. 3–20.

Mathews, G.M. 1934. John Eliot Thayer. *Ibis*, pp. 179–180.

Phillips, J.C. 1934. John Eliot Thayer, 1862–1933. *Auk* **51**: 46–51.

Thayer, J.E. and Bangs, O. 1908. The Present State of the Ornis of Guadaloupe Island. *Condor* **10**: 101–106.

Thomson

Dictionary of National Biography 1885–1900. Vol. XIX, pp. 717–718. London.

Linklater, E. 1974. *The Voyage of the Challenger*. Cardinal edition, Sphere Books Ltd., London.

Sclater, P.L. 1878. Reports on the Collections of Birds made during the Voyage of H.M.S. 'Challenger'.—No. VIII. On the Birds of the Sandwich Islands. *Proceedings of the Zoological Society*, pp. 346–351.

1. Baptized Wyville Thomas Charles Thomson he formally changed his name when gazetted as knight.
2. Sclater, P.L. 1878. *op cit.*, p. 346.
3. *Quoted in* Linklater, E. 1974. *op cit.*, p. 270.

Tolmie

Dictionary of Canadian Biography 1966–91. Vol. XI, pp. 885–888. Toronto.
Ewan, J. and Ewan, N.D. 1981. *Biographical Dictionary of Rocky Mountain Naturalists, 1682–1932*, p. 81 [Gairdner]. Frans A. Stafleu, Utrecht.
Palmer, T.S. 1928. Notes on Persons Whose Names Appear in the Nomenclature of California Birds. *Condor* **30**: 278 [Gairdner], 298 [Tolmie].
Tolmie, W.F. 1963. *The Journals of William Fraser Tolmie. Physician and Fur Trader.* [Includes memoir: *My Father: William Fraser Tolmie*, by S.F.Tolmie]. Edited by R.G. Large. Mitchell Press Ltd., Vancouver.
1,2. Tolmie, W.F. 1963. *op. cit.*, pp. 29, 178.
3. Tolmie's son asserts that his father had "secured the skins of many birds, including that of the MacGillivray's Warbler" on Mt Rainier in August–September 1833 (Tolmie, S.F. *in* Tolmie, W.F. 1963. *op. cit.*, p. 388). This is unconfirmed. No more details are given in the published journal though they could have been in the original. Townsend does not say that Tolmie discovered the warbler but since it is quite common near the Columbia River it is likely that each of them discovered it independently, Tolmie before Townsend since he was there earlier.
4. Tolmie, W.F. 1963. *op. cit.*, p. 274.
5. Townsend, J.K. 1839. *Narrative of a Journey across the Rocky Mountains and a Visit to the Sandwich Islands, Chili, &c.*, p. 343. Henry Perkins, Philadelphia; and *Journal of the Academy of Natural Sciences of Philadelphia* 1839, **8**: 149–50.
6. Audubon, J.J. 1839. *Ornithological Biography*, Vol. V, p. 75. Edinburgh.
7. Tolmie, W.F. 1963. *op. cit.*, p. 386.
8. *Dictionary of Canadian Biography*, *op. cit.*, pp. 887–888.

C.H. Townsend

Bent, A.C. 1922. *Life Histories of North American Petrels and Pelicans and Their Allies*, pp. 84–85. *Bulletin 121, United States National Museum*, Washington, D.C.
Palmer, T.S. 1947. Charles Haskins Townsend. *Auk* **64**: 349–350.
Townsend, C.H. 1887. Notes on the Natural History and Ethnology of Northern Alaska. Introductory letter, p. 83, *in* Healy, M.A. *Report of the Cruise of the Revenue Marine Steamer* Corwin *in the Arctic Ocean in 1885*. Government Printing Office, Washington, D.C.
Townsend, C.H. 1887. List of the Midsummer Birds of the Kowak River, Northern Alaska. *Auk* **4**: 11–13.
Townsend, C.H. 1890. Birds from the Coasts of Western North America and Adjacent Islands, Collected in 1888–'89, with Descriptions of New Species. *Proceedings of the U.S. National Museum* **13**: 131–142.
Townsend, C.H. 1927. Old Times With The Birds: Autobiographical. *Condor* **29**: 224–232.
1,2,3,4,5. Townsend, C.H. 1927. *op. cit.*, pp. 224, 229, 232, 230, 232.

J.K. Townsend

Auk 1930, **47**: 414–415; *Auk* 1931, **48**: 106–109; *Auk* 1934, **51**: 225–226. [Witmer Stone—Frank L. Burns, notes and correspondence on Townsend's Oregon Tubinares etc.]

Dictionary of American Biography 1928–36. Vol. XVIII, pp. 617–618. New York.

Ewan, J. and Ewan, N.D. 1981. *Biographical Dictionary of Rocky Mountain Naturalists, 1682–1932*, pp. 222–223. Frans A. Stafleu, Utrecht.

Ford, A. 1964. *John James Audubon*, pp. 339–340, 342, 346, 348–350, 352–353, 364, 408. [A pro Audubon slant on the acquisition of the Townsend specimens] University of Oklahoma Press, Norman.

Hall, F.S. 1937–38. Studies in the History of Ornithology in the State of Washington (1792–1932), with special reference to the discovery of new species. Part IV. The overland journey of the naturalists Thomas Nuttall and John Kirk Townsend. *Murrelet* **18**: 2–13. [Two portraits, the one labelled Townsend is also of Nuttall]

Stone, W. 1903. John Kirk Townsend. *Cassinia* **7**: 1–5.

Stone, W. 1906. A bibliography and nomenclator of the ornithological works of John James Audubon. *Auk* **23**: 298–312.

Stone, W. 1916. Philadelphia to the coast in early days, and the development of western ornithology prior to 1850. *Condor* **18**: 3–14.

Thwaites, R.G. 1905. *Early Western Travels*, Vol. XXI. [Townsend's narrative is shortened and without the appendix]. Arthur H. Clark Co., Cleveland.

Townsend, J.K. 1837. Description of Twelve New Species of Birds, chiefly from the vicinity of the Columbia river. *Journal of the Academy of Natural Sciences of Philadelphia* **7**: 187–193.

Townsend, J.K. 1839. List of the Birds Inhabiting the Region of the Rocky Mountains, the Territory of the Oregon, and the North West Coast of America. *Journal of the Academy of Natural Sciences of Philadelphia* **8**: 151–158.

Townsend, J.K. 1839. *Narrative of a Journey Across the Rocky Mountains and a Visit to the Sandwich Islands, Chili, &c.* [includes scientific appendix]. Henry Perkins, Philadelphia.

1. Townsend wanted to call it Audubon's Bunting *Emberiza Auduboni*. See Deane, R. 1909. Some original manuscript relating to the history of Townsend's Bunting. *Auk* **26**: 269–272.

2,3,4. Thwaites, R.G. 1905. *op. cit.*, pp. 180, 289, 306.

5. *Quoted in* Wilson, S.B. and Evans, A.H. 1890–99. *Aves Hawaiienses. The Birds of the Sandwich Islands*, p. xiv, note 1. R.H. Porter, London.

6. Audubon, J.J. 1838. *Ornithological Biography*, Vol. IV, pp. xi–xii. Edinburgh.

7. Audubon to Bachman 23 Oct. 1836. *Quoted in* Hanley, W. 1977. *Natural History in America*, pp. 111–112. Quadrangle/The New York Times Book Co., New York.

8. Townsend, J.K. 1839. *Journal of the Academy of Natural Sciences of Philadelphia* **8**: 159.

9. *Quoted in* Dall, W.H. 1915. *Spencer Fullerton Baird. A Biography*, pp. 75–76. J.B. Lippincott Co., Philadelphia and London.

10. Stone, W. 1916. *op. cit.*, pp. 10–11.

11. *Quoted in* Stone, W. 1903. *op. cit.*, p. 5.

12. Street, P.B. 1948. The Edward Harris collection of birds. *Wilson Bulletin* **60**: 167–184.

13. Wilson, S.B. and Evans, A.H. 1890–99. *Aves Hawaiienses. The Birds of the Sandwich Islands*, p. xiv. R.H. Porter, London.

Traill

Audubon, M.R. (Ed.) 1960. *Audubon and His Journals,* Vol. 1. Dover Publications Inc., New York.
Dictionary of National Biography 1885–1900. Vol. LVII, p. 151. London.
Ford, A. 1965. *John James Audubon.* University of Oklahoma Press, Norman.
1. *Quoted in* Audubon, M.R. (Ed.) 1960. *op. cit.,* Vol. 1, p. 163.

Tristram

Fleming, E.M. [unpublished MS, no date] *Recollections of my Grandparents—Canon and Mrs Tristram,* 106 pp.
Dictionary of National Biography, Second Supplement 1901–1911. pp. 535–536. London.
Ibis 1906, p. 602. Obituary.
Ibis 1908. Jubilee Supplement, pp. 153–156. Memoir of Tristram.
Tristram, H.B. 1889. *Catalogue of a Collection of Birds belonging to H.B. Tristram.* Durham.
Additional information and further references in Mearns, B. and Mearns, R. 1988. *Biographies for Birdwatchers. The Lives of Those Commemorated in Western Palearctic Bird Names,* pp. 384–393, 394–402, 453. Academic Press, London.
1. Moreau, R.E. 1959. The Centenarian 'Ibis.' *Ibis,* pp. 19–38, *see* p. 30; Wollaston, A.F.R. 1921. *A Life of Alfred Newton,* p. 67. J. Murray, London.

Vaux

Appleton's Cyclopoedia 1899. Vol. VI, p. 270. D. Appleton & Co., New York.
Miller, R.F. 1966. *Biographies of people for whom birds have been named,* booklet No. 6, unpaged *see* Vaux. Baldwin Bird Club, Baldwin City, Kansas.
Proceedings of the Academy of Natural Sciences of Philadelphia 1883 (1882) **34**: 111–112. [Notice of death].
Vaux, G. 1981. The Vaux family's scientific pursuits. *Frontiers* [Annual of the Academy of Natural Sciences of Philadelphia] **3**: 56–63.

J.P. and J.B.E. Verreaux

Dictionary of South African Biography 1968–81. Vol. II, pp. 811–812 [J.P. Verreaux]. Cape Town.
Günther, A. (Ed.) 1904–12. *The History of the Collections Contained in the Natural History Departments of the British Museum,* Vol. II, p. 503. BM publication, London.
Ibis 1874, pp. 467–468 [Obituary. J.P. Verreaux]
Iredale, T. 1945. Jules Verreaux. *Australian Zoologist,* pp. 71–72.
Lucas, F.A. 1914. The Story of Museum Groups. *American Museum Journal* **14**: 3–14.
Stresemann, E. 1975. *Ornithology, from Aristotle to the Present,* pp. 162–163 [footnote]. Harvard University Press, Cambridge.

Whittell, H.M. 1954. *The Literature of Australian Birds*, pp. 729–730. Paterson Brokenska Pty, Perth.
1. *Quoted in* Günther, A. (Ed.) 1904–12. *op. cit.*, p. 503.
2. Gambel, W. 1847. Remarks on the Birds observed in Upper California. *Journal of the Academy of Natural Sciences of Philadelphia*, 2nd series, **1**: 43.
3. The type is a mounted specimen; there is no mention of any of the Verreaux brothers with it. The bird once belonged to Bonaparte but was bought by the Paris Natural History Museum in 1858.

Whitney

Brewster, E.T. 1909. *Life and Letters of Josiah Dwight Whitney*. Houghton Mifflin Co., Boston and New York.
Dictionary of American Biography 1928–36. Vol. XX, pp. 161–163. New York.
Emerson, W.O. 1899. Dr. James G. Cooper. A Sketch. *Bulletin of the Cooper Ornithological Club* **1**: 1–5.
Emerson, W.O. 1902. In Memoriam: Dr. James G. Cooper. *Condor* **4**: 101–103.
Grinnell, J. 1902. The Ornithological Writings of Dr. J.G. Cooper. *Condor* **4**: 103–105.
Whitney, J.D. 1870. *Geological Survey of California. Ornithology. Volume I. Land Birds*. Edited by S.F. Baird, from the Manuscript and Notes of J.G. Cooper. Published by authority of the Legislature, Cambridge, Massachusetts.
1,2. *Quoted in* Browning, P. 1986. *Place Names of the Sierra Nevada. From Abbot to Zumwalt*, p. 237, 237. Wilderness Press, Berkeley.
3. *Quoted in* Whitney, J.D. 1870. *op. cit.*, p. 443.
4. *Quoted in* Brewster, E.T. 1909. *op. cit.*, p. 277.

Williamson

Appleton's Cyclopoedia 1899, Vol. VI, pp. 537–538. D. Appleton & Co., New York.
Goetzmann, W.H. 1959. *Army Exploration in the American West, 1803–1863*, pp. 250–253, 276, 289–295. Yale University Press, New Haven.
Historical Register and Dictionary of the United States Army 1903, Vol. 1, p. 1044. Government Printing Office, Washington, D.C.
San Francisco Call 14 November 1882, p. 2, column 2.
San Francisco Bulletin 13 November 1882, p. 2, column 4.
Warren, G.K. 1855. Memoir. [Explorations from A.D. 1843 to A.D. 1852 and from A.D. 1852 to A.D. 1857], pp. 55–56, 78–79, 81–82. *Reports of Explorations and Surveys, to ascertain the most practicable route for a Railroad from the Mississippi River to the Pacific Ocean*, Vol. IX. Washington, D.C.
Williamson, R.S. 1943. The Removal of Blossom Rock [with Introductory Note by A.S. Blake]. *Publication of the Society of California Pioneers*, pp. 9–27. San Francisco.
1. Day and month of birth untraced.
2. Newberry, J.S. 1857. Zoology, p. 89, pl. 34. *in* Abbot, H.L. Report upon Explorations for a Route from the Sacramento Valley to the Columbia River. Part IV, No. 2. *Reports of Explorations and Surveys, to ascertain the most practicable route for a Railroad from the Mississippi River to the Pacific Ocean*, Vol. VI. Washington, D.C.

3. Henshaw, H.W. 1920. Autobiographical Notes. Pt. V. *Condor* **22**: 4.
4. *Quoted in* Bent, A.C. 1939. Life Histories of North American Woodpeckers, pp. 154–156. *Bulletin 174, United States National Museum*. Washington, D.C.

Wilson

Cantwell, R. 1961. *Alexander Wilson.* J.B. Lippincott, Philadelphia and New York.
Dictionary of National Biography 1885–1900. Vol. LXII, pp. 75–76. London.
Hunter, C. (Ed.) 1983. *The Life and Letters of Alexander Wilson. Memoirs* of the American Philosophical Society, Vol. 154. Philadelphia.
Mearns, B. and Mearns, R. 1988. *Biographies for Birdwatchers. The Lives of Those Commemorated in Western Palearctic Bird Names*, pp. 419–429, 454–455. Academic Press, London.
Wilson, A. and Bonaparte, C.L. 1831. *American Ornithology; or the Natural History of the Birds of the United States*, 4 vols. Edited by R. Jameson. Constable and Co., Edinburgh.
1,2,3. Hunter, C. (Ed.) 1983. *op. cit.*, pp. 203, 274, 307–308.
4. Wilson, A. and Bonaparte, C.L. 1831. *op. cit.*, Vol. 1, pp. 202–203.
5. Hunter, C. (Ed.) 1983. *op. cit.*, p. 359.
6. Wilson, A. and Bonaparte, C.L. 1831. *op. cit.*, Vol. 3, p. 167.
7. Hunter, C. (Ed.) 1983. *op. cit.*, p. 112.
8. Wilson, A. and Bonaparte, C.L. 1831. *op. cit.*, Vol. 3, pp. 7–8.

Wollweber

No biographical information traced. Details in text are from: Bonaparte, C.L. 1850. Zoologie.—Sur deux espèces nouvelles de Paridæ. *Comptes Rendus de l'Académie des Sciences de Paris* **31**: 478–479.

Wright

Dupree, A.H. 1959. *Asa Gray*, pp. 165–166, 183, 204–211, 324. Harvard University Press, Cambridge.
Ewan, J. and Ewan, N.D. 1981. *Biographical Dictionary of Rocky Mountain Naturalists, 1682–1932*, p. 244. Hans A. Stafleu, Utrecht.
Geiser, S.W. 1958. Chapter 9. Charles Wright. *Naturalists of the Frontier*, 2nd revised edn, pp. 172–198. Southern Methodist University Press, Dallas.
[Sargent, C.S.] 1889. *Scientific Papers of Asa Gray*, Vol. II. Essays; biographical sketches. Charles Wright, pp. 468–474. MacMillan and Co., London.
Urban, I. 1898–1913. *Symbolae Antillanae*, Vol. 3, pp. 141–144. Berolini.
1,2. *Quoted in* Dupree, A.H. 1959. *op. cit.*, pp. 183, 166.
3,4. *Quoted in* [Sargent, C.S.] 1889. *op. cit.*, pp. 474, 468.
5. *Quoted in* Dall, W.H. 1915. *Spencer Fullerton Baird. A Biography*, p. 277. J.B. Lippincott Co., Philadelphia and London.
6. *Annals of the Lyceum of Natural History of New York* 1851, pp. 118–119.

Xántus

Baird, S.F. 1860. Notes on a collection of Birds made by Mr John Xantus. *Proceedings of the Academy of Natural Sciences of Philadelphia* (1859) **2**: 299–306.

Harris, H. 1934. Notes on the Xantus tradition. *Condor* 36: 191–201.

Hume, E.E. 1942. Chapter XXXV. John Xántus (1825–1894). *Ornithologists of the U.S. Army Medical Corps*, pp. 510–532. John Hopkins University Press, Baltimore.

Madden, H.M. 1949. *Xántus, Hungarian Naturalist in the Pioneer West.* Oberösterreichischer Landesverlag, Linz, Austria.

Zwinger, A.H. 1986a. *John Xántus: The Fort Tejon Letters, 1857–1859.* University of Arizona Press, Tucson.

Zwinger, A.H. 1986b. *Xántus. The Letters of John Xántus to Spencer Fullerton Baird from San Francisco and Cabo San Lucas 1859–1861.* Dawson's Bookshop, Los Angeles.

1,2. *Quoted in* Madden, H.M. 1949. *op. cit.*, pp. 23, 22.

3–8. *Quoted in* Zwinger, A.H. 1986a. *op. cit.*, pp. 8–9, 25, 30, 81, 187, 203.

9,10,11. *Quoted in* Zwinger, A.H. 1986b. *op. cit.*, pp. 314, 121, 359.

12. Steinbeck, J. 1960. *The Log from the Sea of Cortez*, pp. 124–125. Pan Books Ltd, London and Sydney.

Appendix

A miscellaneous selection of naturalists commemorated by well-known races or hybrids, by recently obsolete names, or by species that are only accidental visitors to the United States and Canada.

AIKEN, Charles Edward Howard
　　b. Benson, Vermont: 7 September 1850
　　d. Colorado Springs, Colorado: 15 January 1936
Pioneer ornithologist of Colorado. Aiken's father moved to the vicinity of Colorado Springs to begin sheep ranching after losing his business in the 1871 Chicago fire. Aiken developed his early interest in birds by collecting locally and then farther afield. In 1874 he joined one of the Wheeler Survey parties in the southern Colorado mountains and in the following season collected in the region again, in the San Luis valley. In 1876 he went by wagon through New Mexico into Arizona for about four months. When not on field trips he worked on various enterprises including his taxidermy shop in Colorado Springs where he excelled at preparing small birds as well as the routine mounting of deer and game for hunters. In 1907 William Lutley Sclater, Director of the Museum of Colorado College, arranged the purchase of Aiken's bird skins, which eventually amounted to 6800 specimens (now at the University of Colorado Museum). Aiken published little about birds but his notes were used by several other ornithologists including Robert Ridgway in his 'Birds of Colorado', a paper for the *Bulletin of the Essex Institute* (1873). Aiken was held in high regard by Henry W. Henshaw, Edward W. Nelson, Elliott Coues, William Brewster, J.A. Allen and Frank Stephens who all visited him.
　　Dark-eyed Junco *Junco hyemalis aikeni* Ridgway, 1873. Formerly given full specific status as the White-winged or Aiken's Junco.

ARCHILOCHUS, or ARKHILOKHOS
　　b. Paros, Greece: c.720 BC.
　　d. Paros, Greece: c.680 BC.
One of the first Greek lyric poets, famed for his savage satires. As a young man he moved to Thasos where he disgraced himself by running away in battle, afterwards versifying his cowardice. When he eventually returned to his native island he was killed in battle against the neighbouring Naxians.
　　Archilochus is one of several hummingbird genera created by H.G.L. Reichenbach in 1854 to commemorate classical Greeks. The Ruby-throated and Black-chinned Hummingbirds are usually the only species in the genus, though some authors merge *Calypte* with *Archilochus*.

ARMINJON, Vittorio
  b. Chambéry, France: 9 October 1830
  d. Genoa, Italy: 5 February 1897

Italian naval commander. Arminjon served during the Crimean campaign when his organizational and administrative abilities brought him distinction. He afterwards took the corvette *Magenta* on a round the world mission of diplomacy and scientific discovery (1865–68), many of the results being written up by the naturalists H.H. Giglioli and F. de Filippi. Arminjon reformed the Italian naval school and at the age of forty-seven rose to the rank of vice-admiral.

Herald Petrel *Pterodroma arminjoniana* (Giglioli and Salvadori, 1869). First taken during the voyage of the *Magenta*. Henry Giglioli wrote: "I found this species pretty common near Trinadad[e] Island, in the South Atlantic, in lat. 20° S. or thereabouts. On the 23rd of January, 1868, as we lay becalmed about eight miles off the island, many specimens were shot; unfortunately, believing it a well-known species, I had only two skins prepared." Casual in North Atlantic off North Carolina. Accidental New York. Also recorded off the Hawaiian Islands where the race *P.a.heraldica* occurs.

ATTWATER, Henry Philemon
  b. London, England: 28 April 1854
  d. Houston, Texas: 25 September 1931

Agriculturalist and pioneer conservationist in Texas. Attwater emigrated in 1873 to Canada and was married there in 1885. Moving to San Antonio, Texas, in 1889, he played a long and significant role in the agricultural development of his adopted State. He also pursued his lifelong interest in birds and at various times sent collections to the Washington and New York museums as well as to private collectors. He strongly urged the protection of birds through his lectures and newspaper articles and was chiefly responsible for the passing of the Texas bird law. His few ornithological works include a sixty-four page paper on the 'Use and Value of Wild Birds to Texas Farmers and Stockmen and Fruit and Truck Growers' for *Bulletin No.27, Texas Department of Agriculture* (1914). His 'List of Birds Observed in the Vicinity of San Antonio' appeared in *The Auk* (1892).

Greater Prairie-Chicken *Tympanuchus cupido attwateri* Bendire, 1893. Attwater's Prairie-Chicken is found only in south-eastern Texas where its once abundant populations are now isolated and much reduced. In 1972 the Attwater Prairie Chicken National Wildlife Refuge was established near Eagle Lake, Houston, where about 8000 acres are managed for their benefit.

BALDWIN, Lilian Converse Hanna
  b. Cleveland, Ohio: 3 December 1852
  d. Cleveland, Ohio: 27 December 1948

Lilian Hanna was the youngest of seven children, one of whom became a U.S. Senator. At the age of forty-five she married Samuel Prentiss Baldwin and although he was sixteen years her junior she outlived him by ten years. A former geologist turned lawyer, he gave up active practise because of ill-health and devoted the last part of his life to bird study. Every year from 1901, most of each summer was spent on their estate at Gates Mills, near Cleveland, where the Baldwin Bird Research Laboratory was founded. He carried out photographic studies of bird embryos, developed bird catching techniques, made an intensive banding study of House Wrens, and for many years had a late-winter trapping site on another estate at Thomasville, Georgia. Baldwin's early papers on banding demonstrated its wide scientific potential and were so influential that he has been considered "one of the noteworthy ornithologists of all time". They had no children and were prominent in Cleveland social life. Lilian was a generous donor to many institutions including the Medical and Biological Departments of the Western Reserve University and the Metropolitan Opera. She made many trips abroad, using these opportunities to add to her collection of rare glassware, particularly paperweights. In the last year of her life she gave sixty-four acres of her estate to Gates Mills for a park and bird sanctuary.

Lilian's Meadowlark *Sturnella magna lilianae* Oberholser 1930. The type specimen was collected by Wilmot Brown on 2 November 1929 in the Huachuca Mountains, one of 512 birds from southern Arizona and New Mexico purchased from him by Lilian Baldwin. Harry Oberholser worked through the collection and wrote: "It gives us great pleasure to dedicate this beautiful new race of meadowlark to Mrs Lilian Hanna Baldwin (Mrs. S. Prentiss Baldwin), who ... presented the Cleveland Museum of Natural History with the present collection of birds, including the type of this new subspecies." Lilian's Meadowlark may differ sufficiently in vocalizations, morphology and genetics to be separated from the Eastern Meadowlark as a full species (Sibley and Munroe, 1990).

BALTIMORE, 1st Lord
   b. Kipling, Yorkshire, England: 1582
   d. London, England: 15 April 1632

George Calvert was one of the principal secretaries of state under James I. In 1625 he was made Baron Baltimore, in the county of Longford, Ireland. He set up a colony in Newfoundland but because of the harsh climate and attacks by the French he asked permission to move it to Maryland. He died before confirmation was received and his son Cecil inherited the charter of Maryland and sent out twenty colonists and three hundred labourers. Cecil's heir, Charles Calvert, became Proprietary Governor of Maryland 1661–75 and Proprietor of the Province 1675–88. The city of Baltimore takes its name from the family title.

Northern Oriole *Icterus galbula* (Linnaeus, 1758). Linnaeus's description is based on 'The Baltimore Bird' of Mark Catesby's *Natural History of Carolina* (1754). According to Catesby: "This Gold-coloured Bird I have only seen in *Virginia* and *Maryland*; there being none of them in *Carolina*. It is said to have its name from the Lord *Baltimore's* Coat of Arms, which are Paly of six, Topaz and Diamond, a Bend, counterchang'd; his Lordship being a proprietor of those Countries." This indicates that the name had been in use for some time but, though perhaps not referring to the first Lord Baltimore, it was he who began to lay the foundations for one of the most successful colonies. The Baltimore Oriole, named because its plumage was similar to the Baltimore family colours, is now lumped together with Bullock's Oriole as the Northern Oriole.

BELCHER, Edward
   b. Nova Scotia: 1799
   d. London, England: 18 March 1877

British Rear-Admiral. Grandson of the Governor of Halifax and great grandson of the Governor of Massachusetts, New Hampshire and New Jersey. He entered the Royal Navy in 1812 and served at home and abroad. In 1825 he went with Captain Beechey on the *Blossom* to explore for a North-west Passage east of the Bering Straits (at the same time as Franklin's second overland Arctic trip and Parry's approach from Lancaster Sound). The *Blossom* left England in May and was away for three years calling at San Francisco, November 1826, and Monterey, January 1827. Lieutenant Belcher formed mineral collections and assisted Alexander Collie, the surgeon, and George T. Lay, the naturalist. *The Zoology of Captain Beechey's Voyage ... 1825–1828* (1839) had sections on mammals by John Richardson, fishes by G.T. Lay and E.T. Bennett, reptiles by

J.E. Gray and birds by N.A. Vigors. Belcher later explored again in the Pacific as well as in the East Indies; he was in action against China in 1841 and received a knighthood for his services. In 1843 he published a *Narrative of a Voyage Round the World performed in H.M.S. Sulphur ... 1836–42*. Captain Belcher led a small fleet to the Arctic in 1852 to search for Sir John Franklin but abandoned four of his five ships in the pack ice—at a subsequent court-martial he was reluctantly acquitted.

Band-tailed Gull *Larus belcheri* Vigors, 1829. Breeds along the coasts of Peru, north-western Chile and northern Argentina. Casual in Panama. Some Florida records.

BELDING, Lyman
  b. West Farms, Northampton, Massachusetts: 12 June 1829
  d. Stockton, California: 22 November 1917
Ornithologist in California and Baja California. Belding spent much of his boyhood hunting in the wooded hills of Pennsylvania. After an attack of typhoid he was advised in 1851 to take a sea voyage so he signed up for three and a half years on a whaling vessel. He moved to Stockton in 1856 but it was not until 1876, when he received a copy of J.G. Cooper's 'Ornithology of California' that his dormant interest in birds emerged. Encouraged by Baird and Ridgway he collected at Cerros Island, Scammon's Lagoon and other parts of Baja California, 1881–82. He was at Cape San Lucas, 1882–83, and made several more trips to the region. At La Paz in 1881 he took the first nests and eggs of Costa's Hummingbird. Belding was one of the first ornithologists to collect in the Sierras of central California. Between 1878 and 1905 he contributed nearly fifty papers and notes to scientific journals, the most important of which concerned his collections from Baja California and his 'Land Birds of the Pacific District' which appeared as an occasional paper of the California Academy of Sciences in 1890.

Belding's Sparrow *Passerculus sandwichensis beldingi* Ridgway, 1885. This race of Savannah Sparrow was obtained by Belding at San Quentin Bay, Baja California, and occurs as far north as the Santa Barbara region of California. (Belding's Yellowthroat *Geothlypis beldingi* is a resident of southern Baja California and was also named for him by Ridgway.)

BICKNELL, Eugene Pintard
  b. Riverdale on Hudson, New York: 23 September 1859
  d. Hewlett, Long Island, New York: 9 February 1925
New York banker and youngest founder member of the A.O.U. His 'Review of the Summer Birds of a Part of the Catskill Mountains' for the *Transactions of the Linnean Society of New York* (1882) long remained the best ornithological study of that region. A quiet and reserved man, more inclined to watch and listen to birds than shoot them, he also published a series of six articles in the first two volumes of *The Auk* (1884–85) entitled 'A Study of the Singing of our Birds'. Between 1895 and 1917 his attention was diverted to botany but after he took up birds again in his later years some of his notes from Long Island found their way into Ludlow Griscom's *Birds of the New York City Region* (1923). For portrait see p. 156.

Bicknell's Thrush *Catharus minimus bicknelli* (Ridgway, 1882). Bicknell startled his fellow ornithologists in 1881 by discovering this new race of the Gray-cheeked Thrush breeding in the Catskills, an area thought to have been well worked. It was subsequently found to inhabit the 3000 ft to timberline portions of the mountains of New England and eastern New York State and may be a distinct species.

BREWSTER, William

b. Wakefield, Massachusetts: 5 July 1851
d. Cambridge, Massachusetts: 11 July 1919

One of the great American ornithologists. Principal founder and first president of the Nuttall Ornithological Club. Founder member of the A.O.U., President 1895–98. Son of a successful banker, as a youth he had Henry W. Henshaw as a classmate and Ruthven Deane as a neighbour. Brewster loved the area around Cambridge where he remained based for the whole of his life. Among his works were *The Birds of the Cambridge Region of Massachusetts* (1906) and his revised versions of H.D. Minot's *The Land Birds and Game Birds of New England* (1895 and 1903). *Birds of Lake Umbagog* (1924) commemorates his frequent summer visits to the New Hampshire–Maine border. By directing collectors to north-western Mexico, Baja California and the southwestern United States he was also able to specialize in the birds of those regions. These men included Frank Stephens (California and Arizona, 1881, 1884), R.R. McLeod (Mexico, 1883, 1884, 1885), Abbott Frazar (Baja California, 1887), and J.C. Cahoon (Arizona and north Mexico, 1887). His extensive bird collection from these and other parts of North America went to the Cambridge Museum of Comparative Zoology at Harvard. Though Brewster preferred his home State he enjoyed trips to Britain (1891, 1909, 1911), West Virginia, with Ruthven Deane (1874), Illinois, to see Robert Ridgway (1878), the Gulf of St Lawrence (1881), Colorado, with J.A. Allen (1882), South Carolina (1883, 1884, 1885), Florida and Trinidad, with Frank Chapman (1890, 1893). In 1896 Brewster helped organize the Massachusetts Audubon Society and agreed to become its President. The success of this first Audubon Society quickly led to the formation of others. For portrait see p. 156.

Brewster's Linnet *Aegiothus* (*flavirostris* var.) *Brewsterii* Ridgway, 1872. Known only from the type specimen taken at Waltham, Massachusetts in 1870. It is possibly a hybrid between Common Redpoll and Pine Siskin.

Brewster's Warbler *Helminthophaga leucobronchialis* Brewster, 1874. Brewster collected the first specimen at Newtonville, Massachusetts, on 18 May 1870, describing it later as a new species. But in 1881 he published a paper correctly asserting that both Brewster's and Lawrence's Warblers are hybrids of Golden-winged and Blue-winged Warblers.

BULLOCK, William

d. Chelsea, England: 7 March 1849—"at 76 years of age"

British antiquarian, showman, traveller, naturalist and bird collector. He became active in the museum world about 1799 in the north of England. His museum of curios and natural history subjects was a popular attraction in London from 1809 until 1819 when he sold off all the items—acting as his own auctioneer. He set off for Mexico in December 1822 to invest in silver mines there, afterwards publishing *Six Months Residence and Travels in Mexico* (1824). His son, also William Bullock, remained in Mexico sometimes collecting birds with Ferdinand Deppe, many of their birds being worked through by William Swainson. The elder Bullock went back to Mexico for a short period, returning to Britain through the United States where he was particularly taken with the country around Cincinnati. In 1827 he produced a *Sketch of a Journey through the Western States of North America*. The details of his death at 14 Halsey Terrace, Chelsea, from "fatty degeneration of the heart", were rediscovered in 1978 by Edward P. Alexander of the University of Delaware.

Bullock's Oriole *Icterus galbula bullockii* Swainson, 1827. "This, the most beautiful of the group yet discovered in Mexico, will record the name of those ornithologists who have thrown so much light on the birds of that country." When naming this bird

Swainson was obviously thinking of the two William Bullocks, father and son. Though long known as Bullock's Oriole it has recently been reclassified as a race of the Northern Oriole.

CABOT, Samuel
   d. Boston, Massachusetts: 13 April 1885—"in his seventieth year"
Medical doctor and curator of ornithology for the Boston Society of Natural History. He graduated from Harvard in 1836 and the Harvard Medical School in 1839. He made an important collecting trip to Yucatan (1841–42) and published numerous short papers on his discoveries there. From 1850 his professional commitments prevented much further natural history work. His collection went to the Boston Society of Natural History.
   Sandwich Tern *Sterna sandvicensis acuflavida* Cabot, 1847. The American race of the Sandwich Tern, of the east coast of the United States south to Yucatan, is sometimes known as Cabot's Tern.

CANUTE, or CNUT the GREAT
   b. Denmark?: c.995
   d. Shaftesbury, England: 12 November 1035
King of England, Norway and Denmark. Invaded England 1015, established more efficient administration and codified law. Became King of Denmark from 1018, and of Norway after invading it in 1028.
   Red Knot *Calidris canutus* (Linnaeus, 1758). The connection between King Canute and the shorebird can be traced back to the 1607 edition of William Camden's *Britannia*. Camden thought that the word 'knot' alluded to the well-known story concerning Canute, who rebuked his flattering courtiers by physically demonstrating that despite his huge empire he had no power to stop the incoming tide. Camden was only expressing his opinion that the Red Knot, which often frequents the tide edge, could be connected with Canute, but later authors determinedly repeated the idea and Linnaeus compounded the theory by calling the bird *Tringa Canutus*. The word 'knot' is more likely to be derived from the sound of the bird's grunting call; or from the idea that the dense wintering flocks of the birds, at a great distance, can resemble clouds of gnats. Either way, the scientific name refers to Canute, but 'knot' does not.

CONSTANT, Charles
   b. Oignies, Belgium: 13 April 1820
   d. Hénin-Liétard, France: 1905
Licensed medical practitioner who worked in northern France, near Arras. Best known as a poet for his *Chansons inédites* (1845), he was also a horticulturalist, manufacturer of remedies, painter, musician and taxidermist. Though a likely candidate, there is no clear evidence to link him with the scientific name of the Plain-capped Starthroat.
   Plain-capped Starthroat *Heliomaster constantii* (De Lattre, 1843). A Central American species, also known as Constant's Starthroat. Casual in south-eastern Arizona.

COUES, Elliott
   b. Portsmouth, New Hampshire: 9 September 1842
   d. Baltimore, Maryland: 25 December 1899
One of the greatest of all ornithologists. He came under the influence of Baird at an early age and gained a place on an expedition to Labrador where he studied and collected

seabirds. After medical school and the horrific experience of treating Civil War wounded, he was posted to Arizona. "The Apaches are so hostile and daring" he wrote to Baird, "that considerable *caution* will have to tinge my collecting enthusiasm, if I want to save my scalp." In fact his spell in the West was remarkable for the size and quality of his collection, the discovery of birds and mammals new to science and the large volume of published material that ensued. Afterwards he served on the east coast and wrote his *Key to North American Birds* (1872), possibly his most important work; it certainly influenced a whole generation of American ornithologists. His *Check List of North American Birds* (1873) appeared when Coues was back in Indian country at Fort Randall in South Dakota. Four years later Custer was killed at the nearby valley of Little Big Horn, but Coues was then working on the Northern Boundary Survey. In 1876 he was transferred to the U.S. Geological Survey of the Territories to Colorado, and remained connected with it until 1880. Most of this last period was spent at Washington editing survey publications but he also produced *Birds of the Northwest* (1874) and *Birds of the Colorado Valley* (1878). He retired from the army in 1881 and spent much of the rest of his life writing, one of his self-appointed tasks being the documentation of the lives and travels of the earlier frontier explorers—to which he added zoological notations. Coues was one of the most prolific ornithological writers that America has produced: in one year he had over fifty publications. He was one of the three principal founders of the A.O.U. and was a friend and advisor to the bird artist Fuertes. His outspoken views on ornithology and other subjects often made him a controversial character but even those he had antagonized were willing to acknowledge the value of his work, much of which was considered to be brilliant. For portrait see p. 156.

Greater Pewee *Contopus pertinax* Cabanis and Heine, 1859. Long known as Coues's Flycatcher because of specimens obtained by Coues at Fort Whipple and described by him in the *Proceedings of the Academy of Natural Sciences of Philadelphia* (1866).

COX, John Brian
b. Isleworth, England: 15 March 1945.

Cox began birding as a boy in England and travelled throughout the British Isles and much of Europe. At the age of twenty-three he met an Australian girl, was married at Uxbridge, near London, and then emigrated to South Australia where they now live with their four children. Most of his employment has been centred around birds and he has variously conducted environmental impact surveys, participated in expeditions to remote areas, led tours and worked as a National Parks Ranger. He is currently employed by local government to develop a wetland site and nature trail. Most of his forty-seven scientific papers concern seabirds and shorebirds.

Cox's Sandpiper *Calidris paramelanotos* Parker, 1982. In South Australia, on 16 February 1975, Cox and two friends saw an unusual *Calidris* which was later collected. On 5 March 1977 Cox collected a second specimen, which became the type for Cox's Sandpiper, named by Shane Parker, the Curator of Birds at the South Australia Museum. Cox's Sandpiper is very similar to the unique specimen of Cooper's Sandpiper *Calidris cooperi* from Long Island, New York, named by S.F. Baird in 1858. The status of these birds is uncertain: if they are the same species, Baird's name takes priority; if they are both hybrids then both scientific names are invalid. A bird claimed to have been a Cox's Sandpiper was banded and photographed at Duxbury Beach, Plymouth County, Massachusetts, on 15–21 September 1987. Since then Cox has collected two more specimens, which have been deposited in the Australian National Wildlife Collection at Canberra. He expects that biochemical testing will confirm his opinion that Cox's Sandpipers are hybrids.

DEGLAND, Côme Damien
  b. Armentières, France: 6 July 1787
  d. Lille, France: 1 January 1856

Director of the natural history museum at Lille. Author of *Ornithologie Européene*, 2 vols (1849). Bonaparte gave it some well-deserved criticism in his *Revue Critique de l'Ornithologie Européene de M. le Docteur Degland (de Lille)* (1850) saying that "the author had performed a miracle since he had worked without a collection and without a library". A second improved edition by Z. Gerbe appeared in 1867, but again there were no illustrations.

  White-winged Scoter *Melanitta fusca deglandi* (Bonaparte, 1850). Named by Bonaparte despite the foregoing. It is sometimes separated as a full species from the European race, which is known as the Velvet Scoter.

DIAZ, Augustin
  b. Mexico City, Mexico: 1829
  d. Mexico City, Mexico: 1893

Mexican engineer and geographer. Diaz graduated as a lieutenant-engineer in 1847. He served with distinction in the war against the United States, taking part in the construction of the defence works of Mexico City and Chapultepec. From 1850 to 1857 he served with the U.S. and Mexican Boundary Commission. Between 1861–64 and 1869–77 he was Professor of Engineering at the Mexican Military College. He took part in the scientific exploration of Yucatan and was director of the Mexican Geographical and Exploring Commission which made natural history collections in various parts of the country.

  Mexican Duck *Anas diazi* Ridgway, 1886. The A.O.U. *Check-list* advocates the merger of the Mexican Duck with the Mallard because of extensive hybridization between the two groups in south-eastern Arizona, southern New Mexico and west-central Texas.

DRESSER, Henry Eeles
  b. Thirsk Bank, Yorkshire, England: 9 May 1838
  d. Cannes, France: 28 November 1915

Timber trader and ornithologist. Educated in England and Germany, his family timber business required him to travel widely in the Baltic region so he lived for a while in Finland to learn the language. In 1859 he went to New Brunswick to manage a timber estate. Coming home in 1860 he went to Russia, Finland and Germany for two years before returning to New Brunswick in 1862. In 1863 he took a cargo to the Confederate States, remaining in Texas for eighteen months (see Heermann). From 1864 to 1870 he travelled throughout Europe, then set up in the metal business in London. In 1871 he started his famous *A History of the Birds of Europe*, completed it in 1881 and published a supplement in 1896. He also produced *Eggs of the Birds of Europe* (1905–10) and monographs on the bee-eaters and rollers. He used every opportunity to collect during his business journeys and over 11,000 bird skins went eventually to the Manchester University Museum.

  Common Eider *Somateria mollissima dresseri* (Sharpe, 1871). The Atlantic or eastern North American race has been known in the past as the American or Dresser's Eider.

ELLIS, William W.
  b. England: date unknown
  d. Ostend, Belgium: June 1785

Artist–naturalist with Cook's last expedition (see Cook). He joined the *Discovery* as a surgeon's second mate but is best known for his bird notes and bird paintings (including eleven of Hawaiian species). In 1783 he published *An Authentic Narrative of a Voyage performed by Captain Cook [1776–80]*. "His death was occasioned by a fall from the main mast of a ship at Ostend when he was on his way to Germany where the Emperor had engaged him on advantageous terms, to undertake a voyage of discovery."

Greater Akialoa *Hemignathus ellisianus* (Gray, 1859). This scientific name is used if the akialoas are separated into two species. The Greater Akialoa was last seen in the Alakai Swamp on Kauai in 1967. The Lesser Akialoa has not been recorded since 1940. The akialoas are now probably extinct everywhere.

EVELYN. Identity unknown.
Bahama Woodstar *Calliphlox evelynae* (Bourcier, 1847). Resident throughout the Bahama Islands. Casual in southern Florida. It was first described in the 1847 *Proceedings of the Zoological Society of London*, from a specimen from Nassau, but no dedication was given. Bourcier's wife was named Aline, so Evelyn may or may not have been a member of his family. Bourcier was French Consul General in Ecuador from 1849 to 1850 and was subsequently author or co-author of over twenty species of hummingbirds.

FISHER, Albert Kenrick
    b. Sing Sing (Ossining), New York: 21 March 1856
    d. Washington, D.C.: 12 June 1948
A founder member and President of the A.O.U. (1914–16). In 1879 he graduated from the College of Physicians and Surgeons, New York, where C. Hart Merriam was a classmate. They worked together for two years on migration and the geographical distribution of birds, both subsequently giving up medicine to pioneer the Division of Economic Ornithology of the U.S. Government which began in 1885. Fisher served as ornithologist with the Death Valley Expedition in 1891, his report also including the birds of the adjacent areas. His earlier important work on 'The Hawks and Owls of the United States in Their Relation to Agriculture' came out in 1893 and helped to reduce the destruction of these birds by presenting the results of the critical examination of 2700 raptor stomachs. He joined the famous Harriman Alaska Expedition in 1899 with Fuertes as his official assistant. When the U.S. Biological Survey was organized in 1905 Fisher proved his worth by his ability to deal with members of Congress on conservation matters. He went to the South Seas on the 1929 Pinchot Expedition, only retiring in 1931 after forty-six years service with the Federal Government. His son, Walter Kenrick Fisher, was a noted biologist and author of *Birds of Laysan and the Leeward Islands* (1903). For portrait of A.K. Fisher see p. 156.

Seaside Sparrow *Ammodramus maritimus fisheri* Chapman, 1899. The Gulf Coast form of this highly variable sparrow. The type was collected by A.K. Fisher at Grand Isle, Louisiana, on 9 June 1886.

HARCOURT, Edward William Vernon
    b. York, England: 26 June 1825
    d. Nuneham Park, Abingdon, England: 19 December 1891
Descended from the distinguished Vernon family; the additional name Harcourt was assumed only in the 1830s. His father William was Canon of York, and founder, first Secretary and afterwards President of the British Association. Edward Harcourt was M.P. for Oxford, 1878–85, but he mainly led the life of a country gentleman. He made at least two trips abroad, publishing *Sketch of Madeira* (1851) and *Sporting in Algeria* (1859). The former contains his description of a new species of petrel; the latter has several appendices on birds and mammals but they are summaries of the work of Malherbe and Loche. Five of his papers on the birds of Madeira are listed in David Bannerman's *Birds of the Atlantic Islands* (1963–68), vol. 2, pp. xiii–xiv.

Band-rumped Storm-Petrel *Oceanodroma castro* (Harcourt, 1851). This species has two distinct breeding ranges in the tropical regions. The old name Harcourt's Storm-Petrel is now falling into disuse.

HARLAN, Richard
b. Philadelphia, Pennsylvania: 19 September 1796
d. New Orleans, Louisiana: 30 September 1843

Prominent nineteenth century physician and naturalist. Harlan graduated M.D. from the University of Pennsylvania in 1818, practised at Philadelphia then became Professor of Comparative Anatomy at the Philadelphia Museum. His 318-page *Fauna Americana; being a description of the mammiferous animals inhabiting North America* (1825) was an important early work, though he was later more concerned with studies of fossils and reptiles.

Harlan's Hawk *Buteo jamaicensis harlani* (Audubon, 1831). This dark phase of the Red-tailed Hawk was formerly considered a separate species. When naming it, Audubon said that he had "waited until a species should occur, which in its size and importance should bear some proportion to my gratitude toward that learned and accomplished friend". Audubon also gave it the splendid name Black Warrior.

HELOISE. Identity unknown.
Bumblebee Hummingbird *Atthis heloisa* (Lesson and De Lattre, 1839). Also known as Heloise's Hummingbird, this Mexican species was first described from specimens from Jalapa and Coátepec by two French naturalists, R.P. Lesson and Adolphe De Lattre. No dedication was given and the identity of Heloise remains unknown. Accidental in the Huachuca Mountains of Arizona.

HEPBURN, James Edward
b. London, England: 1810 or 1811
d. Victoria, Vancouver Island: 16 April 1869

A little-known west-coast bird collector; his travels were worked out after the re-discovery of his notebooks in the 1920s at the University of California. Educated in Sussex and at Cambridge University, he practised as a barrister in London for a few years. He emigrated to California sometime prior to 6 May 1852—the earliest entry in his notebooks. Over the next eight years he collected primarily in the region of San Francisco, also making trips into the Sierras. In November 1860 he was in British Columbia and in subsequent years collected at San Francisco and up and down the coast of Washington and British Columbia to as far north as Sitka, Alaska. Hepburn was an accurate and reliable observer and assisted Robert Brown in his 'Synopsis of the Birds of Vancouver Island' for *The Ibis* (1868). Hepburn is often quoted in Baird, Brewer and Ridgway's *A History of North American Birds* (1874) but he never published anything himself. His bird collection of 1500 skins of over 300 species was sent back to Cambridge University by his relatives, together with an assortment of mammals, reptiles, fish, insects and American Indian artefacts.

Hepburn's Rosy Finch *Leucosticte arctoa littoralis* Baird, 1869. Hepburn collected this bird in British Columbia at Fort Simpson or in the nearby mountains, in the month of September. The rosy finches are sometimes split into several different species. Hepburn's Rosy Finch is part of the Gray-crowned group.

HODGSON, Brian Houghton
b. Prestbury, Cheshire, England: 1 February 1800
d. Alderley, Gloucester, England: 23 May 1894

Self-taught naturalist who made an enormous contribution to Indian ornithology while pursuing a brilliant career with the Civil Service. He donated his large bird collection from Katmandu and Darjeeling to the British Museum (Natural History).

Olive Tree-Pipit *Anthus hodgsoni* Richmond, 1907. Also known as Olive-backed Pipit. Breeds north-eastern Russia and central Siberia south to the Himalayas, western China and Japan. Casual to western Aleutian Islands.

HOLBÖLL, Carl Peter
b. Copenhagen, Denmark: 31 December 1795
d. At sea, between Denmark and Greenland: Spring 1856

Danish zoologist, son of Frederik Ludvig Holböll (1765–1829) the head gardener at the Copenhagen Botanical Gardens. In 1818 he was appointed 2nd Lieutenant in the Danish navy but obtained leave of absence on full pay to undertake an expedition up the west coast of Greenland to study whaling and collect for the Royal Natural History Museum of Copenhagen. He returned to Denmark in 1824, resigned from the navy and went again to Greenland where he had been appointed Royal Inspector for Trade and Whale Fishery in North Greenland. In 1827 he took up a comparable position in the south of the island and remained involved in Greenland's affairs for the rest of his life. His most important written work was his 'Ornithological Contributions to the Fauna of Greenland' which appeared in Danish in *Naturhistorisk Tidsskrift* (1842–43). He was lost at sea when the brig *Baldur* disappeared without trace on passage from Denmark to Greenland.

Red-necked Grebe *Podiceps grisegena holboellii* Reinhardt, 1854. This is the North American and east Siberian race. It has been known as Holböll's Grebe.

HUTCHINS, Thomas
b. Place and date of birth unknown
d. London, England: 7 July 1790

Surgeon who collected birds and plants in the Hudson Bay area. He entered the Hudson's Bay Company service in 1766 and was employed at York Factory until 1773, during this period becoming friendly with Andrew Graham, the master of Fort Severn who later became acting chief at York Factory. After a year of leave in England Hutchins was appointed chief at Fort Albany and remained in this position until 1782, when he returned to Britain. His observations from around Fort Albany were given to Thomas Pennant who incorporated much of the material into the second volume of his *Arctic Zoology* (1784–85). In this way Hutchins won the reputation of being an authority on Hudson Bay wildlife that was mostly undeserved because he had appropriated as his own large portions of Graham's notes.

Hutchins's Goose *Branta canadensis hutchinsii* (Richardson, 1832). Dr John Richardson described a male goose killed on 19 June 1822 on the Melville Peninsula (during Parry's second Arctic voyage) as a variety of the Brant. He changed his mind in *Fauna Boreali-Americana* (1832), giving it full specific status with the following dedication: "We have designated the [goose] by the name of Hutchinsii, in honour of a gentleman from whom Pennant and Latham derived most of their information respecting the Hudson's Bay birds." This race of Canada Goose is sometimes also called Richardson's Goose.

JOUANIN, Christian
b. Paris, France: 10 July 1925

French pharmacist, ornithologist and conservationist who lives near Paris. Though now retired from pharmacy he maintains his long association with the ornithology department at the Paris Natural History Museum. He made an expedition to Réunion in 1964 and has been many times to the Salvagems, near the Canary Islands, to study seabirds. He still visits the Paris Natural History Museum and has helped with this volume by providing information about many of the French subjects.

Jouanin's Petrel *Bulweria fallax* Jouanin, 1955. A north-west Indian Ocean species whose breeding grounds are still unknown. A single specimen collected on Lisianski (Hawaiian Islands) in 1967 is the only record for the Pacific.

KRIDER, John

d. Philadelphia, Pennsylvania: 1886—"in his 74th year"

Philadelphia gunsmith, taxidermist and sportsman. He collected as far west as Colorado and as far north as the Bay of Fundy but produced no scientific work. He went to Florida with A.L. Heermann, March–April 1848; and took charge of Heermann's manuscript field notes, drawings and eggs when Heermann moved to Texas. In the 1870s Krider frequently went to Iowa. Unfortunately his *Forty Years Notes of a Field Ornithologist* (1879) was written from memory with no anecdotal material about his travels or friends; the ornithological content is considered very unreliable.

Krider's Hawk *Buteo kriderii* Hoopes, 1873. This pale phase of the Red-tailed Hawk from the Great Plains was described as a distinct species by Krider's friend Bernard A. Hoopes from a couple of birds collected by Krider in Winnebago County, Iowa.

KUMLIEN, Ludwig

b. Sumner, Wisconsin: 15 March 1853

d. Milton, Wisconsin: 4 December 1902

Son of Thure Kumlien—a pioneer naturalist of Wisconsin (of Swedish birth) whose observations were included in Baird, Brewer and Ridgway's *A History of North American Birds* (1874–84). Ludwig Kumlien was a graduate from the University of Wisconsin and is best known to ornithologists for his two years in the Arctic. His 'Contributions to the Natural History of Arctic America made in connection with the Howgate Polar Expedition 1877–78' formed *Bulletin of the United States National Museum* No 15 (1879). At other times he was an assistant with the U.S. Fish Commission and a special agent of Fisheries for the Tenth Census. From 1891 until his death he was Professor of Physics and Natural History at Milton College, Wisconsin. He wrote occasionally for *The Auk*, *The Nidologist* and *Forest and Stream*.

Kumlien's Gull *Larus glaucoides kumlieni* Brewster, 1883. This race of Iceland Gull was first described from a bird taken in the Canadian Arctic by Ludwig Kumlien. He found his specimen in Cumberland Sound on 14 June 1878, and also collected an egg—in such poor condition that it had to be held together with thread, being all that was left of the contents of two nests destroyed by Ravens.

LA SAGRA, Ramón de

b. Corunna, Spain: 1801

d. Corunna, Spain: 25 May 1871

Historian and writer on the political economy of Cuba. Shortly after completing his studies at Madrid he went to Cuba and became Director of the Botanical Garden at Havana. He returned to Europe in 1835 and took his natural history collection to Paris where it was worked on by various specialists. He edited the thirteen-volume *Historia física, política y natural de la Isla de Cuba* (1839–61) which also appeared at the same time in French. The bird section was by Alcide d'Orbigny (1839); the mammals by La Sagra and F.L.P. Gervais (1845).

La Sagra's Flycatcher *Myiarchus sagrae* (Gundlach, 1852). Johannes Gundlach pointed out that d'Orbigny had wrongly described this bird under the name of *Tyrannus Phoebe* so Gundlach gave it "the name of M. la Sagra, as he was the first to make it known to the public". Resident on the Bahamas, Cuba and Grand Cayman. Accidental Alabama and Florida.

LICHTENSTEIN, Martin Hinrich Carl
b. Hamburg, Germany: 10 June 1780
d. At sea, Baltic: 3 September 1857

German naturalist who collected and explored in southern Africa, 1802–06. Professor of Zoology at Berlin University, 1811. Made Director of the Berlin Zoological Museum in 1815, he helped to develop it into one of the finest in Europe. His written works include zoological commentaries on the American collections of the explorers Marcgrave, Piso and Hernandez (1823); and an edited version of J.R. Forster's *Descriptiones animalium* (1884) from Cook's second round-the-world voyage.

Altamira Oriole *Icterus gularis* (Wagler, 1829). North of the border this Mexican species is restricted to the lower Rio Grande valley. Long known as Lichtenstein's Oriole, it was named in his honour by Johann Wagler, Director of the Munich Zoological Museum, when describing birds from Ferdinand Deppe's collection.

LILIAN—see BALDWIN

MASSÉNA, François Victor. 2nd Duke of Rivoli and 3rd Prince of Essling
b. Antibes, France: 2 April 1799
d. Paris, France: 16 April 1863

Son of André Masséna, the illustrious Marshal of France. Husband of Anna Masséna. He built up an important bird collection of some 12,500 specimens from all parts of the world. It was bought by Dr J.E. Gray in 1846 on behalf of Dr T.B. Wilson who gifted it to the Philadelphia Academy of Natural Sciences. Masséna's *Catalogue de la magnifique collection d'oiseaux de M. le Prince d'Essling* (1846) extended to forty-one quarto pages. His only other publication appears to be a 'Description de quelque nouvelles espèce d'oiseaux de la famille de Psittacides' for the *Revue et Magasin de Zoologie* (1854) which he co-wrote with his nephew Baron Charles de Souancé; the paper describes a number of new parrots. [Charles de Souancé (1823–96) was the son of Suzanne DeBelle and Gabrielle Guillier, Baron de Souancé. Suzanne was the sister of Anna Masséna, *née* DeBelle.]

Magnificent Hummingbird *Eugenes fulgens* (Swainson, 1827). The alternative name of Rivoli's Hummingbird followed on from Lesson's 1829 dedication *Ornismya Rivoli* and continued in general use until recently. The scientific name proposed by Lesson changed much earlier when it was realized that Swainson had described it first.

Masséna Partridge is an old name for the Montezuma Quail.

MAXIMILIAN, Alexander Philipp. Prince of Wied-Neuwied
b. Neuwied, Germany: 23 September 1782
d. Neuwied, Germany: 3 February 1867

The second son of the ruling Prince Friedrich Karl, he was the last of the Princes of Wied-Neuwied. In 1802 he joined the Prussian army to fight against Napoleon and by 1814 had risen to the rank of Major General, commanding a division during the capture of Paris. Immediately after the fall of Napoleon he retired from the army and led an expedition to south-east Brazil, 1815–17. Back home at his castle on the Rhine, he spent several of the subsequent years writing his story and working through his discoveries for his *Reise nach Brasilien* (1820–21) and *Beiträge zur Naturgeschicte von Brasilien* (1825–33). In 1832 he went to North America for about two years, taking with him the Swiss artist Karl Bodmer. He took a steamboat from St Louis up the Missouri (ten years before Audubon) travelling 2500 miles to Fort Mackenzie, where he turned back because of the general hostility of the tribes that he so wanted to study. His two volume *Reise in das innere Nord-America* [Travels in the interior of North America] was published 1839–41, an English edition appearing in 1843. His works showed him to be a man ahead of his time in the comprehensiveness and thoughtfulness of his studies, though the drawings of the Indians, now so well known, have made his artist more famous. When well into his eighties Maximilian was still contributing ornithological papers to scientific journals. At the time of his death his bird collection numbered 4000 specimens, mainly Brazilian species since much of his North American material perished in a fire on board a steamboat. In 1870 his collection was bought by the American Museum of Natural History at New York.

Brown-crested Flycatcher *Myiarchus tyrannulus* (P.L.S. Müller, 1776). This bird is still sometimes known as Wied's Crested Flycatcher.

Pinyon Jay *Gymnorhinus cyanocephalus* Wied, 1841. Collected by Maximilian between the Marias and Yellowstone rivers, Montana, it used to be known as Maximilian's Jay.

MEARNS, Edgar Alexander

b. Highland Falls, New York: 11 September 1856
d. Washington, D.C.: 1 November 1916

Founder member of the A.O.U. One of the best American field naturalists and a prolific collector. Interested in birds from early childhood he began collecting seriously at the age of sixteen. He entered the army in 1883 two years after graduating from the College of Physicians and Surgeons of New York. Arriving at Fort Verde, Arizona, early in 1884 he spent nearly four years there making collecting trips locally as well as further afield to Texas. From 1888 to 1891 he was at Fort Snelling, Minnesota. From 1892 to 1894 he served with the U.S. and Mexican Boundary Commission, helping to collect some 30,000 natural history specimens between El Paso and the Pacific Coast. In between postings in Virginia, Texas and Rhode Island he served in the Spanish–American War, and at other times worked at the American Museum of Natural History at New York. Military service in the Philippines (1903–04, 1905–07) allowed him to collect Asiatic birds for the first time and led to his discovery of several new bird genera— including *Mearnsia*, a genus of swifts. In 1909 he was retired with the rank of Lieutenant Colonel but given orders to report to the President, Theodore Roosevelt, who wanted him for a year-long hunting expedition to East Africa. He returned to Africa with the Childs Frick expedition (1911–12), exploring and collecting in Ethiopia and Kenya. Though he completed well over a hundred scientific papers ill-health prevented him from writing up all the results of his African travels. At the time of his death more than one-tenth of the bird specimens in the U.S. National Museum had been contributed by Dr Mearns. He also collected an immense number of mammals, reptiles, fish and insects; and at one time there were more plants in the National Herbarium gathered by him than by any other person. For portrait see p. 156.

Montezuma Quail *Cyrtonyx montezumae* (Vigors, 1830). This species has had a variety of common names including Harlequin Quail, Masséna Partridge and Mearns's Quail.

Mearns's Woodpecker and Mearns's Gilded Flicker were names given to races of the Acorn Woodpecker and Northern Flicker found in Arizona and New Mexico.

MIDDENDORFF, Alexander Theodore von

b. St Petersburg, Russia: 6 August 1815
d. Hellenorm, Livonia: 16 (28) January 1894

Russian naturalist and explorer. Early education at St Petersburg. Graduated from Dorpat University, 1837. Also studied at Berlin, Erlangen, Vienna and Breslau. Became

Assistant Professor of Zoology at Kiev, 1839. Shortly after this he joined Karl Ernst von Baer's expedition to the Kola Peninsula. In 1843 he was the first naturalist sent by the St Petersburg Academy of Sciences to explore the Taimyr Peninsula and spent many years working through the zoological, botanical, geographical and anthropological results of this expedition. He subsequently became best known for his *Reise in den äussersten Norden und Osten Sibiriens* [Travels in the extreme North and East of Siberia] (1847–75); his descriptions of the nests of Steller's Eider and Black-bellied Plover were the first ever published. He also wrote 'Die Isepiptesen Russlands', an account of bird migration in Russia prepared for the *Mémoires de l'Académie des Sciences de St. Petersbourg* (1855). He taught natural history to some of the Russian Grand-Dukes and accompanied them on foreign travels.

Middendorff's Grasshopper-Warbler *Locustella ochotensis* (Middendorff, 1853). Breeds in eastern Siberia. A casual migrant to Alaska—Nunivak Island, Near Island, St Lawrence Island, Attu.

Bean Goose *Anser fabalis middendorffii* Severtzov, 1872. This race breeds in Siberia east of the Yenisey. Regular in the western Aleutians and casual St Lawrence Island, the Pribilofs and Seward Peninsula.

MONTEZUMA, II (c.1480–1520)

Aztec Emperor of Mexico (1502–20). Captured by Hernando Cortés in 1519 during the Spanish conquest and held hostage in Mexico City until forced to acknowledge himself a vassal of Charles V. He was restored to liberty after paying a huge ransom in gold and jewels but was killed soon afterwards by his own subjects.

Montezuma Quail *Cyrtonyx montezumae* (Vigors, 1830). Nicholas Vigors published the first description of this bird in the *Zoological Journal* from a specimen in the collection of the Zoological Society of London. He reported that it lived in Mexico but gave no other details. No common name was given and he did not explain why the female genitive suffix was used in the scientific name. It is generally assumed to have been dedicated to Montezuma II.

MURPHY, Robert Cushman

b. New York City: 29 April 1887
d. New York City: 20 March 1973

Chairman of the Department of Birds at the American Museum of Natural History at New York. President of the A.O.U., 1948–50. He grew up at Mount Sinai, Long Island, and met his wife while at Brown University, Providence. Shortly after his marriage in 1911 his wife insisted that he accept an opportunity to act as naturalist for two years on one of the last whaling cruises under sail to the south Atlantic; he eventually produced an account of his experiences for her—*Logbook for Grace* (1947). In 1906, before he went to college, he had spent a year at the American Museum. At the end of his voyage he returned to the New York area and worked at the Brooklyn Museum, 1911–20. From 1921 until his death he was associated with the American Museum, working at first under Frank Chapman. Murphy continued his seabird studies by visiting the coastal islands of Peru and writing a semi-popular account entitled *Bird Islands of Peru* (1925) but he is better known for his highly acclaimed two-volume *Oceanic Birds of South America* (1936). He was the land-based general manager of the well known Whitney South Sea Expedition, and he personally supervised the packing and transportation of the immense Rothschild bird collection from England to New York.

Murphy's Petrel *Pterodroma ultima* Murphy, 1949. "For twenty years" wrote Murphy in 1949, "I have been well aware of the distinctness of this bird, which was obtained in considerable numbers in the central tropical and subtropical Pacific by Mr. Rollo H. Beck and his associates on the Whitney Expedition. Naming has been deferred because of the apparent unlikelihood that such a large, striking, and abundant petrel could fail to turn up in the collections of museums, or at least in the literature of Pacific exploration." Casual off the Hawaiian Islands. Also recorded Oregon and at sea about 350 miles west of Santa Barbara, California.

NAUMANN, Johann Andreas
    b. Ziebigk, Köthen, Germany: 3 April 1744
    d. Ziebigk, Köthen, Germany: 16 May 1826
Farmer without scientific training who devoted most of his life to the study of birds. He produced an important and influential four-volume guide to the birds of middle and northern Europe: *Naturgeschichte der Land u. Wasservögel des nördlichen Deutschlands* (1795–1803). His son Johann Friedrich Naumann (1780–1857) became equally famous as an ornithologist.
    Naumann's Thrush *Turdus naumanni* Temminck, 1820. A Siberian species. Casual in spring to Alaska—St Lawrence Island, Point Barrow and western Aleutians.

NELSON, Edward William
    b. Manchester, New Hampshire: 8 May 1855
    d. Washington, D.C.: 19 May 1934
President of A.O.U., 1908–10. Chief of the U.S. Biological Survey, 1916–27. Nelson lost his father in the Civil War. His mother set up a dressmaking business in Chicago and started again from scratch after the great fire of 1871. When seventeen years old he made a journey to the Rocky Mountains with schoolfriends and Professor E.D. Cope. In 1877 he went to Alaska exploring and collecting at many Bering Sea locations until 1881, often with the U.S. revenue cutter *Corwin*. Extreme privations during his land-based trips weakened his health, and soon after returning to Washington he contracted tuberculosis, which meant that his friend Henry Henshaw had to edit and see through the press his 'Report upon Natural History Collections made in Alaska' (1887). Nelson's mother took him to Arizona to recuperate and most of the remainder of his collecting experiences were in hot dry places—he took part in the Death Valley Expedition (1890–91) as a Special Field Agent. For the next fourteen years he was directed to collect in California and Mexico. He later became more involved in the administrative work of the Biological Survey and he was especially proud of negotiating the Migratory Bird Treaty that helped protect wildfowl migration between the United States and Canada.
    Nelson's Gull *Larus nelsoni* Henshaw, 1884. The name was given to a specimen from St Michael, Alaska, but it is now considered a hybrid between a Glaucous and Herring or Iceland Gull.
    A number of North American bird races are named after Nelson including the Sharp-tailed Sparrow *Ammodramus caudacutus nelsoni* J.A. Allen, 1875. The first specimens of this sparrow were taken by Nelson in the Calumet marshes of south Chicago when he was eighteen years old.

NEWELL, Matthias
    b. Bavaria: 1854
    d. Dayton, Ohio: 12 October 1939
Roman Catholic teaching missionary. He came to America in boyhood and entered the Marianist Brothers' Order in Dayton in 1868. For ten years he worked as a teacher moving from Maryland to Louisiana to Texas to California. He then spent thirty-eight years in the Hawaiian Islands, serving at schools on Maui, Oahu and Hawaii. He was

also involved with the Hawaiian Board of Commissioners of Agriculture and Forestry, Division of Entomology. In 1923 he fell out with the Superintendant of Forestry, resigned and in the following year returned to the mainland. In 1927 he retired to the Mother House of the Marianist Brothers at Dayton where he spent the remainder of his life. According to Henry Henshaw, Brother Matthias had a strong interest in birds and was a keen observer, always ready to pass on any knowledge he had gained. His collection of plants, birds and insects went to the Bernice P. Bishop Museum at Honolulu.

Newell's Shearwater *Puffinus auricularis newelli* Henshaw, 1900. Henry Henshaw was in Hawaii from 1895–1904; the only specimen of this shearwater that he saw was the one presented to him by Brother Matthias. Pratt, Bruner and Berrett in their *Field Guide to the Birds of Hawaii and the Tropical Pacific* (1987) give Newell's Shearwater full specific status but it is often considered to be a race of Townsend's Shearwater.

NUTTING, Charles Cleveland
b. Jacksonville, Illinois: 25 May 1858
d. Iowa City, Iowa: 23 January 1927

American zoologist. Collected birds and antiquities for the Smithsonian Institution in Nicaragua and Costa Rica, 1881–82. He married in 1886 and moved to Iowa City as Professor of Zoology and curator of the museum in the State University of Iowa. His wife died early in 1891 and that same year he led a collecting expedition to Saskatchewan. In 1893 he took some of his students to the Bahamas for marine specimens. He married again in 1897 and continued collecting in foreign parts. In 1902 he went to the Hawaiian Islands on the U.S. Fish Commission steamer *Albatross* as a civilian member of the scientific staff; he was in the West Indies in 1918, and Fiji and New Zealand in 1922. Nutting's energetic, forceful character, his organizing abilities and his enthusiasm for collecting helped to ensure the success of his trips.

Nutting's Flycatcher *Myiarchus nuttingi* Ridgway, 1883. A Central American species. Accidental Roosevelt, Arizona.

PALLAS, Peter Simon
b. Berlin, Germany: 22 September 1741
d. Berlin, Germany: 8 September 1811

The most eminent explorer–naturalist of his day. After moving to St Petersburg he was sent on a six-year expedition across Russia to Lake Baykal and beyond (1768–74). He also explored in the Crimea (1793–94). He published a number of important zoological and botanical works.

Pallas's Cormorant *Phalacrocorax perspicillatus* Pallas, 1811. This extinct species, also known as the Spectacled Cormorant, used to breed on Bering Island at the Soviet end of the Aleutian Chain. Although it may have bred on some of the nearby Alaskan islands there is no conclusive evidence that it did so.

PRÉVOST, Florent
b. place and date of birth untraced
d. Paris: 1870

Assistant naturalist at the Paris Natural History Museum, 1832–70. He was the author of numerous zoological works including volume two (2nd edn) of *Les Pigeons par Madame Knip* (1843), and, with C.L. Lemaire, *Histoire Naturelle des Oiseaux d'Europe* (1845). With O. Des Murs he worked on the birds from the 1836–39 voyage of *La Vénus* (see Néboux) and on the mammals and birds brought back by the 1839–43 French expedition to Abyssinia.

Green-breasted Mango *Anthracothorax prevostii* (Lesson). Also known as Prevost's Mango. The species was given its scientific name by R.P. Lesson in his *Histoire Naturelle des Colibris* (1832). A South and Central American species of hummingbird; one was trapped and banded at Corpus Christi, Texas, winter 1991–92.

RICHARDSON, John

b. Dumfries, Scotland: 5 November 1787
d. Grasmere, England: 5 June 1865

Scottish naval surgeon, Arctic explorer, zoologist, botanist and geologist. He obtained his licence from the Royal College of Surgeons, Edinburgh, in 1807 and was appointed Assistant Surgeon with the Royal Navy in the latter part of the Napoleonic Wars. He then served in eastern Canada during the War of 1812. In 1816 he graduated M.D. at Edinburgh and was soon after appointed to Franklin's first overland expedition to the Arctic, 1819–22. Despite the extreme hardships endured he also took part in Franklin's second expedition to the region, 1825–27. Knighted in 1846 he returned to the Arctic in 1848, with Dr John Rae, to search in the region between the mouths of the Mackenzie and Coppermine Rivers for the missing Franklin expedition. He wrote the volumes on quadrupeds and fishes for the *Fauna-Boreali Americana* (1829–37) and was assisted by Swainson with the volume on birds. Richardson contributed to the zoological appendices of a number of Arctic expeditions including some of those led by Beechey, Belcher, Franklin, Kellett, Parry and James Ross.

Merlin *Falco columbarius richardsonii* Ridgway, 1870. Richardson described a Merlin taken near Carlton House in 1827 as a subspecies. Ridgway redescribed it as a full species (using a specimen from South Dakota) and named it after Richardson but he later relegated it to sub-specific level. The name Richardson's Merlin still persists for this pale prairie form.

Boreal Owl *Aegolius funereus richardsoni* (Bonaparte, 1838). The name Richardson's Owl was once in common use but is now almost obsolete.

Richardson's Goose is another name for Hutchins's Goose.

RICORD, Alexandre

b. Baltimore, Maryland: 1798
d. Place and date of death untraced

Naval surgeon and naturalist who spent eight years in the West Indies and on the east coasts of North and South America, 1826–38. He collected especially in the Antilles, sending his collections to the Paris Natural History Museum. His specimens indicate that he collected on Haiti (1827), Guadeloupe (1829), Haiti (1830) and Cuba (1833–34, 1838). In 1838 he contributed a short bird paper to the *Revue Zoologique* and was probably living in Paris by then because no more foreign specimens were contributed after that date. In 1839 he proposed Jules Verreaux as a member of the Société Cuvierienne.

Cuban Emerald *Chlorostilbon ricordii* (Gervais, 1835). Resident in the Bahamas and on Cuba. There are some unsubstantiated sight records for southern and east-central Florida.

RIEFFER. Identity unknown.

An obscure collector or dealer of the 1830s and 1840s who worked in or around Bogotá, Colombia, and in other parts of South America; the Paris Museum lists specimens purchased in 1841, 1843 and 1844. Many of his birds were described by the French naturalist Boissonneau.

Rufous-tailed Hummingbird *Amazilia tzacatl* (De la Llave, 1833). A Central and

South American species. Accidental Brownsville, southern Texas. Also known as Rieffer's Hummingbird.

RIVOLI—see MASSÉNA

SCHLEGEL, Gustaaf
b. Oegstgêest, Netherlands: 30 September 1840
d. Leyden, Netherlands: 15 October 1903

Son of Hermann Schlegel the Director of the Leyden Museum. He spent eighteen years in the Far East and became an eminent Sinologist. While at Amoy, China, he became friendly with the English ornithologist Robert Swinhoe and collected a few fish and birds amongst which was the type of *Anthus gustavi*.

Pechora Pipit *Anthus gustavi* Swinhoe, 1863. Breeds north-eastern Siberia south to southern Ussuriland and on the Commander Islands. Casual visitor to the western Aleutians, the Pribilofs and St Lawrence Island.

SENNETT, George Burritt
b. Sinclairville, New York: 28 July 1840
d. Youngstown, Ohio: 18 March 1900

Businessman involved in the manufacture of oil-well machinery who was based for most of his life in Erie and Crawford Counties, Pennsylvania. He did not begin to study birds seriously until he was about thirty-four years old but he subsequently proved to be a thorough and intelligent observer, showing sound judgement in his few published papers. His first expedition was to western Minnesota, spring 1876, followed by trips to Texas in 1877, 1878 and 1882. He continued to specialize in the birds of Texas but did not return again, instead he sent William Lloyd to collect there for him in 1887. J.M. Priour collected for him in the adjacent parts of northern Mexico in 1887, 1888 and 1889. As early as 1883 Sennett presented his 8000 bird skins, and nearly as many nests and eggs, to the American Museum of Natural History at New York. From around this time until 1896 he spent his winters in New York and by studying his own collection and those of others he hoped to produce an important monograph on the birds of the lower Rio Grande. Colour plates were produced by Ernest Thompson Seton but the text was delayed by Sennett's business commitments and the book never appeared because of his sudden death following an attack of pneumonia.

Common Nighthawk *Chordeiles minor sennetti* Coues, 1888. Sennett's Nighthawk is the grey form of the northern Great Plains.

Races of White-tailed Hawk, Long-billed Thrasher and Hooded Oriole were also named for Sennett.

SOLANDER, Daniel Carl
b. Norrland, Sweden: 28 February 1736
d. London, England: 16 May 1782

Swedish botanist and favourite pupil of Linnaeus. He went to England in 1760 and joined Joseph Banks on Cook's first Pacific voyage (1768–71). He became Banks's secretary and librarian, travelling with him to Iceland in 1772. His sudden death (in Banks's house in Soho Square) prevented the publication of his plant descriptions from the Pacific.

Solander's Petrel *Pterodroma solandri* (Gould, 1844). Breeds in southern Pacific. Sight record fifty miles north-east of Hawaii in 1964.

STEJNEGER, Leonhard Hess
b. Bergen, Norway: 30 October 1851
d. Washington, D.C.: 28 February 1943

Long-serving employee of the U.S. National Museum at Washington. Educated at Frederic University, Oslo, he studied birds in Norway and the Tyrol. In 1881 he moved to the United States and was sent by the National Museum to collect birds in the Bering Sea, especially on Bering Island. On his return he was appointed Assistant Curator of Birds and in 1885 he completed *Bulletin No. 29* of the U.S. National Museum on his 'Results of ornithological explorations in the Commander Islands and Kamtschatka' in which 400 species are described or noticed. In 1889 he was put in charge of the reptile and amphibian section, producing many papers on these subjects as well as others on the birds of Japan and the northern Pacific islands. He revisited Alaska several times in connection with the Fur Seal investigations and spent many years researching for his exhaustive biography of Steller. Stejneger was exempted from mandatory retirement at the age of seventy and worked on at the museum until a few days before his death at the age of ninety-one. For portrait see p. 442.

Stejneger's Petrel *Pterodroma longirostris* (Stejneger, 1893). Breeds New Zealand and San Fernandez Group, Chile. Recorded at sea in northern Pacific between Hawaiian Islands and North America.

Kauai Amakihi *Hemignathus stejnegeri* (S.B. Wilson, 1890). Endemic to Kauai, where it is still abundant in certain localities. The first specimens were collected by Valdemar Knudsen and afterwards examined by Stejneger and then by Wilson. It is sometimes separated from the Common Amakihi and given full specific status.

STEPHENS, Frank
b. Portage Falls, New York: 2 April 1849
d. San Diego, California: 5 October 1937

Field naturalist and collector active in the south-western States. His family moved to Michigan when he was thirteen and his schooling finished shortly afterwards as he had to take over much of the farmwork because of the shortage of labour during and after the Civil War. He first took up taxidermy when the family transferred to Illinois. At the age of twenty-four he married and went to Kansas but soon continued further west with a mule-drawn wagon, wintering at Colorado Springs where Charles Aiken showed him how to make better bird skins. In the following years Stephens and his wife travelled about between New Mexico, Arizona and California, sometimes collecting for William Brewster and also selling skins to Aiken, Donald R. Dickey, C. Hart Merriam, Charles K. Worthen and the University of California. Stephens was a collector in 1891 for the Death Valley Expedition and in several subsequent years collected birds for the U.S. Biological Survey. Early in 1898 his wife died but he married again later in the year and in 1907 they went together with Miss Annie M. Alexander's party to Alaska. In 1910 he went with Joseph Grinnell on an expedition to the Colorado River. From 1885 he began to spend more time on mammals and published *California Mammals* (1906). Latterly he helped build up the collections for the San Diego Natural History Museum and was still making field trips when in his late seventies, going three times to Baja California with Laurence M. Huey.

Fox Sparrow *Passerella iliaca stephensi* Anthony, 1895. Stephens's Fox Sparrow is a large-billed California form of this highly variable species.

TAVERNER, Percy Algernon
b. Guelph, Ontario, Canada: 10 June 1875
d. Ottawa, Canada: 9 May 1947

The foremost Canadian ornithologist during the first half of the twentieth century. He was the son of Edwin and Emily Fowler but when their marriage was dissolved he took the name of Taverner from his step-father with whom he spent an unsettled childhood travelling the eastern stage circuit. In his teens and early twenties, between studying and practising architecture, he developed his skills in fieldwork and taxidermy around Port Huron and Chicago. On moving to Detroit he fell in with B.H. Swales, A.B. Klugh and W.E. Saunders and (together with J.S. Wallace and J.H. Fleming) they formed the Great Lakes Ornithological Club. Bird work was naturally concentrated at Point Pelee and their findings were summarized in a report by Taverner and Swales for the *Wilson Bulletin* (1907–08). Taverner was later instrumental in winning park status for this important migrant trap. In 1911 the post of Assistant Naturalist and Curator was set up at the National Museum and Taverner was appointed on the recommendation of Fleming, Saunders and Ernest T. Seton. Taverner then set about making a thorough investigation of the ornithology of Canada through field surveying parties and by encouraging correspondents. Conditions at the museum were discouraging but because he was funded through the Geological Survey of Canada there was always money for exploration. He made expeditions to most of the Canadian provinces, travelling to British Columbia and Manitoba several times as well as visiting Labrador, Ellesmere Island and Greenland. During his thirty-one years in office the bird collection increased from about 5000 to 30,000 skins but he failed to sort through a few important collections because of the pressure of paper work. He set up a nationwide survey program which he coordinated through his unique mapping and card index system (still in use today). Taverner wrote on a variety of ornithological subjects but his monuments are his *Birds of Eastern Canada* (1919), *Birds of Western Canada* (1926) and *Birds of Canada* (1934) which appeared at a time when there were no comparable works of such high quality for the United States. Their popularity was enhanced by the numerous colour plates by Allan Brooks and Frank C. Hennessey. Taverner himself executed many of the black and white illustrations and indeed seems to have been capable of turning his hand to anything of a practical nature. Unfortunately he suffered from a stammer which prevented him from seeking more prominent positions in the many bird societies of which he was a member. Among friends, especially when in the field, his speech impediment was less noticeable. One of his colleagues commented that whatever he had to say was usually worth waiting for. His early years with his step-father's repertory company meant that he knew many of the popular and operatic songs and would sometimes sing them around the camp fire after a good day's birding. He was married a month before his fifty-fifth birthday to Martha Hohly Wiest of Detroit.

Timberline Sparrow *Spizella taverneri* Swarth and Brooks, 1925. Breeds locally in montane scrub and dwarf birch in the mountains of south-west Yukon and in British Columbia and Alberta. Usually considered to be a race of Brewer's Sparrow but separated from it by Sibley and Munroe (1990).

TEMMINCK, Coenraad Jacob

  b. Amsterdam, Netherlands: 31 March 1778

  d. Lisse, Netherlands: 30 January 1858

Director of the Leyden Natural History Museum for almost forty years, he made it the best in Europe. Author of *Manuel d'Ornithologie* (1st edn 1815, 2nd edn 1820–40), a most important handbook for European ornithologists. He is also noted for his *Nouveau Recueil de Planches coloriées d'Oiseaux* (1820–39) with over 600 colour plates of birds.

Temminck's Stint *Calidris temminckii* (Leisler, 1812). A Palearctic species. Casual to western Alaska, with records from Cape Prince of Wales, St Matthew Island, St Lawrence Island, the Pribilofs and western Aleutians.

TRUDEAU, James de Bertz
b. On a plantation near New Orleans, Louisiana: 14 September 1817
d. New Orleans, Louisiana: 25 May 1887

Medical practitioner, naturalist, painter and sculptor. While at New York he became a friend and supporter of Audubon. In 1845 Baird visited him in the city and examined his splendid collection of eggs and drawings of eggs. Trudeau went back to Louisiana in 1858 and served with the Confederates as an artillery officer. There is a painting of him as a young man, dressed in Osage costume, by John Woodhouse Audubon.

Trudeau's Tern *Sterna trudeaui* Audubon, 1838. A South American species which breeds in Chile and Argentina and winters along the coasts of Chile and Peru. The type specimen shot by Trudeau was said to have been seen with a few others of the same species at Great Egg Harbor, New Jersey. Trudeau also claimed that these terns had appeared off Long Island, New York. The 1983 A.O.U. *Check-list* considers the natural occurrence of this species in North America to be highly questionable.

VALLISNERI, Antonio
b. Tresilico, Modena, Italy: 3 May 1661
d. Padua, Italy: 18 January 1730

Physician and naturalist. Appointed Professor of Practical Medicine at the University of Padua in 1700. Apart from his medical work he studied the reproduction of insects and mammals and wrote monographs on the chameleon and ostriches. His observations on the reproductive mechanism of an aquatic plant led to its genus being named *Vallisneria*. The results of his researches made him oppose the doctrine of spontaneous generation.

Canvasback *Aythya valisineria* (Wilson, 1814). Alexander Wilson gave this duck its (misspelled) scientific name because of its fondness for Wild Celery *Vallisneria americana*. The bird is not therefore named directly after Vallisneri but after the plant genus that commemorates him.

WIED—see MAXIMILIAN

WOODHOUSE, Samuel Washington
b. Philadelphia, Pennsylvania: 27 June 1821
d. Philadelphia, Pennsylvania: 23 October 1904

Physician and naturalist in the 1850s. In his youth he got to know Townsend, Nuttall, Gambel and Cassin at the Academy of Natural Sciences at Philadelphia. After graduating in medicine from the University of Pennsylvania in 1847 he was appointed (April 1849) surgeon and naturalist to the Creek and Cherokee Boundary survey under Lieutenant Sitgreaves, and then Lieutenant Woodruff. In 1851 he again joined Sitgreaves, on the Zuni River Expedition down the Zuni to the Little Colorado River then across the mountains to the west. A few days before crossing the Colorado River he was wounded in the leg by an arrow as he warmed himself by the camp fire in the early hours of the morning. He recovered and returned home via San Francisco and Nicaragua, publishing his account of the natural history in Sitgreaves's official report (1853). In 1854 he was posted to Fort Delaware but he resigned after two years to go into private practice and around this time took part in expeditions to Central America. During 1859 and 1860 he was surgeon for a shipping line on the Philadelphia–Liverpool run and in the Civil War was surgeon at the Eastern Penitentiary in Philadelphia. He gradually lost touch with ornithologists but in his old age he was pleased to renew his association with the Philadelphia Academy and became a Corresponding Member of the A.O.U.

Scrub Jay *Aphelocoma coerulescens woodhouseii* (Baird, 1858). Based on the *Cyanocitta californica* of Woodhouse in the Sitgreaves Report. This race ranges from the southern parts of Oregon, Idaho and Wyoming to southern California (east of the Sierras), Arizona and New Mexico. It was once considered a distinct species and was known as Woodhouse's Jay.

WORTHEN, Charles Kimball
  b. Warsaw, Illinois: 6 September 1850
  d. Warsaw, Illinois: funeral 29 May 1909
Son of the State Geologist of Illinois, Amos H. Worthen. For ten years from 1867 Worthen was employed by his father as a draughtsman illustrating the reports of the Illinois Geological Survey. He also illustrated some of the reports of the Wheeler Survey West of the Hundredth Meridian and spent a winter at the Agassiz Museum in Cambridge, Massachusetts, making drawings of the teeth of fossil sharks. Worthen is said to have known much about natural history but he published very little. He concentrated on collecting and selling specimens and became well known to a great many naturalists, particularly museum curators.

  Worthen's Sparrow *Spizella wortheni* Ridgway, 1884. Breeds in north-eastern Mexico. Accidental Silver City, New Mexico, where the type specimen was collected on 16 June 1884. It was presented to the U.S. National Museum by Worthen.

WÜRDEMANN, Gustavus Wilhelm
  d. Swedesboro, New Jersey: 29 September 1859—"aged 41 years"
Served with the U.S. Coast Survey from 1837. For last twelve years served as a tidal and meteorological observer in Florida and the Gulf of Mexico. From 1853 to 1854 he was in charge of five parties making hourly tide observations on the Texas coast from the mouth of Matagorda Bay to the mouth of the Rio Grande. He collected birds, reptiles, fish and invertebrates for the Smithsonian Institution from Texas, Louisiana and Florida. His 'Letter relative to the obtaining of Specimens of Flamingos and other Birds from South Florida' in the *Annual Report of the Smithsonian Institution* for 1860 was published posthumously (in 1861).

  Great Blue Heron *Ardea herodias* Linnaeus, 1758. Würdemann's Heron is a colour phase with an all-white head chiefly found in the Florida Keys. Named *Ardea wuerdemanni* by Baird in 1858, it was many years before its true identity was realized.

References

AIKEN—*Auk* 1936, pp. 371–372; *Auk* 1964, p. 123; *Condor* 1936, pp.234–238; Ewan, J. and Ewan, N.D. 1981. *Biographical Dictionary of Rocky Mountain Naturalists, 1682–1932*, p. 2. Frans A. Stafleu, Utrecht.

ARCHILOCHUS—*Encyclopaedia Britannica* 1875, 9th edn., vol. II, pp. 379–380. Edinburgh.

ARMINJON—*Dizionario Biografica degli Italiani* 1960–. Vol.4, pp. 241–242. Rome; Gruson, E.S. 1972. *Words for Birds*, pp. 17–18. New York; quote from *Ibis*, 1869, p. 62.

ATTWATER—*Auk* 1932, pp. 144–145.

BALDWIN—*Auk* 1940, pp. 1–12 [S.P. Baldwin]; *Cleveland Plain Dealer* 28 December 1948; *Cleveland Press* 27 December 1948; and information supplied by the Cleveland Museum of Natural History; quote from *Scientific Publications of Cleveland Museum of Natural History* (1930) **1**: 104.

BALTIMORE—*Complete Peerage of the United Kingdom* 1887. Vol. 1, pp. 226–227. George Bell and Sons, London; quote from Catesby, M. 1754. *The Natural History of Carolina*, Vol. 1, p. 48. London.

574   *Audubon to Xántus*

BELCHER—*Condor* 1928, p. 268; O'Byrne, W.R. 1849. *A Naval Biographical Dictionary*, pp. 68–69. J. Murray, London.
BELDING—*Auk* 1920, pp. 32–45; *Condor* 1928, p. 268.
BICKNELL—*Auk* 1926, pp. 142–149.
BREWSTER—*Auk* 1920, pp. 1–32.
BULLOCK—*Dictionary of National Biography* 1885–1900, Vol.VII, p. 256. London; *Museums Journal* 1917–18, pp. 51–56, 132–137, 180–187; Günther, A. (Ed.) 1904–12. *The History of the Collections Contained in the Natural History Departments of the British Museum*, Vol. II, pp. 208–245 [The Bullock Collection]. BM publication, London; Bullock's death was registered on 16 March 1849 in the North East District of Chelsea (St Luke's, Chelsea).
CABOT—*Auk* 1886, p. 144.
CANUTE—*Encyclopaedia Britannica* 1882, 9th edn, Vol. V, pp. 39–40; Vol. XIV, p. 129. Edinburgh. [Camden's statement is: *Knotts, i Canuti aves, vt opinor e Dania enim aduolare credunter.*]
CONSTANT—*Dictionnaire de Biographie Française* 1961. Vol. IX, p. 499. Paris.
COUES—Cutright, P.R. and Brodhead, M.J. 1981. *Elliott Coues. Naturalist and Frontier Historian*. University of Illinois Press, Chicago.
COX—Biographical material supplied by J.B. Cox; Cox, J.B. 1989. The Story Behind the Naming of Cox's Sandpiper. *Australian Bird Watcher* **13**: 50–57; Cox, J.B. 1990. The enigmatic Cooper's and Cox's Sandpipers. *Dutch Birding* **12**: 53–64.
DEGLAND—*Encyclopaedia Britannica* 1875, 9th edn, vol. XVIII, p. 17. Edinburgh; *Condor* 1928, p. 276.
DIAZ—Gruson, E.S. 1972. *Words for Birds*, pp. 47–48. New York; *Condor* 1928, p. 276; *Diccionario Porrúa, de Historia, Biografía, y Geografía de México* (1964) vol. 1, p. 891. Mexico, D.F.
DRESSER—*Ibis* 1916, pp. 340–342; Mullens, W.H. and Swann, H.K. 1917. *A Bibliography of British Ornithology*, pp. 178–180. London.
ELLIS—Pearse, T. 1968. *Birds of the Early Explorers in the Northern Pacific*, pp. 181–203. T. Pearse, Comox, British Columbia.
EVELYN—*Proceedings of the Zoological Society of London* 1847, p. 44.
FISHER—*Auk* 1951, pp. 210–213; *Condor* 1928, p. 277.
HARCOURT—Boase, F. 1912. *Modern English Biography*, Vol. V (supplement, vol. II), p. 567. Newton and Worth, Truro; *Burke's Peerage* 1959, p. 2284. Burke's Peerage Ltd., London.
HARLAN—Audubon, J.J. 1831. *Ornithological Biography*, Vol. I, p. 441. Edinburgh; *Condor* 1928, p. 281; *Dictionary of American Biography* 1928–36. Vol. VIII, p. 273–274. New York.
HELOISE—*Revue Zoologique* 1839, p. 15.
HEPBURN—*Condor* 1926, pp. 249–253; *Condor* 1931, pp. 169–171.
HODGSON—Hunter, W.W. 1896. *Life of Brian Houghton Hodgson*. J. Murray, London; Mearns, B. and Mearns, R. 1988. *Biographies for Birdwatchers. The Lives of Those Commemorated in Western Palearctic Bird Names*, pp. 191–196, 442. Academic Press, London.
HOLBÖLL—*Condor* 1928, p. 282; *Dansk biografisk Leksikon* 1933–44. Vol. X, pp. 368–369. Copenhagen.
HUTCHINS—*Dictionary of Canadian Biography*, 1966–91. Vol. IV, pp. 377–378. Toronto; Richardson, J. and Swainson, W. 1832. *Fauna Boreali-Americana*, Vol. II, p. 470. London.
JOUANIN—Mearns, B. and Mearns, R. 1988. *Biographies for Birdwatchers. The Lives of Those Commemorated in Western Palearctic Bird Names*, pp. 208–211, 443. Academic Press, London.

KRIDER—Pierce, F.J. 1939. John Krider's Book. *Iowa Birdlife* **9**: 50–51.

KUMLIEN—*Auk* 1903, pp. 93–94. [Thure Kumlien: *Auk* 1889, pp. 204–205.]

LA SAGRA—Barbour, T. 1923. *The Birds of Cuba*, pp. 6–7. *Memoirs of the Nuttall Ornithological Club No. 6.* Cambridge, Massachusetts; *Enciclopedia Universal Ilustrada Europeo-Americana* [c.1920–1958.] Vol. 29, pp. 901–902. Espasa Calpe, S.A., Madrid.

LICHTENSTEIN—*Dictionary of South African Biography*, 1968–81. Vol. III, pp. 520–523. Cape Town; Mearns, B. and Mearns, R. 1988. *Biographies for Birdwatchers. The Lives of Those Commemorated in Western Palearctic Bird Names*, pp. 240–243, 445. Academic Press, London.

MASSÉNA—*Condor* 1928, p. 293; Prestwich, A.A. 1963. *"I Name This Parrot...,"* pp. 64, 89. Edenbridge, Kent.

MAXIMILIAN—*Condor* 1928, p. 287; Gebhardt, L. 1964. *Die Ornithologen Mitteleuropas*, p. 234. Giessen; Prestwich, A.A. 1963. *"I Name This Parrot...,"* p. 65. Edenbridge, Kent.

MEARNS—Hume, E.E. 1942. *Ornithologists of the United States Army Medical Corps*, pp. 295–323. John Hopkins University Press, Baltimore; *The Smithsonian Report for 1917*, pp. 649–662 [Reprinted from *The Auk* 1918].

MIDDENDORFF—*Auk* 1894, p. 264; *Ibis* 1894, p.458; Vaughan, R. 1992. *In Search of Arctic Birds*, pp. 207–208. T. & A.D. Poyser, London.

MONTEZUMA—*Encyclopaedia Britannica* 1878, 9th edn, Vol. XVI, pp. 206–210. Edinburgh.

MURPHY—*Auk* 1974, pp. 1–9.

NAUMANN—Gebhardt, L. 1964. *Die Ornithologen Mitteleuropas*, pp. 256–258. Giessen; Mearns, B. and Mearns, R. 1988. *Biographies for Birdwatchers. The Lives of Those Commemorated in Western Palearctic Bird Names*, pp. 277–281, 448. Academic Press, London.

NELSON—*Auk* 1935, pp. 134–148; *Condor* 1928, p. 289.

NEWELL—Gruson, E.S. 1972. *Words for Birds*, pp. 13–14. New York; *Honolulu Advertiser* 20 October 1939, p. 2, col. 7.

NUTTING—*Dictionary of American Biography* 1928–36, vol. XIII, pp.597–598. New York; Ewan, J. and Ewan, N.D. 1981. *Biographical Dictionary of Rocky Mountain Naturalists, 1682–1932*, p. 163. Frans A. Stafleu, Utrecht.

PALLAS—*Condor* 1928, p. 290; Mearns, B. and Mearns, R. 1988. *Biographies for Birdwatchers. The Lives of Those Commemorated in Western Palearctic Bird Names*, pp. 288–297, 449. Academic Press, London.

PRÉVOST—Ronsil, R. 1948. *Bibliographie Ornithologique Française*, Vol. 1, pp. 406–407. Le Chevalier, Paris.

RICHARDSON—Houston, C.S. (Ed.) 1984. *Arctic Ordeal. The Journal of John Richardson, surgeon-naturalist with Franklin, 1820–1822.* McGill-Queen's University Press, Kingston and Montreal; Johnson, R.E. 1976. *Sir John Richardson.* Taylor & Francis, London.

RICORD—Chardon, C.E. 1949. *Los naturalistas en la América Latina*, p. 213. Ciudad Trujillo; Ricord, A. 1838, Note sur le Gros-bec Père-Noir. *Revue Zoologique*, pp. 167–168; *Revue Zoologique* 1839, p. 96 (footnote); Archives of the Paris Natural History Museum.

RIEFFER—Rounds, R.S. 1990. *Men and Birds in South America*, p. 152. Q.E.D. Press, Fort Bragg. California.

SCHLEGEL—Mearns, B. and Mearns, R. 1988. *Biographies for Birdwatchers. The Lives of Those Commemorated in Western Palearctic Bird Names*, pp. 336–339, 451. Academic Press, London.

SENNETT—*Auk* 1901, pp. 11–23; *Condor* 1928, p. 295.

SOLANDER—*Dictionary of National Biography* 1885–1900, Vol. LIII, p. 212. London.

STEJNEGER—*Dictionary of American Biography. Supplement Three, 1941–1945*, pp. 732–733. New York; *Ibis* 1943, pp. 354–355.

STEPHENS—*Condor* 1918, pp. 164–166; *Condor* 1928, p. 297; *Condor* 1938, pp. 101–110.

TAVERNER—*Auk* 1948, pp. 85–106.

TEMMINCK—Mearns, B. and Mearns, R. 1988. *Biographies for Birdwatchers. The Lives of Those Commemorated in Western Palearctic Bird Names*, pp. 372–376, 452. Academic Press, London.

TRUDEAU—*Condor* 1928, p. 300.

VALLISNERI—*A Biographical Dictionary of Scientists*, 1974 (2nd edn) pp. 523–524. London; *Condor* 1928, pp. 300–301.

WOODHOUSE—*Auk* 1905, pp. 104–106; *Cassinia* 1904, pp. 1–5; Hume, E.E. 1942. *Ornithologists of the United States Army Medical Corps*, pp. 496–509. John Hopkins University Press, Baltimore.

WORTHEN—*Auk* 1909, p. 332.

WÜRDEMANN—*American Journal of Science* 1860, **29**: 304; Geiser, S. W. 1959. Men of Science in Texas. *Field and Laboratory* **27**: 252.

Index of English and Scientific Names of Birds

Index of People